Advances in Optimization and Approximation

Nonconvex Optimization and Its Applications

Volume 1

Advances in Optimization and Approximation

Edited by

Ding-Zhu Du
University of Minnesota, U.S.A.
and Institute of Applied Mathematics, Beijing, P. R. China

and

Jie Sun
National University of Singapore, Singapore

KLUWER ACADEMIC PUBLISHERS
DORDRECHT / BOSTON / LONDON

Library of Congress Cataloging-in-Publication Data

Advances in opcimization and approximation / edited by Ding-Zhu Du, Jie Sun.
p. cm. -- (Nonconvex optimization and its applications ; v. 1)
Includes bibliographical references.
ISBN 0-7923-2785-3
1. Mathematical optimization. 2. Approximation theory. I. Du, Dingzhu. II. Sun, Jie. III. Series.
QA402.5.A374 1994
519.3--dc20 94-9739

ISBN 0-7923-2785-3

Published by Kluwer Academic Publishers,
P.O. Box 17, 3300 AA Dordrecht, The Netherlands.

Kluwer Academic Publishers incorporates
the publishing programmes of
D. Reidel, Martinus Nijhoff, Dr W. Junk and MTP Press.

Sold and distributed in the U.S.A. and Canada
by Kluwer Academic Publishers,
101 Philip Drive, Norwell, MA 02061, U.S.A.

In all other countries, sold and distributed
by Kluwer Academic Publishers Group,
P.O. Box 322, 3300 AH Dordrecht, The Netherlands.

Printed on acid-free paper

Printed in the Netherlands

This book is dedicated to

Professor Minyi Yue

Contents

Preface

The primary target of this volume is two fold. First, we dedicate this book to our teacher, Professor Minyi Yue of the Institute of Applied Mathematics, Beijing, China, to celebrate his fifty years career in mathematical research. Second, we hope to bring to our reader's attention some recent developments in optimization and approximation which we think would possibly point to new research directions in those fields.

Professor Yue is a prominent figure in modern Chinese mathematics. He started his mathematical career in real analysis with Professor Jiangong Chen at Zhejiang University fifty years ago. Later on, by the recommendation of Professor Chen, he joined the research group of Professor Loo-Keng Hua at the Institute of Mathematics, Chinese Academy of Sciences. During that time, he wrote several important papers in number theory while working as a chief assistant of Professor Hua, the director of the Institute. In later 50s, there was a great movement on applied mathematics in China. It was the time that the Chinese postman problem was solved and linear programming and queueing theory were introduced in China. It was also the time that Professor Yue began his study on operations research. He organized various research groups in optimization and queueing theory and extended his interests to nonlinear programming, convex analysis, computational fixed point theory, mathematical economics, and combinatorial optimization. Since then, in a period of thirty years, he has maintained a high level of research, has educated generations of Chinese scholars in applied mathematics, and has taken important administrative responsibilities in the applied mathematics and operations research community of China. He was the deputy director of the Institute of Applied Mathematics and was a founder of the Operations Research Society of China. His collaborators and students now spread a large number of universities in China and around the world.

We express our sincere thanks to the authors of the papers, the anonymous referees, and the publisher for helping us to produce this book.

Ding-Zhu Du and Jie Sun

University of Minnesota and National University of Singapore
November 1993

SCHEDULING MULTIPROCESSOR FLOW SHOPS

BO CHEN
Econometric Institute, Erasmus University
3000 DR Rotterdam, The Netherlands

Abstract. We address the well-known problem of scheduling multiprocessor flow shops. For a given k-stage processing facility, where at each stage one or more identical machines are available, and a given collection of independent jobs, each comprising k tasks to be processed in order, one per stage, find a schedule that minimizes the makespan. Owing to the intractability of the problem, approximation has been the subject of extensive research. In this paper we review the main advances in this aspect, and establish a new result.

1. Introduction

Since Johnson's pioneering work [18], an impressive amount of literature has been created in the theory of flow shop scheduling. In his paper, Johnson considers the special case $k = 2$ of the following k machine flow shop problem: For a given collection of independent and simultaneously available jobs, each to be processed on k continuously available machines sequentially in the same order, assuming that one machine can handle no more than one job at a time, find a schedule that minimizes the overall finishing time.

With the developments in adapting various practical environments comes the corresponding generalization of the traditional flow shop setting. In a general model, which is called the *multiprocessor flow shop scheduling*, each of the k machines is replaced by a number of parallel identical machines, that is, each of these replacing machines is capable of processing the same tasks as the replaced one.

Multiprocessor flow shop scheduling has been frequently called upon in practice. To support a variety of on-line services in addition to carrying an ever increasing computing load, Buten and Shen [1] considered the scheduling model for computer systems with two classes of processors. In [25] Salvador studied such a production system with a successful application in the synthetic fibers industry. Wittrock [30] dealt with the scheduling problem for flexible flow lines which is applicable to flexible manufacturing systems.

However, even simple versions, or even subcomponents, of scheduling multiprocessor flow shops quickly become intractable in practice for searching for the optimal solutions. Therefore, most of the attention has been devoted to devising fast algorithms that find *near* optimal solutions. Addressing the main advances along this line is the purpose of this paper, which is organized as follows. Section 2 provides a brief survey for the advances on the approximations to the problem. A new result is established in Section 3. Finally in Section 4, we make some remarks for further research.

Ding-Zhu Du and Jie Sun (eds.), Advances in Optimization and Approximation, 1–8.

2. A Brief Survey

Let us begin with a formal description of the model under consideration. We are given a k-stage processing facility. The ith stage of the facility, $1 \leq i \leq k$, consists of m_i parallel identical machines. We are also given a set of jobs, denoted by $\{1, 2, \ldots, N\}$. Each job j should be processed at some stages, and its processing at stage i, which will be called its ith *task*, needs p_{ij} time units, and should be performed by at most one of the m_i machines at a time. All machines are continuously available, and can handle at most one task at a time. The order in which all the jobs should go through the stages is $(1, 2, \ldots, k)$, although some jobs may skip some stages. Our objective is to find a schedule that minimize the time (often called *makespan*) by which all the jobs are finished. We consider both of the two job characteristics: *non-preemptive*, which requires that a task, once started, cannot be interrupted until it is completed, and *preemptive*, which allows that the processing of a task can be interrupted and resumed later on any of the machines at the same stage.

The problem as described in the above general setting, when reduced to its most elementary and restrictive forms, are two very common scheduling problems that have received substantial attentions. Firstly, if $k = 1$ then it is the *multiprocessor scheduling* problem. If preemption is allowed, then the problem is solvable in linear time by McNaughton's algorithm [21]. Otherwise, it is strongly NP-hard [9], and is NP-hard even when the number of machines $m_1 = 2$ (see, e.g., [10]). Secondly, if $m_i = 1$ for all $i = 1, \ldots, k$, then we have the traditional *flow shop scheduling* problem. This problem is optimally solved by Johnson's algorithm [18] if $k = 2$. It becomes strongly NP-hard either preemption is allowed or not even when $k = 3$ [11, 12]. For a comprehensive overview on the developments of multiprocessor scheduling, or of flow shop scheduling, one is referred to a recent paper by Lawler et al. [20].

Recently it is proved that our problem remains intractable when it is simplified to the remaining extreme case. Hoogeveen et al. [17] prove that if $k = 2$ and $\max\{m_1, m_2\} = 2$ and $\min\{m_1, m_2\} = 1$, the problem is strongly NP-hard for both cases of preemption and non-preemption.

We have seen that even simple versions, or even subcomponents, of our problem quickly make it impossible to always find an optimal solution efficiently. However, it may still be possible to devise a fast *approximation algorithm*, or *heuristic*, which finds solutions that are provably close to the optimum. Usually there are two criteria for evaluating the quality of a heuristic. The valuation of the *average performance* of a heuristic is carried out by empirical studies, which involves running the heuristic on a large set of problem instances with known or estimated optimal solutions and comparing these to the heuristically obtained results. Another approach is a rigorous mathematical analysis on the maximum deviations of the heuristic solutions from the optimum, or the evaluation of the *worst-case performance* of a heuristics. This stream of research is the one to which this paper is addressed.

Given a heuristic. Let S be a schedule created by the heuristic for some problem instance, and let $C_{\max}(S)$ be its makespan and $C^*_{\max}$ be the corresponding optimal makespan. Define the *worst-case performance ratio* of the heuristic to be the supremum of $C_{\max}(S)/C^*_{\max}$ over all problem instances.

Along the line of devising fast heuristic algorithms such that their worst-case performance ratios are as close to one as possible, an impressive amount of effort has been made in the environments of two special cases, namely, multiprocessor scheduling and flow shop scheduling. For these developments, the reader is referred to [13, 14, 6, 2, 5, 7, 31, 8, 16] and [12, 24, 23, 26, 22, 28, 4], respectively.

Developments in approximations to the multiprocessor flow shop scheduling have been relatively slow. Longston [19] studies the problem with two stages in which the number of machines at each stage are the same. He proposes a heuristic called SORTB, which constructs a priority list of jobs based on a sequence of nonincreasing processing times of the second stage, and then applies *list scheduling* at the first stage, where list scheduling (LS for short) is such a scheduling strategy that, given a list of jobs, when a machine becomes available, it begins work on the list's next unprocessed job. As tasks are completed at the first stage, jobs are queued at the second stage on a first-come, first-serve basis. Longston proves that the worst-case performance ratio of SORTB is 2.

Sriskandarajah and Sethi [27] have established worst-case bounds for some heuristics for the problem also with two stages in which either the number of machines at one stage is one or the two stages have the same number of machines. All the bounds established are not less than 2.

Gupta [15] examines the two-stage multiprocessor flow shop with a single machine at the second stage. Based on the observation that the main issue is to minimize the idle time at the second stage, he develops a heuristic that combines an idle-time-minimization strategy and Johnson's procedure. This algorithm is empirically evaluated and found to be efficient. Recently, Chen [3] has provided an analysis of the worst-case performance of Gupta's algorithm, and finds that its worst-case performance ratio is $3 - 2/m$ where m is the number of machines at the first stage.

As one of the earliest studies on the multiprocessor flow shop scheduling, Buten and Shen's work [1] results in a heuristic algorithm called MJO for the problem with two stages. Based on an auxiliary system, algorithm MJO creates a modified Johnson's order for the jobs (we will discuss it in detail in the next section) and then proceeds with scheduling as Longston's SORTB. The authors prove that, if the jobs satisfy a certain *loading constraint*, which basically requires that the size of each job should not be relatively too big, then the worst-case performance ratio of MJO is $2 - 1/\max\{m_1, m_2\}$. Unfortunately, scrutinizing their proof, one may find that the proof is not free from mistakes and errors.

Note that all the approximation algorithms for the multiprocessor flow shop scheduling previously described allow no preemption. Little research has been done if preemption *is* allowed. In the next section, we propose a new efficient heuristic algorithm for scheduling the two-stage multiprocessor flow shop, and prove that the algorithm has the worst-case performance ratio of $2 - 1/\max\{m_1, m_2\}$ for *both* cases of preemption and non-preemption.

3. A New Heuristic

Throughout this section, a *schedule*, if not specified, should be understood a feasible schedule, either preemptive or non-preemptive, for the two-stage multiprocessor flow shop. The two stages will be called stages A and B, and the number of identical machines at each stage is m and n, respectively. We denote this processing facility by F. Let the processing times of job $j \in \{1, 2, \ldots, N\}$ be a_j and b_j, respectively, at stages A and B.

The fact that if $m = n = 1$ then facility F can be easily handled by Johnson's algorithm [18] invites us to treat stages A and B as two special machines that have "equivalent" power. It is then natural and intuitive to introduce the auxiliary facility F' which is a flow shop consisting of two machines, A' and B', comparable to stages A and B, respectively, in the sense that the processing speeds of machines A' and B' are m and n times, respectively, those of any stage-A and stage-B machine. Formally, the processing times of any job j are a_j/m on machine A' and b_j/n on machine B'. This approach was first adopted by Buten and Shen [1]. We begin with presenting our new heuristic algorithm CLS, which stands for *Concatenating List Schedules of Johnson's Orders*.

Algorithm CLS

Step 1. Renumber the jobs so that, for $i = 1, \ldots, N-1$,

$$\min\{a_i/m, b_{i+1}/n\} \le \min\{b_i/n, a_{i+1}/m\}. \tag{1}$$

Step 2. Apply LS for job list $(1, \ldots, N)$ with processing times $(a_1, \ldots, a_N)$ to m identical machines to create schedule S_1. Apply LS for job list $(N, \ldots, 1)$ with processing times $(b_N, \ldots, b_1)$ to n identical machines to create schedule S_2. Let the finishing times of job j in S_1 and S_2 be r_j and s_j, respectively.

Step 3. Perform the mirror transformation on schedule S_2 with respect to the vertical line of time zero to create schedule $\bar{S}_2$, that is, if job j is processed on machine i in schedule S_2, then the job is processed on the same machine and the processing begins at $-s_j$.

Step 4. Form schedule S for the two-stage flow shop by accepting S_1 as its first-stage schedule and accepting $\bar{S}_2 + \tau$ as its second-stage schedule, where $\bar{S}_2 + \tau$ denotes the schedule in which the processing of each job is postponed from schedule $\bar{S}_2$ by time τ, and $\tau = \max_{1 \le j \le N}\{r_j + s_j\}$.

The choice of τ in Step 4 of algorithm CLS guarantees that the second task of each job *follows* its first one. The first step of algorithm CLS is in fact an application of Johnson's algorithm to facility F', and the following steps are, to some extend, transplantation of the optimal schedule on F' to the original facility F. Therefore, to evaluate the performance of algorithm CLS, we need to (a) Compare the powers of facilities F and F' and (b) Estimate the gain and loss during the transplantation. The first goal can be reached by the following lemma, which shows that facility F' is more powerful than F.

Lemma 3.1 *For any schedule S (preemptive or non-preemptive) for facility F with makespan $C_{\max}(S)$, there exists a non-preemptive schedule S' for facility F' with makespan $C_{\max}(S') \le C_{\max}(S)$.*

Proof. Let r_j and s_j be the finishing time and the starting time of job j at stage A and stage B of schedule S, respectively. Suppose π' and π'' are such two N-permutations that $r_{\pi'(1)} \le r_{\pi'(2)} \le \cdots \le r_{\pi'(N)}$ and $s_{\pi''(1)} \ge s_{\pi''(2)} \ge \cdots \ge s_{\pi''(N)}$.

We claim that, if we schedule all the jobs on machine A' starting from time zero without idle according to sequence π' and schedule all the jobs on machine B' starting from $C_{\max}(S) - \sum_{j=1}^{N} b_j/n$ without idle according to sequence π'', then the resulting schedule S' is a feasible schedule for facility F', and hence the lemma is implied.

We will show that the finishing time of any job on machine A' of facility F' is not later than its finishing time at stage A of facility F, which, by symmetry, implies that the starting time of any job on machine B' of F' is not earlier than its starting time at stage B of F, and thus S' is feasible since so is S.

Consider an arbitrary job j and suppose $j = \pi'(k)$. Since all the jobs of $\{\pi'(1), \ldots, \pi'(k)\}$ are finished at stage A of F in schedule S by time s_j, their total amount of processing $\sum_{i=1}^{k} a_{\pi'(i)}$ is at most ms_j. The fact that this amount of processing can be accomplished by machine A' in s_j time units implies that all these jobs are finished on machine A' by time s_j in schedule S'. □

Our second goal of analyzing the transplantation from a schedule on facility F' to a schedule on facility F is accomplished by the following lemma.

Lemma 3.2 *If S is a schedule found by algorithm* CLS *(hence S is non-preemptive), then*

$$C_{\max}(S) \leq C_{\max}(S^*) + \left(1 - \frac{1}{\max\{m, n\}}\right) \max_{1 \leq j \leq N} \{a_j + b_j\}, \tag{2}$$

for any optimal preemptive schedule S^*.

Proof. Let us introduce for facility F' an auxiliary schedule $\bar{S}$, in which all the jobs are processed without interruption and idle time on both machines A' and B' according to the natural sequence $(1, 2, \ldots, N)$ found by the first step of CLS, and machines A' and B' commence processing at time 0 and $C_{\max}(S) - \sum_{i=1}^{N} b_i/n$, respectively (see Figure 1).

Then in schedule $\bar{S}$ the starting time of job j is $\sum_{i=1}^{j-1} a_i/m$ for any $j = 1, \ldots, N$. On the other hand, from the greediness of LS we see that the starting time of job j, for any $j = 1, \ldots, N$, is not later than $(\sum_{i=1}^{j-1} a_i)/m$. Therefore, the starting time of any job in schedule S is no later than its starting time in schedule $\bar{S}$. By symmetry of algorithm CLS, we assert that the finishing time of any job in schedule S is no earlier than its finishing time in schedule $\bar{S}$ (see Figure 1).

Let k be such that $r_k + s_k = \tau$ where r_k, s_k and τ is defined as in Step 4 of algorithm CLS. From Figure 1 we see that

$$C_{\max}(S) = \tau \leq \sum_{i=1}^{k} a_i/m + \sum_{i=k}^{N} b_i/n + (a_k + b_k) - (a_k/m + b_k/n). \tag{3}$$

From relation (1) we see that the job sequence $(1, \ldots, N)$ is a Johnson's order in facility F' [18]. Hence, the *semiactive* schedule corresponding to schedule $\bar{S}$, which is obtained from $\bar{S}$ by starting all jobs on machine B' as early as possible while keeping the same sequence, is optimal (in both cases of preemption and non-preemption). Therefore, for any schedule S' on facility F', we have

$$C_{\max}(S') \geq \max_{1 \leq j \leq N} \{\sum_{i=1}^{j} a_i/m + \sum_{i=j}^{N} b_i/n\}.$$

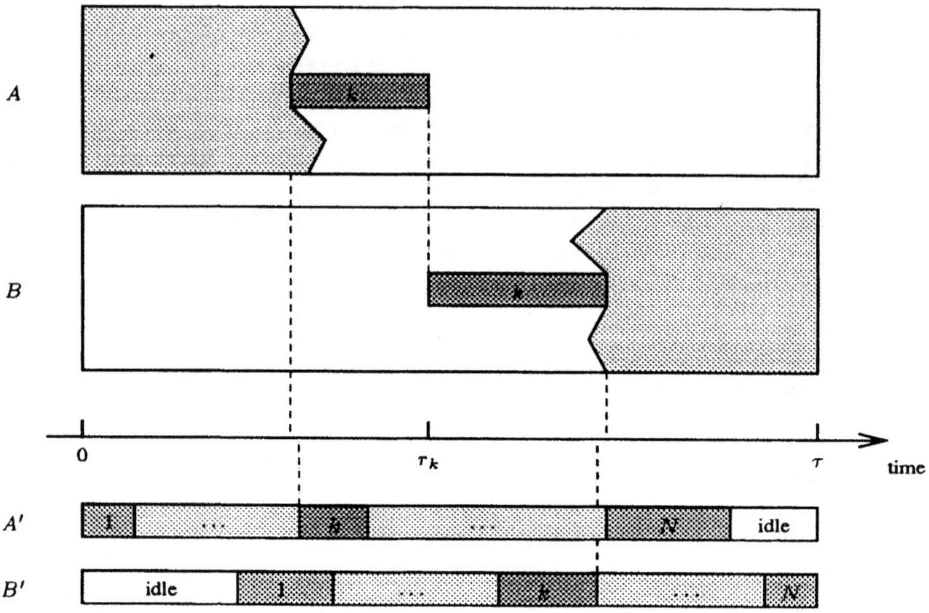

Fig. 1. Heuristic and auxiliary schedules

By (3) we get

$$C_{\max}(S) \le C_{\max}(S') + \left(1 - \frac{1}{\max\{m,n\}}\right)(a_k + b_k),$$

which, together with Lemma 3.1, implies (2). □

Considering that $\max_{1\le j\le N}\{a_j + b_j\} \le C_{\max}(S^*)$ in (2) and the fact that the makespan of any optimal preemptive schedule for a job set is no more than that of any optimal non-preemptive schedule for the same job set, we can draw from Lemma 3.2 the following conclusion, which completes our evaluation of algorithm CLS.

Theorem 3.3 *The worst-case performance ratio of algorithm* CLS *is* $2 - \dfrac{1}{\max\{m,n\}}$.

Proof. We need only to show that the bound is reachable. Suppose without loss of generality that $m \ge n$. Consider the following example. There are $N = 2m - 1$ jobs. Their processing times are as follows: $a_j = m - 1$ for $j = 1, \ldots, m-1$, $a_j = 1$ for $j = m, \ldots, N-1$ and $a_N = m$; $b_j = (N - j + 1)\varepsilon$ for $j = 1, \ldots, N$, where $0 < \varepsilon < 1/(mN)$. Noticing that the natural sequence $(1, 2, \ldots, N)$ is *the* sequence found by algorithm CLS after Step 1, one can easily verify that the makespan of a heuristic schedule is $2m - 1 + \varepsilon$ and that the optimum is at most $m + m(m+1)\varepsilon/2$. Therefore, the ratio between the two makespans tends to $2 - 1/m$ as ε tends to zero. □

4. Concluding Remarks

One clear message that emerges from the previous sections is that approximation for the multiprocessor flow shop scheduling remains an active subject of inquiry.

For one of the two well-known special cases of our problem—approximation for multiprocessor scheduling, successful achievements have been made. For example, Hochbaum and Shmoys [16] have developed a *polynomial approximation scheme* that provides a family of polynomial approximation algorithms with worst-case performance ratios arbitrarily close to one. On the other hand, for the other well-known special case—approximation for flow shop scheduling, a rather negative fact is discovered, although a new positive step has been recently made by Chen et al. [4], which provides an $O(N \log N)$ time heuristic for scheduling three-machine flow shop with the worst-case performance ratio of 5/3. Williamson et al. [29] prove that, for flow shop scheduling, finding a polynomial approximation algorithm that has a worst-case performance ratio less than 5/4 is as difficult as the problem itself!

Coming back from the two special cases to general multiprocessor flow shop scheduling, we are facing such a problem: Although generally there does not exist any polynomial approximation scheme unless P=NP, the existence might be possible if the number of stages is fixed (or even simpler, is fixed at 2). Any answer to the problem, either positive or negative, would be valuable.

Acknowledgements

The author wishes to thank André van Vliet for suggesting the simpler proof of Lemma 3.1.

References

1. R.E. Buten & V.Y. Shen. A scheduling model for computer systems with two classes of processors, *Proceedings of Sagamore Computer Conference on Parallel Processing* (1973), 130–138.
2. B. Chen. A note on LPT scheduling. *Oper. Res. Lett.* 14/1 (1993).
3. B. Chen. Analysis of a heuristic for scheduling two-stage flow shop with parallel machines. *Report 9338/A.* Econometric Institute, Erasmus University, Rotterdam, The Netherlands, 1993.
4. B. Chen, C.A. Glass, C.N. Potts & V.A. Strusevich. A new heuristic for three-machine flow shop scheduling. *Report 9327/A.* Econometric Institute, Erasmus University, Rotterdam, The Netherlands, 1993.
5. E.G. Coffman, Jr., M.R. Garey & D.S. Johnson. An application of bin-packing to multiprocessor scheduling. *SIAM J. Comput.* 7 (1978), 1–17.
6. E.G. Coffman, Jr. & R. Sethi. A generalized bound on LPT sequencing. *Revue Française d' Automatique Informatique, Recherche Operationnelle* 10 (1976), 17–25.
7. D.K. Friesen. Tighter bounds for multifit processor scheduling algorithm. *SIAM J. Comput.* 13 (1984), 170–181.
8. D.K. Friesen & M.A. Longston. Evaluation of a multifit-based scheduling algorithm. *J. Algorithms* 7 (1986), 35–59.
9. M.R. Garey and D.S. Johnson. Strong NP-completeness results: Motivation, examples and implications. *J. Assoc. Comput. Mach.* 25 (1978), 499–508.

10. M.R. Garey and D.S. Johnson. *Computers and Intractability: A Guide to the Theory of NP-Completeness*. Freeman, San Francisco, 1979.
11. M.R. Garey, D.S. Johnson and R. Sethi. The complexity of flowshop and jobshop scheduling. *Math. Oper. Res.* 1 (1976), 117–129.
12. T. Gonzalez & S. Sahni. Flowshop and jobshop schedules: Complexity and approximation. *Oper. Res.* 26 (1978), 36–52.
13. R.L. Graham. Bounds for certain multiprocessor anomalies. *Bell System Tech. J.* 45 (1966), 1563–1581.
14. R.L. Graham. Bounds on multiprocessing timing anomalies. *SIAM J. Appl. Math.* 17 (1969), 416–429.
15. J.N.D. Gupta. Two-stage, hybrid flowshop scheduling problem. *J. Opl. Res. Soc.* 39 (1988), 359–364.
16. D.S. Hochbaum & D.B. Shmoys. Using dual approximation algorithms for scheduling problems: Theoretical and practical results. *J. Assoc. Comput. Mach.* 34 (1987), 144–162.
17. J.A. Hoogeveen, J.K. Lenstra & B. Veltman. Minimizing makespan in a multiprocessor flow shop is strongly NP-hard. *Unpublished manuscript*, 1993. See also: B. Veltman. *Multiprocessor Scheduling with Communication Delays*. Ph.D. Thesis, CWI, Amsterdam, The Netherlands, 1993.
18. S.M. Johnson. Optimal two- and three-stage production schedules with setup times included. *Naval Res. Logist. Quart.* 1 (1954), 61–68.
19. M.A. Langston. Interstage transportation planning in the deterministic flow-shop environment. *Oper. Res.* 35 (1987), 556–564.
20. E.L. Lawler, J.K. Lenstra, A.H.G. Rinnooy Kan & D.B. Shmoys. Sequencing and scheduling: algorithms and complexity. *Handbooks in Oper. Res. and Management Sci., Vol. 4: Logistics of Production and Inventory* (S.C. Graves el al., Eds.) North-Holland, Amsterdam, 1993, 445–522.
21. R. McNaughton. Scheduling with deadlines and loss functions. *Management Sci.* 6 (1959), 1–12.
22. M. Nawaz, E.E. Enscore, Jr. & I. Ham. A heuristic algorithm for the m-machine n-job flowshop sequencing problem. *OMEGA* 11 (1983), 91–95.
23. E. Nowicki & C. Smutnicki. Worst-case analysis of an approximation algorithm for flow-shop scheduling. *Oper. Res. Lett.* 8 (1989), 171–177.
24. H. Röck & G. Schmidt. Machine aggregation heuristics in shop scheduling. *Methods Oper. Res.* 45 (1983), 303–314.
25. M.S. Salvador. A solution to a special class of flow shop scheduling problems. *Proceedings of Symposium on the Theory of Scheduling and its Applications*. Springer-Verlag, Berlin, 1973, 83–91.
26. D.B. Shmoys, C. Stein & J. Wein. Improved approximation algorithms for shop scheduling problems. *Proceedings of the 2nd ACM-SIAM Symposium on Discrete Algorithms* (1991), 148–157.
27. C. Sriskandarajah & S.P. Sethi. Scheduling algorithms for flexible flowshops: worst and average case performance. *European J. Opl. Res.* 43 (1989), 143–160.
28. E. Taillard. Some efficient heuristics methods for the flow shop sequencing problem. *European J. Opl. Res.* 47 (1990), 65–74.
29. D. Williamson, L.A. Hall, J.A. Hoogeveen, C.A.J. Hurkens, J.K. Lenstra, & D.B. Shmoys. Short shop schedules. *Unpublished manuscript*.
30. R.J. Wittrock. An adaptable scheduling algorithm for flexible flow lines. *Oper. Res.* 36 (1988), 445–453.
31. M. Yue. On the exact upper bound for the multifit processor scheduling algorithm. *Ann. Oper. Res.* 24 (1990), 233–259.

THE K-WALK POLYHEDRON

COLLETTE R. COULLARD*
Dept. of Industrial Engineering and Management Sciences

A. BRUCE GAMBLE
J.L. Kellogg Graduate School of Management

and

JIN LIU
Dept. of Industrial Engineering and Management Sciences
Northwestern University
Evanston, Illinois 60208-3118, USA

Abstract. Given a directed graph $G = (V, E)$, with distinguished nodes s and t, a *k-walk* from s to t is a walk with exactly k arcs. In this paper we consider polyhedral aspects of the problem of finding a minimum-weight k-walk from s to t. We describe an *extended* linear programming formulation, in which the number of inequalities and variables is polynomial in the size of G (for fixed k), and all the coefficients have values $0, -1, +1$. We apply the *projection* method on the extended formulation and give two *natural* linear programming formulations, each having one variable for each arc of G. These correspond to two *natural* polyhedra: the convex hull of the (s,t)-k-*walk* incidence vectors, and the dominant of that polytope. We focus on the dominant. We give three classes of facet constraints, showing that the dominant has an exponential number of facet constraints and that some facet constraints have coefficients as large as the size of V.

1. Introduction

A fundamental problem of polyhedral combinatorics is to find a linear inequality system whose solution set is a polyhedron specified by its vertices and extreme rays. For a combinatorial optimization problem, we normally use the incidence vector to represent a combinatorial object, such as a path, tree, matching, etc. Hence we could define the polyhedron associated with the combinatorial optimization problem to be the convex hull of the corresponding incidence vectors, and this is the polyhedron we want to describe by some linear inequality system. Usually the incidence vectors are indexed on some set that is part of the problem input (eg., the node-set or edge-set of a graph). In that case, the linear inequality system has been called a *natural formulation*. If instead the variables are indexed on sets derived from the problem input, the inequality system has been called an *extended formulation*. A formulation is called *compact* if its size is bounded by a polynomial in the problem input size.

The issues of extended versus natural formulations for combinatorial optimization problems have been the subject of several papers. Balas and Pulleyblank, [1983] and [1987], presented extended formulations for the perfectly matchable set polyhedra of bipartite and general graphs, respectively. Barahona and Mahjoub [1987] derived a compact extended formulation for the stable set polytope of series-parallel

* SUPPORTED BY NATIONAL SCIENCE FOUNDATION GRANT NO. DDM-91-96083.

Ding-Zhu Du and Jie Sun (eds.), Advances in Optimization and Approximation, 9–29.

graphs. Martin, Rardin and Campbell [1987] developed methods of obtaining compact extended formulations from dynamic programming. However, few successful conversions from extended formulations to natural formulations have been reported. The work of Ball, Liu, and Pulleyblank [1987] is very significant. They present a compact formulation for the two terminal Steiner tree polyhedron of directed graphs, and then obtain a natural formulation by using the projection method.

In this article we focus on extended and natural formulations for the *Shortest k-Walk Problem*. We assume familiarity with graphs, digraphs, and polyhedra. For a complete reference on polyhedral theory, see Schrijver [1986]. We begin with some preliminaries to establish notation.

A *digraph* is denoted by $G = (V, E)$, where V is the *node set* and E is the *arc set*. Arc $e = uv \in E$ joins its *tail* node u to its *head* node v. A *polyhedron* is the solution set of some finite set of linear inequalities in a finite number of variables. A *polytope* is a bounded polyhedron. Given a polytope P, the *dominant* of P, denoted $dom(P)$, is defined to be the set $Q + R^n_+$, where $R^n_+ = \{x \in R^n : x \geq 0\}$. Given a digraph $G = (V, E)$, and two nodes $s,t \in V$, an (s,t)-*walk* is defined to be a sequence of nodes $a_0, a_1, \ldots, a_n$ in V such that $s = a_0$, $a_n = t$, and there is an arc $e_i = a_{i-1}a_i \in E$ for $1 \leq i \leq n$. Here n is the *length* of the walk, a_i is the *i-th node* of the walk, and e_i is the *i-th arc* of the walk. An (s,t)-*k-walk* is a (s,t)-*walk* of length k. If the nodes a_i, $0 \leq i \leq n$, are distinct, then we have an (s,t)-*path*. Given arc weights $(c_e : e \in E)$ the total weight of a walk is the sum of the weightsof the arcs of the walk, where the weight of an arc is included as many times as the arc is used.

Now we state the *Shortest k-Walk Problem* $(SW_k P)$:

GIVEN A DIGRAPH G=(V,E) WITH DISTINGUISHED NODES s AND t AND WEIGHTS ON THE ARCS, FIND AN (s,t)-k-WALK HAVING MINIMUM TOTAL WEIGHT.

The *incidence vector* of a given (s,t)-k-walk is a vector x indexed on E, where for each arc e, x_e is the number of times the (s,t)-k-*walk* uses arc e. The *k-walk polytope*, P, is the convex hull of all the incidence vectors of (s,t)-k-walks.

In Section 2 we give an efficient algorithm for $(SW_k P)$. This leads to the problem of finding a linear description for the k-walk polytope, P, as well as its dominant, P_d. In Section 2 we give an extended linear programming formulation for $(SW_k P)$, which provides a polytope P', with extra variables. The natural polytope P is a projection of the extended polytope P'. The dominant P_d is a projection of a polyhedron obtained by modifying P' slightly. Explicitly describing P or P_d involves explicitly describing how to project P' onto just the variables of interest (indexed on E), that is, how to project away the extra variables. This involves describing the so-called *projection cone*. We focus on the dominant, P_d, and thus focus on its projection cone. The details of this *projection method* are given in Section 3. We characterize, in a certain sense, the extreme rays of the projection cone for P_d in Sections 4 and 5, and we state a complete finite natural formulation for P_d in Section 5. This description is not tight, in the sense that we have not given a minimal defining list of inequalities; moreover, some of the coefficients are quite large. However, we can say that such a minimal list would be exponential in length, and some of the coefficients would be fairly large. Indeed, in Section 6 we give three interesting classes of constraints and prove they are facet constraints for complete graphs. These classes show there are, in general, exponentially many facet constraints; they also show some of the

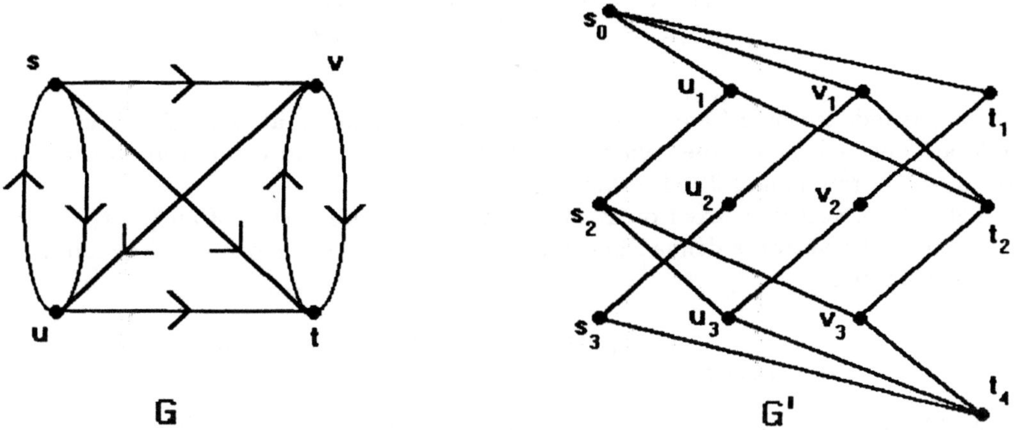

Fig. 1. Example of G and G' with $k = 4$

coefficients can be as large as $|V|$.

2. Polynomial Algorithm and Extended Formulations

In this section we give a polynomial algorithm to solve the Shortest k-Walk Problem. In addition, we give the extended linear programming formulation for this problem that is polynomial in k and the size of the input graph.

Let $G = (V, E)$ be a graph with distinguished nodes s and t. We define an auxiliary graph G' as follows. Let $V^i = \{v_i : v \in V$ is the i-th node of some (s,t)-*k-walk*$\}$ for $0 \leq i \leq k$, and $E^i = \{e' = u_{i-1}v_i : u_{i-1} \in V^{i-1}, v_i \in V^i, e = uv \in E$ is the i-th arc of some (s,t)-k-walk.$\}$ for $1 \leq i \leq k$, V^i is called the *i-th level node set* and E^i is called the *i-th level arc set*. Now $G' = (V', E')$, where $V' = \cup_{i=0}^{k} V^i$, $E' = \cup_{i=1}^{k} E^i$. Notice that $V^0 = \{s_0\}$ and $V^k = \{t_k\}$. See Figure 1 for an example.

For each edge $e' = u_{i-1}v_i \in E'$ (where $e = (u,v) \in E$), let $c_{e'} = c_e$. Now it is straightforward to see that that solving SW_kP on G is equivalent to finding the shortest (s_0, t_k)-*path* on G'. Note that the size of graph G' is a polynomial function of the size of graph G as long as k is so, in which case we have an polynomial algorithm to solve SW_kP by solving a shortest path problem on G'. Furthermore because the multilevel structure of graph G', we can solve the shortest path problem on G' dynamically. We have the following Proposition:

Proposition 2.1 *SW_kP can be solved in $O(k|E|)$ time.*

The shortest dipath problem on the auxiliary graph $G' = (V', E')$ can be formulated as a linear programming problem as follows:

$$\text{Minimize} \quad c'z$$
$$\text{subject to}: \quad Nz = \begin{cases} 1 & \text{if } v = s_0, \\ -1 & \text{if } v = t_k, \\ 0 & \text{otherwise,} \end{cases}$$
$$z \geq 0,$$

where z is indexed on E' and N is the node-arc incidence matrix of G'. (The rows of N are indexed on V', the columns are indexed on E', and column uv has $+1$ in row u, -1 in row v, and 0's elsewhere.)

Let x be a vector indexed on E. If z is the incidence vector of an (s_0, t_k)-path in G', then the vector x defined below is the incidence vector of the corresponding *(s,t)-k*-walk in G.

$$\sum_{(i:\ u_i v_{i+1} \in E^i)} z_{u_i v_{i+1}} = x_e \quad \text{for every arc } e = uv \in E.$$

Moreover, $c'z = cx$. Combining these facts, we get an extended linear programming formulation for $SW_k P$:

$$\text{Minimize} \quad cx$$
$$\text{subject to}: \quad Nz = \begin{cases} 1 & \text{if } v = s_0, \\ -1 & \text{if } v = t_k, \\ 0 & \text{otherwise,} \end{cases} \tag{1}$$
$$z \geq 0$$
$$\sum_{(i:\ u_i v_{i+1} \in E^i)} z_{u_i v_{i+1}} = x_e \quad \text{for every arc } e = uv \in E. \tag{2}$$

Alternately, we can write it as:

$$\text{Minimize}\{cx : (z, x) \in P', \text{ for some } z\},$$

where polytope

$$P' = \{(z, x) \geq 0 : (z, x) \text{ satisfies (1) and (2)}\}.$$

(Note that P' is indeed a polytope, since G' is acyclic.) System (1) corresponds to solving the shortest (s_0, t_k)-path problem on G'. Equations (2) project the solution (z, x) into the space of the x variables, X, indexed on E. Let

$$P = \{x \in R^E : \exists z \ s.t. \ (z, x) \in P'\}.$$

The polytope P is the *projection* of P' into X. This polytope P is the convex hull of incidence vectors of *(s,t)-k*-walks of G; P is the natural polytope we seek to describe via a linear system.

Replacing the equalities in (2) by "$\leq$", we obtain

$$\sum_{(i:\ u_i v_{i+1} \in E^i)} z_{u_i v_{i+1}} \leq x_e \quad \text{for every arc } e = uv \in E. \tag{3}$$

Let polyhedron

$$P'_d = \{(z,x) \geq 0 : (z,x) \text{ satisfies (1) and (3)}\}.$$

Then the set of feasible x,

$$P_d = \{x : \exists\ z\ s.t.\ (z,x) \in P'_d\}$$

is the dominant of polyhedron P; that is

$$P_d = dom(P) = \{x \in R^E :\ \exists\ x' \in P\ s.t.\ x' \leq x\}.$$

Since for nonnegative objective functions, optimizing over $dom(P)$ is equivalent to optimizing over P, the polyhedron P_d is of interest. Because the technical work is simpler, we focus primarily on the dominant P_d.

Note that both P and P_d are presented in terms of the z variables or by P' and P'_d. Now we want a natural formulation in only the x variables to describe P and P_d. This is obtained using the projection method discussed in the next section.

3. Projection Method

The following Projection Theorem was introduced by Balas and Pulleyblank [1983] as a method for obtaining a formulation of a polyhedron associated with a combinatorial optimization problem. Ball, Liu, Pulleyblank [1987] also used it to get a natural formulation for the two-terminal steiner tree polyhedron.

Theorem 3.1 (Balas and Pulleyblank [1983]) *Suppose that a polyhedron $\tilde{P}$ has an extended formulation*

$$C = \{x \in R^m : \exists z \in R^n \text{ such that } Ax + Bz = b,\ z \geq 0\}. \tag{4}$$

Then $dx \leq d_0$ is a valid inequality constraint for $\tilde{P}$ if and only if $d = wA$ and $d_0 = wb$ for some member w in the projection cone $W = \{w : wB \geq 0\}$. Therefore if H is a generator set of W, then

$$hAx \leq hb, \text{ for every } h \in H \tag{5}$$

is a natural formulation for $\tilde{P}$.

For convenience we write equations (1), (2) and (3) in the following matrix forms:

$$\begin{aligned} Nz &= b \\ Dz &= x \\ Dz &\leq x \end{aligned}$$

Now we use Theorem 3.1 to get natural formulations for P and P_d.

Theorem 3.2 *Let $M' = \{(w,u) : wN + uD \geq 0\}$ and $M = \{(w,u) : wN + uD \geq 0,\ u \geq 0\}$. If H (or H') is a generator set of M (or M'), then*

$$wb + ux \geq 0 \text{ for every } (w,u) \in H \text{ (or } H'\text{)} \tag{6}$$

is a natural formulation for P_d (or P).

Proof. We prove the theorem for P_d. The proof for P is similar. By adding surplus variables y to constraints (3), constraints (1) and (3) can be rewritten as

$$\begin{pmatrix} N & 0 & 0 \\ D & -I & I \end{pmatrix} \begin{pmatrix} z \\ x \\ y \end{pmatrix} = \begin{pmatrix} b \\ 0 \end{pmatrix},$$

where $(y, z) \geq 0$. Applying Theorem 3.1, we have

$$W = \{(w,u) : (w,u) \begin{pmatrix} N & 0 \\ D & I \end{pmatrix} \geq 0\},$$

or

$$W = \{(w,u) : u \geq 0, wN + uD \geq 0\} = M.$$

Thus, a complete natural formulation for P_d is

$$P_d = \{x : wb + ux \geq 0, \text{for every } (\mathrm{w}, \mathrm{u}) \in H\}. \square$$

From now on we investigate the projection cone M, for the dominant P_d.

We let w_{ia} denote the w value at node $a_i \in V^i$. Because $wb = w_{0s} - w_{kt}$, inequality (6) can be written as $ux \geq w_{kt} - w_{0s}$. This constraint depends only on the difference of w_{0s} and w_{kt}. Since the rank of N is one less than the number of nodes of G', one equation of system 1 is redundant. Thus, we can redefine

$$M = \{(w,u) : u \geq 0, wN + uD \geq 0, w_{0s} = 0\}. \tag{7}$$

For a generator set H of M, we now have

$$P_d = \{x : ux \geq w_{kt}, \text{ for every } (w,u) \in H\}.$$

Since G' is connected, it follows that *lin.space*$(M) = \{0\}$, and therefore M is a pointed cone (when the variable w_{0s} is projected out). It follows that M is generated by its extreme rays. Next we investigate the set of extreme rays of M and the constraints of P_d they induce. In particular, we want to characterize those extreme rays that induce facet constraints.

4. Projection Cone M

In this section, we give several propositions to illustrate the features of the extreme rays of M. Here we follow closely the approach of Ball, Liu and Pulleyblank [1987].

Proposition 4.1 *If (w,u) is an extreme ray of M and $w = 0$, then there exists an arc $e_0 \in E$ such that*

$$u_e = \begin{cases} 1 & \text{if } e = e_0, \\ 0 & \text{otherwise.} \end{cases}$$

These extreme rays induce the nonnegativity constraints $x_e \geq 0$.

Proposition 4.2 *If (w,u) is an extreme ray of M and $w \neq 0$, then for every arc $e = ab \in E$, we have*

$$u_{ab} = \max_{i:a_i b_i \in E^i} \{0, w_{(i+1)b} - w_{ia}\} \tag{8}$$

Proof. By (4) we can write M as:

$$\begin{aligned} M \;=\; & \{(w,u) : w_{ia} - w_{(i+1)b} + u_{ab} \geq 0, u_{ab} \geq 0, w_{0s} = 0; \\ & \forall\, e = ab \in E, \forall\, 0 \leq i \leq k-1 \;s.t.\; a_i b_{i+1} \in E^i\}. \end{aligned} \tag{9}$$

Thus, $u_{ab} \geq \max_{a_i b_{i+1} \in E^i, 0 \leq i \leq k-1}\{0, w_{(i+1)b} - w_{(ia)}\}$, for every arc $ab \in E$. If there exists an arc $e = ab \in E$, such that $u_{ab} > \max_{a_i b_{i+1} \in E^i, 1 \leq i \leq k-1}\{0, w_{(i+1)b} - w_{ia}\}$, then we could decrease u_{ab} by a positive amount until either $u_{ab} = 0$ or there exists an index i such that $w_{ia} - w_{(i+1)b} + u_{ab} = 0$. This contradicts the assumption that (w,u) is an extreme ray.□

Given $(w,u) \in M$, we define $F(w,u)$ to be the face of P_d induced by constraint $ux \leq w_{kt}$. We say a k-walk *belongs to that face*, or $p^k \in F(w,u)$, if the incidence vector of p^k belongs to $F(w,u)$.

Proposition 4.3 *Let $(w,u) \in M$, and let $p^k = a(0), a(1), \ldots, a(k)$ be a k-walk of G, where $a(i) \in V$, for $0 \leq i \leq k$, $a(0) = s$, and $a(k) = t$. If $p^k \in F(w,u)$, then we have*

$$w_{ia(i)} - w_{(i+1)a(i+1)} + u_{a(i)a(i+1)} = 0 \;\; for \;\; 0 \leq i \leq k-1,$$

and $w_{ia(i)} \geq 0$ for $0 \leq i \leq k$.

Proof. Let x_{p^k} be the incidence vector of k-walk p^k. Since $p^k \in F(w,u)$, we have $ux_{p^k} = w_{kt}$ or $\sum_{i=0}^{k-1} u_{a(i)a(i+1)} = w_{kt}$. On the other hand, from $(w,u) \in M$ we have

$$w_{ia(i)} - w_{(i+1)a(i+1)} + u_{a(i)a(i+1)} \geq 0 \quad for \quad 0 \leq i \leq k-1. \tag{10}$$

Summing these inequalities over i we have $\sum_{i=0}^{k-1} u_{a(i)a(i+1)} \geq w_{kt} - w_{0s} = w_{kt}$. Thus equality must hold for every inequality in (10). Finally, summing these equations from 0 to $i-1$ we get $w_{ia(i)} = \sum_{j=1}^{j=i-1} u_{a(j)a(j+1)} \geq 0$.□

Proposition 4.3 says that if a k-walk is on face $F(w,u)$, then the cone constraint for each of its corresponding arcs in G' must be tight on (w,u); and every node on G' that is used by this walk has a nonnegative w value. Indeed, if we restrict w to be nonnegative, then we still get all facet inducing constraints of P_d:

Proposition 4.4 *For each $(w,u) \in M$ there exists $(\tilde{w},u) \in M$ such that $\tilde{w} \geq 0$ and $\tilde{w}_{kt} \geq w_{kt}$. Thus, constraint $ux \geq w_{kt}$ is dominated by constraint $ux \geq \tilde{w}_{kt}$.*

Proof. For each node $a_i \in V'$, let $\tilde{w}_{ia} = \max\{0, w_{ia}\}$. Now for any constraint in M: $w_{ia} - w_{(i+1)b} + u_{ab} \geq 0$ implies $\max\{0, w_{ia}\} - w_{(i+1)b} + u_{ab} \geq 0$, which implies $\max\{0, w_{ia}\} - \max\{0, w_{(i+1)b}\} + u_{ab} \geq 0$. Thus, $\tilde{w}_{ia} - \tilde{w}_{(i+1)b} + u_{ab} \geq 0$, implying $(\tilde{w},u) \in M$, as desired. Note that $\tilde{w}_{kt} \geq w_{kt}$ by the definition of $\tilde{w}$.□

By Proposition 4.4, restricting w to be nonnegative in the definition of M still yields a complete linear system for P_d. Therefore, we redefine M as:

$$M = \{(w,u) : w \geq 0, \; u \geq 0, wN + uD \geq 0, w_{0s} = 0\} \tag{11}$$

Note that Propositions 4.1, 4.2, and 4.3 all hold for M as it is now defined.

5. More on Projection Cone M

In this section we show how to test whether a given (w,u) is an extreme ray of a given M. (This is an alternative to the straightforward checking via the definition of M.) Let $S(w,u)$ be the set of constraints of M that (w,u) satisfies with equality. (Set $S(w,u)$ is called the *tight set* of (w,u). We say (w,u) *is dominated by* $(\tilde{w},\tilde{u})$ if $S(w,u) \subseteq S(\tilde{w},\tilde{u})$. The following proposition characterizes this dominance.

Proposition 5.1 *Let $(w,u) \in M$. Then (w,u) is dominated by extreme ray $(\tilde{w},\tilde{u}) \in M$ if and only if the following three conditions are met:*

(i) *If $u_{ab} = 0$, then $\tilde{w}_{ia} - \tilde{w}_{(i+1)b} \geq 0$, for $a_i b_{i+1} \in E^i$.*

(ii) *If $u_{ab} = w_{(i+1)b} - w_{ia}$, for some i, then $\tilde{w}_{(i+1)b} - \tilde{w}_{ia} \geq 0$, and $\tilde{w}_{(i+1)b} - \tilde{w}_{ia} \geq \tilde{w}_{(j+1)b} - \tilde{w}_{ja}$, for $a_j b_{j+1} \in E^j$.*

(iii) *If $w_{ia} = 0$, then $\tilde{w}_{ia} = 0$.*

Proof.
("$\Rightarrow$") Assume that $S(w,u) \subseteq S(\tilde{w},\tilde{u})$. Then we have for **(i)**: Since $u_{ab} = 0$, we have that $\tilde{u}_{ab} = 0$, and thus $\tilde{w}_{ja} - \tilde{w}_{(j+1)b} \geq 0$, for $a_j b_{j+1} \in E^j$, as desired. For **(ii)** : Since $u_{ab} = w_{(i+1)b} - w_{ia}$, we have that $\tilde{w}_{(i+1)b} - \tilde{w}_{ia} = \tilde{u}_{ab}$. Now since $\tilde{u}_{ab} \geq \tilde{w}_{(j+1)b} - \tilde{w}_{ja}$, for $a_j b_{j+1} \in E^j$, we have that $\tilde{w}_{(i+1)b} - \tilde{w}_{ia} \geq \tilde{w}_{(j+1)b} - \tilde{w}_{ja}$, for $a_j b_{j+1} \in E^j$, as desired. Finally, **(iii)** is immediate from $S(w,u) \subseteq S(\tilde{w},\tilde{u})$.
("$\Leftarrow$") Assume that conditions **(i)**,**(ii)**,**(iii)** hold. Now suppose $u_{ab} = 0$. Then by **(i)** we have $\tilde{w}_{ja} - \tilde{w}_{(j+1)b} \geq 0$, for $a_j b_{j+1} \in E^j$. By Proposition 4.2, $\tilde{u}_{ab} = 0$, as desired. Suppose $w_{ia} - w_{(i+1)b} + u_{ab} = 0$. Then by **(ii)** we have $\tilde{w}_{(i+1)b} - \tilde{w}_{ia} \geq 0$ and $\tilde{w}_{(i+1)b} - \tilde{w}_{ia} \geq \tilde{w}_{(j+1)b} - \tilde{w}_{ja}$ for $a_j b_{j+1} \in E^j$. Thus, by Proposition 4.2, $\tilde{w}_{ia} - \tilde{w}_{(i+1)b} + \tilde{u}_{ab} = 0$, as desired. Finally suppose $w_{ia} = 0$, then by **(iii)** $\tilde{w}_{ia} = 0$.□

Next we define a new cone Λ; there is one cone for each point in M. We will characterize the extreme rays of M in terms of their associated cones Λ. Given $(w,u) \in M$, $\Lambda(w,u)$ is a set of vectors λ indexed on the nodes V' of auxiliary digraph G', and is the solution set of the following linear system:

$\lambda_{ia}w_{ia} - \lambda_{(i+1)b}w_{(i+1)b}$	$\geq$	$0,$	for all i and all $a_i b_{i+1} \in E^i$ with $u_{ab} = 0$;
$\lambda_{(i+1)b}w_{(i+1)b} - \lambda_{ia}w_{ia}$	$\geq$	$0,$	for all i and all $a_i b_{i+1} \in E^i$ with $w_{ia} - w_{(i+1)b} + u_{ab} = 0$;
$\lambda_{(i+1)b}w_{(i+1)b} - \lambda_{ia}w_{ia}$	$\geq$	$\lambda_{(j+1)b}w_{(j+1)b} - \lambda_{ja}w_{ja},$	for all i and all $a_i b_{i+1} \in E^i$, and all j with $a_j b_{j+1} \in E^j$, and $w_{ia} - w_{(i+1)b} + u_{ab} = 0$;
λ_{ia}	$=$	$0,$	for all i, a with $w_{ia} = 0$;
λ_{ia}	$\geq$	$0,$	for all $a_i \in V'$.

By the definition of Λ we have the following proposition.

Proposition 5.2 *$S(w,u) \subseteq S(\tilde{w},\tilde{u})$ if and only if there exists a $\lambda \in \Lambda(w,u)$ such that $\tilde{w}_{ia} = \lambda_{ia} w_{ia}$ for all $a_i \in V'$.*

Proof.
("$\Leftarrow$") By the definition of Λ, it follows that $\tilde{w}$ satisfies the three conditions of Proposition 5.1.
("$\Rightarrow$") Assume $S(w,u) \subseteq S(\tilde{w},\tilde{u})$. Then all three conditions of Proposition 5.1 hold. It follows that letting $\lambda_{ia} = \begin{cases} 0 & \text{if } w_{ia} = 0, \\ \tilde{w}_{ia}/w_{ia} & \text{if } w_{ia} > 0, \end{cases}$
we have that $\lambda \in \Lambda(w,u)$.□

Given $(w,u) \in M$, let $I_{ia}(w) = \begin{cases} 0 & \text{if } w_{ia} = 0, \\ 1 & \text{if } w_{ia} > 0. \end{cases}$

Obviously $I(w) \in \Lambda(w,u)$. We call $I(w)$ the *trivial member* of $\Lambda(w,u)$. Now we can write $\Lambda(w,u)$ as $\Lambda(w,u) = \{\lambda \geq 0 : \Lambda^{=}\lambda = 0, \Lambda^{\neq}\lambda \geq 0\}$, where $\Lambda^{=}$ are all the constraints which hold equality when $\lambda = I(w)$ and rest constraints are $\Lambda^{\neq}\lambda \geq 0$.

Proposition 5.3 *$(w,u) \in M$ is an extreme ray of M if and only if emery member of Λ is a scalar multiple of $I(w)$.*

Proof.
("$\Leftarrow$") By assumption and Proposition 5.2, every member of M that dominates (w,u) is a scalar multiple of (w,u). Thus, (w,u) is extreme.
("$\Rightarrow$")

Let λ be a member of Λ. By Proposition 5.2, the vector $(\tilde{w},\tilde{u})$ obtained by letting $\tilde{w}_{ia} = \lambda_{ia} w_{ia}$, for all $a_i \in V'$, dominates (w,u). Since (w,u) is extreme, every member of M that dominates (w,u) is a scalar multiple of (w,u). Thus, λ is the trivial member of $\Lambda(w,u)$. □

By Proposition 5.3 we have

Proposition 5.4 *Vector (w,u) in M is an extreme ray if and only if rank($\Lambda_{=}$) is equal to one less than the number of nodes of G'.*

We can use Proposition 5.3 and Proposition 5.4 to check whether a given (w,u) in M is an extreme ray.

We are now ready to give a complete natural formulation for P_d. Let E be an integer matrix and W the cone $\{Ex \geq 0\}$. If U is the maximum determinant of a square submatrix of E, then (by Cramer's Rule) there exists an integral generating set for W such that no component of any member of this generating set is greater than U^2. Consider the projection cone $M = \{(w,u) : wN + uD \geq 0, w \geq 0, u \geq 0\}$. All the entries of matrix $\begin{pmatrix} N & I & 0 \\ D & 0 & I \end{pmatrix}$ are $0, +1, -1$; moreover, no column has more than three nonzero entries. Thus, all its subdeterminants are bounded in absolute value by $3^{|V'|+|E|}$. An interesting open question is whether or not this bound can be improved.

Now we can give a finite description of P_d based on Λ and this determinant bound.

Theorem 5.5 *Given a digraph $G = (V, E)$ with distinguished nodes s and t, the dominant P_d of the convex hull of incidence vectors of $(r,t)-k$ walks of G is described by the following system:*

$$ux \geq w_{kr}, \text{ for every } (w,u) \text{ in } H,$$

where H consists of the vectors (w,u) such that

- *Vector w is indexed on the nodes V' of auxiliary digraph G', and each component satisfies $0 \leq w_{ia} \leq 3^{|V'|+|E|}$.*
- *Vector u is indexed on E, and satisfies, for each $ab \in E$:*

$$u_{ab} = \max\{0, \max\{w_{(i+1)b} - w_{ia} : 0 \leq i \leq k, a_i b_{i+1} \in E'\}\}.$$

- *Every member of the cone $\Lambda(w,u)$ is a scalar multiple of the trivial member.*

Note that the number of constraints in this description is exponential in the size of the graph G. However we will see from the following sections that P_d does have exponentially many facet constraints. Theorem 5.5 presents constraints with exponentially large coefficients. We have no reason to believe a minimal system requires such large coefficients, but we will show that P_d does have facet constraints with coefficients as large as $|V|$.

Theorem 5.5 leaves open some questions:

- Given a digraph $G = (V, E)$, can a linear description for P_d be obtained in time polynomial in that description?
- What is the tight bound on the size of the coefficients?
- What is the structure of the set of tight arcs

$$\{e = a_i b_{i+1} \in E' : w_{ia} - w_{(i+1)b} + u_{ab} = 0\}.$$

  as a subgraph of G', for a vector (w,u) that is an extreme point of M? (or facet-inducing for P_d?) We know this set contains a spanning tree of G', but can we give a structural characterization?

These questions remain for future work.

6. Facet Constraints

We now turn our attention to determining which of the constraints we have identified are facet inducing. The following proposition provides a useful tool for identifying facet inducing inequalities. Given $(w,u) \in M$, we will let $U_+ = \{e \in E : u_e > 0\}$ and $U_0 = \{e \in E : u_e = 0\}$. For each $x \in P$ we define the *u-positive vector of x* to be the vector $x_+ = (x_e : e \in U_+)$ and the *u-zero vector of x* to be the vector $x_0 = (x_e : e \in U_0)$. We can rewrite $x = (x_+, x_0)$. Also, for all $e \in E$, let I_e be the unit vector (indexed on E) corresponding to e. (That is, $I_e(e) = 1$ and $I_e(a) = 0$ for all $a \neq e$.) Recall that given $(w,u) \in P_d$, $F(w,u)$ is the face of P_d induced by constraint $ux \leq w_{kt}$.

Proposition 6.1 *Suppose $p^1, \ldots, p^m$ are all the k-walks on $F(w,u)$, and $w_{kt} > 0$; then $ux \geq w_{kt}$ is a facet constraint if and only if rank$(p^1_+, \ldots, p^m_+) = | U_+ |$.*

Proof.
("$\Leftarrow$") Assume that $\text{rank}(p_+^1, \ldots, p_+^m) = |U_+|$. Let $p^e = p^1 + I_e$ for each arc $e \in U_0$. Then we have $p^e \in F(w, u)$. It is easy to check that there are $(|U_+| + |U_0|)$ linearly independent vectors in set $\{p^1, \ldots, p^m\} \cup \{p^e : e \in U_0\}$; therefore $ux \geq w_{kt}$ is a facet constraint.
("$\Rightarrow$") Assume that $ux \geq w_t$ is a facet constraint. We can write

$$F(w, u) = \{\sum_{i=1}^{m} \lambda_i p^i + \sum_{e \in U_0} \mu_e I_e \ : \ \sum_{i=1}^{m} \lambda_i = 1, \lambda \geq 0, \mu \geq 0\}.$$

Let $\{q^j : 1 \leq j \leq (|U_+| + |U_0|)\}$ be a set of linearly independent vectors from the set $F(w, u)$, and we write $q^j = \sum_{i=1}^{m} \lambda_i^j p^i + \sum_{e \in U_0} \mu_e^j I_e$, as above. Consider the set of independent vectors $I = \{p^e : e \in U_0\} \cup p^1$. By the basis substitution theory of linear algebra, we know that there are $(|U_+| - 1)$ many vectors from $\{q^j : 1 \leq j \leq (|U_+| + |U_0|)\}$, say they are $\{q^j : 1 \leq j \leq |U_+| - 1\}$, which, together with I, are independent. By the special structure of I, it follows that $\{q_+^j : 1 \leq j \leq |U_+| - 1\} \cup \{p_+^1\}$ are linearly independent vectors; since each of these vectors is spanned by the set $\{p_+^1, \ldots, p_+^m\}$, it follows that that $\text{rank}(p_+^1, \ldots, p_+^m) = |U_+|$.□

We are now ready to give some classes of facet constraints. From now on we will assume that G is a complete directed graph.

6.1. Bottom Constraints

The first class of facets to be considered we will refer to as *bottom facets*. Given $S \subseteq V$ (with $s \in S$), we let

$$b_S = \begin{cases} k & \text{if } t \in S, \\ k-1 & \text{if } t \notin S. \end{cases}$$

The bottom constraint corresponding to S is:

$$2 \sum_{(i \notin S, j \in S)} x_{ij} + \sum_{(i,j \in S \text{ or } i,j \notin S)} x_{ij} \geq b_S. \tag{12}$$

Figure 2 displays the coefficients of a bottom constraint.

Now we want to show the constraints given by 12 are facet constraints. We first show that a given facet inducing bottom constraint for a small graph can be extended to a facet inducing bottom constraint on a larger graph by adding nodes. Proposition 6.2

Proposition 6.2 *Let $G = (V, E)$ be a complete directed graph. Suppose we are given $k \geq 4$ and $S \subseteq V$ satisfying $min\{|S|, |V - S|\} \geq 2, s \in S, t \notin S$. Suppose that the constraint*

$$2(\sum_{(i \notin S, j \in S)} x_{ij}) + \sum_{(i,j \in S \text{ or } i,j \notin S)} x_{ij} \geq k - 1 \tag{13}$$

is facet inducing for P_d. Then for the complete directed graph $\tilde{G} = (\tilde{V}, \tilde{E})$ where $\tilde{V} = V \cup \{r\}$, we have that

$$2(\sum_{(i \notin \tilde{S}, j \in \tilde{S})} x_{ij}) + \sum_{(i,j \in \tilde{S} \text{ or } i,j \notin \tilde{S})} x_{ij} \geq k - 1 \tag{14}$$

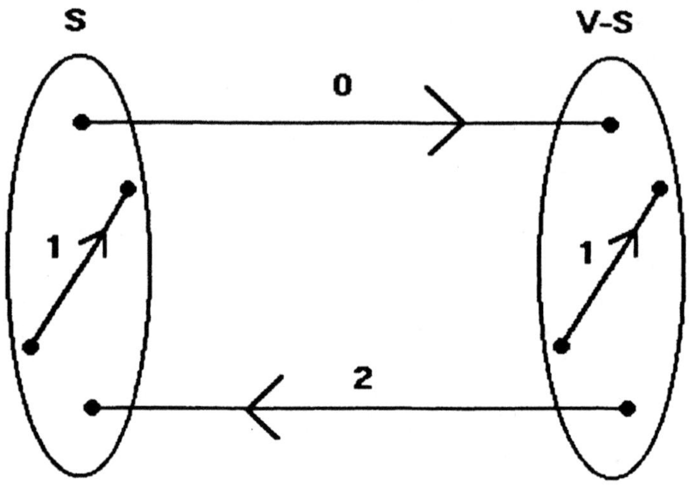

Fig. 2. Coefficients of a Bottom Constraint

is a facet inducing constraint (for P_d of $\tilde{G}$), where $\tilde{S} = S$.

Proof. Assume that constraint (13) is a facet inducing constraint for P_d with respect to G. By Proposition 6.1 there exist a sequence of k-walks $\{p^i : 1 \leq i \leq | U_+ |\}$ such that $\{p^i_+ : 1 \leq i \leq | U_+ |\}$ is a set of independent vectors. By adding the new node r to V, we get the following three types of new arcs in $\tilde{G}$ with nonzero coefficients. They are

type 1: arc $e = rv,\ v \in V - S$,

type 2: arc $e = rv,\ v \in S$.

type 3: arc $e = ur,\ u \notin V - S$,

It is straightforward to check that (since $k \geq 4$) for each new arc e of type 1 or 2, there is a k-walk p^e in $\tilde{G}$ that uses arc e exactly once and does not use any other new arcs; in addition, for each type 3 new arc e, there is a k-walk p^e that uses e and no other type 3 new arc. The vectors $\{p^i_+ : 1 \leq i \leq | U_+ |\}$ (with respect to $\tilde{G}$) and vectors $\{p^e_+ : e \text{ is a new arc }\}$ are independent, and so by Proposition 6.1, (14) is a facet constraint.□

Proposition 6.3 *Let G be the complete directed graph on three nodes and let $k = 2m$ for some $m \geq 2$. Then the constraint*

$$2(x_{21} + x_{31}) + (x_{23} + x_{32}) \geq 2m - 1 \tag{15}$$

is a facet inducing constraint for P_d. (Note that this constraint is the bottom constraint where $S = \{1\}$.)

Proof. There are four variables having nonzero coefficients, so by Proposition 6.1 it suffices to find 4 k-walks such that their u-positive vectors are independent. Here are these walks:

$p^1 : s = 1 \rightarrow 2 \rightarrow 3 \rightarrow 2 \rightarrow 3 \rightarrow 2 \rightarrow \ldots \rightarrow 3 \rightarrow 2 \rightarrow 3$

$p^1_+ = (0, 0, m, m-1)$

$p^2 : s = 1 \rightarrow 3 \rightarrow 2 \rightarrow 3 \rightarrow 2 \rightarrow 3 \rightarrow \ldots \rightarrow 3 \rightarrow 2 \rightarrow 1 \rightarrow 3$

$p^2_+ = (1, 0, m-2, m-1)$

$p^3 : s = 1 \rightarrow 2 \rightarrow 1 \rightarrow 2 \rightarrow 1 \rightarrow 2 \rightarrow \ldots \rightarrow 1 \rightarrow 2 \rightarrow 3$

$p^3_+ = (m-1, 0, 1, 0)$

$p^4 : s = 1 \rightarrow 3 \rightarrow 1 \rightarrow 3 \rightarrow 1 \rightarrow 3 \rightarrow \ldots \rightarrow 1 \rightarrow 3 \rightarrow 1 \rightarrow 2 \rightarrow 3$

$p^4_+ = (1, m-1, 1, 0)$

It is not difficult to verify that $\{p^i_+ : 1 \leq i \leq 4\}$ are independent.□

One implication of the above propositions is that P_d has exponentially many facets. Another property of these bottom constraints is that every k-walk in G satisfies them with equality. That is, each facet induced by a bottom constraint contains all extreme points of P_d. In fact, as we next show, these are the only such facets.

Proposition 6.4 *Let $G = (V, E)$ be a complete directed graph and let (w, u) be an extreme ray of M such that the constraint $ux \geq w_{kt}$ is satisfied with equality by all k-walks of G. Then $ux \geq w_{kt}$ must be a bottom constraint.*

Proof. For the given (w, u), the elements of w correspond to the nodes of the auxiliary graph G' as shown in Figure 3. For the sake of clarity, the arcs of G' are not shown in this figure. We claim that there exists a positive number α such that for node $s = 1$, and for each $0 \leq j \leq k-1$, we have that $w_{j1} = j\alpha$; moreover, for each node i, $2 \leq i \leq n$, we have, for $1 \leq j \leq k$, $w_{ji} = (j-1)\alpha$ or $j\alpha$. Given the claim, if we let $S = 1 \cup \{i : w_{1i} = \alpha\}$, then we get the desired constraint.

To prove the claim, we consider only three nodes node $s = 1$, node 2, and node $t = n$. (The argument for arbitrary node i is similar.) Because every k-walk is on $F(w, u)$, Proposition 4.3 implies every arc of G' is tight.

For arc $e = (12)$ we have

$$
\begin{aligned}
w_{12} - w_{01} &= w_{32} - w_{21} \\
&= w_{42} - w_{31} \\
&\;\;\vdots \\
&= w_{(k-2)2} - w_{(k-3)1} \\
&= w_{(k-1)2} - w_{(k-2)1};
\end{aligned}
$$

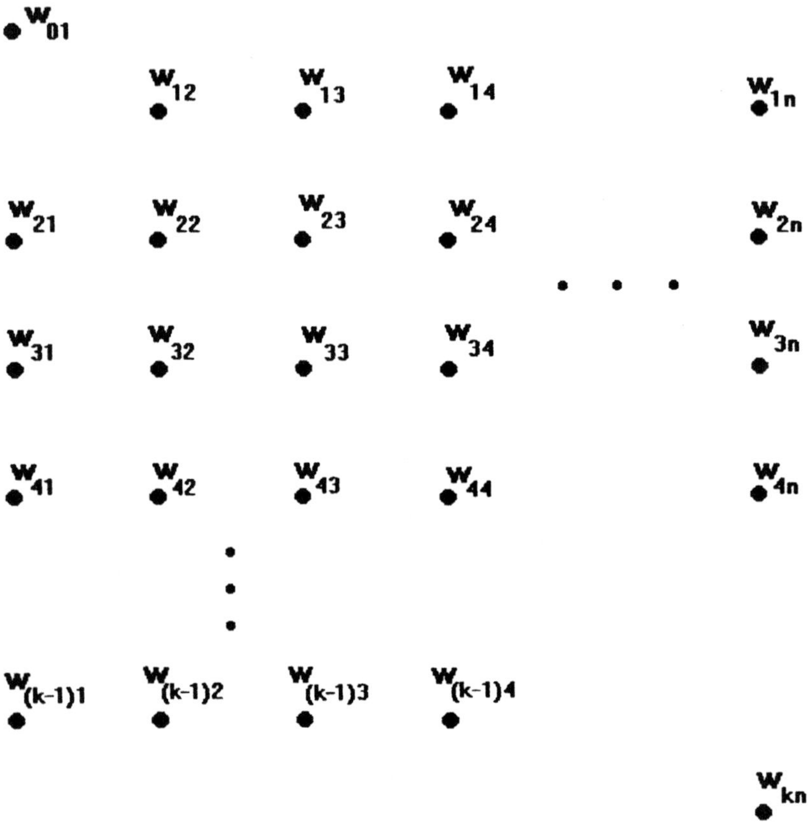

Fig. 3. The nodes of G'

so we have

$$\begin{aligned} w_{42} - w_{32} &= w_{31} - w_{21} \\ w_{52} - w_{42} &= w_{41} - w_{31} \\ &\vdots \\ w_{(k-2)2} - w_{(k-3)2} &= w_{(k-3)1} - w_{(k-4)1} \\ w_{(k-1)2} - w_{(k-2)2} &= w_{(k-2)1} - w_{(k-3)1}. \end{aligned} \tag{16}$$

For arc $e = (21)$ we have

$$\begin{aligned} w_{21} - w_{12} &= w_{31} - w_{22} \\ &= w_{41} - w_{32} \end{aligned}$$

$$\begin{aligned} &\vdots \\ &= w_{(k-2)1} - w_{(k-3)2} \\ &= w_{(k-1)1} - w_{(k-2)2}; \end{aligned}$$

so we have

$$\begin{aligned} w_{31} - w_{21} &= w_{22} - w_{12} \\ w_{41} - w_{31} &= w_{32} - w_{22} \\ &\vdots \\ w_{(k-2)1} - w_{(k-3)1} &= w_{(k-3)2} - w_{(k-4)2} \\ w_{(k-1)1} - w_{(k-2)1} &= w_{(k-2)2} - w_{(k-3)2}. \end{aligned} \tag{17}$$

Compare (16) and (17). Let $w_{22} - w_{12} = \alpha, and w_{32} - w_{22} = \beta$; then we have

$$w_{22} - w_{12} = w_{42} - w_{32} = w_{62} - w_{52} = \ldots = \alpha,$$

$$w_{32} - w_{22} = w_{52} - w_{42} = w_{72} - w_{62} = \ldots = \beta,$$

$$w_{31} - w_{21} = w_{51} - w_{41} = w_{71} - w_{61} = \ldots = \alpha,$$

$$w_{41} - w_{31} = w_{61} - w_{51} = w_{81} - w_{71} = \ldots = \beta.$$

From the above equations we have

$$\begin{aligned} &(w_{12}, w_{22}, w_{32}, w_{42}, w_{52}, \ldots) \\ = &(w_{12}, w_{12} + \alpha, w_{12} + \alpha + \beta, w_{12} + 2\alpha + \beta, w_{12} + 2\alpha + 2\beta, \ldots), \end{aligned} \tag{18}$$

$$\begin{aligned} &(w_{21}, w_{31}, w_{41}, w_{51}, w_{61}, \ldots) \\ = &(w_{21}, w_{21} + \alpha, w_{21} + \alpha + \beta, w_{21} + 2\alpha + \beta, w_{21} + 2\alpha + 2\beta, \ldots). \end{aligned} \tag{19}$$

For arc $e = (1n)$ we have

$$\begin{aligned} w_{1n} - w_{01} &= w_{3n} - w_{21} \\ &= w_{4n} - w_{31} \\ &\vdots \\ &= w_{(k-2)n} - w_{(k-3)1} \\ &= w_{kn} - w_{(k-1)1}; \end{aligned}$$

so we have

$$\begin{aligned} w_{4n} - w_{3n} &= w_{31} - w_{21} \\ w_{5n} - w_{4n} &= w_{41} - w_{31} \\ &\vdots \\ w_{(k-2)n} - w_{(k-3)n} &= w_{(k-3)1} - w_{(k-4)1} \\ w_{kn} - w_{(k-2)n} &= w_{(k-1)1} - w_{(k-3)1}. \end{aligned} \tag{20}$$

For arc$e = (n1)$ we have

$$\begin{aligned} w_{21} - w_{1n} &= w_{31} - w_{2n} \\ &= w_{41} - w_{3n} \\ &\vdots \\ &= w_{(k-2)1} - w_{(k-3)n} \\ &= w_{(k-1)1} - w_{(k-2)n}; \end{aligned}$$

so we have

$$\begin{aligned} w_{2n} - w_{1n} &= w_{31} - w_{21} \\ w_{3n} - w_{2n} &= w_{41} - w_{31} \\ &\vdots \\ w_{(k-3)n} - w_{(k-4)n} &= w_{(k-2)1} - w_{(k-3)1} \\ w_{(k-2)n} - w_{(k-3)n} &= w_{(k-1)1} - w_{(k-2)1}. \end{aligned} \tag{21}$$

Comparing equations(19), (20), (20) and (21) we get

$$\begin{aligned} &(w_{1n}, w_{2n}, w_{3n}, w_{4n}, w_{5n}, \ldots, , w_{(k-2)n}, , w_{kn}) \\ = &(w_{1n}, w_{1n} + \alpha, w_{1n} + \alpha + \beta, w_{1n} + 2\alpha + \beta, w_{1n} + 2\alpha + 2\beta, \ldots, \\ &w_{(k-2)n}, w_{(k-2)n} + \alpha + \beta). \end{aligned} \tag{22}$$

Now from arc $e = (2n)$ we have $w_{2n} - w_{12} = w_{3n} - w_{22}$, and so $w_{3n} - w_{2n} = w_{22} - w_{12}$; notice that this implies $\alpha = \beta$.

By equations (19),(20) and (22)

$$\begin{aligned} w_{21} &= w_{32} - w_{12} = 2\alpha; \\ w_{j1} &= w_{21} + (j-2)\alpha = j\alpha, \text{ for } 2 \leq j \leq k-1; \\ w_{j2} &= w_{12} + (j-1)\alpha, \text{ for } 1 \leq j \leq k-1; \\ w_{jn} &= w_{1n} + (j-1)\alpha, \text{ for } 1 \leq j \leq k, j \neq k-1. \end{aligned}$$

Because (w, u) is an extreme ray of M, by Proposition 5.3 we know that $\Lambda(w, u)$ has only the trivial member $I(w)$, which is true if and only if $w_{12} = \alpha$ or 0 and $w_{1n} = \alpha$ or 0. Notice that similar results hold for all other nodes, so we are done.□

6.2. South-East Constraints

Another interesting class of constraints are the *South-East* facets. Let us arbitrarily order the nodes of G from 1 to n with $s = 1$ and $t = n$. Then the corresponding South-East constraint will be:

$$\sum_{i<j} (j - i + 1)x_{ij} \geq k + n - 1.$$

Note that we get a new constraint from each different ordering of the nodes. Note also that the coefficients of this constraint range in magnitude from 1 to n.

To see that this constraint is valid for P_d, consider the following two constraints that are valid for P:

$$\sum_{e \in E} x_e = k;$$

$$\sum_{ij} (j - i) x_{ij} = n - 1.$$

The first constraint says simply that there are k arcs in any k-walk. For the second constraint, think of the nodes of G laid out in sequence from 1 to n. The coefficient for each arc is simply the number of "steps" the arc moves along the sequence. The constraint says that any k-walk must move a total of $n - 1$ steps (from $s = 1$ to $t = n$). If we add together these two constraints and delete those x_{ij}'s with negative coefficients, we obtain the South-East constraint.

Alternatively we can obtain this constraint from the (w, u) shown on the auxiliary graph G' in Figure 4. For clarity, only the nodes of G' are illustrated in this figure. The number shown at each node is the w_{ji} value corresponding to that node.

The constraint seems to say that on the auxiliary graph, in order to go from s to t, you have to proceed right (east) at least $n - 1$ and go (south) down at least k. It is easy to see that an (s,t)-k-walk is in $F(w, u)$ if and only if it uses only arcs from set $\{e : e = ij,\ i < j, \text{ or } i = j + 1\}$.

Proposition 6.5 *The South-East constraints are facet inducing for* $k \geq 5$.

Proof. We prove it by induction on n the number of nodes of G.

For $n = 3$, the South-East constraint will be

$$2x_{12} + 3x_{13} + 2x_{23} \geq 3 + k - 1 = k + 2.$$

Now we prove this is a facet constraint by giving three k-walks p^1, p^2, p^3 such that p^1_+, p^2_+, p^3_+ are independent vectors; hence by Proposition 6.1 it is a facet constraint.

k even: Suppose that $k = 2m$; then we have

$p^1 : 1 \to 2 \to 1 \to 2 \to \ldots 1 \to 2 \to 3;$

$p^2 : 1 \to 3 \to 2 \to 3 \to \ldots 3 \to 2 \to 1 \to 3;$

$p^3 : 1 \to 2 \to 3 \to 2 \to \ldots 3 \to 2 \to 3;$

and we have

$$(p^1_+,\ p^2_+,\ p^3_+) = \begin{pmatrix} m & 0 & 1 \\ 0 & 2 & 0 \\ 1 & m-2 & m-1 \end{pmatrix}.$$

k odd: Suppose that $k = 4m + 3$; then we have

$p^1 : 1 \to 2 \to 1 \to 2 \to \ldots 1 \to 2 \to 3;$

$p^2 : 1 \to 3 \to 2 \to 3 \to \ldots 3 \to 2 \to 3;$

$p^3 : 1 \to 2 \to 3 \to 2 \to 1 \to 2 \to 3 \to 2 \to 1 \ldots 1 \to 2 \to 3;$

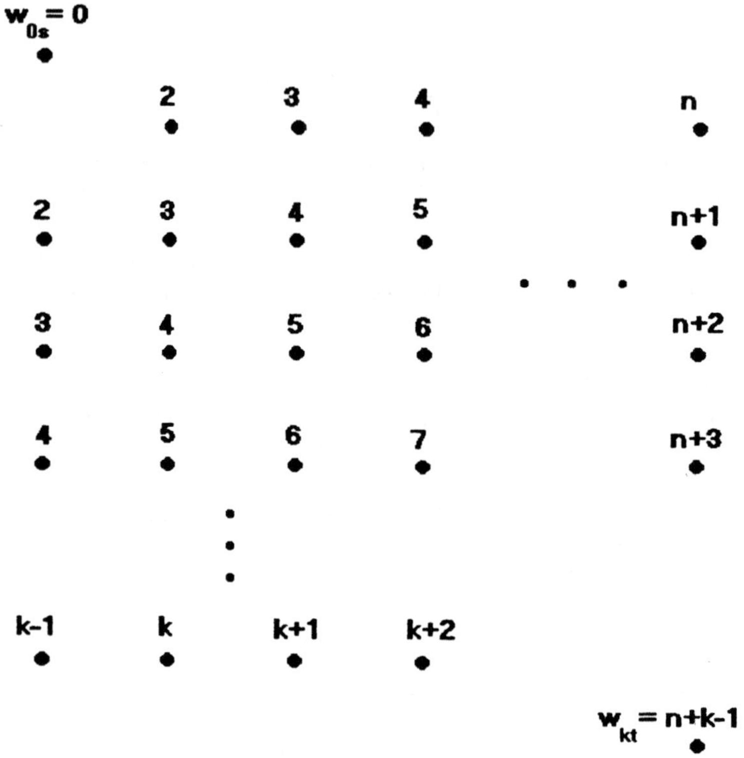

Fig. 4. The nodes of G' for a South-East Constraint

and we have

$$(p_+^1,\ p_+^2,\ p_+^3) = \begin{pmatrix} 2m+1 & 0 & m+1 \\ 0 & 1 & 1 \\ 1 & 2m+1 & m \end{pmatrix}.$$

Suppose that $k = 4m+1$; then we have

$p^1 :\ 1 \to 2 \to 1 \to 2 \to \ldots 1 \to 2 \to 1 \to 3;$

$p^2 :\ 1 \to 3 \to 2 \to 3 \to \ldots 3 \to 2 \to 3;$

$p^3 :\ 1 \to 3 \to 2 \to 1 \to 3 \to \ldots 3 \to 2 \to 3;$

and we have

$$(p_+^1,\ p_+^2,\ p_+^3) = \begin{pmatrix} 2m & 0 & 1 \\ 1 & 1 & m-2 \\ 0 & 2m & m \end{pmatrix}.$$

It is easily verified that p^1_+, p^2_+, p^3_+ are independent vectors, so the proposition is true when n=3.

Suppose it is true for all graphs with n nodes and let G be a graph with $n+1$ nodes ($s = 1$ and $t = n+1$). Let $m = n(n-1)/2$. Now consider the graph induced by the nodes $1, \ldots, n-1, n+1$. Let B be the set of arcs with nonzero coefficients in the South-East constraint for this induced graph. Note that $| B |= m$. By induction, there are m k-walks $p^1, \ldots, p^m$ such that $p^1_+, \ldots, p^m_+$ are linearly independent vectors.

We now construct $(n+1)(n)/2$ independent k-walks on G that are in $F(w,u)$. For each node i, $2 \le i \le (n-1)$, we can find a k-walk q^i such that the first 3 steps of q^i are: $1 \rightarrow i \rightarrow n \rightarrow (n-1)$ and for the last $k-3$ steps we use only arcs from B; so $q^i \in F(w,u)$. For arc $(1n)$ we can find a k-walk q^1 such that the first 2 steps of q^i are: $1 \rightarrow n \rightarrow n-1$ and for the last $k-2$ steps we use only arcs from B; so $q^1 \in F(w,u)$. For arc $n(n+1)$ we can find a k-walk q^n such that the first 4 steps of q^n are: $1 \rightarrow n \rightarrow n+1 \rightarrow n \rightarrow n-1$ and for the last $k-4$ steps we use only arcs from B; so $q^n \in F(w,u)$. It is not difficult to verify that $p^1_+, \ldots, p^m_+, q^1_+, \ldots, q^n_+$ are independent vectors, and so by Proposition 6.1 we know that (6.2) is a facet constraint for G.□

6.3. Parity Constraints

In this section, we give another class of facet inducing constraints, called *parity constraints.* Let S_1, S_2, S_3, S_4 be a partition of the node set V such that S_4 is not empty, starting node $s = 1 \in S_1$, and terminal node $t = n \in S_1$ if k is odd, $t \in S_4$ if k is even. The parity constraint is:

$$\sum_{i,j \in S_1} x_{ij} + \sum_{i,j \in S_4} x_{ij} + \sum_{i \in S_1, j \in S_3} x_{ij} + \sum_{i \in S_2, j \in S_1} x_{ij} \\ + \sum_{i \in S_2, j \in S_3} x_{ij} + \sum_{i \in S_2, j \in S_4} x_{ij} \sum_{i \in S_4, j \in S_3} x_{ij} \ge 1 \qquad (23)$$

To see that these constraints are valid, consider first the special case where $S_2, S_3 = \emptyset$. The constraint coefficients will then be as illustrated in Figure 5.

It is readily apparent that there is no (s,t)-walk with the same parity as k that uses only arcs with coefficient zero. More generally, if S_2 and/or S_3 are nonempty note that any walk that enters S_3 or leaves S_2 will satisfy (23). Any k-walk that includes no nodes from S_2 and S_3 must satisfy (23) by the parity argument, so the constraint is valid.

Parity constraints are induced by (w,u) as shown on the auxiliary graph G' in Figure 6. Again the number shown with each node is the w_{ji} value corresponding to that node. For the displayed value of w, there are four types of columns in G'. Columns that have alternating w_{ij} values (starting with w_{1j}) of $1,0,1,\ldots$ correspond to nodes that belong to S_1. Columns that have alternating w_{ij} values of $0,1,0,\ldots$ correspond to nodes that belong to S_4. The columns with all w_{ij} values equal to zero belong to S_2 and the columns with all w_{ij} values equal to one belong to S_3. A k-walk is in $F(w,u)$ if and only if it uses exactly one of the arcs appearing in the constraint.

Proposition 6.6 *The parity constraints are facet inducing if* $k \ge 6$.

Proof. We prove it for the case k is odd (when k is even, the proof is similar). Let A be the set of the arcs appeared in the above constraint. If we can show that for

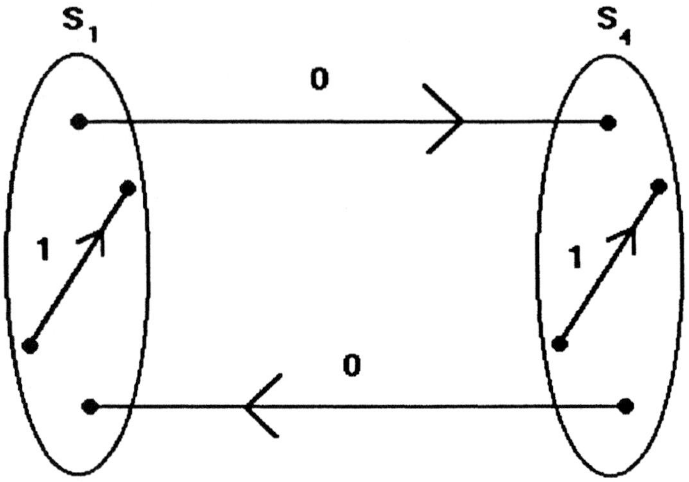

Fig. 5. Parity Constraint

each arc $e \in A$, there is a k-walk p^e using arc e exactly once and arc e is the only arc from A being used by p^e, then we know that $\{p^e_+ : e \in A\}$ are independent vectors, and hence by Proposition 6.1, constraint (23) is a facet inducing constraint. Here are the seven types of arcs appeared in the parity constraint, and for each arc we do find the required k-walk.

type1: arc $e = uv, u \in S_1, v \in S_3$. Pick a node x from S_4, the k-walk we find is: $s \rightarrow x \rightarrow u \Rightarrow v \rightarrow x \rightarrow t \rightarrow x \rightarrow t \ldots \rightarrow x \rightarrow t$.

type2: arc $e = uv, u \in S_2, v \in S_1$. Pick a node x from S_4, the k-walk we find is: $s \rightarrow x \rightarrow u \Rightarrow v \rightarrow x \rightarrow t \rightarrow x \rightarrow t \ldots \rightarrow x \rightarrow t$.

type3: arc $e = uv, u \in S_2, v \in S_3$. Pick a node x from S_4, the k-walk we find is: $s \rightarrow u \Rightarrow v \rightarrow t \rightarrow x \rightarrow t \rightarrow x \rightarrow t \ldots \rightarrow x \rightarrow t$.

type4: arc $e = uv, u \in S_2, v \in S_4$. The k-walk we find is: $s \rightarrow u \Rightarrow v \rightarrow t \rightarrow v \rightarrow t \rightarrow v \rightarrow t \ldots \rightarrow v \rightarrow t$.

type5: arc $e = uv, u \in S_4, v \in S_3$. The k-walk we find is: $s \rightarrow u \Rightarrow v \rightarrow t \rightarrow u \rightarrow t \rightarrow u \rightarrow t \ldots \rightarrow u \rightarrow t$.

type6: arc $e = uv, u \in S_4, v \in S_4$. The k-walk we find is: $s \rightarrow u \Rightarrow v \rightarrow t \rightarrow u \rightarrow t \rightarrow u \rightarrow t \ldots \rightarrow u \rightarrow t$.

type7: arc $e = uv, u \in S_1, v \in S_1$. Pick a node x from S_4, the k-walk we find is: $s \rightarrow x \rightarrow u \Rightarrow v \rightarrow x \rightarrow t \rightarrow x \rightarrow t \ldots \rightarrow x \rightarrow t$;

It is straightforward to check that the k-walks we found above are the required k-walks, and so we are done.□

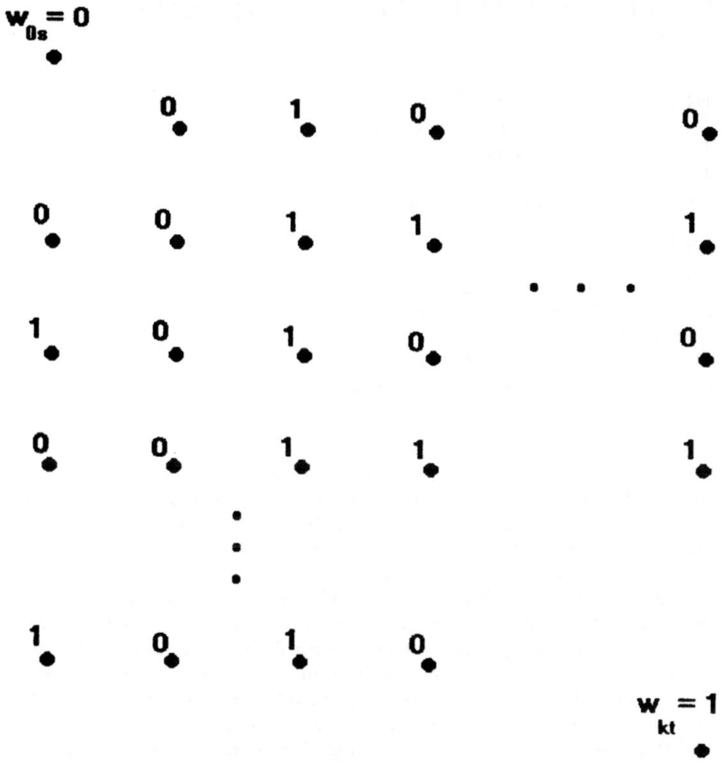

Fig. 6. G' for Parity Constraint

References

1. E. Balas and W.R. Pulleyblank [1983], The perfect matchable subgraph polytope of a bipartite graph, *Networks,* Vol. 13 (1983) 495-516.
2. E. Balas and W.R. Pulleyblank [1987], The perfect matchable subgraph polytope of an arbitrary graph, *Research Report CORR87-33(1987),* Department of Combinatorics and Optimization, University of Waterloo, to appear in Combinatorica.
3. M. O. Ball, W-G. Liu and W.R. Pullyblank [1987], Two terminal Steiner tree polyhedra, *Research Report CORR87-33(1987),* Department of Combinatorics and Optimization, University of Waterloo. Appeared In *Proceedings of CORE 20th Anniversary Conference.*
4. A. Schrijver [1986], *Theory of Linear and Integer Programming,* John Wiley and Sons Ltd., Chichester, 1986.
5. F. Barahona and A.R Mahjoub [1987], Compositions of graphs and polyhedra II: stablesets, Research Report CORR 87-47 1987, Department of Combinatorics and Optimization, University of Waterloo.
6. R.K. Martin, R.L. Rardin and B.A. Campbell [1987], Polyhedral characterization of discrete dynamic programming, *Research Report* CC-87-24, School of Industrial Engineering, Purdue University.

TWO GEOMETRIC OPTIMIZATION PROBLEMS

BHASKAR DASGUPTA
Department of Computer Science
University of Minnesota
Minneapolis, MN 55455-0159
email: dasgupta@cs.umn.edu

and

VWANI ROYCHOWDHURY
School of Electrical Engineering
Purdue University
West Lafayette, IN 47907-1285
email: vwani@drum.ecn.purdue.edu

Abstract. We consider two optimization problems with geometric structures. The first one concerns the following minimization problem, termed as the *rectilinear polygon cover* problem: "Cover certain features of a given rectilinear polygon (possibly with rectilinear holes) with the minimum number of rectangles included in the polygon." Depending upon whether one wants to cover the interior, boundary or corners of the polygon, the problem is termed as the *interior, boundary* or *corner* cover problem, respectively. Most of these problems are known to be NP-complete. In this chapter we survey some of the important previous results for these problems and provide a proof of impossibility of a polynomial-time approximation scheme for the interior and boundary cover problems. The second problem concerns routing in a segmented routing channel. The related problems are fundamental to routing and design automation for Field Programmable Gate Arrays (FPGAs), a new type of electrically programmable VLSI. In this chapter we survey the theoretical results on the combinatorial complexity and algorithm design for segmented channel routing. It is known that the segmented channel routing problem is in general NP-Complete. Efficient polynomial time algorithms for a number of important special cases are presented.

Key words: Rectilinear Polygons, Segmented Channel Routing, Approximation Heuristics

1. Introduction

The problem of covering a certain class of features of a rectilinear polygons with the minimum number of rectangles belongs to a more general class of geometric covering and decomposition problems. Depending upon whether one wants to cover the interior, boundary or corners of the polygon, such a problem is termed as *interior, boundary or corner* cover problem, respectively. The rectilinear cover problem has received particular attention, partly because very little progress has been made in finding efficient algorithms for covering arbitrary polygons with primitive shapes, and also partly because the rectilinear cover problem has important applications in storing images[19], and in the manufacture of integrated circuits[20]. Also, an investigation of this problem has given rise to special kinds of perfect graphs of interest[21]. Unfortunately, most of these problems are NP-complete in general, hence there is need to develop efficient heuristics for these problems. In this chapter we survey some previous results for these problems and provide proofs of some very

Ding-Zhu Du and Jie Sun (eds.), Advances in Optimization and Approximation, 30–57.

recent results in this area.

Conventional channel routing [16] concerns the assignment of a set of connections to tracks within a rectangular region. The tracks are freely customized by the appropriate mask layers. Even though the channel routing problem is in general NP-Complete [26], efficient heuristic algorithms exist and are in common use in many placement and routing systems. In this chapter we study the more restricted channel routing problem (see Fig. 13), where the routing is constrained to use fixed wiring segments of predetermined lengths and positions within the channel. Such segmented channels are incorporated in channeled Field Programmable Gate Arrays (FPGAs) [12]. In [11, 25] it is demonstrated that a well designed segmented channel needs only a few tracks more than a freely customized channel. The problem of segmented channel routing is shown to be NP-Complete in [11, 25]. In this chapter we survey some of the polynomial time algorithms for cases with special geometrical structures.

The rest of the chapter is organized as follows:

- In Section 2.1 we state the basic definitions and provide a precise statement of the rectilinear polygon cover problem.
- In Section 2.2 we state results about the complexity of an exact solution for most of the above problems.
- In Section 2.3 we state various approximation heuristics for approximating these problems.
- In Section 2.4 we state some results which gives polynomial time solutions for some special cases of the rectilinear polygon cover problem.
- In Section 2.5 we prove a result showing the impossibility of having a polynomial-time approximation scheme for the interior and boundary cover problem under the assumption of P≠NP.
- In Sections 3.1 and 3.2 we provide a description of segmented channel routing, introduce the related optimization problems and survey some of the previous results.
- In Section 3.3 we introduce some of the polynomial time algorithms for segmented channel routing.
- We conclude in Section 4 with some open questions about these problems.

2. The Rectilinear Polygon Cover Problems

2.1. Definitions and Preliminaries

A *rectilinear polygon* P is a polygon with its sides parallel to the coordinate axes. Such a polygon may or may not have holes, but if the holes are present they are also rectilinear. The polygon P is said to be in *general position* if no three of its vertices are on the same horizontal or vertical line. In all subsequent discussions we assume that the given polygon is *simple*, *i. e.*, no two non-consecutive edges of the polygon cross each other.

The corners of the given polygon can be classified into *convex*, *degenerate convex* and *concave* types. A *convex* corner is a corner produced by the intersection of two consecutive sides of the polygon which form a 90° angle inside the interior of the polygon. A *degenerate convex* corner is produced by the intersection of two pairs of edges forming two 90° angles. The remaining corners are the *concave* corners, produced by the intersection of two consecutive edges of the polygon which form a 270° angle inside the interior of the polygon.

The interior (*resp.* boundary, corner) cover problem for a rectilinear polygon of n vertices is the following minimization problem: find a set of (possibly overlapping) rectangles of minimum cardinality so that the union of these rectangles covers the interior (*resp.* boundary, corners) of the given polygon. For the corner cover, it is sufficient that each corner is on the boundary (possibly a corner) of one of the rectangles in the given set.

An ***anti-rectangle*** set is a set of points inside the given rectilinear polygon such that no two of them can be covered jointly by one rectangle which does not contain a part of the exterior of the polygon. Depending upon whether it is an interior, boundary or corner cover problem, these points can be placed in the interior, boundary or corners only of the given polygon, respectively. If θ is the size of a cover for one of the cover problems, and α is the size of an anti-rectangle set for this cover, then it is obvious that $\theta \geq \alpha$. When the cover size is minimum and the size of the anti-rectangle set is maximum, $\theta = \alpha$ holds for some special cases of the cover problems. However, the equality is not true in general for either the interior, boundary or corner cover problems. Erdös asked if $\theta/\alpha \leq c$ (for some positive constant c) for the interior cover problem for arbitrary rectilinear polygons (mentioned by Chaiken et. al.[4]) and the answer is not known yet (the best known bound is $\frac{\theta}{\alpha} \leq \log \alpha$[8]).

Let A be an NP-complete minimization problem and H be a polynomial-time approximation heuristic for A. We say that H has a performance ratio of c iff for any instance I of A,

$$cost_H(I) \leq c \;.\; cost_{opt}(I)$$

where $cost_H(I)$ (resp. $cost_{opt}(I)$) denotes the cost of the solution of the instance I as obtained by H (resp., by the optimal algorithm).

For any minimization problem A let c_{opt} be the cost of the optimal solution and c_{approx} be the cost of an approximate solution produced by a heuristic. Let $\epsilon_n = \frac{|c_{opt} - c_{approx}|}{c_{opt}}$ be the relative error of the approximate solution for input size n. A polynomial-time approximation scheme for A is an algorithm that takes as input an instance of the problem and a constant $\epsilon > 0$, and produces a solution with $\epsilon_n \leq \epsilon$ in time polynomial in n[6]. A more detailed discussion of the related concepts is available in [6, 13].

2.2. Complexities of Exact Solution

In this section we summarize the previous results about the complexities of exact solutions for the interior, boundary and corner cover problems.

2.2.1. *Interior Cover Problem*

Masek[19] was the first to show that the interior cover problem is NP-complete for rectilinear polygons with holes. For a long time the complexity of this problem was unknown for polygons without holes, until Culberson and Reckhow[7] showed the interior cover problem is NP-complete even if the polygon has no holes, and even if

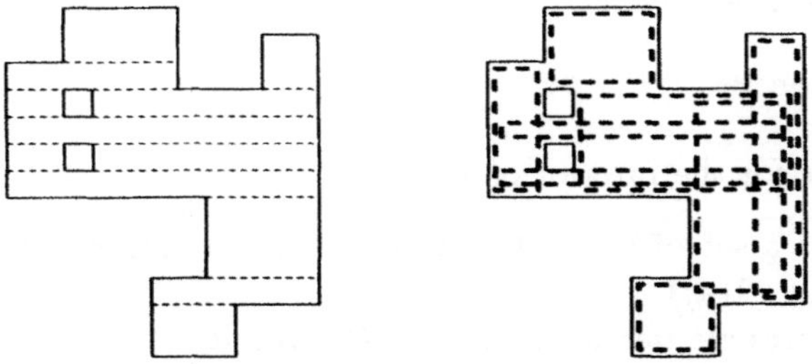

Fig. 1. *The sweepline heuristic for the interior cover problem. (a) The partition phase (b) The Extend/Delete phase. The rectangles are shown slightly offset for clarity.*

the polygon is required to be in general position. The NP-completeness reduction of Culberson and Reckhow is quite involved. They reduce the satisfiability problem to this problem. Given an instance of the satisfiability problem, they construct an instance of the interior cover problem for polygons without holes such that the interior of the constructed polygon can be covered with a specified number of rectangles if and only if the given formula is satisfiable.

2.2.2. *Boundary Cover Problem*

Conn and O'Rourke[5] showed that the boundary cover problem is NP complete for polygons with holes, even if the polygon is in general position. As before, the complexity of the problem for polygons without holes was not known for quite some time until Culberson and Reckhow[7] showed the boundary cover problem is NP-complete even if the polygon has no holes, and even if the polygon is required to be in general position.

2.2.3. *Corner Cover Problem*

Conn and O'Rourke[5] showed that the following version of the corner cover problem is NP-complete: cover each concave corner by two rectangles along both the perimeter segments defining this corner. Using similar techniques, Berman and DasGupta[2] showed that the corner problem is NP-complete for polygons with holes. The complexity of the corner cover problem for polygons without holes is still unknown, although it is conjectured to be NP-complete by Conn and O'Rourke[5].

2.3. Approximation Heuristics

Because of the hardness of the rectilinear cover problems as stated in the previous section, it is of importance to consider efficient heuristics for the problem. Below we summarize some of the known heuristics for these problems.

2.3.1. *Interior Cover*

Franzblau[8] proposed the following sweep-line heuristic for the interior cover problem.

ALGORITHM Partition/Extend.

Input: Rectilinear polygon P.

Output: Rectangle cover for the interior of P.

- Partition P into a set of disjoint rectangles by extending each horizontal edge of P (see fig. 1(a)).
- Extend each rectangle vertically inside P until it is vertically maximal. Delete any repeated rectangles (see fig. 1(b)).

It is possible to implement the above heuristic so that it runs in $O(n \log n)$ time using standard data structures.

Let $c_{PE}(I)$ be the number of rectangles used by the above heuristic and $c_{opt}(I)$ be the optimal number of rectangles needed for the cover for an instance I. The following theorem was proved by Franzblau[8].

Theorem 2.3.1 *[8]* $c_{PE}(I) = O(c_{opt}(I). \log(c_{opt}(I)))$.

However, if the given polygon has no holes, then the heuristic performs considerably better as shown by the following theorem.

Theorem 2.3.2 *[8] If the given polygon has no holes then*

$$c_{PE}(I) \leq 2.c_{opt}(I) - 1$$

2.3.2. *Boundary Cover*.
Berman and DasGupta[2] suggested a very simple heuristic for the boundary cover problem. The following theorem is proved in [2].

Theorem 2.3.3 *[2] It is possible to design a heuristic for the boundary cover problem which which runs in $O(n \log n)$ time and has a performance ratio of* 4.

2.3.3. *Corner Cover*
Berman and Dasgupta[2] suggested the following heuristic for the corner cover problem for polygons with holes.

ALGORITHM Corner Cover.

Input: A rectilinear polygon P, possibly with holes.

Output: A set of rectangles which together cover the corners of P.

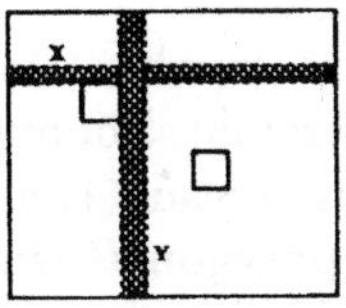

Fig. 2. *X and Y are two vertices of the graph and (X,Y) is an edge*

- Form the two sets S and T; S contains, for each horizontal segment e of P, an adjacent rectangle of width 1 of maximal horizontal extent (this is the ***principal side*** for this rectangle), and T contains, for each vertical segment e of P, an adjacent rectangle of width 1 of maximal vertical extent (the principal side is defined similarly). Now, form a bipartite graph $G = (S \cup T, E)$, where $E = \{(y, z) \in S \times T \mid$ principal sides of y and z share a corner $\}$ (see *fig.* 2 for an example). Construct a minimum vertex cover R of G using maximum matching. The set of rectangles R constitutes our approximate cover.

The following theorem was proved in [2].

Theorem 2.3.4 *[2] The performance ratio of the above heuristic is* 4. *It runs in* $O(n \log n)$ *time.*

When the given polygon has no holes, Berman and DasGupta[2] designed a new heuristic for the corner cover problem which runs in time $O(n \log n)$ and has a performance ratio of 2. Details of this heuristic are quite involved and available in [2].

2.4. Polynomial Time Solutions for Special Cases

In this section we survey some of the results which provide polynomial time solution for these problems when the given polygon is restricted.

2.4.1. *Interior Cover*

A rectilinear polygon P is called x-convex (resp. y-convex) if the intersection of any horizontal (resp. vertical) line segment with the interior of P is a connected (possibly empty) segment. A rectilinear polygon is *rectilinearly convex* iff it is both x-convex and y-convex. Chaiken et. al.[4] showed that for the interior cover problem for a rectilinearly convex polygon $\alpha = \theta$ (where α and θ are the sizes of maximum cardinality anti-rectangle set and minimum cardinality interior cover, respectively), and use this to provide a polynomial time solution to the interior cover problem for this special case. The result was further extended by Franzblau and Kleitman[9] who proved a similar result when the polygon is just y-convex. Lubiw[17, 18] gives polynomial time algorithms for the interior cover problem for a slightly more general class of polygons.

2.4.2. *Boundary Cover*
Using matching techniques in graphs, Conn and O'Rourke[5] gives an $O(n^{\frac{5}{2}})$ time algorithm for covering the horizontal boundary segments of a polygon P provided P is in general position. The same result can be used to devise an $O(n^{\frac{5}{2}})$ time heuristic for the boundary cover of a polygon P which has a performance ratio of 2 *provided* P is in general position.

2.4.3. *Corner Cover*
Conn and O'Rourke[5] presented an $O(n^{\frac{5}{2}})$ time algorithm for covering the convex vertices of a given polygon P optimally. However, this algorithm does not generalize to covering the concave corners of P.

2.5. Impossibility of Approximation Schemes

In this section we show that the vertex cover problem for graphs in which the degree of any vertex is bounded by a *constant* B can be reduced to the interior or boundary cover problems preserving the nature of approximation. Due to a recent result of Arora et al[1], this shows that a polynomial time approximation scheme for these covering problems is impossible, unless $P = NP$.

Let (F, P) and (G, Q) be two combinatorial optimization problems where F and G are the cost functions and P and Q are the feasibility predicates. Let $opt_{F,P}(x)$ and $opt_{G,Q}(x)$ be the optimal cost values for an instance x and $quality(a, b) = \max\left(\frac{a-b}{b}, \frac{b-a}{a}\right)$ for positive integers a and b. We say (F, P) can be reduced to (G, Q) preserving approximation[3] with amplification c if and only if there exists two deterministic polynomial time algorithms T_1 and T_2 and a positive constant c such that for all x and y, if $\bar{x} = T_1(x)$ then

(1) $Q(\bar{x}, y) \Rightarrow P(x, T_2(\bar{x}, y))$, and

(2) if $Q(\bar{x}, y)$ then $quality(opt_{F,P}(x), F(T_2(\bar{x}, y))) \leq c(quality(opt_{G,Q}(\bar{x}), G(y)))$.

This reduction ensures that an approximation G will result in an equally good approximation of F.

The *bounded-degree* vertex cover problem for graphs is as follows:

INSTANCE: An undirected graph $G = (V, E)$ with every vertex of degree at most some constant B, and a positive integer K.

QUESTION: Is there a subset $V' \subseteq V$, $\mid V' \mid \leq K$, such that for every edge $\{u, v\} \in E$, either $u \in V'$ or $v \in V'$?

Culberson and Reckhow[7] showed the NP-hardness of the rectilinear cover problem without holes by reducing the 3-SAT problem to this problem. However, their reduction is not approximation preserving since it introduces quadratically many rectangles in the optimal solution and hence does not preserve the approximation quality in the sense described above. Here, we show how to reduce the bounded-degree vertex cover problem to the interior or boundary cover problems *preserving* the approximation nature. We first consider the interior cover problem.

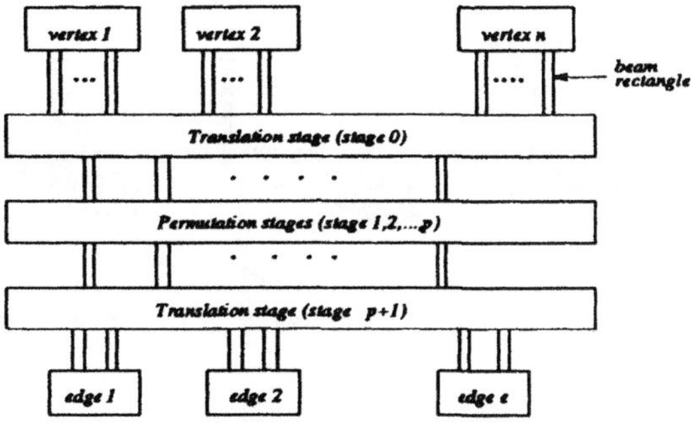

Fig. 3. *The overall scheme*

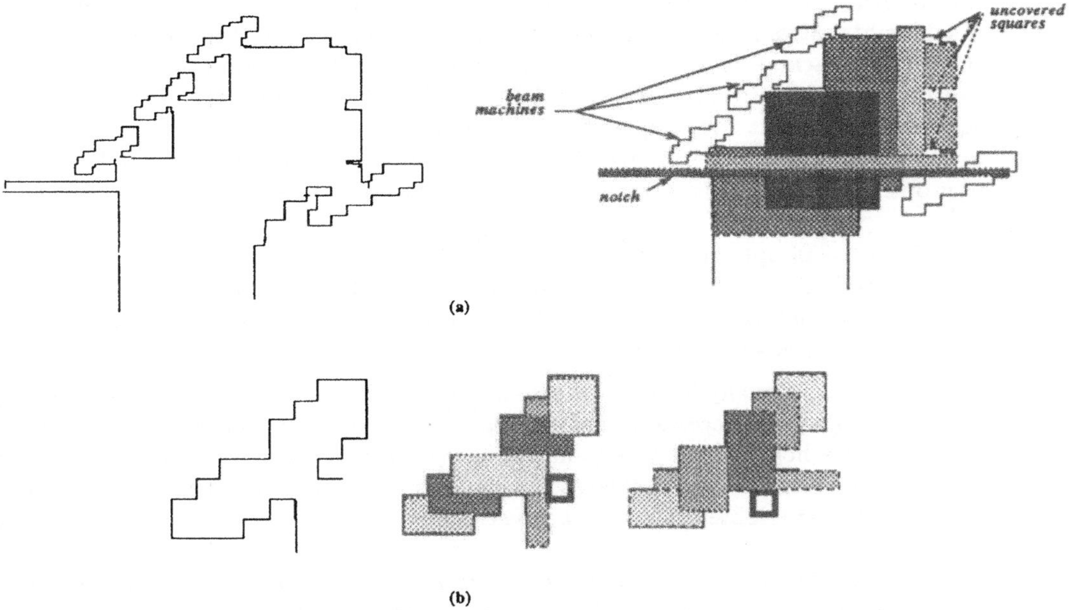

Fig. 4. *(a) A vertex gadget and its background covers and uncovered squares. (b) A beam machine and its two optimal covers. The uncovered square near its mouth is shown by thick lines.*

The overall scheme of our approach is shown in *fig.* 3. We use a *gadget* for every vertex. Beams (rectangles) coming out of a gadget indicate that this vertex participates in vertex cover. The beams are first translated, then permuted appropriately, again translated and finally enter the edge-gadgets. Each edge gadget is coupled with two beams and represents an edge between the two vertices which correspond to the two beams.

Now, we describe each component in details.

Vertex gadget: The vertex gadget is shown in full details in *fig.* 4(a). It consists of B beam machines when B is the degree of this vertex ($B = 3$ in the figure). Each beam machine can be covered optimally with 6 rectangles with only one

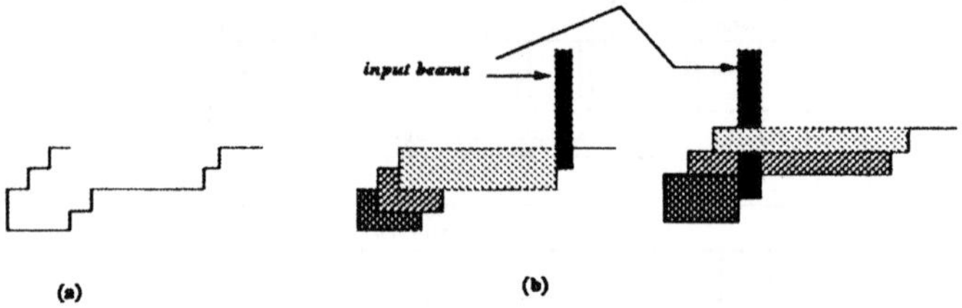

Fig. 5. *(a) An edge gadget. (b) Its two optimal covers.*

rectangle extending through its mouth in horizontal or vertical direction (see *fig.* 4(b)), and in each of these two optimal covers there is an uncovered square near the mouth of the machine (shown in *fig.* 4(b)). There is one additional beam machine at bottom right, and a notch at the extreme left bottom, which forces this beam machine to use its horizontal beam. The background of this gadget can be covered optimally with $2B + 1$ rectangles, thus leaving out B uncovered squares (*fig.* 4(a)). These squares can be covered by the horizontal beams of B beam machines in an optimal cover, or by one more additional rectangle in a non-optimal cover. This structure has the following properties.

(a) There is an optimal cover of this gadget with $8B + 1$ rectangles when no beam from any of the beam machines is "used" (*i. e.,* extends vertically downwards). This corresponds to the case when this vertex does not participate in a vertex cover.

(b) If a beam from any beam machine extends vertically downwards then $8B+2$ rectangles are necessary and sufficient to cover this gadget. The same property holds when more than one beam from one or more beam machines extends vertically downwards. This corresponds to the case when this vertex participates in a vertex cover.

For each of the covers described above, it is also possible to place an equal number of anti-rectangle points in the interior of the polygon.

Edge gadget: This gadget is shown in *fig.* 5. If either of its two input beams are used then it can be covered optimally with 4 rectangles, otherwise it requires at least 5 rectangles. This is same as the *inverter* structure in[7].

Translation stage: It consists of e pairs of *joints*[7] where e is the number of edges in the graph. A pair of joint rectangles is an aligned pair of beam machines (*fig.* 6). If the "incoming" beam for the left polygon of the joint is present (*i. e.* , the corresponding edge is present) then the "outgoing" beam from the right polygon of the joint should be used for optimal cover, otherwise the common horizontal rectangle between should be used. The unique background cover for this stage (covering the staircases with the uncovered squares at the mouth of the joint polygons (refer to *fig.* 4(b) for such uncovered squares)) involves $2e$ rectangles and is shown in *fig.* 6(b). This is a slightly simplified version of the joint structure in Culberson and Reckhow[7]. There are two translation stages.

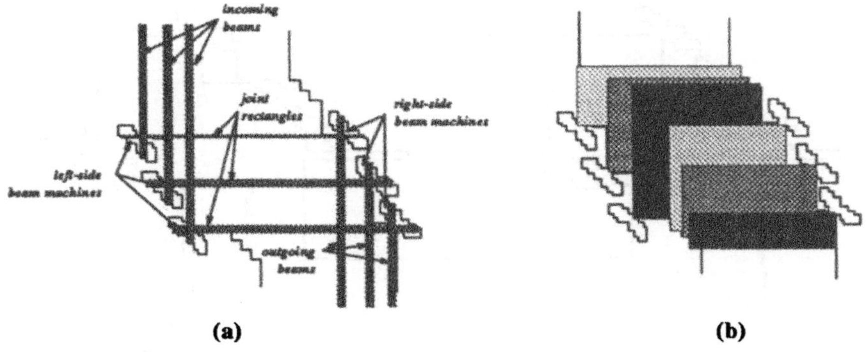

Fig. 6. *(a) A translation stage for 3 beams. (b) Its background covers.*

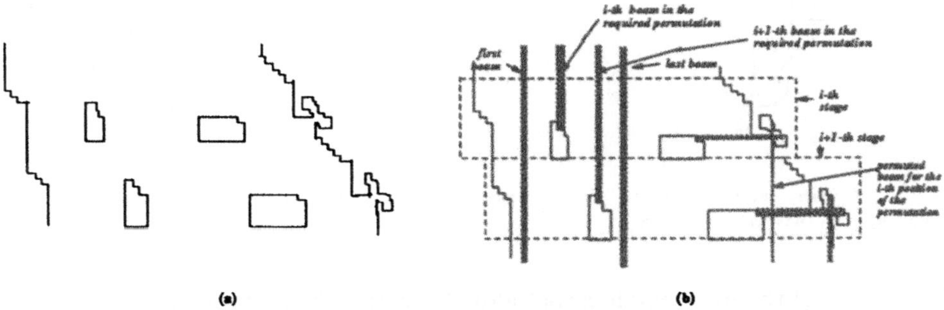

Fig. 7. *(a) A permutation stage. (b) The beam which is permuted is shown.*

They are used to allow the optimal covering of permutation stages not to affect the vertex gadgets and the edge gadgets.

Permutation stage: There are at most $p \leq 2e$ permutation stages when e is the number of edges in the graph. These are needed because the order in which the beams come out of the vertex gadgets is not necessarily the same as they should arrive at the edge gadgets. Each stage consists of staircases and holes as shown in *fig.* 7(a). In stage i we put the ith beam from left "in the required permutation" in its correct place and so the boundary of the polygon is always visible from the right-side hole. The right-side hole is placed so that it makes impossible the right-staircases of the previous stages to be covered by any rectangle which covers optimally the background this or later stages. If a beam is present and covers one notch of the left-side hole, the vertical beam of the beam machine at the right is used for the optimal cover, otherwise, for optimal cover, the horizontal beam of this beam machine covers the notch of the right-side hole (and, so, its vertical beam cannot be used for optimal cover). In all we need 8 rectangles to cover each stage, because, there are 8 anti-rectangle points, and none of these can be covered by rectangles covering those of previous or later stages. The left-side hole can be merged with the boundary for stage n. The details are given in a lemma below.

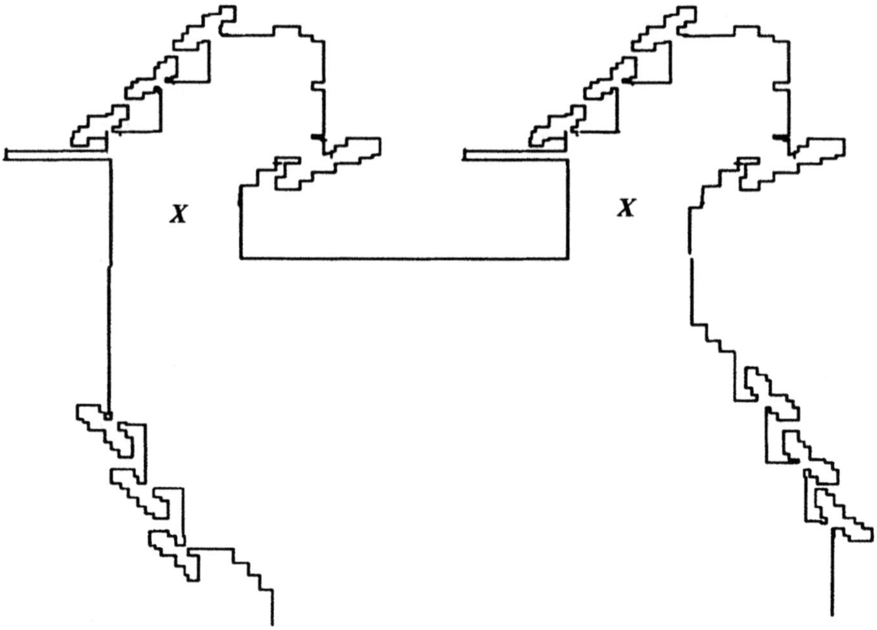

Fig. 8. *Joining vertex gadgets (for a graph having 2 vertices) to the translation stage (anti-rectangle points marked by **X**).*

The proofs of the following lemma follow from the above discussion.

Lemma 2.5.1 *A vertex gadget can be covered optimally only if all of its beam rectangles are not "used".*

Lemma 2.5.2 *The part of the polygon connecting the vertex gadgets to the translation stage needs v rectangles to cover it independent of the cover of any other stage or gadgets, where v is the number of vertices of the graph and these rectangles do not influence the optimal cover of the total polygon.*

Proof. We can place anti-rectangle points at each joints of vertex gadgets to the translation stage, and none of these anti-rectangle points can be covered together with those in the covers of the vertex gadgets, the translation stage or any other stages (see *fig.* 8).□

The proofs of the following two lemmas are straightforward from the discussion above.

Lemma 2.5.3 *For each translation stage the following are true:*

(a) *It requires 2e rectangles to cover its staircases along with the mouth of the joint polygons (i. e., its background).*

(b) *If the incoming beam to the left-side polygon is present, then for optimal cover the outgoing beam of the right-side polygon should be present. However, if the*

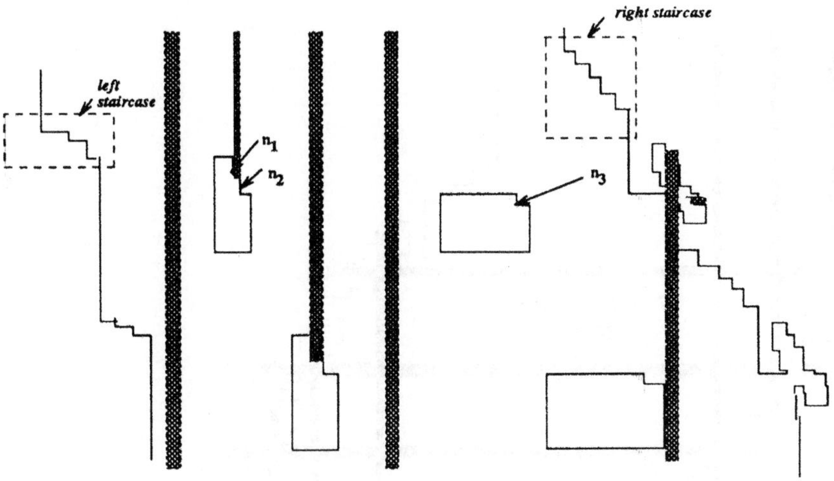

Fig. 9. *Permutation stages. The incoming beam is shown to be present.*

input beam is not present, then for optimal cover the joint rectangle must be present.

(c) *The joint polygons can always be covered optimally, if the rule of (b) above is followed.*

(d) *The optimal cover of the background of this stage cannot be affected by a rectangle that participates in an optimal cover of vertex or edge gadgets or other stages.*

Lemma 2.5.4 *Consider the ith permutation stage (stage i). Then the following are true (fig. 9):*

(a) *Stage i has* 8 *anti-rectangle points (for its background cover) which cannot be covered together with those of any other stage. Also,* 8 *rectangles are sufficient to cover background of stage i.*

(b) *Let the two notches of the left hole be* $n1$ *and* n_2 *and that of right hole be* n_3. *Then,*

(i) *if* n_1 *is covered by the incoming beam, then for optimal cover* n_2, n_3 *and the three left staircases must be covered by rectangles which cover the* 5 *right staircases and hence the right-side beam machine can send out its vertical beam.*

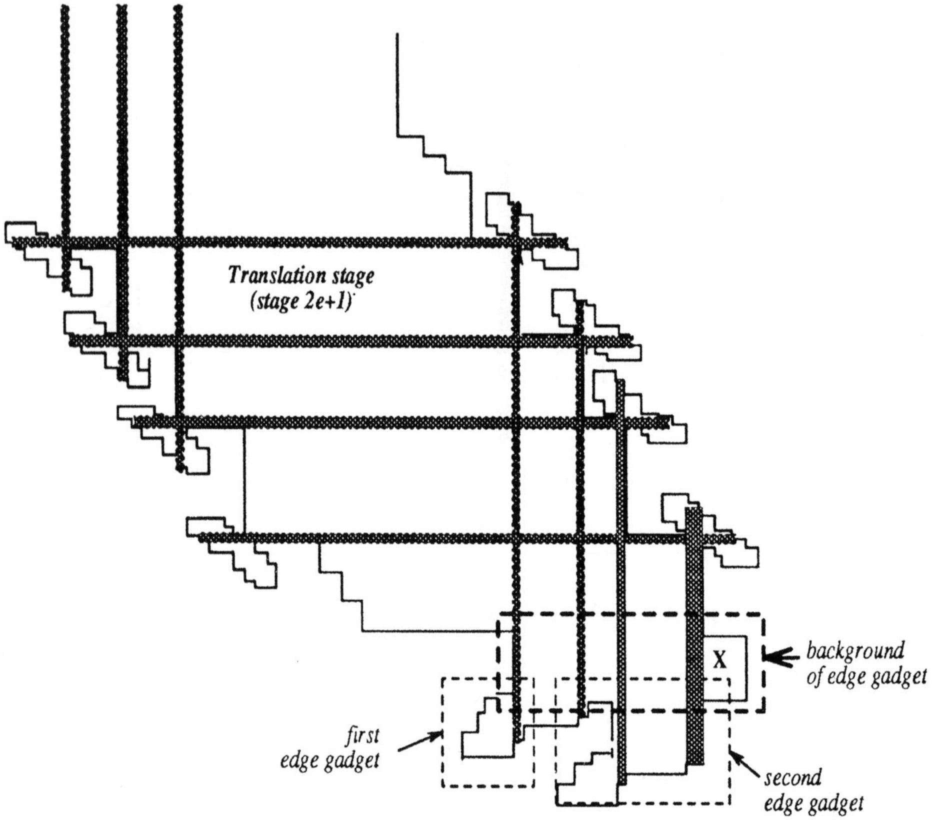

Fig. 10. *Edge gadgets for a graph having 2 edges. The anti-rectangle point in the background cover of the edge is shown by* **X**.

(ii) *if n_1 is not covered by the incoming beam, then for optimal cover notches n_1, n_2 and the three left staircases must be covered by the rectangles which cover the* 5 *right staircases and n_3 is covered by the horizontal beam of the beam machine (and hence the outgoing beam is absent).*

(c) *Anti-rectangle points for covering the background of stage i cannot be covered together with those of any other stage.*

(d) *The stage i places the ith beam from the left in the required permutation into its correct position.*

Lemma 2.5.5 *The background of the edge gadget needs 1 rectangle to cover optimally independent of any other stage. This optimal cover of the background of the edge gadgets is not influenced by the covering rectangles of the previous stages. If* *an edge gadget gets one or two incoming beams, it needs only* 4 *more rectangles to cover it.*

Proof. Rectangles covering vertex gadgets or stages 1 to $2e$ cannot extend to the edge gadget. The beam rectangles in stage $2e+1$ can extend to cover edge gadgets (see *fig.* 10), but these rectangles cannot take part in the optimal cover of the background of stage $2e+1$. One rectangle is sufficient to cover the background of the edge gadget, and the corresponding anti-rectangle point is shown in *fig.* 10. □

Lemma 2.5.6 *Any covering of the polygon (corresponding to a given graph) can be transformed in polynomial time (in the size of the given graph) to another cover with the following properties without increasing the number of covering rectangles:*

(a) *Either all the beams of a vertex gadget are used or none is used.*

(b) *Each edge gadget uses at least one of its two beams from the previous translation stage (and hence needs* 4 *additional rectangles for its optimal cover).*

(c) *The background of the vertex gadgets, translation stages, permutation stages and edge gadgets are covered with the optimal number of rectangles.*

Proof.

(a) If only some beams are used for a vertex gadget we can as well use all the beam rectangles without increasing the number of rectangles for the vertex gadget. Then, we keep moving from one stage to another till we arrive at the edge gadget, turning the joint rectangle of each stage off and using the incoming and outgoing beam rectangles (if this was not the case, the cover for a particular stage was not optimal and we actually improve, refer to lemma 2.5.3, 2.5.4). Surely we do not use more rectangles at the edge gadget (we will improve if it had no incoming beam rectangles).

(b) Assume this is not the case. We keep moving from the edge gadget through successive stages towards the vertex gadgets in a manner similar to as described in (a) above (we arbitrarily turn on one of the two beams of the edge gadget). We may use one more rectangle at the vertex gadget, if it was not using its beams, but the gain of one rectangle at the corresponding edge gadget compensates.

(c) Once we have done the transformations needed for parts (a) and (b) above, we can select the necessary background covers depending on whether the incoming beams of a particular stage are used or not as outlined in lemma 2.5.2, 2.5.3 (part (a)), 2.5.4 and 2.5.5, since the optimal covers of these parts of the polygon are independent of each other and depends only on the presence or absence of the beam rectangles.

It is obvious that traversal in (a), (b) or (c) above takes polynomial time.□

Using lemmas 2.5.2, 2.5.3, 2.5.4, 2.5.5 and 2.5.6, and the properties of the gadgets, it is possible to prove the following lemma (see [2] for a proof).

Lemma 2.5.7 *Let $G = (V, E)$ be a graph in which the degree of any vertex is at most a constant $B > 0$ and P be the polygon constructed from it by the above procedure. Let the number of permutation stages needed be p ($p \leq B\cdot \mid V \mid$). Then, the following holds:*

(a) *A minimal vertex cover of G of size m corresponds to a rectilinear cover of P of size at most* $30\cdot \mid E \mid +14\cdot p + (8\cdot B + 2)\cdot \mid V \mid +m + 1$.

(b) *A rectilinear cover of P, after the transformation as outlined in lemma 2.5.6, of size θ corresponds to a minimal vertex cover of size at least* $\theta - (30\cdot \mid E \mid +14\cdot p + (8\cdot B + 2)\cdot \mid V \mid +1)$.

Theorem 2.5.1 *The reduction of the vertex cover problem to the rectilinear cover problem as outlined above is approximation preserving.*

Proof. T_1 is the transformation needed to construct the polygon using gadgets as outlined above. The transformation T_2 consists of the following:

(a) Modify the cover as outlined in lemma 2.5.6,

(b) Select a vertex in the vertex cover if and only if the corresponding gadget in the polygon uses all its beam rectangles.

The quality constraint for the reduction follows from lemma 2.5.7 above, and the fact that $p \leq 2\cdot \mid E \mid \leq B\cdot \mid V \mid$. □

Hence, we have proved the following result.

Theorem 2.5.2 *No polynomial-time approximation scheme exists for the interior cover problem, unless P=NP.*

A careful examination of our construction shows that all the results hold for boundary cover also, in particular we can always place all the anti-rectangle points on the boundary of the polygon. Hence, we also prove the following result.

Theorem 2.5.3 *No polynomial-time approximation scheme exists for the boundary cover problem, unless P=NP.*

3. Segmented Channel Routing

3.1. Motivation

The architecture of channeled FPGAs [12] is similar to that of conventional (mask programmed) gate arrays: comprising rows of logic cells separated by segmented routing channels (Fig. 11). The inputs and outputs of the cells each connect to a dedicated vertical segment. Programmable switches are located at each crossing of vertical and horizontal segments and also between pairs of adjacent horizontal segments in the same track. By programming a switch, a low resistance path is created between the two crossing or adjoining segments.

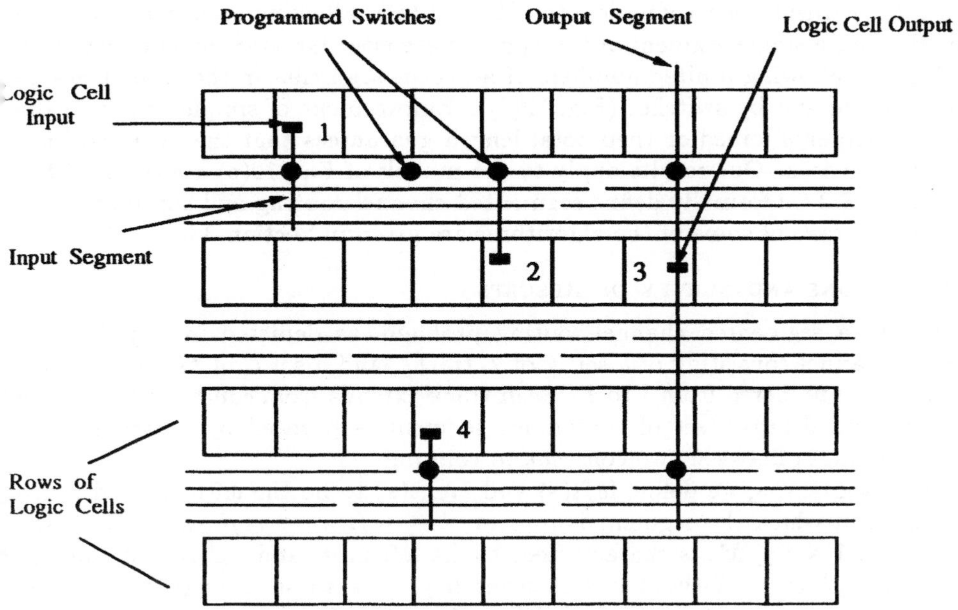

Fig. 11. **FPGA routing architecture. • denotes a programmed switch; unprogrammed switches are omitted for clarity.**

A typical example of routing in a channeled FPGA is shown in Fig. 11. The vertical segment connected to the output of cell 3 is connected by a programmed switch to a horizontal segment, which in turn is connected to the input of cell 4 through another programmed switch. In order to reach the inputs of cells 1 and 2, two adjacent horizontal segments are connected to form a longer one.

The choice of the wiring segment lengths in a segmented channel is driven by tradeoffs involving the number of tracks, the resistance of the switches, and the capacitances of the segments. These tradeoffs are illustrated in Fig. 12.

Fig. 12(a) shows a set of connections to be routed. With the complete freedom to configure the wiring afforded by mask programming, the *left edge* algorithm [14] will always find a routing using a number of tracks equal to the density of the connections (Fig. 12(b)). This is the case since there are no 'vertical constraints' in the problems we consider.

In an FPGA, achieving this complete freedom would require switches at every cross point. Furthermore, switches would be needed between each two cross points along a wiring track so that the track could be subdivided into segments of arbitrary length (Fig. 12(c)). Since all present technologies offer switches with significant resistance and capacitance, this would cause unacceptable delays through the routing. Another alternative would be to provide a number of continuous tracks large enough to accommodate all nets (Fig. 12(d)). Though the resistance is limited, the capacitance problem is only compounded, and the area is excessive.

A segmented routing channel offers an intermediate approach. The tracks are divided into segments of varying lengths (Fig. 12(e)), allowing each connection to be routed using a single segment of the appropriate size. Greater routing flexibility is obtained by allowing limited numbers of adjacent segments in the same track to be joined end-to-end by switches (Fig. 12(f)). Enforcement of simple limits on the number of segments joined or their total length guarantees that the delay will not be unduly increased. Our results apply to the models of Fig. 12(e) and Fig. 12(f).

In Section 3.2 we formally define segmented channel routing and summarize the key results. Details of some of the algorithms are given in Section 3.3.

3.2. Definitions and Survey of Results

The input to a segmented channel routing problem, as depicted in Fig. 13, is a segmented channel consisting of a set $\mathcal{T}$, of T tracks, and a set $\mathcal{C}$ of M connections. The tracks are numbered from 1 to T. Each track extends from column 1 to column N, and is divided into a set of contiguous segments separated by switches. The switches are placed between two consecutive columns.

For each segment s, we define $left(s)$ and $right(s)$ to be the leftmost and rightmost column in which the segment is present, $1 \leq left(s) \leq right(s) \leq N$. Each connection c_i, $1 \leq i \leq M$, is characterized by its left-most and right-most column: $left(c_i)$ and $right(c_i)$. Without loss of generality, we assume throughout that the connections have been sorted so that $left(c_i) \leq left(c_j)$ for $i < j$.

A connection c may be ***assigned*** to a track t, in which case the segments in track t that are present in the columns spanned by the connection are considered *occupied*. More precisely, a segment s in track t is occupied by the connection c if $right(s) \geq left(c)$ and $left(s) \leq right(c)$. In Fig. 13 for example, connection c_3 would occupy segments s_{21} and s_{22} in track 2 or segment s_{31} in track 3.

Definition 3.2.1 [Routing]
A routing, R, of a set of connections is an assignment of each connection to a track such that no segment is occupied by more than one connection.
A K**-segment** routing is a routing that satisfies the additional requirement that each connection occupies at most K segments.

We can now define the following segmented channel routing problems:

Problem 3.2.1 [Unlimited Segment Routing] Given a set of connections and a segmented channel, find a routing.

To reduce the delay through assigned connections, it may be desirable to limit the number of segments used for each connection.

Problem 3.2.2 [K-Segment Routing] Given a set of connections and a segmented channel, find a K-segment routing.

It is often desirable to determine a routing that is optimal with respect to some criterion. We may thus specify a weight $w(c, t)$ for the assignment of connection c to track t, and define:

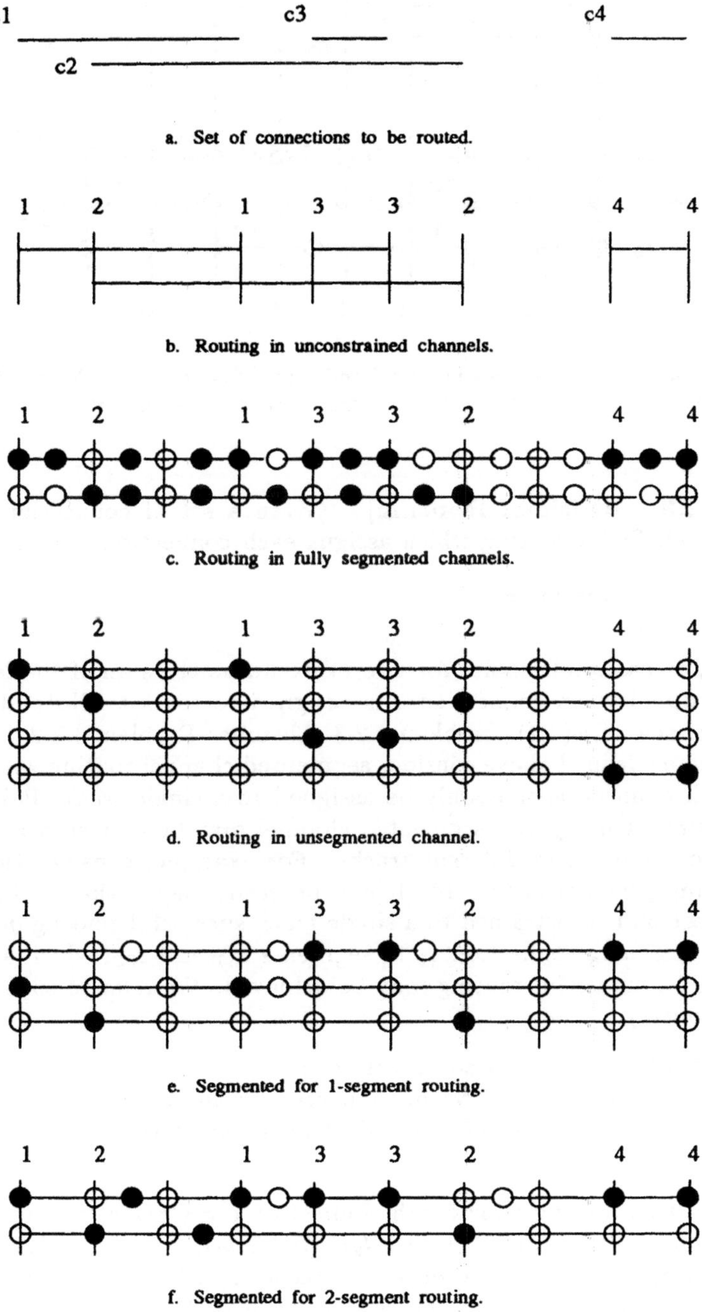

Fig. 12. Examples of channel routing. o denotes open switch; • closed switch.

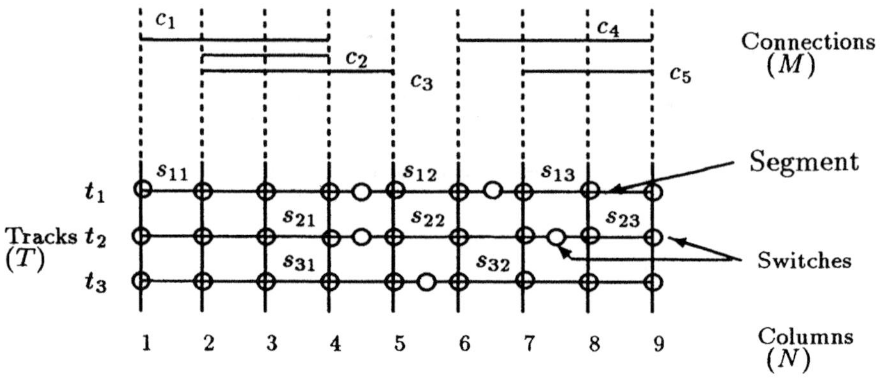

Fig. 13. **An example of a segmented channel and a set of connections.** $M = 5$, $T = 3$, $N = 9$. Connections: c_1, c_2, c_3, c_4, c_5. **Segments:** s_{11}, s_{12}, s_{13}, s_{21}, s_{22}, s_{23}, s_{31}, s_{32}.

Problem 3.2.3 **[Optimal Routing]** Given a set of connections and a segmented channel, find a routing which assigns each connection c_i to a track t_i such that $\sum_{i=1}^{M} w(c_i, t_i)$ is minimized.

For example, a reasonable choice for $w(c,t)$ would be the sum of the lengths of the segments occupied when connection c is assigned to track t. Note also that with appropriate choice of $w(c,t)$, Problem 3.2.3 subsumes Problem 3.2.2.

The problems defined above consider segmented channel routing with the restriction that each connection may only be assigned to a single track. It is easy to see that the routing capacity of a segmented channel may be increased if a connection is assigned to segments in different tracks. For example, consider the segmented channel routing problem in Fig. 14. It can be easily shown that if the assignment of each connection is constrained to a single track successful routing does not exist. However, by assigning connection c_2 to segments s_{11} and s_{33}, which are located in tracks t_1 and t_3, successful routing may be achieved. We refer to such a routing as generalized routing.

Definition 3.2.2 **[Generalized Routing]**
A generalized routing, R_G, of a set of connections consists of an assignment of each connection to one or more tracks such that no segment is occupied by more than one connection.

Thus a generalized routing allows each connection $c = (left(c), right(c))$ to be *split* into p ($p \geq 1$) parts: $(left(c), l_1)$, (l_1+1, l_2), (l_2+1, l_3), $\cdots$, $(l_{p-1}+1, right(c))$, such that each part can be assigned to different tracks. A column l_i, where a connection is split, is referred to as a column where the connection c *changes* tracks.

Detailed hardware implementations may be developed to support generalized routing. For example, vertical wire segments may be added to facilitate track changing. In this case if a connection changes tracks, two switches must be programmed

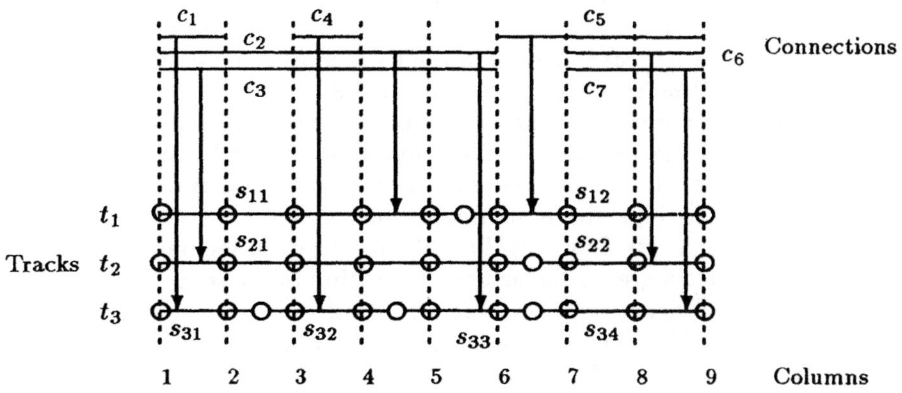

Fig. 14. An example where generalized routing is necessary for successful assignment.

compared to only one if the connection is assigned to two contiguous segments in the same track. Thus allowing connections to occupy multiple tracks might lead to increase in area and to greater delays.

Motivated by such penalties, constraints may be imposed on the generalized segmented channel routing problem leading to the following potentially important special cases:

1. Determine a generalized routing that uses at most k segments for routing any particular connection.
2. Determine a generalized routing that uses at most l different tracks for routing any connection.
3. Determine a generalized routing where connections can switch tracks only at predetermined columns.

Preliminary results on the following unconstrained version of generalized segmented channel routing problem is presented in [25].

Problem 3.2.4 [Generalized Segmented Channel Routing] Given a set of connections and a segmented channel, find a generalized routing.

In [25] the following results are presented.

Theorem 3.2.1 Determining a solution to Problem 3.2.1 is strongly NP-complete.

Theorem 3.2.2 Determining a solution to Problem 3.2.2 is strongly NP-complete even when $K = 2$.

The reductions used to prove these theorems are rather tricky, and may have applications to problems in the area of task-scheduling on non-uniform processors.

Although Theorems 3.2.1 and 3.2.2 show that segmented channel routing is in general NP-complete, several special cases of the problem are tractable. In this chapter we present polynomial-time algorithms for the following special cases:

Identically Segmented Tracks: Two tracks will be defined to be identically segmented if they have switches at the same locations, and hence, segments of the same length. The *left edge* algorithm used for conventional channel routing can be applied to solve Problems 3.2.1, 3.2.2, and 3.2.3.
1-Segment Routing: A routing can be determined by a linear time ($O(MT)$) greedy algorithm that exploits the geometry of the problem. The corresponding optimization problem can be also solved in polynomial time by reducing it to a weighted maximum bipartite matching problem.
At Most 2-Segments Per Track: If each track is segmented into at most two segments then also a greedy linear time algorithm (similar to the one for 1-Segment routing) can be designed to determine a routing.

We have also developed a general $O(T!M)$-time algorithm using dynamic programming for solving Problems 3.2.1, 3.2.2, and 3.2.3. This general algorithm can be adapted to yield more efficient algorithms for the following cases:
Fixed Number of Tracks: If the number of tracks is fixed then the general algorithm directly yields a polynomial time algorithm.
K-segment Routing: The general algorithm can be modified to yield an $O((K+1)^T M)$-time algorithm. Note that for small values of K the modified algorithm performs better than the general one.
Fixed types of Tracks: If the number of tracks is *unbounded* but the tracks are chosen from a fixed set, where T_i is the number of tracks of type i, then an $O((\prod_1^l T_i^{K+2})M)$ time (hence, a polynomial-time) algorithm can be designed.

3.3. Algorithms

In this section we present algorithms for various special cases of Problems 3.2.1-3.2.3. We first discuss algorithms that exploit the geometry of the segmented channels. We then discuss a general algorithm based on dynamic programming. Finally we discuss a heuristic algorithm (based on linear programming) that appears to work surprisingly well in practice.

3.3.1. *Geometrical Algorithms*

Identically Segmented Tracks

If all tracks are identically segmented (i.e., the locations of the switches are the same in every track), then Problems 3.2.1 and 3.2.2 can be solved by the left edge algorithm [14] in time $O(MT)$. Assign the connections in order of increasing left ends as follows: assign each connection to the first track in which none of the segments it would occupy are yet occupied.

Note that the density of the connections does not provide an upper bound on the number of tracks required for routing (as is the case for conventional routing when the left edge algorithm is used in the absence of vertical constraints). However, if prior to computing the density, the ends of each connection are extended until a column adjacent to a switch is reached, then the density would be a valid upper bound.

1-Segment Routing

If we restrict consideration to 1-segment routings, Problem 3.2.2 can be solved by the following greedy algorithm.

> The connections are assigned in order of increasing left ends as follows. For each connection, find the set of tracks in which the connection would occupy one segment. Eliminate any tracks where this segment is already occupied. From among the remaining tracks, choose one where the unoccupied segment's right end is closest to the left (i.e. the right end coordinate of the segment in the chosen track is the smallest) and assign the connection to it. If there is a tie, then it is broken arbitrarily. In the example of Fig. 13, the algorithm assigns c_1 to s_{11}, c_2 to s_{21}, c_3 to s_{31}, c_4 to s_{32}, and c_5 to s_{13}. The time required is $O(MT)$.

Next we show that if some connection cannot be assigned to any track, then no complete routing is possible. The proof of the following theorem can be found in [25].

Theorem 3.3.1 The above algorithm solves Problem 3.2.2 if $K = 1$.

For 1-segment routing, Problem 3.2.3 may be solved efficiently by reducing it to a bipartite matching problem. The underlying bipartite graph can be constructed as follows: the left side has a node for each connection and the right side a node for each segment. An edge is present between a connection and a segment if the connection can be assigned to the segment's track. The weight $w(c,\ t)$ is assigned to the edge between connection c and a segment in track t. A minimum-weight matching indicates an optimal routing. The time required using the best known matching algorithm (see [24]) is $O(V^3)$, where $V \leq M + NT$ is the number of nodes.

At Most 2-Segments Per Track

In a track with 2-segments, the first segment from the left will be referred to as the *initial* segment and the next one will be referred to as the *end* segment. If the track is unsegmented, i.e. it has only one segment, then for our purposes we will refer to the only segment as an end segment.

The following greedy algorithm, which is similar to the one for 1-Segment routing, can be used to determine a solution to Problem 3.2.1:

> The connections are assigned in order of increasing left ends (ties are resolved arbitrarily). During the execution of the algorithm a track will be considered as *unoccupied* if no connection has been assigned to it.
>
> Now for each connection, determine the set of tracks in which the connection would occupy a single segment. Eliminate any track where this segment is already *occupied*. Now consider the following two cases:
> Case 1. If no track is available (i.e. after the above mentioned elimination of tracks) then append the connection to the pool, P, of unassigned (but already examined) connections.
> Case 2. If tracks are available then assign the connection to a track where the unoccupied segment's right end is closest to the left (i.e. the right end coordinate of the segment in the chosen track is the smallest). If more than one track qualifies then the tie is broken arbitrarily.

Next, if $|P|$ (i.e. the number of unassigned, but already examined, connections) equals the number of tracks unoccupied by any connection, then assign the connections in P to these unoccupied tracks in any order; mark these tracks as occupied, and remove the assigned connections from P. Else, if $|P|$ is greater than the number of such unoccupied tracks then stop, and signal that no valid routing is possible.
Continue with the next connection.
When, all the connections are examined and pool P is non-empty then assign the connections in P to unoccupied tracks.

Theorem 3.3.2 The above mentioned algorithm determines a routing, if there exists one, for the case where every track has at most two segments.

A proof for the above theorem is outlined in [25].

3.3.2. *A General Algorithm for Determining Routing*
Although the problem of determining a routing for a given segmented channel and a set of connections is in general NP-complete, we describe below an algorithm that finds a routing in time linear in M (the number of connections) when T (the number of tracks) is fixed. This is of interest since T is often substantially less than M. The algorithm may also be quite efficient when there are many tracks, but they are segmented in a limited number of ways (see Theorem 7 below). The algorithm first constructs a data structure called an ***assignment graph*** and then reads a valid routing from it. The same algorithm applies to both Problems 3.2.1 and 3.2.2, though with different time and memory bounds. It can also be extended to Problem 3.2.3.

Frontiers and the Assignment Graph

Given a valid routing for connections c_1 through c_i, it is possible to define a *frontier* which constitutes sufficient information to determine how the routing of $c_1 ... c_i$ may be extended to include an assignment of c_{i+1} to a track such that no segment occupied by any of c_1 through c_i will also be occupied by c_{i+1}. Fig. 15 shows an example of a frontier. It will be apparent that c_{i+1} may be assigned to any track t in which the frontier has not advanced past the left end of c_{i+1}. For example, in Fig. 15 connection c_4 can be assigned to track t_2 but not to track t_1.

More precisely, given a valid routing of $c_1, \cdots, c_i$, $1 \le i < M$, define the frontier $\mathbf{x}$ to be a T-tuple ($\mathbf{x}[1]$, $\mathbf{x}[2]$, ..., $\mathbf{x}[T]$) where $\mathbf{x}[j]$ is the leftmost unoccupied column in track t_j at or to the right of column $left(c_{i+1})$. (A column in track t_j is considered unoccupied if the segment present in the column is not occupied.) The frontier is thus a function $\mathbf{x} = F_i(t_{c_1}, \cdots, t_{c_i})$ of the tracks $t_{c_1}, \cdots, t_{c_i}$ to which $c_1 ... c_i$ are respectively assigned. For $i = 0$, let $\mathbf{x} = F_0$, where $F_0[t] = left(c_1)$ for all t. For $i = M$, let $\mathbf{x} = F_M$, where $F_M[t] = N + 1$ for all t.

Next, we describe a graph called the assignment graph which is used to keep track of partial routings and the corresponding frontiers. A node at level i, $1 \le i < M$, of the assignment graph corresponds to a frontier resulting from some valid routing of c_1 through c_i. Level 0 of the graph contains the root node, which corresponds to F_0. If a complete valid routing for $c_1, \cdots, c_M$ exists, then level M of the graph contains a single node corresponding to F_M. Otherwise, level M is empty.

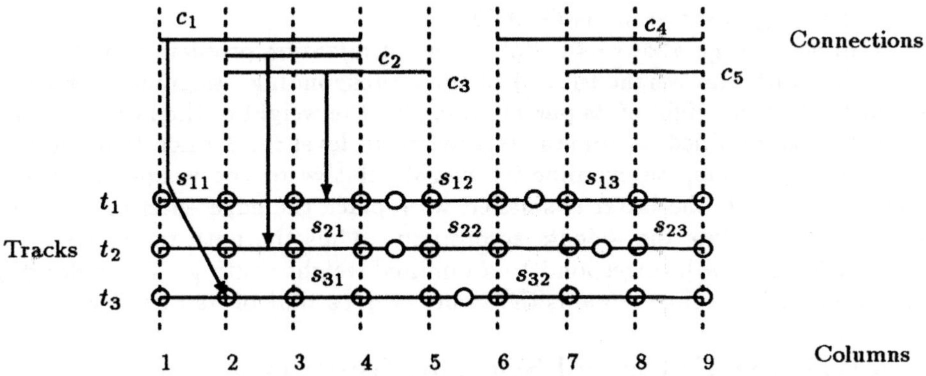

Fig. 15. A frontier for the example of Fig. 13. Connections c_1, c_2, and c_3 are assigned to segments s_{31}, s_{21}, and $\{s_{11}, s_{12}\}$ respectively. The frontier is $\mathbf{x} = [7, 6, 6]$.

The assignment graph is constructed inductively. Given level $i \geq 0$ of the graph, construct level $i+1$ as follows. (For convenience, we identify the node by the corresponding frontier.)

For each node $\mathbf{x}_i$ in level i {
    For each track t_j, $1 \leq j \leq T$ {
        If $\mathbf{x}_i[j] = left(c_{i+1})$ {
            /* c_{i+1} can be assigned to track t_j. */
            Let $\mathbf{x}_{i+1}$ be the new frontier after c_{i+1} is assigned to track t.
            If $\mathbf{x}_{i+1}$ is not yet in level $i+1$ {
                Add node $\mathbf{x}_{i+1}$ to level $i+1$.
                Add an edge from node $\mathbf{x}_i$ to node $\mathbf{x}_{i+1}$. Label it with t_j.
            }
        }
        Else {
            /* $\mathbf{x}_i[j] > left(c_{i+1})$ so c_{i+1} cannot be assigned to track t_j. */
        Continue to next track t_{j+1}.
        }
    }
}

If there are no nodes added at level $i+1$, then there is no valid assignment of c_1 through c_{i+1}.

Searching for the node $\mathbf{x}_{i+1}$ in level $i+1$ can be done in $O(T)$ time using a hash table. Insertion of a new node in the table likewise requires time $O(T)$.

If there are a maximum of L nodes at each level, then construction of the entire assignment graph requires time $O(MLT^2)$. Once the assignment graph has been constructed, a valid routing may be found by tracing a path from the node at level M back to the root, reading the track assignment from the edge labels. (If there is no node at level M, then no complete valid assignment exists.) This takes only $O(M)$

time, so the overall time for the algorithm is $O(MLT^2)$. The memory required to store the assignment graph is $O(MLT)$.

A minor change allows us to solve the optimization problem as well. Each edge is labeled with the weight $w(c, t_c)$ of the corresponding assignment. Each node is labeled with the weight of its parent node plus the weight of the incoming edge. The algorithm is modified as follows. If a search in level $i+1$ finds that the new node $\mathbf{x}_{i+1}$ already exists, we examine its weight relative to the weight of node $\mathbf{x}_i$ plus $w(c_{i+1}, t_{c_{i+1}})$. If the latter is smaller, we replace the edge entering $\mathbf{x}_{i+1}$ with one from $\mathbf{x}_i$ and update the weights accordingly. Thus the path traced back from the node at level M will correspond to a minimal weight routing. The order of growth of the algorithm's time remains the same, as does that of its memory.

Analysis for Unlimited Segment Routing

The following theorem shows that for unlimited segment routing, $L \leq 2\ T!$, so that the time to construct the assignment graph and find an optimal routing is $O(M\ T^2\ T!)$ and the memory required is $O(M\ T\ T!)$. A proof of the theorem can be found in [25].

Theorem 3.3.3 For unlimited segment routing, the number of distinct frontiers that may occur for some valid assignment of c_1 through c_i is at most $2\ T!$.

Analysis for K-Segment Routing

The following theorem shows that for K-segment routing, $L \leq (K+1)^T$, so that the time to construct the assignment graph and find an optimal routing is $O(MT^2(K+1)^T)$ and the memory required is $O(MT(K+1)^T)$ (see [25] for a proof).

Theorem 3.3.4 for K-segment routing, the number of distinct frontiers that may occur for some valid routing of c_1 through c_i is at most $(K+1)^T$.

Case of Many Tracks of a Few Types

Suppose the T tracks fall into two types, with all tracks of each type segmented identically. Then two frontiers that differ only by a permutation among the tracks of each type may be considered equivalent for our purposes in that one frontier can be a precursor of a complete routing if and only if the other can. Thus we can restrict consideration to only one of each set of equivalent frontiers, and strengthen the result of Theorem 6 as follows.

Theorem 3.3.5 Suppose there are T_1 tracks segmented in one way, and $T_2 = T - T_1$ segmented another way. The number of distinct frontiers $\mathbf{x}$ that may occur for some valid K-segment routing of c_1 through c_i, and that satisfy $\mathbf{x}[i] \leq \mathbf{x}[j]$ for all $i < j$ with tracks t_i and t_j of the same type, is $O((T_1T_2)^K)$.

A proof for the above theorem can be found in [25]. It follows that a K-segment routing may be found in time $O(M(T_1T_2)^K T^2)$, and memory $O(M(T_1T_2)^K T)$. The result of Theorem 3.3.5 may easily be generalized to the case of l types of tracks, in which case the time is $O(M(\prod_1^l T_i^K))$, and the memory is $O(M(\prod_1^l T_i^K)T)$.

3.3.3. *A Linear Programming Approach*
Problems 3.2.1 and 3.2.2 can be reduced to 0–1 Linear Programming (LP) problems via a straight-forward reduction procedure. The 0–1 LP is in general NP-Complete. For our purposes, however, such a reduction is interesting because simulations in [25] showed that for almost all cases the corresponding 0–1 LP problems could be solved by viewing them as ordinary LP problems for which efficient algorithms are known.

We now briefly describe the reduction procedure for Problem 3.2.1. The corresponding reduction for Problem 3.2.2 follows after minor modifications. Let us define binary variables x_{ij}, for $1 \leq i \leq M$, and $1 \leq j \leq T$, as follows: if $x_{ij} = 1$, then connection c_i is assigned to track t_j, else if $x_{ij} = 0$, then connection c_i is not assigned to track t_j. Since in a routing each connection is assigned to at most one track, one has the following constraints:

$$\sum_{j=1}^{T} x_{ij} \leq 1; \quad \forall\ 1 \leq i \leq M$$

One also has to make sure that in any routing two connections assigned to the same track must not share a segment. Consider a track t_j; one can then easily determine sets of connections $P_{j1}, \cdots, P_{jl_j}$ (not necessarily disjoint) such that at most one from each set can be assigned to the track t_j. Hence for each such set P_{jk}, one must satisfy

$$\sum_{c_i \in P_{jk}} x_{ij} \leq 1$$

Finally, one must make sure that all the connections are routed, this can be ensured by maximizing the following objective function:

$$\sum_{i=1}^{M} \sum_{j=1}^{T} x_{ij}$$

One can now easily verify that the above $0-1$ LP's objective function achieves the value of M if and only if there is a solution to Problem 3.2.1 [25].

4. Conclusion and Open Problems

We have discussed some of the important results for the rectilinear polygon cover problems. However, the following problems still remain open and may be worth investigating further:

- Can we prove a better upper bound for the performance ratio of the sweepline heuristic (or any other heuristic) for the interior cover problem for polygons with holes? There is currently no example known in which the sweepline heuristic has a performance ratio more than 3.
- Prove or disprove if $\frac{\theta}{\alpha} \leq c$ for some positive constant c for the interior cover problem.
- Prove or disprove that the corner cover problem is NP-complete for rectilinear polygons without holes.

We have also introduced problems concerning the design and routing for segmented channels. There are several open issues in this new area of routing. For example, although efficient algorithms for many special cases of the routing problem have been developed, several other interesting cases are yet to be solved; following are some relevant ones: 1. connection lengths are bounded; and 2. connections are non-overlapping. An important open problem is to develop efficient algorithms for approximate segmented channel routing. Also, efficient algorithms for the generalized routing problems are not known.

Acknowledgements

The first author wishes to thank Prof. Piotr Berman with whom part of the research work on this problem was carried out and Prof. Georg Schnitger for suggesting the problem of non-approximability of the interior cover problem. The second author would like to thank Prof. Abbas El Gamal and Dr. Jonathan Greene with whom parts of the research on segmented channel routing was carried out.

References

1. S. Arora, C. Lund, R. Motwani, M. Sudan and M. Szegedy. Proof Verification and Hardness of Approximation Problems. *Proc. 33rd IEEE Symp. Found. of Comp. Sc.(FOCS)* (1992), pp. 14-23.
2. P. Berman and B. DasGupta, Results on Approximation of the Rectilinear Polygon Cover Problems, Tech Rep. # CS-92-07, Department of Computer Science, Penn State, 1992.
3. P. Berman and G. Schnitger. On the Complexity of Approximating the Independent Set Problem. *Symp. on Theo. Aspects of Computing(STACS)* (1989), pp. 256-268.
4. S. Chaiken, D. J. Kleitman, M. Saks and J. Shearer. Covering Regions by Rectangles. *SIAM J. Algebraic Discrete methods*, 2 (1981), pp. 394-410.
5. H. E. Conn, and J. O'Rourke. Some restricted rectangle covering problems. Tech Rep. JHU-87/13, The Johns Hopkins University, Baltimore, MD, 1987, also appeared in *Proc. of the 1987 Allerton Conference* (1987), pp. 898-907.
6. T. H. Corman, C. E. Leiserson and R. L. Rivest. *Introduction to Algorithms.* The MIT Press, Cambridge, Massachusetts, 1990.
7. J. Culberson and R. A. Reckhow. Covering Polygon is Hard. *Proc. 29th IEEE Symp. Found. of Comp. Sc.(FOCS)* (1988), pp. 601-611.
8. D. S. Franzblau. Performance Guarantees on a Sweep-line Heuristic for Covering Rectilinear Polygons with Rectangles. *SIAM J. Disc. Math.*, 2 (1989), pp. 307-321.
9. D. S. Franzblau and D. J. Kleitman. An Algorithm for Constructing Regions with Rectangles. *Proc. 16th ACM Symp. Theory of Comput.(STOC)*, 1984, pp. 167-174.
10. A. El Gamal. Two Dimensional Stochastic Model for Interconnections in Master Slice Integrated Circuits. *IEEE Trans. on Circuits and Systems*, CAS-28, 127-138, February, 1981.
11. A. El Gamal, J. Greene and V. P. Roychowdhury. Segmented Channel Routing is as Efficient as Conventional Routing (and Just as Hard). *Proc. 13th Conference on Advanced Research in VLSI*, UC Santa Cruz, pp. 192-211, March 1991.
12. A. El Gamal, J. Greene, J. Reyneri, E. Rogoyski, K. El-Ayat, and A. Mohsen. An Architecture for Electrically Configurable Gate Arrays. *IEEE J. Solid-State Circuits*, Vol. 24, No. 2, April, 1989, pp. 394-398.
13. M. R. Garey and D. S. Johnson. *Computers and Intractability - A Guide to the Theory of NP-Completeness.* W. H. Freeman & Co., 1979.
14. A. Hashimoto and J. Stevens. Wire Routing by Optimizing Channel Assignment Within Large Apertures. *Proc. 8th IEEE Design Automation Workshop*, 1971.
15. H. Hsieh, et al. A Second Generation User Programmable Gate Array. *Proc. Custom Integrated Circuits Conf.*, May 1987, pp. 515-521.

16. M. Lorenzetti and D. S. Baeder. Routing. Chapter 5 in *Physical Design Automation of VLSI Systems*, B. Preas and M. Lorenzetti, eds. Benjamin/Cummings, 1988.
17. A. Lubiw. Ordering and some Combinatorial Optimization Problems with some Geometric Applications. Ph. D. Thesis, University of Waterloo, Ontario, Canada, 1985.
18. A. Lubiw. The Boolean Basis Problem and how to Cover Polygons by Rectangles. University of Waterloo, Ontario, Canada, Manuscript, 1988. Referenced in [8].
19. W. J. Masek. Some NP-complete Set Covering Problems. MIT, Cambridge, MA, unpublished manuscript. Referenced in [13].
20. C. Mead and L. Conway. *Introduction to VLSI Systems*. Addision-Wesley, Reading, MA., 1980, Chapters 2 and 4.
21. R. Motwani, A. Raghunathan and H. Saran. Perfect Graphs and Orthogonally Convex Covers. *4th SIAM Conf. on Discr. Math.* (1988).
22. T. Ohtsuki. Minimum Dissection of Rectilinear Polygons. *Proc. IEEE Symp. on Circuits and Systems* (1982), pp. 1210-1213.
23. L. Pagli, E. Lodi, F. Lucçio, C. Mugnai and W. Lipski. On Two Dimensional Data Organization 2. *Fund. Inform.*, 2 (1979), pp. 211-226.
24. C. H. Papadimitrou and K. Steiglitz. Combinatorial Optimization: Algorithms and Complexity. Prentice Hall, Inc., New Jersey, 1982.
25. V. P. Roychowdhury, J. Greene and A. El Gamal. Segmented Channel Routing. *IEEE Trans. on Computer-Aided Design of Integrated Circuits and Systems*, vol. 12, No. 1, pp. 75-95, January 1993.
26. T. Szymanski. Dogleg Channel Routing is NP-Complete. *IEEE Trans. CAD*, CAD-4(1), pp. 31-41, Jan. 1985.

A SCALED GRADIENT PROJECTION ALGORITHM FOR LINEAR COMPLEMENTARITY PROBLEMS

JIU DING
Department of Mathematics
The University of Southern Mississippi
Hattiesburg, MS 39406-5045, U.S.A.

Abstract. A new interior point algorithm is presented for linear complementarity problems. Unlike the homotopy method that follows the path of centers or the potential reduction method that minimizes the potential function, our method minimizes the corresponding equivalent quadratic programming problem directly, using the scaled gradient projection technique.

Key words: LCP's, Interior point methods

1. Introduction

Let $R^{n\times n}$ denote the space of all $n \times n$ real matrices and R^n the n-dimensional Euclidean space. The linear complementarity problem (abbreviated as the LCP), that of finding $(x, y) \in R^n \times R^n$ which satisfies

$$y = Mx + q, \quad (x, y) \geq 0, \quad x^T y = 0, \tag{1}$$

where $M \in R^{n\times n}$ and $q \in R^n$, has been studied over decades because of its many applications in linear and convex quadratic programs, bimatrix games, and the others [1].

Interior point methods have attracted tremendous attention in mathematical programming since Karmarkar published his seminal paper [4] in 1984. Recently, several new polynomial-time algorithms have been proposed for solving LCP's. Kojima et al. [7] gave the first polynomial-time algorithm for a class of LCP's with positive semidefinite matrices, based on the theoretical formulation given by Megiddo [9]. A generalization of their method was presented in [3] using the Euler prediction. A potential-reduction method was proposed in [5]. For some other recent work, see [2] [6] [8] [10] [11] [12].

In this paper we propose a new interior point algorithm for solving general linear complementarity problems. Unlike the homotopy continuation method which follows the path of "centers" resulting from the perturbed Karush-Kuhn-Tucker condition, and unlike the potential reduction method which employs the concept of "potential functions" to guarantee the sufficient decrease in the objective function, our method just minimizes directly the quadratic programming problem which is equivalent to the original problem. Starting from a strictly feasible point, the algorithm generates a sequence of iterates in the interior of the feasible set, with the help of the technique of projected scaled gradients.

Ding-Zhu Du and Jie Sun (eds.), Advances in Optimization and Approximation, 58–67.

The next section will give the description of the algorithm. The convergence analysis of the method is presented in Section 3. We explore the polynomial-time complexity of the algorithm in Section 4, and some conclusions and final remarks are given in the last section.

2. The Algorithm

Denote by R^n_+ and R^n_{++} the nonnegative and positive orthants of R^n, respectively. Let

$$S = \{(x, y) \in R^n_+ \times R^n_+ : y = Mx + q\}$$

be the feasible set of the LCP (1), and let

$$S_{int} = \{(x, y) \in R^n_{++} \times R^n_{++} : y = Mx + q\}$$

be the (relative) interior of S. We assume that $S_{int} \neq \emptyset$ and that the LCP has a solution.

Throughout the paper $\|x\| = (\sum_{i=1}^n x_i^2)^{1/2}$ denotes the usual Euclidean spectral norm for $x \in R^n$, and $\|A\|$ stands for the corresponding matrix norm of $A \in R^{n\times n}$. Thus, $\|A\| = \sup\{\|Ax\| : \|x\| \leq 1\}$ is the square root of the maximal eigenvalue of A^TA. For $v \in R^m, v > 0$ means $v_i > 0$ for $i = 1, \ldots, m$. We write $X = \text{diag}(x_1, \ldots, x_n) \in R^{n\times n}$ for $x = (x_1, \ldots, x_n)^T \in R^n$. We use I to denote an identity matrix of some dimension from the context.

The LCP (1) is equivalent to the following quadratic programming problem (abbreviated as the QP) with the optimal value 0:

$$\min\ x^Ty, \text{ subject to } y = Mx + q,\ (x, y) \in R^n_+ \times R^n_+. \tag{2}$$

In this paper it is the QP (2) that we want to solve to solve the original LCP (1).

Suppose an initial strictly feasible point $(x^1, y^1) \in S_{int}$ is known. Given a precision $\epsilon > 0$. The general algorithm goes as follows.

Step 0: Let $k = 1$.

Step 1: If $(x^k)^Ty^k < \epsilon$, then stop. Otherwise go to Step 2.

Step 2: Let

$$p^k = \begin{bmatrix} p_x^k \\ p_y^k \end{bmatrix} = P_{N([(Y^k)^{-1}MX^k\ -I])}\begin{bmatrix} X^ky^k \\ X^ky^k \end{bmatrix},$$

where P_B is the orthogonal projection into a subspace B and $N(C)$ is the null space of a matrix C. Then define

$$\begin{bmatrix} \delta_x^k \\ \delta_y^k \end{bmatrix} = -\beta_k\begin{bmatrix} X^kp_x^k \\ Y^kp_y^k \end{bmatrix}/\|p^k\|,$$

where $\beta_k = \min\{\frac{\|p^k\|}{\|X^kY^k\|}, \frac{1}{2}\}$.

Step 3: Let

$$x^{k+1} = x^k + \delta_x^k,\quad y^{k+1} = y^k + \delta_y^k,$$

and $k := k+1$. Then go to Step 1.

Remark 1 The direction (δ_x^k, δ_y^k) is actually the solution of the following problem:

$$\min\ (y^k)^T\delta_x + (x^k)^T\delta_y,$$

subject to

$$\delta_y = M\delta_x,\quad \|(X^k)^{-1}\delta_x\|^2 + \|(Y^k)^{-1}\delta_y\|^2 \le (\beta_k)^2.$$

Thus, the searching direction of our method is a scaled gradient projection.

Remark 2 The initial point $(x^1, y^1) \in S_{int}$ can be obtained through the "phase I" procedure by means of the construction of an "artificial LCP" as discussed in [6].

In the following, we shall estimate the extent of the decrease in x^Ty after one iteration. Furthermore, we shall investigate the property of the polynomial-time complexity of the algorithm.

3. Convergence Analysis

Now we give a brief analysis of the previous algorithm. To this purpose, we need to explore the implementation of the algorithm for each cycle of the iteration.

Suppose $(x, y) \in S_{int}$ and

$$(x', y') = (x + \delta_x, y + \delta_y)$$

where

$$\begin{bmatrix}\delta_x\\ \delta_y\end{bmatrix} = -\beta\begin{bmatrix}Xp_x\\ Yp_y\end{bmatrix}/\|p\|$$

for some $\beta \in (0,1)$ and

$$p = \begin{bmatrix}p_x\\ p_y\end{bmatrix} = P_{N([Y^{-1}MX\ -I])}\begin{bmatrix}Xy\\ Xy\end{bmatrix}.$$

It is easy to see that $(x', y') \in S_{int}$. Furthermore,

$$\begin{aligned} x'^T y' &= (x+\delta_x)^T(y+\delta_y) = x^Ty + \delta_x^T\delta_y + \delta_x^T y + \delta_y^T x \\ &= x^Ty - \beta\frac{(Xp_x)^Ty + (Yp_y)^Tx}{\|p\|} + \beta^2\frac{p_x^T XYp_y}{\|p\|^2}. \end{aligned}$$

Denote $A \equiv A(x,y) = Y^{-1}MX$.

Proposition 3.1

$$\begin{aligned} p &= \begin{bmatrix}p_x\\ p_y\end{bmatrix} = \{I - \begin{bmatrix}A^T\\ -I\end{bmatrix}(I + AA^T)^{-1}[A\ \ -I]\}\begin{bmatrix}Xy\\ Xy\end{bmatrix} \\ &= \begin{bmatrix}Xy\\ Xy\end{bmatrix} - \begin{bmatrix}XM^T\\ -Y\end{bmatrix}(Y^2 + MX^2M^T)^{-1}(MX - Y)Xy. \end{aligned}$$

Proof. From the theory of the Moore-Penrose generalized inverse, we have

$$\begin{aligned}
p &= P_{N([A\ -I])}\begin{bmatrix} Xy \\ Xy \end{bmatrix} = (I - P_{R([A\ -I]^T)})\begin{bmatrix} Xy \\ Xy \end{bmatrix} \\
&= (I - \begin{bmatrix} A^T \\ -I \end{bmatrix}\begin{bmatrix} A^T \\ -I \end{bmatrix}^+)\begin{bmatrix} Xy \\ Xy \end{bmatrix} \\
&= \{I - \begin{bmatrix} A^T \\ -I \end{bmatrix}([A\ -I]\begin{bmatrix} A^T \\ -I \end{bmatrix})^{-1}[A\ -I]\}\begin{bmatrix} Xy \\ Xy \end{bmatrix} \\
&= \{I - \begin{bmatrix} A^T \\ -I \end{bmatrix}(I + AA^T)^{-1}[A\ -I]\}\begin{bmatrix} Xy \\ Xy \end{bmatrix} \\
&= \{I - \begin{bmatrix} XM^TY^{-1} \\ -I \end{bmatrix}(I + Y^{-1}MX^2M^TY^{-1})^{-1}[Y^{-1}MX\ -I]\}\begin{bmatrix} Xy \\ Xy \end{bmatrix} \\
&= \{I - \begin{bmatrix} XM^T \\ -Y \end{bmatrix}(Y^2 + MX^2M^T)^{-1}[MX\ -Y]\}\begin{bmatrix} Xy \\ Xy \end{bmatrix} \\
&= \begin{bmatrix} Xy \\ Xy \end{bmatrix} - \begin{bmatrix} XM^T \\ -Y \end{bmatrix}(Y^2 + MX^2M^T)^{-1}(MX - Y)Xy. \text{Q.E.D.}
\end{aligned}$$

Since $\text{Rank}([A\ -I]) = \text{Rank}([I\ A^T]^T) = n$ and

$$[A\ -I]\begin{bmatrix} I \\ A \end{bmatrix} = 0,$$

$N([A\ -I]) = R([I\ A^T]^T)$. Thus we have another expression for p.

Proposition 3.2

$$p = \begin{bmatrix} I \\ A \end{bmatrix}(I + A^TA)^{-1}(I + A)^TXy = \begin{bmatrix} X^{-1} \\ Y^{-1}M \end{bmatrix}(X^{-2} + M^TY^{-2}M)^{-1}(y + M^Tx).$$

Proof.

$$\begin{aligned}
p &= P_{N([A\ -I])}\begin{bmatrix} Xy \\ Xy \end{bmatrix} = P_{R([I\ A^T]^T)}\begin{bmatrix} Xy \\ Xy \end{bmatrix} = \begin{bmatrix} I \\ A \end{bmatrix}\begin{bmatrix} I \\ A \end{bmatrix}^+\begin{bmatrix} Xy \\ Xy \end{bmatrix} \\
&= \begin{bmatrix} I \\ A \end{bmatrix}([I\ A^T]\begin{bmatrix} I \\ A \end{bmatrix})^{-1}[I\ A^T]\begin{bmatrix} Xy \\ Xy \end{bmatrix} \\
&= \begin{bmatrix} I \\ A \end{bmatrix}(I + A^TA)^{-1}[I\ A^T]\begin{bmatrix} Xy \\ Xy \end{bmatrix} \\
&= \begin{bmatrix} I \\ A \end{bmatrix} + (I + A^TA)^{-1}(I + A)^TXy \\
&= \begin{bmatrix} X^{-1} \\ Y^{-1}M \end{bmatrix}(X^{-2} + M^TY^{-2}M)^{-1}(y + M^Tx). \text{Q.E.D.}
\end{aligned}$$

Proposition 3.3 *If $y \neq -M^Tx$, then $p \neq 0$. In particular, if M is a copositive matrix or a P_0-matrix, then $p \neq 0$ at any $(x, y) \in S_{int}$.*

Proof. Suppose $p = 0$. Then there exists a vector $\eta \in R^n$ such that

$$\begin{bmatrix} Xy \\ Xy \end{bmatrix} = \begin{bmatrix} A^T \\ -I \end{bmatrix} \eta$$

which implies that $Xy = -A^T Xy$. It follows that

$$y = -M^T Y^{-1} \eta = -M^T Y^{-1} Xy = -M^T x.$$

If M is copositive, then $y^T x = -x^T M^T x \leq 0$ which is impossible since $x > 0$ and $y > 0$. Now suppose M is a P_0-matrix. Then so is M^T. From Theorem 3.4.2 in [1], there exists an index k such that $x_k (M^T x)_k \geq 0$. On the other hand, we have

$$0 < x_i y_i = -x_i (M^T x)_i, \quad i = 1, \ldots, n,$$

which leads to a contradiction. **Q.E.D.**

Let $H \equiv H(x, y) = 2I - (I - A)^T (I + AA^T)^{-1} (I - A)$ and $||Xy||_H^2 = (Xy)^T H(x, y) Xy$. Note that $H(x, y)$ is symmetric and positive semidefinite.

Proposition 3.4

$$||p|| = ||Xy||_H.$$

Proof.

$$\begin{aligned}
||p||^2 &= ||P_{N([A\ -I])} \begin{bmatrix} I \\ I \end{bmatrix} Xy||^2 = (Xy)^T [I\ I] P_{N([A\ -I])} \begin{bmatrix} I \\ I \end{bmatrix} Xy \\
&= (Xy)^T [I\ I] \{ I - \begin{bmatrix} A^T \\ -I \end{bmatrix} ([A\ -I] \begin{bmatrix} A^T \\ -I \end{bmatrix})^{-1} [A\ -I] \} \begin{bmatrix} I \\ I \end{bmatrix} Xy \\
&= (Xy)^T \{ [I\ I] \begin{bmatrix} I \\ I \end{bmatrix} - [I\ I] \begin{bmatrix} A^T \\ -I \end{bmatrix} \\
&\qquad ([A\ -I] \begin{bmatrix} A^T \\ -I \end{bmatrix})^{-1} [A\ -I] \begin{bmatrix} I \\ I \end{bmatrix} \} Xy \\
&= (Xy)^T [2I - (I - A)^T (I + AA^T)^{-1} (I - A)] Xy \\
&= (Xy)^T H(x, y) Xy = ||Xy||_H^2. \textbf{Q.E.D.}
\end{aligned}$$

Similarly if we let $S \equiv S(x, y) = (I + A)(I + A^T A)^{-1} (I + A)^T$ and $||Xy||_S^2 = (Xy)^T S(x, y) Xy$, we have another expression for $||p||$.

Proposition 3.5

$$||p|| = ||Xy||_S.$$

Proof.

$$||p||^2 = ||P_{N([A\ -I])} \begin{bmatrix} I \\ I \end{bmatrix} Xy||^2 = ||P_{R([I\ A^T]^T)} \begin{bmatrix} I \\ I \end{bmatrix} Xy||^2$$

$$= (Xy)^T[I\ I]P_{R([I\ A^T]^T)}[{I \atop I}]Xy$$

$$= (Xy)^T[I\ I]\{[{I \atop A}]([I\ A^T][{I \atop A}])^{-1}[I\ A^T]\}[{I \atop I}]Xy$$

$$= (Xy)^T(I+A)(I+A^TA)^{-1}(I+A)^TXy$$

$$= (Xy)^TS(x,y)Xy = ||Xy||_S^2.\text{Q.E.D.}$$

Remark 3 We have actually proved that for any $n \times n$ matrix A,

$$[{I_n \atop A}](I_n + A^TA)^{-1}[I_n\ A^T] + [{A^T \atop -I_n}](I_n + AA^T)^{-1}[A\ \ -I_n] = I_{2n}$$

and consequently,

$$(I_n + A)(I_n + A^TA)^{-1}(I_n + A)^T + (I_n - A)^T(I_n + AA^T)^{-1}(I_n - A) = 2I_n.$$

Proposition 3.6

$$(Xp_x)^Ty + (Yp_y)^Tx = ||p||^2.$$

Proof.

$$(Xp_x)^Ty + (Yp_y)^Tx = p_x^TXy + p_y^TXy = [p_x^T\ p_y^T][{Xy \atop Xy}] = ||p||^2.\text{Q.E.D.}$$

Now it is easy to prove the following basic result.

Theorem 3.1

$$(x')^Ty' \leq x^Ty - \beta||p|| + \frac{\beta^2}{2}||XY||.$$

Proof. From the previous lemma, using some inequalities, we obtain

$$\begin{aligned}(x')^Ty' &= x^Ty - \beta\frac{||p||^2}{||p||} + \beta^2\frac{p_x^TXYp_y}{||p||^2} \\ &\leq x^Ty - \beta||p|| + \beta^2\frac{||p_x||||XY||||p_y||}{||p||^2} \\ &\leq x^Ty - \beta||p|| + \beta^2||XY||\frac{||p_x||^2 + ||p_y||^2}{2||p||^2} \\ &= x^Ty - \beta||p|| + \frac{\beta^2}{2}||XY||.\text{Q.E.D.}\end{aligned}$$

4. Complexity Analysis

In this section we analyze the complexity of the algorithm. From Theorem 3.1, it is important to estimate $||p||$ from below in terms of $||XY||$ or x^Ty for the complexity analysis. Proposition 3.4 and Proposition 3.5 give that

$$||p||^2 = (Xy)^TH(x,y)Xy = (Xy)^TS(x,y)Xy,$$

where

$$H(x,y) = 2I - (I - Y^{-1}MX)^T(I + Y^{-1}MX^2M^TY^{-1})^{-1}(I - Y^{-1}MX),$$
$$S(x,y) = (I + Y^{-1}MX)(I + XM^TY^{-2}MX)^{-1}(I + Y^{-1}MX)^T.$$

From Theorem 3.1, the complexity of the algorithm depends on a lower bound of $||p||$ in terms of $||XY||$ where (x,y) vary in a suitable subset of S_{int}. Since $||p||$ is the norm induced by the symmetric positive definite matrices $H(x,y)$ or $S(x,y)$, the lower bound problem of $||p||$ is determined by the linear algebra problem of minimal eigenvalues of $H(x,y) = S(x,y)$.

As defined by Kojima et al. in [6], let $\lambda(x,y)$ be the smallest eigenvalue of $S(x,y)$, let

$$\lambda(M) = \inf\{\lambda(x,y) : (x,y) \in S_{int},\},$$

let

$$\lambda_K(M) = \inf\{\lambda(x,y) : (x,y) \in S_{int}, x_iy_i \leq K,\ i = 1,\cdots,n\}$$

with $K > 0$, and let

$$\lambda_K^* = \min\{\lambda(x,y) : (x,y) \in S_{int},\ x^Ty \geq \epsilon,\ x_iy_i \leq K,\ i = 1,\cdots,n\}$$

where $\epsilon > 0$ is a fixed parameter. Noting that the value of x^Ty is decreased throughout the implementation of the algorithm, we have immediately

Proposition 4.1 *Throughout the execution of the algorithm with initial point* (x^1, y^1),

$$||p|| \geq \sqrt{\lambda(x,y)}||Xy|| \geq \sqrt{\lambda_K(M)}||Xy||$$

with $K = (x^1)^Ty^1$.

By Theorem 3.1,

$$(x')^Ty' \leq x^Ty - \beta||p|| + \frac{\beta^2}{2}||XY||$$

after one iteration. Let $\phi(\beta) = -||p||\beta + ||XY||\frac{\beta^2}{2}$ for $\beta > 0$, then ϕ obtains its minimum value $-\frac{||p||^2}{2||XY||}$ at $\beta = \frac{||p||}{||XY||}$. Let

$$\beta = \min\{\frac{||p||}{||XY||}, \frac{1}{2}\}.$$

Suppose for a class of matrices M, we have $\lambda_K^* = \alpha^2 > 0$ where $\alpha > 0$ and $K = (x^1)^Ty^1$. Then throughout the implementation of the algorithm starting at $(x^1, y^1) \in S_{int}$,

$$||p|| \geq \alpha||Xy||.$$

If $||p|| \leq \frac{1}{2}||XY||$, then $\beta = \frac{||p||}{||XY||}$. Thus,

$$\begin{aligned}(x')^Ty' &\leq x^Ty - \frac{||p||^2}{2||XY||} \leq x^Ty - \frac{\lambda_K^*||Xy||^2}{2||XY||} \\ &\leq x^Ty - \frac{\lambda_K^*}{2}||Xy|| \leq (1 - \frac{\lambda_K^*}{2\sqrt{n}})x^Ty.\end{aligned}$$

If $\|p\| > \frac{1}{2}\|XY\|$, then $\beta = \frac{1}{2}$. Hence,

$$\begin{aligned}(x')^T y' &\leq x^T y - \frac{1}{2}\|p\| + \frac{1}{8}\|XY\| < x^T y - \frac{1}{2}\|p\| + \frac{1}{4}\|p\| = x^T y - \frac{1}{4}\|p\| \\ &\leq x^T y - \frac{\alpha}{4}\|Xy\| \leq x^T y - \frac{\alpha}{4\sqrt{n}} x^T y = (1 - \frac{\alpha}{4\sqrt{n}}) x^T y.\end{aligned}$$

Therefore we obtain

Proposition 4.2 *Let M be in some class of matrices such that $\lambda_K^* > 0$ where $K = (x^1)^T y^1$. Starting at $(x^1, y^1) \in S_{int}$, the algorithm generates a point $(x, y) \in S_{int}$ such that $x^T y \leq \epsilon$ in a polynomial time in terms of L, $|\ln \epsilon|$, and $1/\lambda_K^*$, where L is the input size.*

Now assume M is a P-matrix. Then it was shown in [6] that $\lambda_K^* > 0$. Hence we have the following result.

Corollary 4.1 *Let M be a P-matrix. Given $\epsilon > 0$ and starting at $(x^1, y^1) \in S_{int}$, the algorithm generates a point $(x, y) \in S_{int}$ such that $x^T y \leq \epsilon$ in a polynomial time in terms of L, $|\ln \epsilon|$, and $1/\lambda_K^*$.*

Now we try to explore the complexity of the algorithm for other matrices. First we assume $q = 0$ in the LCP. In this case, M must be an S-matrix for the algorithm to work, that is, there exists a $z > 0$ such that $Mz > 0$. We show that our algorithm solves the special LCP in polynomial time. Then we cast the general LCP to the special one. Let $(x, y) \in S_{int}$ as before, let

$$\pi = (Y^2 + MX^2M^T)^{-1}(Y - MX)Xy,$$

and let $\bar{x} = x - \pi$ and $\bar{y} = y + M^T\pi$. Then

$$p = \begin{bmatrix} X(y + M^T\pi) \\ Y(x - \pi) \end{bmatrix} = \begin{bmatrix} X\bar{y} \\ Y\bar{x} \end{bmatrix}.$$

Rewrite p as

$$p = \begin{bmatrix} X\bar{y} \\ Y\bar{x} \end{bmatrix} - \frac{\bar{\Delta}}{2n} e + \frac{\bar{\Delta}}{2n} e,$$

where $\bar{\Delta} = x^T\bar{y} + y^T\bar{x}$ and e is the $2n$-dimensional vector of all 1's. Then,

$$\begin{aligned}\|p\|^2 &= \|\begin{bmatrix} X\bar{y} \\ Y\bar{x} \end{bmatrix} - \frac{\bar{\Delta}}{2n} e + \frac{\bar{\Delta}}{2n} e\| \\ &= \|\begin{bmatrix} X\bar{y} \\ Y\bar{x} \end{bmatrix} - \frac{\bar{\Delta}}{2n} e\|^2 + \|\frac{\bar{\Delta}}{2n} e\|^2 \\ &\geq \|\frac{\bar{\Delta}}{2n} e\|^2 = \frac{\bar{\Delta}^2}{2n}.\end{aligned}$$

Since $q = 0$, we have

$$\begin{aligned}\bar{\Delta} &= x^T\bar{y} + y^T\bar{x} = x^T(y + M^T\pi) + y^T(x - \pi) \\ &= 2x^T y + \pi^T Mx - \pi^T y = 2x^T y - q^T\pi = 2x^T y.\end{aligned}$$

Thus, we have proved the following proposition.

Proposition 4.3 *If $q = 0$ in the LCP (1), then*

$$\|p\| \geq \frac{\sqrt{2}x^T y}{\sqrt{n}}.$$

Proposition 4.4 *Under the same assumption as above, if $\|p\| \leq \frac{1}{2}\|XY\|$, then*

$$(x')^T y' \leq (1 - \frac{1}{n})x^T y.$$

Otherwise,

$$(x')^T y' \leq (1 - \frac{1}{2\sqrt{2n}})x^T y.$$

Proof. If $\|p\| \leq \frac{1}{2}\|XY\|$, then $\beta = \frac{\|p\|}{\|XY\|}$, and from Proposition 4.3,

$$(x')^T y' \leq x^T y - \frac{\|p\|^2}{2\|XY\|} \leq x^T y - \frac{1}{n}\frac{(x^T y)^2}{\|XY\|} \leq (1 - \frac{1}{n})x^T y.$$

Now suppose $\|p\| > \frac{1}{2}\|XY\|$, then $\beta = \frac{1}{2}$. Hence, from Proposition 4.3,

$$\begin{aligned}(x')^T y' &\leq x^T y - \frac{1}{2}\|p\| + \frac{1}{8}\|XY\| < x^T y - \frac{1}{2}\|p\| + \frac{1}{4}\|p\| = x^T y - \frac{1}{4}\|p\| \\ &\leq x^T y - \frac{1}{2\sqrt{2}\sqrt{n}}x^T y = (1 - \frac{1}{2\sqrt{2n}})x^T y.\text{Q.E.D.}\end{aligned}$$

The following theorem is an easy consequence of the above analysis.

Theorem 4.1 *Suppose $q = 0$, then a solution to the LCP (1) can be obtained by the algorithm in polynomial time and $(x^k)^T y^k$ converge to zero at least linearly.*

Remark 4 It is not known under what conditions the algorithm finds a nonzero solution of the special LCP if it exists.

Now we use our algorithm to solve the general LCP with $q \neq 0$. Given the LCP (1), let

$$\tilde{M} = \begin{bmatrix} M & q \\ d^T & \mu \end{bmatrix}, \ \tilde{x} = \begin{bmatrix} x \\ \eta \end{bmatrix}, \ \tilde{y} = \begin{bmatrix} y \\ \xi \end{bmatrix},$$

where $d \in R^n$ and $\mu \in R$. Consider the following LCP :

$$\tilde{y} = \tilde{M}\tilde{x}, \ (\tilde{x}, \tilde{y}) \geq 0, \ \tilde{x}^T \tilde{y} = 0. \tag{3}$$

Then it is easy to see that if $(\tilde{x}, \tilde{y})$ is a solution of (3) with $\tilde{x}_{n+1} = \eta > 0$, then $(\frac{x}{\eta}, \frac{y}{\eta})$ is a solution of (1).

Theorem 4.2 *Suppose a strictly feasible point $(x^1, y^1) \in S_{int}$ is given in the LCP (1). Then the algorithm can be used to obtain a solution $(\tilde{x^*}, \tilde{y^*})$ to the LCP (3) in polynomial time. If $\tilde{x^*}_{n+1} = \eta^* > 0$, then $(\frac{x^*}{\eta^*}, \frac{y^*}{\eta^*})$ is a solution to the LCP (1).*

Proof. It is easy to see that for the given $(x^1, y^1) \in S_{int}$, we can find a $d \in R^n$ and $\mu \in R^1$ such that

$$(\tilde{x}^1, \tilde{y}^1) = \left(\begin{bmatrix} x^1 \\ 1 \end{bmatrix}, \begin{bmatrix} y^1 \\ 1 \end{bmatrix}\right)$$

is a strictly feasible initial point to the LCP (3). For example, we may choose $\mu = -1$ and $d = \frac{2}{\|x^1\|^2} x^1$. Now Theorem 4.1 shows that our algorithm solves the LCP (3) in polynomial time. The last conclusion of the theorem follows from Proposition 4.1. **Q.E.D.**

5. Conclusions

In this paper, we proposed a polynomial-time algorithm for the general linear complementarity problem. The method is based on the scaled gradient projection technique. We obtained a complexity result for solving the LCP with a P-matrix. Moreover, our algorithm can be applied to a special LCP with $q = 0$ in polynomial time, and can also be used for the general LCP. A problem that remains to be investigated is the case when the solution to the augmented LCP does not give rise to a solution to the original LCP.

References

1. R. W. Cottle, J.-S. Pang, and R. E. Stone, The Linear Complementarity Problem, Academic Press, New York, 1992.
2. J. Ding, "A continuation algorithm for a class of linear complementarity problems using an extrapolation technique," Linear Algb. Appl., 186, 199-214, 1993.
3. J. Ding and T. Y. Li, "A polynomial-time predictor-corrector algorithm for a class of linear complementarity problems," SIAM J. Optimization, Vol. 1, No. 1, 83-92, 1991.
4. N. Karmarkar, "A new polynomial-time algorithm for linear programming," Combinatorica, 4, 373-395, 1984.
5. M. Kojima, N. Megiddo, T. Noma, and A. Yoshise, A Unified Approach to Interior Point Algorithms for Linear Complementarity Problems, Lecture Notes in Computer Science 538, Springer-Verlag, New York, 1991.
6. M. Kojima, N. Megiddo, and Y. Ye, "An interior point potential reduction algorithm for the linear complementarity problem," Math. Programming, 54, 267-279, 1992.
7. M. Kojima, S. Mizuno, and A. Yoshise, "A polynomial-time algorithm for a class of linear complementarity problems," Math. Programming, 44, 1-26, 1989.
8. M. Kojima, S. Mizuno, and A. Yoshise, "An $O(\sqrt{n}L)$ potential reduction algorithm for linear complementarity problems," Research Report, Department of Information Sciences, Tokyo Institute of technology, Tokyo, Japan, 1987.
9. N. Megiddo, "Pathways to the optimal set in linear programming," Proceedings of 6th Mathematical Programming Symposium of Japan, 1-35, Nagoya, Japan, 1986.
10. Y. Ye, "An $O(n^3L)$ potential reduction algorithm for linear programming," Math. Programming, 50, 239-258, 1991.
11. Y. Ye, "A further result on the potential reduction algorithm for the P-matrix linear complementarity problem," Manuscript, Dept. of Management Sciences, Univ. of Iowa, Iowa City, 1988.
12. Y. Ye and P. M. Pardalos, "A class of linear complementarity problems solvable in polynomial time," Linear Algb. Appl., 152, 3-19, 1991.

A SIMPLE PROOF FOR A RESULT OF OLLERENSHAW ON STEINER TREES*

XIUFENG DU
Department of Mathematics, Qiqihaer Teacher's College, Heilongjiang, China.

DING-ZHU DU and BIAO GAO
Department of Computer Science, University of Minnesota
Minneapolis, MN 55455, USA.

and

LIXUE, QŬ
Department of Mathematics, Qiqihaer Teacher's College, Heilongjiang, China.

Abstract. We give a simple proof for a result of Ollerenshaw that if two full Euclidean Steiner trees exist for four points, then the one with a longer edge between two Steiner points is the shorter tree. In addition, we show that the result of Ollerenshawe holds in the rectilinear plane.

1. Introduction

The *minimum Steiner tree* for a set of finitely many points in a metric space is the shortest network interconnecting the point set. In each minimum Steiner tree, every vertex in the point set is called a *regular point* while every vertex not in the point set is called a *Steiner point*. Every minimum Steiner tree has the following properties:

(1) At each Steiner point, the subtree induced by the Steiner point and its neighbors is the minimum Steiner tree for its neighbors.

(2) At each regular point, the subtree induced by the regular point and its neighbors is the minimum Steiner tree for the regular point and its neighbors.

A tree interconnecting the given point set and satisfying (1) and (2) may not be the minimum Steiner tree, however, is called a *Steiner tree*. A Steiner tree with n regular points can have at most $n-2$ Steiner points. The Steiner tree is *full* if it has exactly $n-2$ Steiner points. In a full Steiner tree, every internal vertex is a Steiner point. For a set of four points, there are two topologies of Steiner trees as shown in Figure 1. (The topology of a Steiner tree is the graph structure of the tree or the adjacent relation of the tree.) For each topology, there may exists more than one full Steiner tree. However, in every normed plane, all full Steiner trees with the same topology for the same set of points have the same length. In the case of four points, if full Steiner trees exist for both topologies, then the shorter one is the minimum Steiner tree. How can we know which one is shorter? K. Ollerenshaw [1] gave the following.

Theorem 1 *In the Euclidean plane, suppose that two full Steiner trees for four points exist. Then the one with a longer edge between two Steiner points is the*

* Support in part by the National Science Foundation of USA under grant CCR-9208913.

Ding-Zhu Du and Jie Sun (eds.), Advances in Optimization and Approximation, 68–71.

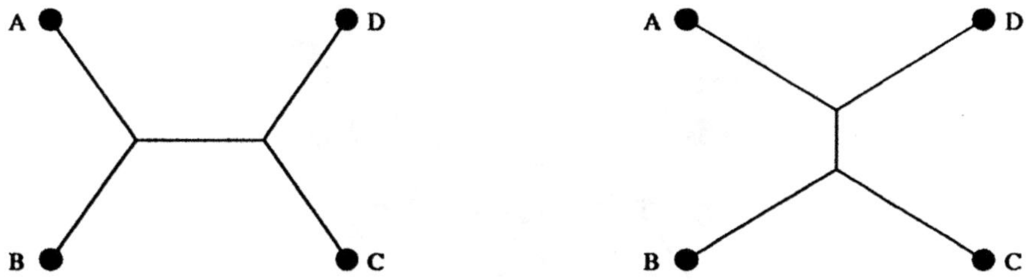

Fig. 1. Full Steiner trees for four points.

shorter tree.

His proof is pretty complicated. In this note, we first present a simple proof for Ollerenshaw's result and then extend his result to the rectilinear plane.

2. In the Euclidean Plane

In the Euclidean plane, a Steiner tree for a point set P must satisfy the following:

(1) All leaves are points in P.

(2) Any two edges meet at an angle of at least 120°.

(3) Every Steiner point has degree exactly three. Every angle at a Steiner point has degree 120°.

We first give a simple proof for Ollerenshaw's result in this section.

Consider a full Steiner tree T for the four points A, B, C, and D as shown in Figure 2. Denote $2a = d(A,C)$, $2b = d(B,D)$, and $2x = d(S_1, S_2)$, where $d(A,C)$ is the Euclidean distance between points A and C and S_1 and S_2 are two Steiner points on T. Move AC parallelly along S_1A and CS_2 to a position $A'C'$ such that $d(S_1, A') = d(C', S_2) = (d(S_1, A) + d(C, S_2))/2$. Clearly, $2a = d(A', C')$ and $A'S_1C'S_2$ is a parallelogram. Thus, $A'C'$ intersects S_1S_2 at their middle point E. Similarly, move BD parallelly along BS_1 and S_2D to a position $B'D'$ such that $d(B', S_1) = d(S_2, D') = (d(C, S_1) + d(S_2, D))/2$. Then $2b = d(B', D')$ and $B'D'$ passes through the point E which is also the middle point of $B'D'$. Using the cosine theorem in triangle ΔS_1EA, we obtain

$$a^2 = x^2 + d(A,E)^2 + x \cdot d(A,E).$$

Thus,

$$d(A,E) = \frac{-x + \sqrt{4a^2 - 3x^2}}{2}.$$

Similarly,

$$d(B,E) = \frac{-x + \sqrt{4b^2 - 3x^2}}{2}.$$

Therefore, the length of T equals

$$2(d(A,E) + d(B,E) + x) = \sqrt{4a^2 - 3x^2} + \sqrt{4b^2 - 3x^2}$$

which is a decreasing function in x. This completes the proof of Ollerenshaw's result.

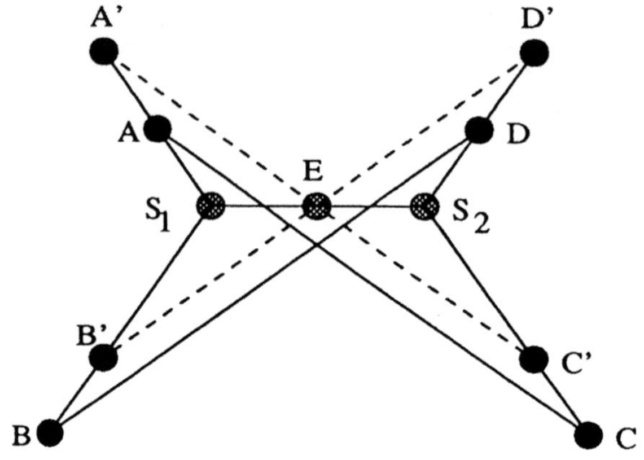

Fig. 2. Move AC and BD respectively to $A'C'$ and $B'D'$.

3. In the Rectilinear Plane

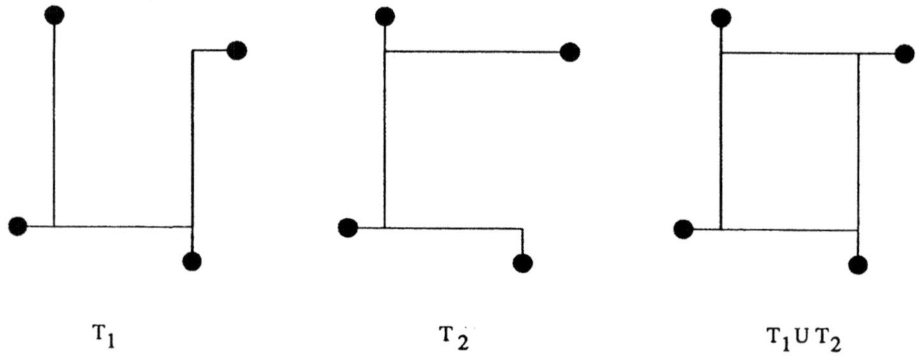

Fig. 3. Ollerenshaw theorem holds in the rectilinear plane.

Ollerenshaw theorem in the rectilinear plane can also be proved very easily but in a different way. As shown in Figure 3, suppose that there exist two full Steiner trees T_1 and T_2 with different topologies for four points. First, note that for each T_i, there may exist other full Steiner trees with the same topologies as that of T_1. However, every such tree has the same total length and the same edge length between two Steiner points as those of T_i. Let s_i denote the length of edge between Steiner points in T_i for $i = 1, 2$. Then we can choose T_1 and T_2 such that the union of T_1 and T_2 contain a rectangle with edge lengths s_1 and s_2, T_1 can be obtained from $T_1 \cup T_2$ by removing an edge of length s_2 and T_2 can be obtained from $T_1 \cup T_2$ by removing an edge of length s_1. Therefore, $s_1 \leq s_2$ if and only if $\ell(T_1) \geq \ell(T_2)$ where $\ell(T_i)$ denotes the length of tree T_i. This completes the proof of Ollerenshaw theorem in the rectilinear plane.

Theorem 2 *In the rectilinear plane, suppose that two full Steiner trees for four points exist. Then the one with a longer edge between two Steiner points is the* *shorter tree.*

4. Discussion

From this result, one might guess that Ollerenshaw theorem holds for all normed planes. However, it is not the case. We give a counterexample in the following.

In the plane with the L_3 norm, consider four points:

$$A(0.943111, 0.544172), \quad B(-2.565462, 3.611771),$$

$$C(-0.943111, -0.544172), \quad D(2.565462, -3.611771).$$

Two full Steiner trees for these four ponits exist. Their lengths, Steiner points, and distances between Steiner points are shown as follows:

full Steiner tree	length	Steiner points	the distance between Steiner points
$AB - CD$	9.870102	(0.267196, 0.355822) (-0.267196, -0.355822)	0.800526
$AD - BC$	9.828168	(-0.306930, 0.134219) (0.306930, -0.134219)	0.6305116

This example shows that Ollerenshaw's result does not hold in the $L - 3$ plane.

References

1. K. Ollerenshaw, Minimum networks linking four points in a plane, *Inst. Math. Appl.* 15 (1978) 208-211.

OPTIMIZATION ALGORITHMS FOR THE SATISFIABILITY (SAT) PROBLEM

JUN GU

Dept. of Electrical and Computer Engineering,
University of Calgary,
Calgary, Alberta T2N 1N4, Canada
gu@enel.UCalgary.CA

March 12, 1993

Abstract. The satisfiability problem (SAT) is a fundamental problem in mathematical logic and computing theory. Methods to solve the satisfiability problem play an important role in the development of computing theory and systems. Traditional methods treat the SAT problem as a discrete, constrained decision problem. In this chapter, we show how to translate the SAT problem into an unconstrained optimization problem and use efficient local search and global optimization methods to solve the transformed optimization problem. This approach gives significant performance improvements for certain classes of conjunctive normal form (CNF) formulas. It offers a complementary approach to the existing SAT algorithms.[1]

Key words: Constraint satisfaction problem (CSP), local search, global optimization, satisfiability (SAT) problem.

1. Introduction

The satisfiability (SAT) problem has three components [53, 103]:

- A set of m variables: x_1, x_2, ..., x_m.
- A set of literals. A literal is a variable ($Q = x$) or a negation of a variable ($Q = \bar{x}$).
- A set of n distinct clauses: C_1, C_2, ..., C_n. Each clause consists of only literals combined by just logical *or* ($\vee$) connectors.

The goal of the satisfiability problem is to determine whether there exists an assignment of truth values to variables that makes the following conjunctive normal form (CNF) formula satisfiable:

$$C_1 \wedge C_2 \wedge \cdots \wedge C_n, \tag{1}$$

where $\wedge$ is a logical *and* connector.

[1] This research was supported in part by 1987-1988 and 1988-1989 ACM/IEEE Academic Scholarship Awards, and is presently supported in part by NSERC Strategic Grant MEF0045793 and NSERC Research Grant OGP0046423. This work was present in part in [119, 121, 122].

Ding-Zhu Du and Jie Sun (eds.), Advances in Optimization and Approximation, 72–154.

The satisfiability problem is fundamental in solving many practical problems in mathematical logic, inference, machine learning, constraint satisfaction, and VLSI engineering (see Section 8). Theoretically, the SAT problem is a *core* of a large family of computationally intractable NP-complete problems [53, 103]. Several such NP-complete problems have been identified as central to a variety of areas in computing theory and engineering. Therefore, methods to solve the satisfiability problem play an important role in the development of computing theory and systems. There has been great interest in designing efficient algorithms to solve the SAT problem.

Traditional methods treat the SAT problem as a constrained decision problem. Many SAT algorithms and techniques have been developed. In this chapter, we show how to translate the SAT problem into a discrete, unconstrained, local optimization problem or a continuous, unconstrained, global optimization problem. We then show how to use existing local search and global optimization methods to solve the transformed optimization problem. Experimental results and theoretical study indicate that these optimization methods achieve significant performance improvements for certain classes of conjunctive normal form (CNF) formulas. The optimization method to the SAT problem offers a complementary approach to the existing SAT algorithms.

The rest of this chapter is organized as follows: In the next section, we give a unified framework, an algorithm space, that combines existing SAT algorithms into a unified framework. This is followed with a brief survey of the existing SAT algorithms. The relationship between SAT problem and the constraint satisfaction problem (CSP) is discussed. Section 3 gives some preliminaries that simplify our discussions. In Section 4, we discuss some basic SAT problem instance models and SAT algorithm's completeness. Section 5 shows the basic principles behind the local search and global optimization methods. Section 6 describeds local search algorithms to SAT problem. Global optimization techniques to the SAT problem are presented in Section 7. In Section 8, some applications of the SAT problem are described. Future work for SAT research is discussed in Section 9. Finally, Section 10 concludes this chapter.

Throughout this chapter, for each optimization algorithm discussed, we give its experimental results and performance comparisons with other existing SAT methods.

2. A Classification of SAT Algorithms

In this section, we first introduce the algorithm space which is a useful tool to develop efficient search algorithms for the SAT problem. Following this, we give a brief survey of the existing SAT algorithms in terms of the algorithm space. The relationship between SAT problem and constraint satisfaction problem (CSP) is briefly discussed which indicates some interacting activities between two areas and the original sources of some SAT algorithms.

2.1. The Algorithm Space

Most search algorithms belong to discrete, combinatorial optimization. How do they relate to the continuous optimization methods in operations research? Can we solve a discrete search problem with a numerical algorithm in real space? For a given problem, how could one predicate the existence and performance of a particular algorithm from an existing algorithm? We believe a unified framework for search and optimization would shed light on developing efficient algorithms for a search problem.

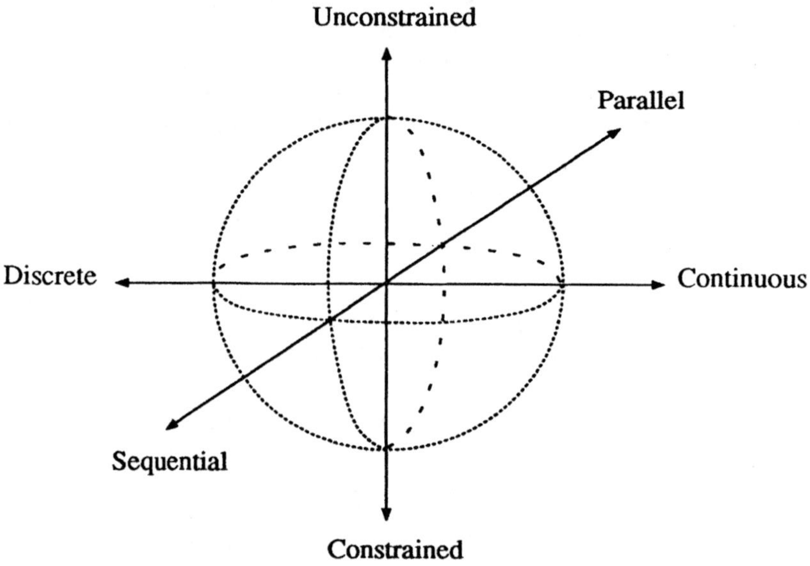

Fig. 1. The algorithm space is a unified framework for discrete search algorithms and continuous optimization algorithms. Eight octants represent eight basic classes of algorithms.

Figure 1 shows a basic *algorithm space.* It gives a unified framework for search and optimization algorithms in terms of variable domain, constraint strength, and the parallelism in the algorithms [131, 134]. The algorithm space is divided into eight octants, representing eight basic algorithm classes of search and optimization methods. This includes four sequential algorithm classes, i.e.,

- discrete constrained algorithms
- discrete unconstrained algorithms
- continuous constrained algorithms
- continuous unconstrained algorithms

and four parallel algorithm classes, i.e.,

- parallel discrete constrained algorithms
- parallel discrete unconstrained algorithms

- parallel continuous constrained algorithms
- parallel continuous unconstrained algorithms

An algorithm defines an ***algorithm ponit*** in the algorithm space. A set of algorithms, i.e., a number of algorithm points, defines an ***algorithm trajectory*** in the algorithm space. A more complicated algorithm space for SAT problem may have more than three dimensions [134].

An important step to design a search algorithm is to examine its physical and philosophical background. A good search algorithm should be physically solid and elegant. An efficient search algorithm must be natural and simple. The algorithm space has been a useful tool to the development of efficient search algorithms along this direction [131, 134].

Figure 2 shows, with some typical examples, four sequential classes of SAT algorithms in the algorithm space. This figure is a 2-dimensional cross section of the algorithm space cut at the sequential side. The left half plane and the right half

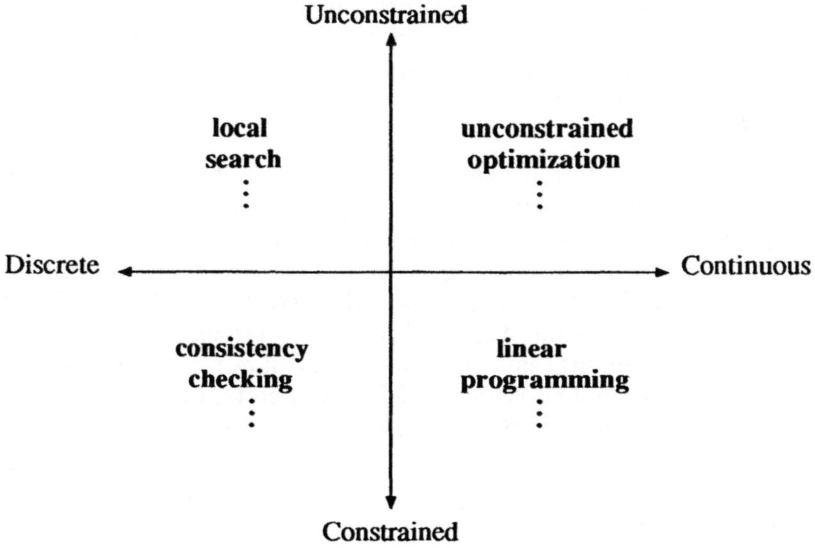

Fig. 2. A 2-dimensional cross section of the algorithm space cut at the sequential side. It indicates a unified framework for some existing discrete search methods and continuous optimization techniques for the SAT problem.

plane in the figure represent a discrete search domain and a continuous search domain, respectively. In discrete search domain, variables, values, constraints, and the objective functions are defined with discrete values. Discrete, constrained search methods include consistency checking algorithms [134, 218, 232, 323]. A typical discrete, unconstrained search technique is local search [119, 126, 130, 295, 297, 299]. In a continuous search domain, variables, values, constraints, and the objective functions are defined quantitatively. Typical constrained and unconstrained optimization techniques include numerical global optimization methods in operations research.

Most constrained search methods have unconstrained versions. Most discrete search algorithms have continuous optimization counterparts. For instance, discrete consistency algorithms [134, 218, 232, 323] are constrained algorithms. If one formulates the amount of "inconsistency" in the problem into an objective function, an unconstrained, local search method often solves the problem efficiently [295, 297, 299, 119, 126, 130]. Furthermore, local search operates in a discrete search space. By extending a search problem into a real search space, constrained and unconstrained global optimization algorithms can be developed to solve SAT problem [19, 119, 127, 131, 155, 178, 177].

The algorithm space provides a unified and global perspective to the development of search and optimization algorithms for the SAT problem. For a given search problem in general, if one can find an algorithm in one octant, then one could possibly find the affined algorithms in other octants. As shown in Figure 2, once we had consistency algorithms and local search algorithms for the SAT problem, it would be natural to think about unconstrained optimization algorithms for the SAT problem since that right-top octant was empty (something must be put there to meet natural symmetry). This was our original incentive to develop unconstrained optimization algorithms for the SAT problem. Similarly, some continuous, constrained algorithms for the SAT problem, such as integer linear programming, would fit into the right-bottom octant in Figure 2.

Some deep and interesting insights into a search problem may be gained from the algorithm space. By changing constraint strength and variable domain in the algorithm space, an algorithm trajectory for the given problem could be drawn. For the given search problem, this trajectory indicates some important properties, such as the asymptotically "poor" algorithms, the asymptotically "good" algorithms, possible algorithms with certain degree of constraint strengths, possible algorithms in different variable domains, and potential performance of these algorithms. For example, in Figure 2, from local search which is a discrete, unconstrained algorithm, we can predict the algorithm structure and performance of continuous, unconstrained algorithms [134].

2.2. Basic SAT Algorithm Classes

Following the algorithm space, a number of major SAT algorithm classes can be identified. They are given in Figure 3 in their chronological order. Most existing SAT algorithms can be grouped into these categories.

– *Discrete, constrained algorithms.* Algorithms in this category treat the SAT problem as a constrained decision problem, solving the problem with discrete search and inference procedures. One straightforward way to solve the SAT problem is to enumerate all possible truth assignments to see if one satisfies the formula. Many improved techniques, such as consistency algorithms [134, 218], backtracking algorithms [18, 30, 34, 201, 257], term-rewriting [68, 163], production system [292], multi-valued logic [288], Binary Decision Diagrams

- Discrete
  - Constrained
    - 1960: Davis-Putnam (DP) algorithm [63]
    - 1965: Resolution [273]
    - 1971: Consistency algorithms [49, 118, 167, 322]
    - 1978: Loveland's Davis-Putnam (DPL) [214]
    - 1986: Parallel consistency chips [118, 117, 132]
    - 1988: Chip and conquer [106]
    - 1990: DPL plus heuristic (DPLH) [177]
    - 1992: Binary decision diagrams [10]
  - Unconstrained
    - 1987: *CNF* Local search (SAT1.0) [119]
    - 1987: *DNF* Local search (SAT1.4) [119]
    - 1987: Parallel local search (SAT1.6) [119]
    - 1987: Local search hardware [119]
    - 1988: Local search for n-queen [118, 294, 295]
    - 1989: Local search & backtracking [120, 121, 130]
    - 1990: *Unison* algorithm [300]
    - 1990: *Unison* hardware [301]
    - 1991: Local search complexity [135, 243]
    - 1992: GSAT [287]
- Continuous
  - Constrained
    - 1986: Branch-and-bound (APEX) [19]
    - 1988: Programming models [19, 178]
    - 1988: Cutting plane [156, 154]
    - 1989: Branch-and-cut [157]
    - 1990: Interior point method [186]
  - Unconstrained
    - 1987: *UniSAT* models [119]
    - 1987: Global optimization (SAT6) [119, 122, 131]
    - 1989: Neural net models [181, 38]
    - 1990: Global optimization & backtracking [122, 124, 125, 131]
    - 1991: SAT14 algorithms [122, 138, 131]

Fig. 3. Some typical algorithms for the SAT problem.

[10, 33], chip and conquer [106], resolution and regular resolution [108, 214, 240, 273, 312, 325], and independent set algorithm [174], have been proposed. Among them, resolution is the most widely used method.

Specific versions of resolution principle include Davis-Putnam algorithm [63, 62], simplified Davis-Putnam algorithms [99, 113, 260], Davis-Putnam algorithm in Loveland's form (DPL) [214, 214], Davis-Putnam-Loveland procedure plus heuristic (DPLH) [177], and a simplified Davis-Putnam algorithm with strict ordering of variables [164]. The Davis-Putnam algorithm improved in certain aspects over Gilmore's proof method [109]. It has been a major practical method for solving the SAT problem.

A number of special SAT problems, such as 2-satisfiability and Horn clauses, are ***solvable*** in polynomial time [3, 53, 240]. Several linear time algorithms [11, 86], polynomial time algorithm [243], and polynomial time bound [281], were proposed.

– *Discrete, unconstrained algorithms.* In this approach, the number of unsatisfiable *CNF* clauses or the *DNF* distance against solution assignment is formulated as the value of the objective function, so the SAT problem is transformed into a discrete, unconstrained minimization problem to the objective function. Local search is a major class of discrete, unconstrained search methods [119, 126, 127, 128, 130, 134, 135]. It can be used to solve the transformed problem (see Section 6). A special case of local search is simulated annealing [144].
– *Constrained programming algorithms.* Methods in this class were developed based on the fact that the SAT problem can be written as integer program and one could try to solve it as a linear programming problem [19, 155, 156, 178, 186, 246, 324]. Many approaches, including branch-and-bound [19], cutting plane method [156, 154], branch-and-cut [157], interior point algorithm [186], and improved interior point algorithm [289], have been proposed to solve the integer program representing the inference problem. Researchers found integer programming methods faster than resolution for certain classes of problems but they do not possess robust convergent property and often fail to solve hard satisfiability problems [19, 155, 156, 178, 186].
– *Unconstrained programming algorithms.* Special models are formulated that transform a discrete SAT problem on Boolean space $\{0,1\}^m$ into a quantitative, unconstrained *UniSAT* problem on real space E^m. Thus, the original decision problem is transformed into an unconstrained global optimization problem which can be solved by existing global optimization methods [119, 122, 126, 127, 129, 131, 134] (see Section 7).

In practice, most sequential search algorithms can be mapped onto parallel computing machines, producing parallel search algorithms. Accordingly, there are four classes of parallel algorithms for the SAT problem:

– *Parallel, discrete, constrained algorithms.* These include parallel consistency algorithms [117, 123, 118, 132], Boolean difference method [130, 300, 301], and parallel PROLOG languages.
– *Parallel, discrete, unconstrained algorithms.* A number of discrete local optimization algorithms was implemented with parallel computing structures [119, 130, 300, 301].
– *Parallel, constrained programming algorithms.* Kamath et al. implemented an interior point zero-one integer programming algorithm on a *KORBX(R)* parallel/vector computer which has 32 MFlops at a full vector concurrent mode [186].

- *Parallel, unconstrained programming algorithms.* Several of them have been implemented [119].

Most existing search and optimization techniques are in one or more of the above categories.

2.3. CSP and SAT Problems

In a constraint satisfaction problem (CSP), one seeks a set of value assignment to variables that satisfies the given constraints, making all variable expressions jointly satisfiable. In the SAT problem, one seeks a truth assignment to Boolean variables that satisfies the given clauses (constraints), i.e., making all clauses jointly satisfiable. Clauses in the SAT problem are a specific form of variable expressions in CSP.

The SAT problem and CSP are a natural "twin." For a CSP with Boolean variables, there is a direct correspondence between CSP and SAT problems. The SAT problem can be viewed as a CSP in which all variables are Boolean and all constraints are represented explicitly in the negative form [165]. For a CSP with multiple-valued variables, using variable transformation and constraint transformation, the CSP can be converted into a SAT problem [165].

Since CSP and SAT problems are interrelated each other, algorithms and architectures developed for CSP are applicable to the SAT problem as well [127, 134, 117, 123, 132, 165]. The relationship between SAT and CSP gives the basic ideas of several local search algorithms to the SAT problem.

The n-queen problem is a typical problem in CSP. According to recent survey [182, 196, 197], the earlest local search algorithms for the n-queen problem were developed in 1988 by Sosič and Gu. Four versions of the algorithms were developed. The *Queen_Search1* ($QS1$) algorithm is a probabilistic local search algorithm that runs in approximately $O(nlogn)$ time [294, 295, 296]. The $QS2$ and $QS3$ are near linear local search algorithms with an efficient random selection [297]. The $QS4$ algorithms is a linear time local search algorithm with an initial search procedure [298, 299].

The local search solutions to the large-size n-queen problem have been enlightening to other CSPs. Recently, Selman, Levesque and Mitchell reported the empirical results of a local search algorithm, i.e., the $GSAT$ algorithm, for the SAT problem [287]. Selman has recently addressed [286] that the local search solutions to the large size n-queen problem was "the original impetus" to the $GSAT$ algorithm [287].

3. Preliminaries

To simplify our discussion, throughout this chapter, let:

- F be a *CNF* or *DNF*[2] formula,
- n be the number of clauses in F,

[2] A *DNF* formula is a Boolean formula in disjunctive normal form. It is a sum of terms each of which is a product of literals, for example, $x_1\bar{x}_2 + x_2x_4$.

- m be the number of variables in F,
- C_i be the ith clause,
- $|C_i|$ be the number of literals in clause C_i,
- Q_{ij} be the jth literal in the ith clause, and
- l be the average number of literals: $\frac{\sum_{i=1}^{n} |C_i|}{n}$,

where $i = 1, ..., n$ and $j = 1, ..., m$.

On Boolean space $\{0, 1\}^m$, let:

- x_j be the jth variable,
- $\mathbf{x}$ be a vector of m variables,
- $C_i(\mathbf{x})$ be the ith clause function,
- $Q_{ij}(\mathbf{x})$ be the jth literal function of the ith clause function, and
- $F(\mathbf{x})$ be a Boolean function from $\{0, 1\}^m$ to $\{0, 1\}$,

where $i = 1, ..., n$ and $j = 1, ..., m$.

On real space E^m, let:

- y_j be the jth variable,
- $\mathbf{y}$ be a vector of m variables,
- $c_i(\mathbf{y})$ be the ith clause function,
- $q_{ij}(\mathbf{y})$ be the jth literal function of the ith clause function, and
- $f(\mathbf{y})$ be a real function from E^m to E,

where $i = 1, ..., n$ and $j \doteq 1, ..., m$.

Following [216], a real-valued function f defined on a subset of E^m is said to be *continuous* at $\mathbf{y}$ if $f(\mathbf{y}_k) \to f(\mathbf{y})$. A set of real-valued functions $f_1, f_2, ..., f_n$ on E^m forms a vector function $\mathbf{f} = (f_1, f_2, ..., f_n)$ whose ith components is f_i. It is *continuous* if each of its component functions is continuous.

If a component of $\mathbf{f}$, i.e., $f = f(\mathbf{y}) = f(y_1, y_2, \ldots, y_m)$, is continuous on E^m, and if f has first partial derivatives which are continuous on the set, we define the *gradient* of f to be the vector

$$\nabla f(\mathbf{y}) = \left(\frac{\partial f(\mathbf{y})}{\partial y_1}, \frac{\partial f(\mathbf{y})}{\partial y_2}, \ldots, \frac{\partial f(\mathbf{y})}{\partial y_m} \right). \tag{2}$$

In matrix calculations, the *gradient* is considered to be a row vector.

If f has second partial derivatives which are continuous on this set, we define the *Hessian* of f at $\mathbf{y}$ to be the $m \times m$ matrix denoted by

$$\mathbf{H}(\mathbf{y}) = \nabla^2 f(\mathbf{y}) = \left[\frac{\partial^2 f(\mathbf{y})}{\partial y_i \partial y_j} \right]. \tag{3}$$

The *norm* of a vector $\mathbf{y}$ on E^m is defined by

$$||\mathbf{y}|| = \sqrt{\mathbf{y}^t \mathbf{y}} = \sqrt{\sum_{i=1}^{m} y_j^2}. \tag{4}$$

We will call a $\mathbf{y} \in E^m$ with $f(\mathbf{y}) = 0$ a solution of f, denoted as y^*.

Two aspects of the iterative optimization algorithms are global convergence and local convergence rate [216]. *Global convergence* concerns, starting from an initial point, whether the sequence of points will converge to the final solution point. *Local convergence rate* is the rate at which the generated sequence of points converges to the solution.

Definition 1 *[216] Let the sequence $\{r_k\}$ converge to r. The order of convergence of $\{r_k\}$ is defined as the supremum of the nonnegative numbers p satisfying*

$$0 \leq \overline{\lim_{k\to\infty}} \frac{|r_{k+1} - r|}{|r_k - r|^p} < \infty.$$

Definition 2 *[216] If a sequence $\{r_k\}$ converges to r in such a way that*

$$\lim_{k\to\infty} \frac{|r_{k+1} - r|}{|r_k - r|} = \beta < 1,$$

the sequence $\{r_k\}$ is said to converges linearly to r with convergence ratio β.

4. Complete Algorithms and Incomplete Algorithms

4.1. SAT Problem Models

Problem instances are used to test the performance of SAT algorithms. The following SAT problem models are often used by researchers in the area. The first two models generate random CNF formulas. The use of randomly generated problem instances is a standard technique in algorithm design [54, 113, 156, 157, 186, 187].

1. **The exact l-SAT model.** In the exact l-SAT model, a randomly generated CNF formula consists of independently generated random clauses. Each clause contains exactly l literals. Each literal Q is uniformly chosen from $X = \{x_1, \ldots, x_m\}$ without replacement, i.e., for $x_i \in X$, $\Pr[Q = x_i] = 1/|X|$, where $|X|$ denotes the number of elements in X, and then Q is negated with probability p. Similar models were used in [95, 119, 127, 130, 231].
2. **The average l-SAT model.** In the average l-SAT model, a randomly generated CNF formula consists of independently generated random clauses. Each of the m variables occurs positively with probability $p/2$, negatively with probability $p/2$, and is absent with probability $1 - p$. The average number of literals in the clauses is approximately l. This model and its variations were used in [113, 119, 127, 126, 130, 156, 157, 186, 289].
3. **The regular SAT models.** Models such as the graph coloring problem and the n-queen problem are used to assess the performance of SAT algorithms [119, 129, 138, 127, 165].
4. **Practical models.** Many people use SAT problem instances generated from the practical application problems (see Section 8).

There are two basic and related concepts, the ***hardness*** of the SAT problem instances and the ***hard-and-easy distributions*** of the SAT problem instances. They are essential to understanding the difficulty of the SAT problem instances.

Informally, the *hardness* of the SAT problem instances refers to the satisfiable and unsatisfiable (sat/unsat) boundary of the SAT problem instances (see Figure 5). The hardness of the SAT problem instances is an intrinsic property of the problem instances and is determined by the SAT problem models. Generally, for the randomly generated CNF formulas, the hardness of the SAT problem increases, up to a point, when n/m increases or the number of literals l ($l > 3$) in each clause decreases. That is, fewer literals and larger number of clauses reduce the possibility of making all clauses jointly satisfiable. The 3-SAT problem is thought to be a difficult problem to solve. An informal indication of hardness can be derived from the percentage of sat/unsat statistics of the SAT problem instances.

The *hard-and-easy distributions* of the specific SAT problem instances refer to the distributions of the SAT problem instances in terms of algorithms. That is, such distributions not only depend on the inherent hardness of the SAT problem models and also on the functionality of the algorithms used to solve the problem. The results of hard-and-easy distributions of an algorithm cannot be generalized to a different algorithm. As an analogy, assume that "the problem instance" to be solve is to go to New York and "the SAT algorithms" available are the alternative tools of traveling. Then, flying to New York is much easier than walking to New York. So, the difficulty of "going to New York" depends on tools of traveling as well. The "running time" of flying to New York can be a few hours but the "running time" of walking to New York could be several thousands hours.

A number of researchers studied various SAT instance distributions, the probabilistic performance, and the average time complexities of certain SAT algorithms [30, 95, 94, 111, 112, 113, 164, 256, 262, 255]. For a random *CNF* formula consisting of n independently generated random clauses. In each clause, each of the m variables occurs positively with probability p, negatively with probability p, and is absent with probability $1 - 2p$. Goldberg, Purdom and Brown [113] showed that, for this random model, the expected time for a backtracking algorithm due to Davis-Putnam to determine whether a random formula is satisfiable is polynomial in n and m (but exponential in $1/p$). Franco and Paull [95] indicated that for random formulas with n clauses, m variables, and constant clause lengths, the expected running time of the Davis-Putnam procedure is exponential. Despite of the worst-case complexity of the SAT problem, algorithms and heuristics with polynomial average time complexities have been constantly reported [42, 43, 92, 93, 257, 259, 260, 261, 326].

Some researchers investigated the number of solutions from the SAT problem instances. Dubois gave a combinatorial formula computing the number of solutions of a set of any clauses [82]. He and Carlier also studied the mathematical expectation of the number of solutions for a probabilistic model [83].

4.2. Complete Algorithms and Incomplete Algorithms

From algorithm functionality point of view, there are basically two classes of SAT algorithms, complete ones and incomplete ones. The class of SAT algorithms is *complete* if they are able to verify satisfiability as well as unsatisfiability, thus determining whether the given CNF formula is satisfiable or unsatisfiable. Algorithms in this category includes resolution, the Davis-Putnam algorithm, and some complete optimization algorithms [63, 119, 120, 124, 125, 127, 129, 130, 273]. The *incomplete* algorithms can only find one random solution for certain CNF formulas and give no answer if the CNF formula is not satisfiable. Algorithms in this class include the local search algorithms [119, 128, 130, 243] and some integer programming algorithms [186, 289].

Depending on the SAT algorithms used, for the same problem instances, the hard-and-easy distributions of SAT problem instances are completely different. We use problem instances generated from the exact 3-SAT model as an example. Figure 4 shows, for 50 variables, the real execution results of the Davis-Putnam algorithm for 100 to 500 clauses. The results of the number of DP operations are shown in a so-called *bell curve* (solid line). It consists of a left region, a peak region, and a right region [119, 130, 231]. The dotted line shows the percentage of the satisfiability of the random problem instances. Clearly, the SAT problem instances generated in the left region of the bell curve are satisfiable and problem instances in the right region of the bell curve are unsatisfiable with the peak region as a sat-to-unsat transition region. Because Davis-Putnam algorithm is a complete algorithm, it is able to verify satisfiability and unsatisfiability. So it gives results for problem instances in all three regions.

The results of the Davis-Putnam algorithm may not hold for a different SAT algorithm. As an example, for an incomplete algorithm, it may find a satisfiable solution much quickly than the Davis-Putnam algorithm but it can verify neither satisfiability nor unsatisfiability. That is, it gives no answer if a CNF formula is not satisfiable. In this case, for most problem instances in the peak region and all problem instances in the right region of the bell curve, an incomplete algorithm will take an *infinite* amount of computing time to "find a solution."

We may consider that why does the Davis-Putnam algorithm produce a bell-like run time curve. The algorithm is able to verify unsatisfiability *quickly* for certain classes of random problem instances but is slow when it comes to verifying satisfiability, as *all* possible resolutions need to be tried out before concluding that the inference relation holds or that the problem is satisfiable. In the left-half of the bell curve, most random problem instances are satisfiable, the computing time of the Davis-Putnam algorithm increases as the number of clauses increases. In the right-half of the bell curve, as the number of clauses increases, more random problem instances become unsatisfiable. DP algorithm verifies unsatisfiability *quickly* for this type of random problems.

The above conclusion can not be generalized to an incomplete algorithm. An incomplete algorithm can not verify unsatisfiability. Thus an unsatisfiable problem instance that could be solve by the Davis-Putnam algorithm in a fraction of a second

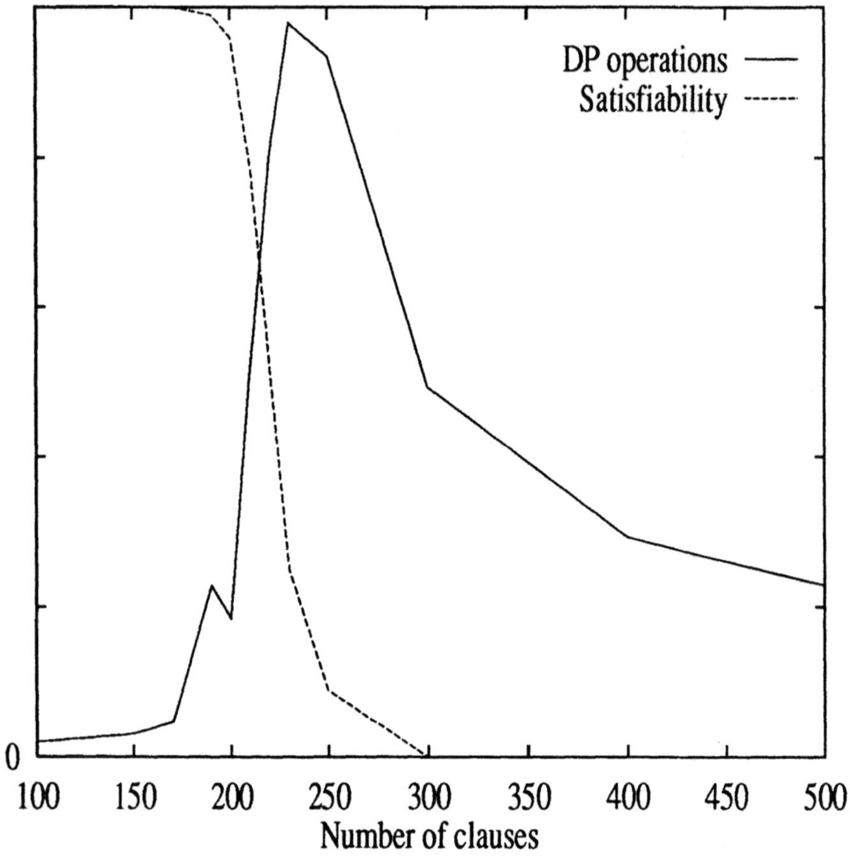

Fig. 4. The execution results of the Davis-Putnam algorithm for exact 3-SAT problem instances (50 variables).

would take an incomplete algorithm an *infinite* amount of computing time.

4.3. The Comparison of SAT Problem Models

Figures 5 and 6 show, for 50 variables, the percentage of satisfiability and real computing time of a complete SAT algorithm for problem instances generated from the exact 3-SAT model and the average 3-SAT model [119, 121, 130]. Figure 5 indicates that, for the same problem sizes, problem instances generated from the average 3-SAT model have much lower percentage of satisfiability, as compared to those generated from the exact 3-SAT model. The sat-and-unsat boundary of the

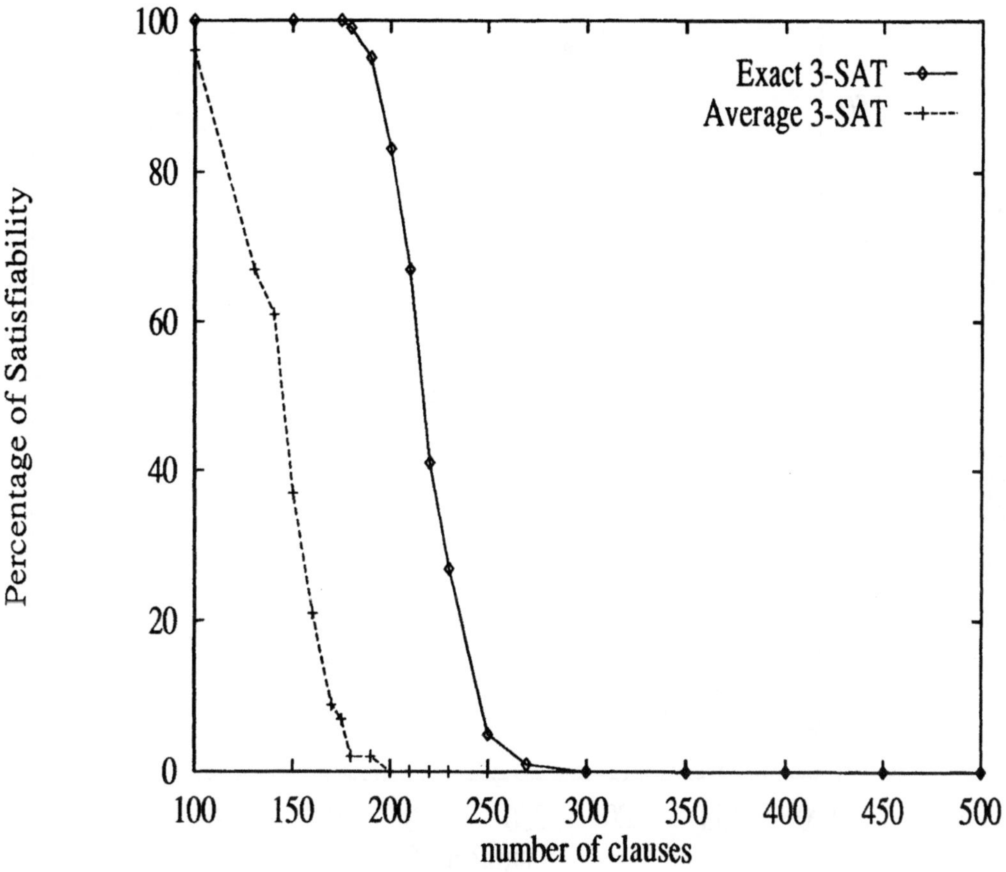

Fig. 5. Percentage of satisfiability for the exact and average 3-SAT problem models with 50 variables.

average 3-SAT model is drawn by smaller n/m values than those for the exact 3-SAT problem model.

In general, for different problem instance models, the n/m values that separate the sat and unsat regions differ considerably. Even for the same problem model, a tiny variation to the parameters of the model will significantly change the n/m values and thus the sat-and-unsat boundary. For the average 3-SAT problem model (see Figure 7), problem instances generated from a model with minimum number of literal being 1 have much lower percentage of satisfiability than that with minimum number of literal being 2. Only at some very restricted cases, the n/m values that separate sat and unsat regions are fixed. This case *hardly* happens in practical

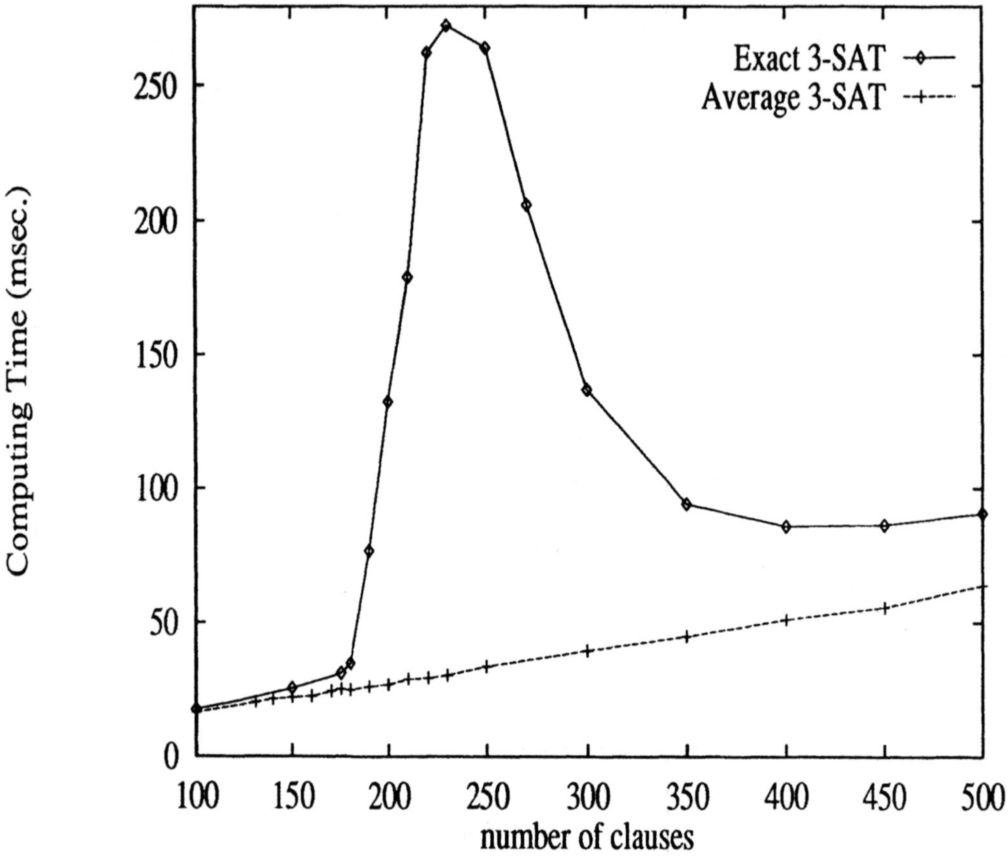

Fig. 6. Computing time for the exact and average 3-SAT problem models (with 50 variables) on a SUN SPARC 1 workstation.

application problems.

The n/m values that determine the sat-and-unsat boundary also depend on the algorithms used to solve the problem. An incomplete algorithm can not decide a sat-and-unsat boundary for any SAT problem instance model.

5. Optimization: An Iterative Refinement Process

Optimization has been an important unifying theme that cuts across many areas in science, engineering, management, industry, transportation, and other areas. The concept of optimization is well rooted as a principle underlying the analysis of many

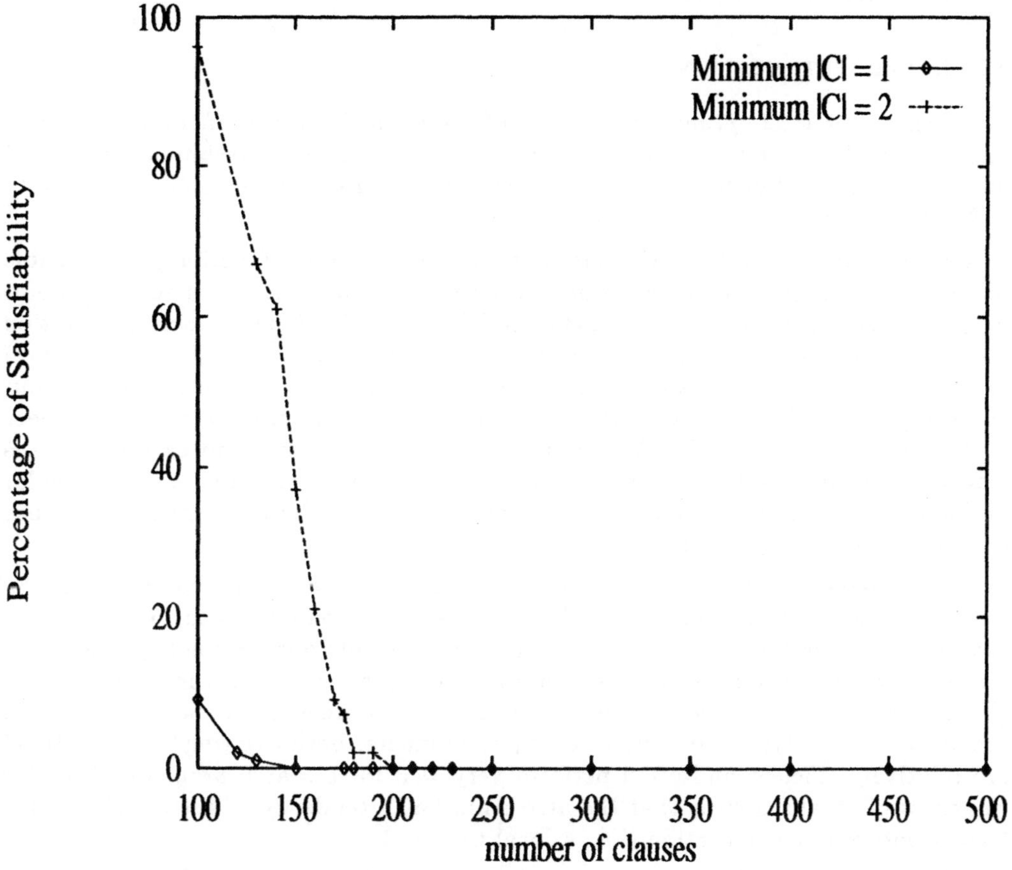

Fig. 7. Percentage of satisfiability for two average 3-SAT problem models with $|C|_{min} = 1$ and $|C|_{min} = 2$ (50 variables). Problem instances generated from a model with smaller minimum number of literal have much lower percentage of satisfiabilty.

complex decision problems. When one deals with a complex decision problem, involving the selection of values to a number of interrelated variables, one should focus on a single objective (or a few objectives) designed to qualify performance and measure the quality of the decision. The *core* of the optimization process is to minimize (or maximize) the objective function subject to the constraints imposed upon the values of the decision variables by the problem.

Most optimization algorithms are designed as an iterative refinement process. Typically, in seeking a vector that solves the optimization problem, an initial vector $\mathbf{y}_0$ is selected and the algorithm generates an improved vector $\mathbf{y}_1$. The process is repeated and an even better solution $\mathbf{y}_2$ is found. Continuing in this fashion, a sequence of ever-improving points $\mathbf{y}_0$, $\mathbf{y}_1$, ..., $\mathbf{y}_k$, ..., is found that approaches a solution point $\mathbf{y}^*$. This is the basic ideas of local search and global optimization algorithms for the SAT problem.

5.1. Local Optimization

Local search, or *local optimization*, is one of the primitive forms of continuous optimization in the discrete search space. It was one of the early techniques proposed during the mid-sixties to cope with the overwhelming computational intractability of NP-hard combinatorial optimization problems.

Given a minimization (maximization) problem with object function f and feasible region R, a typical local search algorithm requires that, with each solution point $x_i \in R$, there is a predefined *neighborhood* $N(x_i) \subset R$. Given a current solution point $x_i \in R$, the set $N(x_i)$ is searched for a point x_{i+1} with $f(x_{i+1}) < f(x_i)$ $(f(x_{i+1}) > f(x_i))$. If such a point exists, it becomes the new current solution point, and the process is iterated. Otherwise, x_i is retained as a *local optimum* with respect to $N(x_i)$. Then, a set of feasible solution points is generated, and each of them is "locally" improved within its neighborhood. To apply local search to a particular problem, one need only to specify the neighborhood and the randomized procedure for obtaining a feasible starting solution.

Local search is very efficient in several aspects. First, at the beginning of the search, using a full assignment of all variables in the search space, it reduces an exponential growth search space to a much manageable one. Secondly, it searches for improvement within its local neighborhood using a testing for improvement and, if there is any improvement, takes an action for improvement. Since the object function has a polynomial number of input numbers, both "testing" and "action" are relatively efficient and often perform very well for a search problem. A major weakness of local search is that the algorithm has a tendency of getting stuck at a locally optimum configuration (i.e., a local minima).

Many search techniques, such as statistical optimization [56, 283], simulated annealing [190], stochastic evolution [279], conflict minimization [228, 295, 299], and genetic algorithms [114], are either local search or variations of local search. Local search is one among very few successful techniques for combinatorial optimization. For most search problems encountered, in terms of computing time and memory space, local search often achieves many orders of magnitude of performance improvement [126, 130, 139, 264, 297, 299].

5.2. Global Optimization

Global optimization is also referred as *global search* in some related area. Global optimization is concerned with the characterization and computation of global minima and maxima of unconstrained nonlinear functions and constrained nonlinear problems. Global optimization problems belong to the class of NP-hard problems.

Both local search and global optimization aim at solving a minimization (maximization) problem using local neighborhood search strategy. They differ only in the way that the objective function is reduced: local search accepts any solution point as long as it reduces the value of the objective function; global optimization only admits a solution point that yields the *minimum* value to the objective function. This remarkable distinction shows that local search is a nondirective, qualitative global optimization and global optimization is a directed, quantitative local search.

Many search problems are discrete in nature. How does one use continuous optimization techniques to solve a discrete search problem? There are many ways to do so. First, we can directly supply the discrete values to the continuous variables as constraints to the given problem. Second, we can encode the discrete values and add them into the objective function. Third, most discrete values can be given as a separate instance of a function with the special interpretation of its values. The last, as described below, we can apply the *carrier principle* and make a mixed mode global optimization.

In a telecommunication system, a speech signal can not be transmitted directly. Before transmission, a modulator adds the speech signal on a high frequency signal called *carrier signal*. The carrier signal carries the speech signal during the transmission. At the receiver side, a demodulator takes the speech signal away from the carrier signal. Using the carrier principle for mixed mode optimization, we simply put a discrete search problem into a real search space with continuous variables. The problem is then solved by global optimization techniques. Finally, the results of continuous variables are decoded back to their discrete counterparts which, after verification, are taken as the solution to the original search problem.

In the next section, we will describe some local search algorithms for the SAT problem. They are the essential basis to the global optimization algorithms for the SAT problem.

6. Local Search Algorithms for SAT

Local search algorithms are a major class of the discrete, unconstrained optimization algorithms in the discrete search space. Using local search for SAT problem, the number of unsatisfiable *CNF* clauses or the *DNF* distance against solution assignment is formulated as the value of an objective function, so the SAT problem is transformed into a discrete, unconstrained minimization problem to the objective function [119, 126, 127, 128, 130, 134, 135].

6.1. A Discrete, Unconstrained Optimization Model for *CNF* Formulas

Most discrete local search algorithms were developed based on a discrete, unconstrained optimization model, the *SAT1 model*. In this model, the truth values assigned to the variables are swapped between "+1" and "−1," that is:

$$x_i = \begin{cases} True & \text{if } x_i = 1 \\ False & \text{if } x_i = -1 \end{cases} \tag{5}$$

The objective function in *SAT1* model is defined as:

$$F(\mathbf{x}) = \sum_{i=1}^{n} C_i(\mathbf{x})|_{C_i=False} \tag{6}$$

where $C_i(\mathbf{x})$ is an unsatisfiable clause, i.e., $C_i(\mathbf{x})$ evaluates to *False*. $F(\mathbf{x})$ gives the number of unsatisfiable clauses as its objective value.

The *SAT1* model is a discrete, unconstrained optimization model for SAT problem expressed in a *CNF* formula. Its dual model, the *SAT4 model* (see Section 6.4), is a discrete, unconstrained optimization model for a *DNF* formula.

6.2. Local Search Algorithms for SAT Problem

Many discrete local search algorithms have been developed for the SAT problem. In this section, we will describe a general local search algorithm for the satisfiability problem. For details of other algorithms, see references [119, 121, 131, 127, 126, 128, 134, 132].

6.2.1. *A General Local Search Algorithm for the SAT Problem*

The general local search algorithm (Figure 8) consists of an initialization stage and a search stage. Several operations, *get_a_SAT_instance*(), *select_an_initial_point*(), *evaluate_object_function*(), *test_swap*(), *perform_swap*(), and *local_handler*() are needed in the algorithm. We will see their operation in the following discussion of the algorithm.

Initialization. Initially, the *SAT*1.0 algorithm requires a practical or a randomly generated SAT problem instance. At the beginning of search, procedure *get_a_SAT_instance*() produces an instance of the satisfiability problem. Procedure *select_an_initial_point*() randomly selects an initial starting point: $\mathbf{x}_0$.. An initial starting point is an initial random truth assignment to the variables of the SAT problem. The objective function, $F(\mathbf{x})$, computes the number of unsatisfiable clauses in the SAT problem. Based on the *SAT1* model, the original SAT problem is transformed into an unconstrained minimization problem to the objective function.

Local Search. The search process is an iterative local optimization process. During the kth iteration, for each variable x_i in vector $\mathbf{x}$, we cyclically search through its neighborhood to select a solution point that minimizes the objective function. That is, during a single search step, a literal x_i is considered for its contribution to the objective value. The test to see if a swap of the truth value of x_i with $\overline{x_i}$ will reduce the total number of unsatisfiable clauses is performed by function *test_swap*$(x_i,\overline{x_i})$. Function *test_swap*$(x_i,\overline{x_i})$ returns *true* if the swap operation would reduce the total number of unsatisfiable clauses; otherwise, it returns *false*. The actual swap operation is performed by procedure *perform_swap*$(x_i,\overline{x_i})$, which produces a new solution point, i.e., $\mathbf{x}_{k+1}$. Procedure *evaluate_object_function*() updates the value of the object function, i.e., the total number of unsatisfiable clauses.

```
procedure SAT1.0 ()
begin
    /* initialization */
    get_a_SAT_instance();
    x_0 := select_an_initial_point();
    F(x) := evaluate_object_function(x_0);

    /* search */
    k := 0;
    while F ≠ 0 do
    begin
        for each variable i := 1 to m do
            /* if swap(x_i, x̄_i) does not increase F */
            if test_swap(x_i, x̄_i) then
            begin
                x_{k+1} := perform_swap(x_i, x̄_i);
                F(x) := evaluate_object_function();
            end;
        if local then local_handler();
        k := k + 1;
    end;
end;
```

Fig. 8. **SAT1.0: a local search algorithm for SAT problem.**

```
procedure local_handler ()
begin
    randomly choose some variable x_t's;
    x_{k+1} := perform_swap(x_i's, x̄_i's);
    conflicts := compute_conflicts(x_{k+1});
end;
```

Fig. 9. **A simple local handler.**

Termination. In practice, before the object function reduces to zero, the algorithm could be stuck at a locally optimum point. To solve this problem, in the $SAT1.0$ algorithm, a simple local handler (see Figure 9) that may negate the truth values of up to m variables is added to improve its convergence performance. The basic idea is to generate 'random exchanges' in some current solution points, when the algorithm was stuck on a local minima, accepting a modified point as a new current solution not only if the value of the cost function is better but also if it is worse.

In the $SAT1.0$ algorithm, due to excessive floating point computation and poorer convergence, the probability given by a Boltzmann-like law as used in simulated annealing was not taken as a decision for accepting a random change in the *local_handler*(). Instead, in this local handler, a definite selection of a new search state is taken. This local handler has effectively improved the convergent performance of the $SAT1.0$ algorithm.

As observed by many researchers, local optima produced by local search have a certain average quality; they are related to the quality of the initial starting points. The use of a random initial starting point is an ***unbiased*** decision that enables one to sample the set of local optima widely, producing an improved starting point. In the local handler, the use of some random exchanges has the similar effect of reassigning the initial feasible starting point which is a fairly general technique applied to tackle the pathological effect of the worst-case instances. The early success of this strategy has been renowned as Lin's important contribution to the solution of the traveling salesman problem (TSP) [211].

In most combinatorial optimization problems, such as the TSP, it is effective to use a completely random starting point [211]. In practical algorithm design, however, the effectiveness of the idea must be justified by additional running time. In the local handler used in $SAT1.0$ algorithm, generating a completely random starting point requires $O(m)$ time, while selecting a few variables to swap takes only a constant time. So we have used a swap more frequently than would a completely random starting point.

Running Time. The running time of the $SAT1.0$ algorithm can be estimated as follows: Procedure *get_a_SAT_instance*() and function *evaluate_object_function*() each take $O(nl)$ time since they go through n clauses and examine, on average, l literals per clause. Procedure *select_an_initial_point*() takes $O(m)$ time to produce an initial solution point. Accordingly, the total time for initialization is $O(nl)$.

The running time of the *while* loop equals the number of *while* loops executed, k, multiplied by the running time of the *for* loop. In the worst case, *perform_swap()* takes $O(m)$ time, *test_swap()*, *evaluate_object_function*(), *local_handler*() each will take $O(nl)$ time. Combining $O(m)$ time cyclic scan of m variables, the running time of the *for* loop is $O(m(nl))$. Summarizing the above, the time complexity of $SAT1.0$ is:

$$O(nl) + O(knml) = O(knml), \tag{7}$$

where the number of iterations, k, is determined by the nature of the problem instance and is therefore nondeterministic. We will discuss the average value of k in Section 6.3.

As is common with any probabilistic local search algorithm [54], $SAT1.0$ does not guarantee the worst case running time for a particular instance of execution, although it exhibits satisfactory performance and robust behavior. Hence, an unsuccessful attempt may occur and the algorithm will call the local handler.

Other SAT1 Algorithms. Many local search algorithms have been developed [119, 126, 127, 128, 130, 134, 135]. $SAT1.1$ algorithm is a greedy local search algorithm. $SAT1.2$ algorithm is a local search algorithm with monotonic conflict minimization. $SAT1.3$ algorithm is a greedy local search with perturbation. $SAT1.4$ is a local search algorithm for *DNF* formulas. $SAT1.5$ uses local tracking to avoid being trapped into a local loop. $SAT1.6$ to $SAT1.10$ are improved local search algorithms developed to speed up the computation.

6.2.2. *Parallel Local Search Algorithms*

A typical parallel local search algorithm, the $SAT1.6$ algorithm, is shown in Figure 10 [119, 121, 130]. Many efficient implementations were developed. Depending on implementations, there are many alternative ways of grouping "variables" or "clauses" together. The $SAT1.6$ algorithm needs not necessarily be run on a parallel computer. By taking advantage of bit-level parallelism, we can pack many Boolean data into the bits of a computer word and evaluate them simultaneously on a sequential computer [119, 121, 130, 134, 300]. $SAT1.6$ algorithm can be used in conjunction with a backtracking/resolution procedure to verify satisfiability and unsatisfiability.

6.2.3. *Complete Local Search Algorithms*

If we put local search algorithms into a variety of complete search framework, for any given CNF formula, they are able to verify satisfiability as well as unsatisfiability efficiently. $SAT1.10$ to $SAT1.20$ include efficient, parallel, and complete SAT algorithms developed to solve hard and practical engineering applications [119, 120, 130, 134]. Some of the algorithms combine parallel local search with resolution/backtracking procedures, achieving computing efficiency and search completeness. For a complete local search algorithm, if at a node of search tree a solution point is found unsatisfiable, then the algorithm backtracks and searches the next solution point until a solution is found or unsatisfiability is proved to be true. Figures 11 and 12 give two typical backtracking local search algorithms.

6.3. Average Time Complexity

A number of theoretical analysis was made on the complexity of the local search procedures for SAT problem. Koutsoupias and Papadimitriou showed that the "greedy" local search algorithm (that flips the variable that improves the most clauses) converges in m steps with a high probability [192, 244]. Gu and Gu studied the average time complexity of some local search algorithms [135, 127]. For the randomly generated CNF formulae with n clauses, m variables, and l literals in each clause, the average run time of the $SAT1.1$ algorithm for verifying the satisfiability is $O(m^{O(1)}n)$ for $l \geq \log m - \log\log m - c$ and $n/m \leq 2^{l-2}/l$, where c is a constant. For $l \geq 3$ and $n/m \leq \alpha 2^l/l$, where $\alpha < l$ is a constant, the average run time of the $SAT1.2$ algorithm for verifying the satisfiability is $O(m^{O(1)}n^2)$. The average run time of the $SAT1.3$ algorithm is $O(ln(m \log m)^2)$ for $l \geq 3$ and $n/m \leq \alpha 2^l/l$, where $\alpha < l$ is

```
procedure SAT1.6 ()
begin
    /* initialization */
    get_a_SAT_instance();
    x0 := select_an_initial_point();
    F(x) := evaluate_object_function(x0);

    /* search */
    k := 0;
    while F ≠ 0 do
    begin
        ||for some x_i(k)'s ∈ x_k do
            /* if swap(x'_i(k)s) does not increase F */
            if test_swap(x'_i(k)s) then
            begin
                x_k+1 := perform_swap(x'_i(k)s);
                F(x) := evaluate_object_function();
            end;
        if local then local_handler();
        k := k + 1;
    end;
end;
```

Fig. 10. **SAT1.6: a parallel local search algorithm for SAT problem.**

a constant. These results improve significantly on the well-known Davis-Putnam algorithm which has an $O(n^{O(m/l)})$ average time complexity for certain classes of *CNF* formulas.

6.4. A Discrete, Unconstrained Optimization Model for *DNF* Formulas

Using a simple transformation — taking the "distance" between the truth value and the solution value — we obtain another unconstrained optimization model, the *SAT4 model* [119]:

$$x_i = \begin{cases} True & \text{if } x_i - 1 = 0 \\ False & \text{if } x_i + 1 = 0 \end{cases} \tag{8}$$

The objective function in *SAT4* model is defined as:

$$F(\mathbf{x}) = \sum_{i=1}^{n} C_i(\mathbf{x}) \tag{9}$$

```
Procedure SAT1.11 ()
begin
    /* initialization */
    get_a_SAT_instance();
    x0 := select_an_initial_point();
    f := evaluate_object_function(x0);

    /* backtracking with local search */
    x* := backtracking(x0);
end;


procedure backtracking(xi)
begin
    /* local search assigns v to xi */
    v := local_search();
    xi := v;
    Vi := Vi - {v};
    /* append variable xi to the partial path */
    path[xi] := i;

    if path broken then backtracking;
    if solution found then return x*;
    else backtracking(next xi);
end;
```

Fig. 11. **SAT1.11: a complete local search algorithm with backtracking.**

where

$$C_i(\mathbf{x}) = \prod_{j=1}^{m} Q_{ij}(x_j) \tag{10}$$

where

$$Q_{ij}(x_j) = \begin{cases} |x_j - 1| & \text{if } x_j \text{ is a literal in } C_i \\ |x_j + 1| & \text{if } \bar{x_j} \text{ is a literal in } C_i \\ 1 & \text{if neither } x_j \text{ nor } \bar{x_j} \text{ is in } C_i \end{cases} \tag{11}$$

A $SAT1.4$ and $SAT14.5$ algorithms were implemented for the $SAT4$ model [119]. The $SAT14.5$ algorithm is discussed in Section 7.

The *SAT4 model* is a dual model to the *SAT1 model*. It is an important model that transforms a *CNF* formula into a *DNF* formula. In Section 7, we extend the $SAT4$ model into real space and develop universal SAT problem model, *UniSAT*, for

```
Procedure SAT1.14 ()
begin
    /* initialization */
    get_a_SAT_instance();
    x0 := select_an_initial_point();
    f := evaluate_object_function(x0);

    /* backtracking with local search */
    x* := backtracking(x0);
end;


procedure backtracking(x_i)
begin
    x_i := v;
    V_i := V_i - {v};
    /* local search with x_i assigned with v */
    local_search(v);
    /* append variable x_i to the partial path */
    path[x_i] := i;

    if path broken then backtracking;
    if solution found then return x*;
    else backtracking(next x_i);
end;
```

Fig. 12. **SAT1.14: a complete local search algorithm with backtracking.**

the SAT problem. This has resulted in many continuous, unconstrained optimization algorithms for the SAT problem.

6.5. Experimental Results

In this section, we give experimental results of a parallel (32-bit) local search algorithm (the $SAT1.7$ algorithm) and a complete local search algorithm (the $SAT1.13$ algorithm) for the SAT problem [119, 126, 130]. We will test $SAT1.7$ and $SAT1.13$ algorithms with various SAT problem instance models including Pehoushek's problem instances [250]. We will also compare the performance of these local search algorithms with those of the Davis-Putnam algorithm [63, 62], the $GSAT$ algorithm [287], and an interior point zero-one integer programming algorithm [186], for the SAT problem.

Due to limited memory space on a workstation computer, in order to experiment with the large-scale SAT problem instances, $SAT1.7$ and $SAT1.13$ algorithm were

TABLE I
Real execution performance of a *SAT*1.7 algorithm on a SUN SPARC 2 workstation (Time Units: seconds)

Problems (n,m,l)			Ten Trials		Execution Time			Iterations (k)		
Clauses	Variables	Literals	Global	Local	Min	Mean	Max	Min	Mean	Max
100	100	3	10	0	0.000	0.001	0.010	1	1.900	7
200	100	3	10	0	0.000	0.004	0.010	2	2.500	4
300	100	3	10	0	0.000	0.008	0.020	2	5.900	12
400	100	3	10	0	0.010	0.027	0.060	6	28.40	76
1000	1000	3	10	0	0.020	0.030	0.050	2	2.600	4
1500	1000	3	10	0	0.040	0.055	0.080	3	3.600	5
2000	1000	3	10	0	0.070	0.084	0.110	4	4.800	6
2500	1000	3	10	0	0.100	0.124	0.180	5	6.500	10
3000	1000	3	10	0	0.130	0.179	0.250	6	9.700	14
3500	1000	3	10	0	0.200	0.636	1.380	12	41.50	91
1000	1000	4	10	0	0.010	0.022	0.040	1	1.700	3
2000	1000	4	10	0	0.050	0.061	0.080	2	2.100	3
3000	1000	4	10	0	0.070	0.094	0.140	2	3.000	4
4000	1000	4	10	0	0.110	0.144	0.180	3	3.900	5
5000	1000	4	10	0	0.160	0.227	0.300	4	5.800	7
6000	1000	4	10	0	0.240	0.383	0.490	6	10.60	14
7000	1000	4	10	0	0.560	0.865	1.120	19	29.10	37
8000	1000	4	10	0	0.890	1.896	5.990	36	84.30	274
8500	1000	4	10	0	1.840	10.786	22.69	86	557.0	1232
10000	1000	5	10	0	0.330	0.451	0.800	4	6.300	12
11000	1000	5	10	0	0.350	0.489	0.600	5	6.700	9
12000	1000	5	10	0	0.420	0.593	0.740	6	9.100	11
13000	1000	5	10	0	0.510	0.761	1.220	8	12.90	22
14000	1000	5	10	0	0.720	1.107	1.760	12	21.40	38
15000	1000	5	10	0	0.810	1.671	2.900	15	37.50	72
10000	400	6	10	0	0.280	0.497	0.930	7	13.30	26
10000	500	6	10	0	0.190	0.377	0.530	3	7.200	11
10000	600	6	10	0	0.220	0.328	0.500	3	4.500	7
10000	700	6	10	0	0.200	0.284	0.430	3	3.600	5
10000	800	6	10	0	0.230	0.277	0.400	3	3.200	4
10000	900	6	10	0	0.220	0.289	0.450	2	2.600	4
10000	1000	6	10	0	0.210	0.264	0.350	2	2.300	3
20000	1000	7	10	0	0.330	0.500	0.820	2	2.400	4
30000	2000	7	10	0	0.690	0.882	1.080	2	2.300	3
40000	3000	7	10	0	1.140	1.289	1.580	2	2.200	3
50000	4000	7	10	0	1.510	1.666	1.890	2	2.000	2
60000	5000	7	10	0	0.890	1.260	3.440	1	1.100	2
10000	1000	10	10	0	0.030	0.050	0.070	1	1.000	1
20000	2000	10	10	0	0.090	0.124	0.180	1	1.000	1
30000	3000	10	10	0	0.190	0.258	0.350	1	1.000	1
40000	4000	10	10	0	0.210	0.305	0.400	1	1.000	1
50000	5000	10	10	0	0.330	0.441	0.590	1	1.000	1

TABLE II
Performance comparison between the Davis-Putnam algorithm and a $SAT1.7$ algorithm on a SUN SPARC 2 workstation (Average of ten algorithm executions. Time Units: seconds. Symbol "S/F" stands for algorithm's *success/failure* of giving a solution within a time limit of $120 \times n/m$ seconds.)

Problems Clause, Variable, Literal			Execution Time Davis-Putnam Algorithm				Execution Time $SAT1.7$ Algorithm			
n	m	l	S/F	Min	Mean	Max	S/F	Min	Mean	Max
500	500	3	10/0	2.066	2.159	2.249	10/0	0.010	0.013	0.020
750	500	3	10/0	2.699	2.916	3.183	10/0	0.010	0.015	0.020
1000	500	3	10/0	3.149	3.657	6.316	10/0	0.020	0.035	0.050
1250	500	3	9/1	3.966	5.797	10.02	10/0	0.030	0.049	0.080
1500	500	3	6/4	7.499	9.147	11.60	10/0	0.050	0.108	0.220
1000	500	4	10/0	4.533	4.684	4.933	10/0	0.020	0.026	0.030
1500	500	4	10/0	6.766	7.960	16.90	10/0	0.030	0.040	0.060
2000	500	4	8/2	8.999	10.27	14.25	10/0	0.040	0.066	0.100
2500	500	4	2/8	13.63	15.96	18.28	10/0	0.060	0.074	0.090
3000	500	4	1/9	46.33	46.33	46.33	10/0	0.070	0.118	0.160
3000	500	5	10/0	16.23	16.90	18.82	10/0	0.070	0.094	0.140
4000	500	5	5/5	21.72	28.39	44.66	10/0	0.090	0.119	0.160
5000	500	5	0/10	> 1200	> 1200	> 1200	10/0	0.110	0.180	0.290
6000	500	5	0/10	> 1440	> 1440	> 1440	10/0	0.210	0.313	0.450
7000	500	5	0/10	> 1680	> 1680	> 1680	10/0	0.380	0.591	0.980
10000	1000	10	10/0	99.71	101.8	103.1	10/0	0.030	0.047	0.070
12000	1000	10	10/0	122.1	124.3	126.7	10/0	0.030	0.073	0.100
14000	1000	10	10/0	140.7	145.2	148.7	10/0	0.040	0.077	0.150
16000	1000	10	10/0	165.1	167.1	168.7	10/0	0.060	0.098	0.140
18000	1000	10	10/0	185.5	188.6	189.9	10/0	0.060	0.136	0.210

implemented in C in a space-efficient manner.

6.5.1. *Performance of the $SAT1.7$ Algorithm*

Real Execution Time of the $SAT1.7$ Algorithm. The initial problem instances were generated based on the exact l-SAT model. The real performance of the $SAT1.7$ algorithm running on a SUN workstation is illustrated in Table I. The number of clauses (n), the number of variables (m), and the number of literals per clause (l), are give in the first three columns. Column 4 (*Global*) indicates the number of times that the algorithm succeeds in finding a solution, i.e., a global minimum point. Column 5 (*Local*) indicates the number of times that the algorithm was stuck at local minimum points.

From these results we can observe that, in terms of global convergence and local convergent rate, the $SAT1.7$ algorithm exhibits good convergent properties and fast computing speed.

As discussed in Section 6.3 and [135], beyond a certain range of hardness, for example, for $n = 8500, m = 1000$, and $l = 4$, the computing time of the $SAT1.7$

TABLE III
Performance comparison between a $SAT1.7$ algorithm running on a SUN SPARC 2 workstation and the $GSAT$ algorithm running on a MIPS computer with comparable computing power (Time Units: seconds)

Problems			Execution Time	
n	m	l	GSAT	SAT1.7
215	50	3	0.400	0.019
301	70	3	0.900	0.054
430	100	3	6.000	0.336
516	120	3	14.00	0.596
602	140	3	14.00	0.260
645	150	3	45.00	0.102
860	200	3	168.0	1.776
1062	250	3	246.0	3.106
1275	300	3	720.0	8.822

TABLE IV
Execution performance of a $SAT1.7$ algorithm with Pehoushek's Benchmarks on a SUN SPARC 2 workstation (Time Units: seconds)

Problems				Ten Trials		Execution Time			Iterations (k)		
Names	n	m	l	Global	Local	Min	Mean	Max	Min	Mean	Max
Prob1	600	180	3	10	0	0.400	3.640	9.990	334	3221.9	8880
Prob2	600	180	3	10	0	0.160	0.501	0.990	116	418.90	879
Prob3	600	180	3	10	0	0.040	0.705	2.030	32	596.70	1745
Prob4	600	180	3	10	0	0.220	1.408	2.770	185	1181.9	2285
Prob5	600	180	3	10	0	0.220	0.727	3.260	154	572.90	2560
Prob6	600	180	3	10	0	0.060	0.703	1.580	40	567.20	1309
Prob7	600	180	3	10	0	0.170	0.997	3.740	135	866.10	3321
Prob8	600	180	3	10	0	0.030	0.814	1.830	16	675.90	1616
Prob9	600	180	3	10	0	0.100	1.134	2.670	76	881.30	2150
Prob10	600	180	3	10	0	0.070	0.539	2.630	51	437.60	2059

algorithm starts to increase.

The Probabilistic Behavior of the $SAT1.7$ Algorithm. The $SAT1.7$ algorithm uses a probabilistic local handler. Table I also shows its probabilistic behavior, the number

TABLE V
Performance comparison between a $SAT1.7$ algorithm running on a SUN SPARC 2 workstation and an interior point zero-one integer programming algorithm running on a $KORBX(R)$ parallel/vector computer (Time Units: seconds)

Problems Clause,Variable,Literal			Execution Time Interior Point Algorithm				Execution Time $SAT1.7$ Algorithm			
n	m	l	S/F	Min	Mean	Max	S/F	Min	Mean	Max
100	50	5	52/0	0.5	0.7	7.7	10/0	0.000	0.001	0.010
200	100	5	70/0	1.0	1.1	1.7	10/0	0.000	0.006	0.020
400	200	7	69/0	2.4	3.5	6.3	10/0	0.000	0.007	0.020
800	400	10	31/0	5.1	5.6	7.7	10/0	0.000	0.009	0.020
800	400	7	20/0	4.7	7.8	16.1	10/0	0.010	0.014	0.040
1000	500	10	49/0	6.5	7.4	9.8	10/0	0.000	0.012	0.020
2000	1000	10	10/0	14.9	18.5	20.5	10/0	0.020	0.032	0.090
2000	1000	7	50/0	18.1	21.5	27.3	10/0	0.030	0.056	0.090
2000	1000	3	49/1	8.7	420.7	9595.5	10/0	0.180	2.657	13.07
4000	1000	4	1/1	8314.3	8314.3	8314.3	10/0	1.080	10.63	26.64
4000	1000	10	10/0	24.0	25.1	25.8	10/0	0.050	0.055	0.070
8000	1000	10	10/0	37.5	38.0	38.8	10/0	0.090	0.219	0.410
16000	1000	10	10/0	46.2	66.4	92.1	10/0	0.490	0.603	0.800
32000	1000	10	10/0	80.1	232.4	311.3	10/0	1.450	1.701	2.780

of iterations required, and its success in finding a solution. The table indicates that the algorithm succeeds for most given SAT problem instances. For many practical problem instances we have encountered, $SAT1.7$'s performance seems fairly robust.

Performance Comparison with the Davis-Putnam Algorithm. The execution results of the Davis-Putnam algorithm and the $SAT1.7$ algorithm for the exact l-SAT problem instances are given in Table II. Ten algorithm executions were made for each algorithm. The minimum, maximum, and average execution times are reported. Because the Davis-Putnam algorithm was slow for hard problem instances, a maximum execution time of $120 \times n/m$ seconds was set as the time limit of its execution. Symbol "S/F" in Column 4 stands for algorithm's *success/failure* of giving an answer within such a time limit. For the Davis-Putnam algorithm, the average execution time does not include the maximum execution time limit if some of the ten executions were successful; the average execution time was taken as the maximum execution time limit only if all ten executions were failed.

From numerous algorithm executions, we have observed that, for the random problem instances listed in the table, the Davis-Putnam algorithm was approximately hundreds to thousands times slower than the $SAT1.7$ algorithm. As the hardness of the problem instances increases, the number of failure, F, increases quickly. For some slightly hard problem instances, such as $n = 5000, m = 500$, and $l = 5$, all ten algorithm executions failed after a reasonably long time limit. Due to its $O(n^{O(m/l)})$ average run time complexity, even for some fairly easy problem instances, such as $n = 10000, m = 1000$, and $l = 10$, the Davis-Putnam algorithm

TABLE VI
Real execution performance of a *SAT*1.13 algorithm on a SUN SPARC 2 workstation (Time Units: seconds)

Problems (n,m,l)			Ten Trials		Execution Time			Iterations (k)		
Clauses	Variables	Literals	Global	Local	Min	Mean	Max	Min	Mean	Max
100	100	3	10	0	0.000	0.001	0.010	1	1	1
200	100	3	10	0	0.000	0.010	0.020	1	2.30	4
300	100	3	10	0	0.000	0.015	0.030	2	5.10	12
400	100	3	10	0	0.010	0.145	0.840	9	145.4	893
1000	1000	3	10	0	0.040	0.048	0.060	1	1.40	2
1500	1000	3	10	0	0.060	0.078	0.090	1	2.30	3
2000	1000	3	10	0	0.070	0.113	0.150	2	3.80	6
2500	1000	3	10	0	0.120	0.158	0.210	4	5.50	8
3000	1000	3	10	0	0.200	0.310	0.480	7	13.3	23
3500	1000	3	10	0	0.490	1.008	2.770	25	55.2	167
1000	1000	4	10	0	0.030	0.045	0.060	1	1	1
2000	1000	4	10	0	0.080	0.103	0.140	1	1.7	3
3000	1000	4	10	0	0.120	0.160	0.210	1	2.2	3
4000	1000	4	10	0	0.190	0.230	0.260	2	3.0	4
5000	1000	4	10	0	0.290	0.338	0.430	4	5.0	7
6000	1000	4	10	0	0.350	0.465	0.680	5	8.3	15
7000	1000	4	10	0	0.430	0.729	1.040	7	17.5	31
8000	1000	4	10	0	0.960	2.088	4.600	27	75.1	183
8500	1000	4	10	0	2.950	7.947	14.63	111	369.2	743
10000	1000	5	10	0	0.500	0.610	0.720	3	4.80	7
11000	1000	5	10	0	0.590	0.800	0.990	4	7.40	11
12000	1000	5	10	0	0.680	0.839	1.110	5	8.30	13
13000	1000	5	10	0	0.780	1.154	1.620	7	14.5	23
14000	1000	5	10	0	0.910	1.308	1.810	10	19.4	33
15000	1000	5	10	0	1.210	2.027	3.250	16	37.0	66
10000	400	6	10	0	0.500	0.625	0.710	7	9.9	12
10000	500	6	10	0	0.480	0.640	0.820	3	6.2	10
10000	600	6	10	0	0.400	0.493	0.570	2	3.0	4
10000	700	6	10	0	0.430	0.550	0.770	2	2.9	5
10000	800	6	10	0	0.430	0.494	0.630	1	1.9	3
10000	900	6	10	0	0.460	0.523	0.700	2	1.9	3
10000	1000	6	10	0	0.430	0.488	0.680	1	1.3	3
20000	1000	7	10	0	0.910	1.124	1.440	1	1.9	4
30000	2000	7	10	0	1.550	1.733	1.950	1	1.3	2
40000	3000	7	10	0	2.160	2.382	2.580	1	1.1	2
50000	4000	7	10	0	2.920	3.036	3.140	1	1	1
60000	5000	7	10	0	3.530	3.719	4.060	1	1	1
10000	1000	10	10	0	0.340	0.377	0.420	1	1	1
20000	2000	10	10	0	0.780	0.821	0.880	1	1	1
30000	3000	10	10	0	1.200	1.311	1.470	1	1	1
40000	4000	10	10	0	1.710	1.826	1.970	1	1	1
50000	5000	10	10	0	2.120	2.372	2.660	1	1	1

TABLE VII
Real performance comparison between the Davis-Putnam algorithm and a $SAT1.13$ algorithm on a SUN SPARC 2 workstation (Average of ten algorithm executions. Time Units: seconds. Symbol "S/F" stands for algorithm's *success/failure* of giving a solution within a time limit of $120 \times n/m$ seconds.)

Problems Clause,Variable,Literal			Execution Time Davis-Putnam Algorithm				Execution Time $SAT1.13$ Algorithm			
n	m	l	S/F	Min	Mean	Max	S/F	Min	Mean	Max
500	500	3	10/0	2.066	2.159	2.249	10/0	0.020	0.021	0.030
750	500	3	10/0	2.699	2.916	3.183	10/0	0.020	0.030	0.040
1000	500	3	10/0	3.149	3.657	6.316	10/0	0.040	0.048	0.060
1250	500	3	9/1	3.966	5.797	10.02	10/0	0.050	0.064	0.090
1500	500	3	6/4	7.499	9.147	11.60	10/0	0.070	0.117	0.190
1000	500	4	10/0	4.533	4.684	4.933	10/0	0.030	0.040	0.050
1500	500	4	10/0	6.766	7.960	16.90	10/0	0.060	0.075	0.100
2000	500	4	8/2	8.999	10.27	14.25	10/0	0.070	0.105	0.180
2500	500	4	2/8	13.63	15.96	18.28	10/0	0.090	0.130	0.160
3000	500	4	1/9	46.33	46.33	46.33	10/0	0.150	0.201	0.320
3000	500	5	10/0	16.23	16.90	18.82	10/0	0.110	0.137	0.190
4000	500	5	5/5	21.72	28.39	44.66	10/0	0.160	0.204	0.240
5000	500	5	0/10	> 1200	> 1200	> 1200	10/0	0.180	0.253	0.320
6000	500	5	0/10	> 1440	> 1440	> 1440	10/0	0.320	0.370	0.500
7000	500	5	0/10	> 1680	> 1680	> 1680	10/0	0.390	0.575	0.960
10000	1000	10	10/0	99.71	101.8	103.1	10/0	0.350	0.382	0.450
12000	1000	10	10/0	122.1	124.3	126.7	10/0	0.420	0.458	0.530
14000	1000	10	10/0	140.7	145.2	148.7	10/0	0.480	0.526	0.560
16000	1000	10	10/0	165.1	167.1	168.7	10/0	0.560	0.596	0.620
18000	1000	10	10/0	185.5	188.6	189.9	10/0	0.650	0.716	0.780

took an excessive amount of time to find a solution. In comparison, the $SAT1.7$ algorithm was successful for all ten executions. It was able to find a solution to the given problem instances efficiently.

Since $SAT1.7$ algorithm is *incomplete*, it can neither find a solution for an unsatisfiable *CNF* formula nor give an answer if a *CNF* formula is not satisfiable.

Performance Comparison with the GSAT Algorithm. Table III gives real performance comparison between a $SAT1.7$ algorithm running on a SUN SPARC 2 workstation and the *GSAT* algorithm [287] running on a MIPS computer with comparable computing power [286]. For the exact 3-SAT problem instances generated from the same problem model used in [231], $SAT1.7$ algorithm is approximately tens to hundreds times faster than the *GSAT* algorithm.

Performance Comparison with Pehoushek's Benchmarks. Pehoushek has recently provided an *i-j-regular 3-SAT* problem model [250]. In his model, one puts i positive and j negative copies of each variable in a set, pull out 3 at a time forming a

TABLE VIII
Performance comparison between a *SAT*1.13 algorithm running on a SUN SPARC 2 workstation and the *GSAT* algorithm running on a MIPS computer with comparable computing power (Time Units: seconds)

Problems			Execution Time	
n	m	l	GSAT	SAT1.13
215	50	3	0.400	0.100
301	70	3	0.900	0.010
430	100	3	6.000	0.040
516	120	3	14.00	0.810
602	140	3	14.00	8.060
645	150	3	45.00	0.190
860	200	3	168.0	0.970
1062	250	3	246.0	20.71
1275	300	3	720.0	19.66

TABLE IX
Execution performance of a *SAT*1.13 algorithm with Pehoushek's Benchmarks on a SUN SPARC 2 workstation (Time Units: seconds)

Problems				Ten Trials		Execution Time			Iterations (k)		
Names	n	m	l	Global	Local	Min	Mean	Max	Min	Mean	Max
Prob1	600	180	3	10	0	0.660	4.509	13.77	527	3945.5	11935
Prob2	600	180	3	10	0	0.080	0.381	1.580	57	307.90	1288
Prob3	600	180	3	10	0	0.040	0.682	1.710	17	556.00	1440
Prob4	600	180	3	10	0	0.050	2.228	6.840	29	1868.0	5816
Prob5	600	180	3	10	0	0.080	0.628	1.910	42	507.80	1590
Prob6	600	180	3	10	0	0.050	0.480	1.580	28	379.10	1392
Prob7	600	180	3	10	0	0.230	6.479	49.78	181	1882.4	5919
Prob8	600	180	3	10	0	0.300	0.718	1.690	214	526.90	1248
Prob9	600	180	3	10	0	0.060	0.839	2.670	36	678.90	2323
Prob10	600	180	3	10	0	0.040	0.753	1.750	17	597.80	1496

clause. Table IV shows real execution performance of the *SAT*1.7 algorithm for ten benchmark SAT problem instances provided by Pehoushek. Presently we are unable to make any performance comparison with the existing SAT algorithms since there are no any execution data with Pehoushek's benchmarks in the public area. According to Pehoushek's performance measure [250], *SAT*1.7 is *extremely* fast on

TABLE X
Performance comparison between a *SAT*1.13 algorithm running on a SUN SPARC 2 workstation and an interior point zero-one integer programming algorithm running on a *KORBX(R)* parallel/vector computer (Time Units: seconds)

Problems Clause, Variable, Literal			Execution Time Interior Point Algorithm				Execution Time *SAT*1.13 Algorithm			
n	*m*	*l*	S/F	Min	Mean	Max	S/F	Min	Mean	Max
100	50	5	52/0	0.5	0.7	7.7	10/0	0.000	0.004	0.010
200	100	5	70/0	1.0	1.1	1.7	10/0	0.010	0.010	0.010
400	200	7	69/0	2.4	3.5	6.3	10/0	0.010	0.014	0.020
800	400	10	31/0	5.1	5.6	7.7	10/0	0.030	0.034	0.050
800	400	7	20/0	4.7	7.8	16.1	10/0	0.020	0.032	0.060
1000	500	10	49/0	6.5	7.4	9.8	10/0	0.030	0.037	0.050
2000	1000	10	10/0	14.9	18.5	20.5	10/0	0.070	0.091	0.110
2000	1000	7	50/0	18.1	21.5	27.3	10/0	0.080	0.099	0.140
2000	1000	3	49/1	8.7	420.7	9595.5	10/0	0.110	0.162	0.290
4000	1000	4	1/1	8314.3	8314.3	8314.3	10/0	0.510	11.07	40.19
4000	1000	10	10/0	24.0	25.1	25.8	10/0	0.160	0.189	0.270
8000	1000	10	10/0	37.5	38.0	38.8	10/0	0.370	0.456	0.610
16000	1000	10	10/0	46.2	66.4	92.1	10/0	0.970	1.042	1.110
32000	1000	10	10/0	80.1	232.4	311.3	10/0	2.230	2.720	3.090

the problem instances.

Performance Comparison with Interior Point Zero-One Integer Programming Algorithm Recently Kamath et al. used an interior point zero-one integer programming algorithm to solve the SAT problem [186]. They implemented their algorithm in FORTRAN and C languages and ran the algorithm on a *KORBX(R)* parallel/vector computer with problem instances generated from the average 3-SAT problem model. The *KORBX(R)* parallel computer operates in scalar mode at approximately 1 MFlops and at 32 MFlops with full vector concurrent mode. Their execution results are given in Table V.

We ran the *SAT*1.7 algorithm for the same problem instances on a sequential SUN SPARC 2 workstation. The results are given in Table V. Apparently, compared to the interior point zero-one integer programming algorithm running on a parallel computer, in addition to improved global convergence, *SAT*1.7 algorithm is much simpler and it achieves many orders of magnitude of performance improvements in terms of computing time.

6.5.2. *Performance of the SAT1.13 Algorithm*

In Tables VI – X, we give the real execution performance of the *SAT*1.13 algorithm. Since *SAT*1.13 combines local search with backtracking, it is a *complete* algorithm. For some problem instances, it takes slightly more execution time than the *SAT*1.7

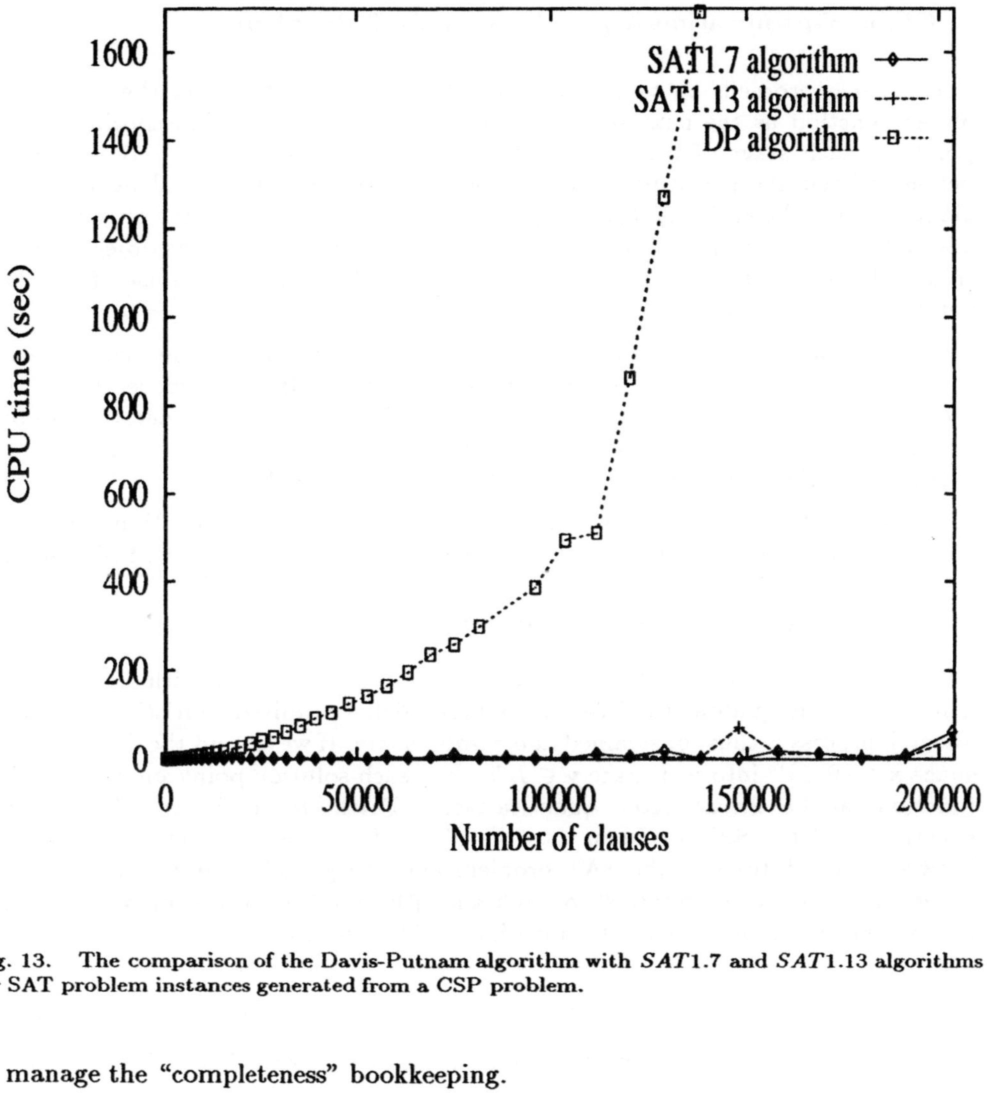

Fig. 13. The comparison of the Davis-Putnam algorithm with $SAT1.7$ and $SAT1.13$ algorithms for SAT problem instances generated from a CSP problem.

to manage the "completeness" bookkeeping.

6.5.3. *Experiments with the structured SAT models*

To assess the performance of the $SAT1$ algorithms with *non-binary* problem instances, we also tested SAT problem instances generated from regularly structured problems. Figure 13 shows performance comparisons between the Davis-Putnam algorithm and the $SAT1.7$ and $SAT1.13$ algorithms for SAT problem instances generated from a regular CSP problem.

In the next section, we extend the discrete local search algorithm on Boolean space into continuous global optimization algorithms on real space.

7. Global Optimization Algorithms for SAT Problem

Local search proceeds by taking *any* solution point that decreases the value of the object function as the next solution point. There are many neighboring solution points at each stage of the search. A local search does not take into account its neighbors' relative performance to the object function. A natural heuristic is to globally select the *best neighbor* that yields the minimum value to the object function and take this *best neighbor's direction* as the descent direction of the object function. In a real search space, many global optimization algorithms make use of this best-neighbor heuristics.

Using a global optimization method to solve the SAT problem, special models are formulated that transform a discrete SAT problem on Boolean space $\{0,1\}^m$ into a continuous SAT problem on real space E^m. Thus, the discrete decision problem is transformed into an unconstrained optimization problem which can be solved by existing global optimization methods [119, 122, 127, 129, 131].

In this section, we give some transformation models, optimization algorithms, and theoretical analyses of some global optimization methods for the SAT problem.

7.1. *UniSAT*: Universal SAT Problem Models

The representation of a solution point and the objective function with discrete values is restricted and qualitative. Discrete variables deliver limited heuristic information, a search process is thus constrained and cumbersome. If we extend the discrete search space $\mathbf{x} \in \{0,1\}^m$ into real space $\mathbf{y} \in E^m$, then each solution point and the objective function can be characterized quantitatively. Furthermore, by encoding solution constraints of the SAT problem into the object function, a direct correspondence between the solutions of the SAT problem and the global minimum points of the object function can be established. Subsequently, the SAT problem is transformed into an unconstrained global optimization problem on E^m.

7.1.1. *The Universal DeMorgan Laws*

DeMorgan Laws are well-known, when applying them to *CNF* and *DNF*, we have:

$$\overline{A \vee B \vee C \cdots} = \overline{A} \wedge \overline{B} \wedge \overline{C} \wedge \cdots \tag{12}$$

$$\overline{A \wedge B \wedge C \wedge \cdots} = \overline{A} \vee \overline{B} \vee \overline{C} \vee \cdots \tag{13}$$

To use global optimization techniques for the SAT problem on a real space, we need the following generalizations and transformations to extend Boolean DeMorgan Laws into the universal DeMorgan Laws on real space E^m.

Operator generalization. We first replace Boolean $\vee$ and $\wedge$ operators with real operators $+$ and $\times$, respectively:

$$\overline{A+B+C\cdots}=\overline{A}\times\overline{B}\times\overline{C}\times\cdots \tag{14}$$

$$\overline{A\times B\times C\times\cdots}=\overline{A}+\overline{B}+\overline{C}+\cdots \tag{15}$$

Variable generalization. Following the *SAT4 model,* we extend Boolean variables $\mathbf{x}$ $(x_1, x_2, ..., x_m)$ on Boolean space $\{0,1\}^m$ to real variables $\mathbf{y}$ $(y_1, y_2, ..., y_m)$ on real space E^m. The correspondence between $\mathbf{x}$ and $\mathbf{y}$ is defined as follows (for $1 \leq i \leq m$):

$$x_i = \begin{cases} True & \text{if } y_i = 1 \\ False & \text{if } y_i = -1 \\ undefined & \text{otherwise} \end{cases} \tag{16}$$

which is equivalent to

$$x_i = \begin{cases} True & \text{if } y_i - 1 = 0 \\ False & \text{if } y_i + 1 = 0 \\ undefined & \text{otherwise} \end{cases} \tag{17}$$

The following transformations are used to derive the objective function in the *UniSAT* models.

Literal transformation. Based on real variables, a literal function is defined for each literal:

$$q_{ij}(y_j) = \begin{cases} |y_j - 1| & \text{if } x_j \text{ is a literal in } C_i \\ |y_j + 1| & \text{if } \bar{x_j} \text{ is a literal in } C_i \\ 1 & \text{if neither } x_j \text{ nor } \bar{x_j} \text{ is in } C_i \end{cases} \tag{18}$$

In this quantitative formulation, the solution assignment is coded explicitly in the function.

Clause transformation. A clause function $c_i(\mathbf{y})$ from E^m to E^1 is defined as a product with at most m literal functions:

$$c_i(\mathbf{y}) = \prod_{j=1}^{m} q_{ij}(y_j). \tag{19}$$

The clause function transforms each clause $C_i(\mathbf{x})$ on Boolean space $\{0,1\}^m$ into a clause function $c_i(\mathbf{y})$ on real space E^m.

The UniSAT Model. Using literal and clause functions, a *CNF* formula $F(\mathbf{x})$ on Boolean space $\{0,1\}^m$ is transformed into an object function $f(\mathbf{y})$ on real space E^m:

$$f(\mathbf{y}) = \sum_{i=1}^{n} c_i(\mathbf{y}) = \sum_{i=1}^{n}\prod_{j=1}^{m} q_{ij}(y_j). \tag{20}$$

The *UniSAT* models transform the SAT problem from a discrete, constrained decision problem into an unconstrained global optimization problem [119, 122, 129, 126, 131]. A good property of the transformation is that *UniSAT* models establish a correspondence between the global minimum points of the objective function and the solutions of the original SAT problem. A *CNF* $F(\mathbf{x})$ is true *if and only if* $f(\mathbf{y})$ takes the global minimum value 0 on the corresponding $\mathbf{y}$.

Following the above formulation, for example, a *CNF* $F(\mathbf{x}) = (x_1 \vee \bar{x}_2) \wedge (\bar{x}_1 \vee x_2 \vee x_3)$ is translated into $f(\mathbf{y}) = |y_1 - 1||y_2 + 1| + |y_1 + 1||y_1 - 1||y_3 - 1|$. The solution to the SAT problem corresponds to a set of global minimum points of the object function. To find a *True* value of $F(\mathbf{x})$ is equivalent to find a *False* value, i.e., 0, of f($\mathbf{y}$).

7.1.2. *Basic UniSAT Problem Models*

A transformation model determines the performance of an optimization algorithm. Many *UniSAT* problem models were developed. A variety of the objective functions with polynomial, logarithm, and exponential formations were studied. For the sake of simplicity, we give some typical *UniSAT* problem models.

The UniSAT5 Model: Given a *CNF* formula $F(\mathbf{x})$ from $\{0,1\}^m$ to $\{0,1\}$ with n clauses $C_1, \ldots, C_n$, we define a real function $f(\mathbf{y})$ from E^m to E that transforms the SAT problem into an unconstrained global optimization problem [119, 122, 129, 126, 131]:

$$\min_{\mathbf{y} \in \mathbf{E}^m} f(\mathbf{y}) \tag{21}$$

where

$$f(\mathbf{y}) = \sum_{i=1}^{n} c_i(\mathbf{y}). \tag{22}$$

A clause function $c_i(\mathbf{y})$ is a product of m *literal functions* $q_{ij}(y_j)$ $(1 \leq j \leq m)$:

$$c_i = \prod_{j=1}^{m} q_{ij}(y_j), \tag{23}$$

where

$$q_{ij}(y_j) = \begin{cases} |y_j - 1| & \text{if literal } x_j \text{ is in clause } C_i \\ |y_j + 1| & \text{if literal } \bar{x}_j \text{ is in clause } C_i \\ 1 & \text{if neither } x_j \text{ nor } \bar{x}_j \text{ is in } C_i \end{cases} \tag{24}$$

The correspondence between $\mathbf{x}$ and $\mathbf{y}$ is defined as follows (for $1 \leq i \leq m$):

$$x_i = \begin{cases} 1 & \text{if } y_i = 1 \\ 0 & \text{if } y_i = -1 \\ undefined & \text{otherwise} \end{cases}$$

Clearly, $F(\mathbf{x})$ is true iff $f(\mathbf{y})$=0 on the corresponding $\mathbf{y} \in \{-1,1\}^m$.

The following theorem gives an important condition for stopping the optimization process for solving the *UniSAT5* problem.

Theorem 3 *For object function f defined in the UniSAT5 problem model, if $f(\mathbf{y}) < 1$ for some $\mathbf{y} \in E^m$, then we can find a solution $\mathbf{y}^* \in \{-1, 1\}^m$ such that $f(\mathbf{y}^*) = 0$, i.e., we can find a solution of CNF formula F.*

Proof: From $f(\mathbf{y}) < 1$, we have that for each clause function c_i $(1 \le i \le n)$ of f, $c_i(\mathbf{y}) < 1$. Therefore, for each clause function c_i $(1 \le i \le n)$, there exists a literal function q_{ij} in c_i such that $q_{ij}(\mathbf{y}) < 1$. Define a *round-off operation* as follows:

$$y_j^* = \begin{cases} 1 & \text{if } y_j \ge 0 \\ -1 & \text{if } y_j < 0 \end{cases} \tag{25}$$

Let $\mathbf{y}^*$ be the vector obtained from a round-off operation on $\mathbf{y}$. Then clearly, the literal function $q_{ij}(\mathbf{y}^*) = 0$. This implies that the clause function $c_i(\mathbf{y}^*) = 0$ and thus $f(\mathbf{y}^*) = 0$. □

The *UniSAT5 model* on real space is a direct extension of the discrete *SAT4 model* on Boolean space. A similar model to *UniSAT5* was proposed independently in neural network area [181].

The UniSAT7 Model: Given a CNF formula $F(\mathbf{x})$ from $\{0,1\}^m$ to $\{0,1\}$ with n clauses $C_1, \ldots, C_n$, we define a real function $f(\mathbf{y})$ from E^m to E that transforms the SAT problem into an unconstrained global optimization problem [119, 122, 129, 126, 131]:

$$\min_{\mathbf{y} \in \mathbf{E}^m} f(\mathbf{y}), \tag{26}$$

where

$$f_1(\mathbf{y}) = \sum_{i=1}^{n} c_i(\mathbf{y}). \tag{27}$$

A clause function $c_i(\mathbf{y})$ is a product of m *literal functions* $q_{ij}(y_j)$ $(1 \le j \le m)$:

$$c_i = \prod_{j=1}^{m} q_{ij}(y_j), \tag{28}$$

where

$$q_{ij}(y_j) = \begin{cases} (y_j - 1)^{2p} & \text{if } x_j \text{ is in clause } C_i \\ (y_j + 1)^{2p} & \text{if } \bar{x}_j \text{ is in clause } C_i \\ 1 & \text{if neither } x_j \text{ nor } \bar{x}_j \text{ is in } C_i \end{cases} \tag{29}$$

where p is a positive integer.

The correspondence between $\mathbf{x}$ and $\mathbf{y}$ is defined as follows (for $1 \le i \le m$):

$$x_i = \begin{cases} 1 & \text{if } y_i = 1 \\ 0 & \text{if } y_i = -1 \\ undefined & \text{otherwise} \end{cases}$$

Clearly, $F(\mathbf{x})$ is true iff $f(\mathbf{y})$=0 on the corresponding $\mathbf{y} \in \{-1, 1\}^m$.

To ensure the theoretical convergence ratios, instead of f_1, the object function

$$f(\mathbf{y}) = f_1(\mathbf{y}) + f_2(\mathbf{y}) \tag{30}$$

will be considered [137], where

$$f_2(\mathbf{y}) = \sum_{j=1}^{m} (y_j - 1)^2 (y_j + 1)^2. \tag{31}$$

For all $\mathbf{y} \in \{-1, 1\}^m$, clearly $f_2(\mathbf{y}) = 0$. From this, for any $\mathbf{y} \in \{-1, 1\}^m$, $f(\mathbf{y}) = 0$ if and only if $f_1(\mathbf{y}) = 0$. Thus, $F(\mathbf{x})$ is true if and only if $f(\mathbf{y}) = 0$ on the corresponding point $\mathbf{y} \in \{-1, 1\}^m$.

For the object function f and the *UniSAT7* problem, we have the following theorem.

Theorem 4 *Let f be an object function of a CNF formula F. For any $\mathbf{y} \in E^m$ with $f(\mathbf{y}) < 1$, we can find a vector $\mathbf{y}^* \in \{-1, 1\}^m$ such that $f(\mathbf{y}^*) = 0$, i.e., we can find a solution of the formula F.*

Proof: Let $\mathbf{y}$ be a vector in E^m such that $f(\mathbf{y}) < 1$. Then from $f = f_1 + f_2$, we have $f_1(\mathbf{y}) < 1$ and $f_2(\mathbf{y}) < 1$. Then we have for each clause function c_i $(1 \leq i \leq n)$ of f_1 defined in (28), $c_i(\mathbf{y}) < 1$. Therefore, for each clause function c_i $(1 \leq i \leq n)$, there exists a literal function q_{ij} in c_i defined in (29) such that $q_{ij}(\mathbf{y}) < 1$. Define a *round-off operation* as follows:

$$y_j^* = \begin{cases} 1 & \text{if } y_j \geq 0 \\ -1 & \text{if } y_j < 0 \end{cases} \tag{32}$$

Let $\mathbf{y}^*$ be the vector obtained from the round-off operation on $\mathbf{y}$. Then clearly, the literal function $q_{ij}(\mathbf{y}^*) = 0$. This implies that the clause function $c_i(\mathbf{y}^*) = 0$ and thus $f_1(\mathbf{y}^*) = 0$.

For each clause function $(y_j - 1)^2(y_j + 1)^2$ $(1 \leq j \leq m)$ in f_2, $f_2 < 1$ implies that either $(y_j - 1)^2 < 1$ or $(y_j + 1)^2 < 1$. Therefore, for each clause function $(y_j - 1)^2(y_j + 1)^2$ $(1 \leq j \leq m)$ in f_2, by the round-off operation, we have $(y_j^* - 1)^2(y_j^* + 1)^2 = 0$. Thus, we have $f_2(\mathbf{y}^*) = 0$. Combining this and $f_1(\mathbf{y}^*) = 0$, $\mathbf{y}^*$ is a solution of f. □

From the above theorem, the optimization process for solving the *UniSAT7* problem can be stopped when a vector $\mathbf{y} \in E^m$ with $f(\mathbf{y}) < 1$ is found.

The UniSAT8 Model: Given a CNF formula $F(\mathbf{x})$ from $\{0, 1\}^m$ to $\{0, 1\}$ with n clauses $C_1, \ldots, C_n$, we define a real function $f(\mathbf{y})$ from E^m to E that transforms the SAT problem into an unconstrained global optimization problem [119, 122, 129, 126, 131]:

$$\min_{\mathbf{y} \in \mathbf{E}^m} f(\mathbf{y}), \tag{33}$$

where

$$f(\mathbf{y}) = \sum_{i=1}^{n} c_i(\mathbf{y}). \tag{34}$$

A clause function $c_i(\mathbf{y})$ is a product of m *literal functions* $q_{ij}(y_j)$ $(1 \leq j \leq m)$:

$$c_i = \prod_{j=1}^{m} q_{ij}(y_j), \tag{35}$$

where

$$q_{ij}(y_j) = \begin{cases} (y_j - 1)^{2p+1} & \text{if } x_j \text{ is in clause } C_i \\ (y_j + 1)^{2p+1} & \text{if } \bar{x}_j \text{ is in clause } C_i \\ 1 & \text{if neither } x_j \text{ nor } \bar{x}_j \text{ is in } C_i \end{cases} \tag{36}$$

where p is a positive integer.

Clearly, the *UniSAT8* problem model must be used with a quadralization procedure.

The translation of SAT problem into a nonlinear program is quite different from the integer program approach. In integer programming approach, one views the SAT problem as a 0/1 integer programming problem and tries solving it as a linear programming problem [19, 155, 156, 178, 186, 324]. If the solution is non-integer, one rounds off the values to the nearest integers and checks whether this corresponds to a solution of the original problem. If the rounded off values do not correspond to a solution, one computes another solution of the linear programming problem. In the *UniSAT* formulations, any solution to the nonlinear programming problem can be shown to correspond to some solution of the SAT problem (assuming a satisfiable assignment exists).

7.2. Global Optimization Algorithms for SAT Problem

Although nonlinear problems are intrinsically more difficult to solve, an unconstrained optimization problem is conceptually simple and easy to handle. Many powerful solution techniques have been developed to solve unconstrained optimization problems, which are based primarily upon calculus, rather than upon algebra and pivoting, as in the simplex method.

Many families of unconstrained global optimization algorithms for the *UniSAT* problem have been developed during the past eight years [119, 122, 127, 129, 131]. In this section, we first describe a general global optimization algorithm for the SAT problem. We then give several typical global optimization algorithms for the SAT problem.

7.2.1. *An Unconstrained Global Optimization Algorithm for SAT Problem*

A general optimization algorithm for the SAT problem, the *SAT*6.0 algorithm, is given in Figure 14 [119, 126, 127, 122, 131]. The algorithm performs the following several procedures.

```
procedure SAT6.0 ()
begin
    /* initialization */
    obtain_a_SAT_instance();
    y0 := select_an_initial_solution();
    f(y0) := evaluate_object_function(y0);

    /* search */
    k := 0;
    while not(solution_testing()) do
        for some y_i(k)s ∈ y_k
        begin
            /* minimizer */
            if test_min(f(y_i(k)s)) then
            begin
                y_k+1 := perform_min(f(y_i(k)s));
                f(y_k+1) := evaluate_object_function();
            end
            if close_to_solution() then x := approximate(y_k+1);
        end;
        /* local handler */
        if local then local_handler();
        k := k + 1;
    end;
end;
```

Fig. 14. **SAT6.0:** A general global optimization algorithm for the satisfiability problem.

Initialization. To start, procedure *obtain_a_SAT_instance*() initializes a SAT problem instance or it reads in a practical SAT problem instance. An object function, f, or a set of object functions, $\mathbf{f}$, is formulated according to a given *UniSAT* model. The SAT problem thus becomes a minimization problem to the object function(s). To begin, procedure *select_an_initial_solution*() selects an initial starting point $\mathbf{y}_0 \in E^m$. The corresponding value of the object function, $f(\mathbf{y}_0)$, is evaluated by function *evaluate_object_function*().

Global Optimization. The optimization process is an iterative minimization process to the object function. Function *test_min*() tests if the value of the object function can be minimized. If this is true, the minimization operation is performed by the procedure *perform_min*(), followed by *evaluate_object_function*() that updates the value of the object function. Often procedures *test_min*(), *perform_min*(), and *evaluate_object_function*() are performed in a whole without clear distinction.

Depending on the global optimization strategy, during a single search phase, i.e., in one iteration, the object function(s) can be minimized in one dimension or in more than one (up to m) dimensions. Methods able to optimize f in one dimension of the problem include line search methods, the coordinate descent methods, and coordinate Newton's methods. Methods that optimize f in more than one dimensions include the steepest descent methods, multi-dimensional Newton's methods, and many others.

If using a line search method, in the kth iteration, only one variable, y_i, is chosen as the descent dimension. Since variables other than y_i, i.e., y_j $(j = 1, ..., i-1, i+1, ..., m)$, remain as constants, $f(\mathbf{y})$ in (27) is a parabolic function having a minimum point. The minimization procedure is performed by *perform_min*() that determines the descent direction of object function and produces a new solution point, $\mathbf{y}_{k+1}$. This can be done by a variety of global optimization methods, such as (in a basic line search method) the evaluation of

$$\frac{\partial f(\mathbf{y}_k)}{\partial y_{i(k)}} \tag{37}$$

if the object function f is differentiable, or the evaluation of

$$\frac{f(y_{1k}, y_{2k}, ..., y_{ik} + \Delta h, ..., y_{mk}) - f(y_{1k}, y_{2k}, ..., y_{ik}, ..., y_{mk})}{\Delta h} \tag{38}$$

if the object function f is continuous but not differentiable.

A typical multi-dimension global optimization method is the steepest descent method. In this case, the descent direction can be found by selecting α_k that minimizes

$$\mathbf{f}(\mathbf{y}_k - \alpha_k \nabla \mathbf{f}(\mathbf{y}_k)^t). \tag{39}$$

The new solution point is then $\mathbf{y}_{k+1} = \mathbf{y}_k - \alpha_k \nabla \mathbf{f}(\mathbf{y}_k)^t$. Often we may directly use gradients to evaluate α_k and then decide the new solution point:

$$\alpha_k = \beta \frac{\mathbf{f}(\mathbf{y}_k)}{||\nabla \mathbf{f}(\mathbf{y}_k)||^2} \tag{40}$$

where β is a convergence control factor.

Termination. As the iterative improvement progresses, a global minimum point may be approached gradually. The *closeness* between the present solution point and the global minimum solution point can be tested by solution point testing or objective value testing. Procedure *close_to_solution*() performs the closeness testing.

If the present solution point is sufficiently close to a global minimum point, procedure *approximate*() performs the *round-off* operation (as defined in (25) and (32)). This converts a solution point $\mathbf{y}$ in real space E^m to a solution point $\mathbf{x}$ in Boolean space $\{0, 1\}^m$ which can be a solution of the original SAT problem. Procedure *solution_testing*() takes the solution generated from procedure *approximate*() and substitutes it into the given *CNF* formula to verify its correctness.

In practice, the search process could be stuck at a locally optimum point. To improve the convergence performance of the algorithm, one or more local handlers may be added. In the basic SAT algorithm, a simple local handler that negates the truth values of up to m variables was used. It has effectively improved the convergence performance of this algorithm.

Running Time. The running time of the $SAT6.0$ algorithm can be estimated as follows. For a SAT problem with n clauses and on average l literals in each clause, the initialization portion takes $O(ln)$ time. The running time of the *while* loop equals the number of *while* loops executed, k, multiplied by the running time of the internal *for* loop. In the worst case, *evaluate_object_function*() and *local_handler*() each will take $O(ln)$ time. Depending on the global optimization scheme used, procedure *close_to_solution*() may take $O(m)$ time or at most $O(ln)$ time. Procedure *approximate*() takes $O(m)$ time. The run time of *perform_min*() is determined by the chosen global optimization method. For a line search approach, procedure *perform_min*() can be done in $O(ln)$ time. So the run time of $SAT6.0$ in this case is:

$$O(nl) + O(knml) = O(knml), \tag{41}$$

where k is the number of *while* loops executed which is determined by the nature of the problem instances.

Space. There is not any large data storage in the $SAT6.0$ algorithm. The major space required is to store n clauses. Most of the SAT algorithms were implemented in a space-efficient way. Literals in each clause are coded using a 2-dimensional linked list array, so the total space taken for the algorithm is $O(nl)$.

Any existing unconstrained global optimization methods can be used to solve the *UniSAT* problems (see textbooks and literature). Depending on global optimization strategies used, we have developed many optimization algorithms for the SAT problem. These include the basic algorithms ($SAT6$ family), steepest descent methods ($SAT7$ family), modified steepest descent methods ($SAT8$ family), newton methods ($SAT10$ family), quasi-newton methods ($SAT11$ family), descent methods ($SAT14$ family), cutting-plane methods ($SAT15$ family), conjugate direction methods ($SAT16$ family), ellipsoid methods ($SAT17$ family), homotopy methods ($SAT18$ family), and linear programming methods ($SAT21$ family). In each algorithm family, different approaches and heuristics can be used to design objective functions, select starting initial points, scramble search space, formulate higher order local handlers, deflect descent directions, utilize parallelism, and implement hardware architectures to speed up computations.

In the following, we give three specific global optimization algorithms on the real space.

7.2.2. *The $SAT14.5$ Algorithm*

Based on a coordinate descent method [216], in Figure 15, we give a simple algorithm, the $SAT14.5$ algorithm, for the *UniSAT5* problem. The kernel of the $SAT14.5$

```
Procedure SAT14.5 ()
begin
    /* initialization*/
    y := initial_vector();
    local := search := 0;
    limit := bm log m;

    /* search */
    while (f(y)≥ 1 and local ≤ limit) do
    begin
        old_f := f(y); search := search + 1;

        /* minimizer */
        for i := 1 to m do
            minimize f(y) with respect to y_i;

        /* local handler */
        if (f(y)=old_f or (search > b' log m and f(y)≥ 1)) then
        begin
            y := initial_vector();
            search := 0; local := local + 1;
        end;
    end;
    if f(y) < 1 then y* := round_off(y) else y* := enumerate();
end;
```

Fig. 15. **SAT14.5: A global optimization algorithm for the *UniSAT5* problem.**

algorithm is a *minimizer* which minimizes the object function f by the coordinate descent method.

Given a function f on E^m, the $SAT14.5$ algorithm initially chooses a vector $\mathbf{y}$ from E^m and then minimize the function f with respect to variables y_j $(1 \leq j \leq m)$ in *minimizer* until $f < 1$. Since each variable y_j appears in one clause function c_i at most once, the function $f(\mathbf{y})$ can be expressed as

$$f(\mathbf{y}) = a_j |y_j - 1| + b_j |y_j + 1| + d_j$$

for $(1 \leq j \leq m)$, where a_j, b_j, and d_j are *local gain factors* which are independent to y_j. They can be computed in $O(ln)$ time. Therefore, $f(\mathbf{y})$ takes its minimum value with respect to y_j at point either $y_j = 1$ or $y_j = -1$. Thus, the *minimizer* optimizes the function f as follows: if $a_j \geq b_j$ then set y_j equal to 1; otherwise set y_j equal to -1.

In practice, before $f < 1$, the algorithm could be stuck at a local minimum point.

To overcome this problem, a simple local handler is added. The local handler simply generates a new initial vector $\mathbf{y}$ to start an independent search. In the $SAT14.5$ algorithm, if the object function f can no longer be reduced *or* after $b' \log m$ (b' is a constant) iterations of the *while* loop f is still at least one, then the *local-handler* is called.

The number of the executions of the *local handler* is recorded in the $SAT14.5$ algorithm. If the algorithm can not find a solution for a CNF formula after calling too many times the *local handler*, then the algorithm calls procedure *enumerate*() which enumerates all possible assignments to see whether there is one that satisfies the formula. Clearly, procedure *enumerate*() always gives an answer in $O(2^m)$ time. The *limit* of calling the *local-handler* in the $SAT14.5$ algorithm is set to $bm \log m$, where b is a constant.

The run time of the $SAT14.5$ algorithm can be estimated as follows: Procedure *initial_vector*() takes $O(m)$ time and compute $f(\mathbf{y})$ takes $O(ln)$ time. In one iteration of the *while* loop, clearly, minimize $f(\mathbf{y})$ with respect to one variable can be computed in $O(ln)$ time, and thus, the *minimizer* takes $O(m(ln))$ time. Clearly one execution of the local handler takes $O(m)$ time. Thus, without calling procedure *enumerate*(), the run time of the $SAT14.5$ algorithm is

$$O(ln) + O(km(ln)) = O(klmn),$$

where k is the iteration number of the *while* loop.

Several versions of the $SAT14.5$ algorithms were implemented. The $SAT14.6$ algorithm is a $SAT14.5$ algorithm with elaborated control structures to local gain factors. $SAT14.6$ complements $SAT14.5$ algorithm well in many practical application problems.

7.2.3. *The SAT14.7 Algorithm*

Based on a coordinate descent method [216], in Figure 16, we give a $SAT14.7$ algorithm for the *UniSAT7* problem on E^m. The kernel of the $SAT14.7$ algorithm is a *minimizer* which minimizes the object function by the coordinate descent method.

In practice, to optimize the function f_1 is more efficient than to optimize the function f for the *UniSAT7* problem. Given the object function f_1 on E^m, the $SAT14.7$ algorithm initially chooses $\mathbf{y}$ from E^m and then the function f_1 is minimized with respect to each variables y_j $(1 \le j \le m)$ in the *minimizer* until $f_1 < 1$. Since for each clause function c_i $(1 \le i \le n)$ in f_1, each variable y_j $(1 \le j \le m)$ appears in c_i at most once, f_1 is a quadratic function with respect to y_j:

$$f_1(\mathbf{y}) = a_j(y_j - 1)^2 + b_j(y_j + 1)^2 + c_j$$

for $(1 \le j \le m)$, where a_j, b_j, and c_j are constants which can be computed in $O(ln + m)$ time. Thus, minimizing f_1 with respect to one variable can be done in $O(nl)$ time. When $f_1 < 1$ then a *round_off* operation defined in (32) is performed to find the solution.

```
Procedure SAT14.7 ()
begin
    /* initialization */
    y := initial_vector();
    local := 0; limit := poly(m);

    /* search */
    while (f1(y) ≥ 1 and local ≤ limit) do
    begin
        old_f := f1(y);

        /* minimizer */
        for i := 1 to m do
            minimize f1(y) with respect to yi;

        /* local handler */
        if f1(y) ≥ old_f then
        begin
            y := initial_vector();
            local := local + 1;
        end;
    end;
    if f1(y) < 1 then y* := round_off(y) else y* := enumerate();
end;
```

Fig. 16. **SAT14.7: A global optimization algorithm for the *UniSAT7* problem.**

In practice, before $f_1 < 1$, the algorithm could be stuck at a local minimum point. To overcome this problem, a *local handler* is added in the SAT14.7 algorithm. In the local handler, a new initial vector $\mathbf{y}$ is generated.

The run time of the SAT14.7 algorithm can be estimated as follows. The initial portion and the computation of $f_1(\mathbf{y})$ take $O(ln)$ time. In one iteration of the *while* loop, minimizing $f_1(\mathbf{y})$ with respect to one variable can be computed in $O(ln)$ time, and thus, the *minimizer* takes $O(lmn)$ time. Clearly, one execution of the local handler takes $O(m)$ time. Summarizing the above, the run time of the SAT14.7 algorithm is $O(klmn)$, where k is the iteration times of the *while* loop. The experimental results show that the iteration times of the *while* loop for optimizing f_1 is less than that for f.

Using the results of the iteration times in Section 7.3.2, the value of k is expected to be $O(\log(m+n))$ and the average time complexity of the SAT14.7 algorithm is expected to be $O(lmn\log(m+n))$ if the SAT14.7 algorithm is not stuck at a local minimum point.

```
Procedure SAT14.11 ()
begin
    /* initialization */
    get_a_SAT_instance();
    x0 := select_an_initial_point();
    f := evaluate_object_function(x0);

    /* backtracking with global optimization */
    x* := backtracking(x0);
end;


procedure backtracking(xi)
begin
    /* global optimization assigns v to xi */
    v := global_optimization();
    xi := v;
    Vi := Vi - {v};
    /* append variable xi to the partial path */
    path[xi] := i;

    if path broken then backtracking;
    if solution found then return x*;
    else backtracking(next xi);
end;
```

Fig. 17. **SAT14.11: a complete global optimization algorithm with backtracking.**

One version of the *SAT*14.7 algorithms was implemented for a weighted *UniSAT7* [138]. The results were reported elsewhere.

7.2.4. *Complete Global Optimization Algorithms*

The *SAT*14.5, *SAT*14.6, and *SAT*14.7 algorithms are *incomplete* algorithms. In order to achieve computing efficiency and search completeness, in *SAT*14.11 to *SAT*14.20 algorithms, we combine global optimization algorithms with backtracking/resolution procedures and make them complete. Therefore, they are able to verify satisfiability as well as unsatisfiability. Figures 17 and 18 give two backtracking global optimization algorithms.

For SAT problem *per se*, backtracking is able to verify unsatisfiability quickly for certain classes of problems but is slow when it comes to verifying satisfiability, as all possible resolutions need to be tried out before concluding that the inference relation holds or that the problem is satisfiable. From our experience, a *combined* global

```
Procedure SAT14.13 ()
begin
    /* initialization */
    get_a_SAT_instance();
    x0 := select_an_initial_point();
    f := evaluate_object_function(x0);

    /* backtracking with global optimization */
    x* := backtracking(x0);
end;


procedure backtracking(xi)
begin
    xi := v;
    Vi := Vi - {v};
    /* global optimization with xi being assigned v */
    global_optimization(v);
    /* append variable xi to the partial path */
    path[xi] := i;

    if path broken then backtracking;
    if solution found then return x*;
    else backtracking(next xi);
end;
```

Fig. 18. **SAT14.13: a complete global optimization algorithm with backtracking.**

optimization algorithm with backtracking/resolution procedures would perform well for certain classes of satisfiable and unsatisfiable problem instances.

Most global optimization SAT algorithms have been built with backtracking/ resolution procedures. The combined algorithms achieve certain efficiency improvements while retaining search completeness.

7.3. Convergence Property

Gu, Gu and Du [137] have analyzed the convergence ratio and efficiency of three basic methods: the steepest descent method, Newton's method, and the coordinate descent method for object function f defined in the *UniSAT7* problem model. In this section, we briefly summarize major results.

7.3.1. *Convergence Analysis*

The main result of this section is that for any Boolean *CNF* formula F, if $\mathbf{y}^*$ is a solution point of f defined in the *UniSAT7* problem model, then the *Hessian* matrix $\mathbf{H}(\mathbf{y}^*)$ of f at $\mathbf{y}^*$ is positive definite. From this result, the convergence ratios of the three methods can be derived.

Definition 5 *[306]. An $m \times m$ real symmetric matrix $\mathbf{H}$ is positive definite if and only if for all nonzero vector $\mathbf{d}$ in E^m, $\mathbf{d}\cdot\mathbf{H}\cdot\mathbf{d}^T > 0$. Or equivalently, $\mathbf{H}$ is positive definite if and only if all the eigenvalues of $\mathbf{H}$ are larger than zero.*

Theorem 6 *Let $\mathbf{y}^* \in \{-1,1\}^m$ be a solution point of f. Then the Hessian matrix $\mathbf{H}(\mathbf{y}^*)$ of f is positive definite.*

Proof: See [137].

Now we give the convergence ratios of the steepest descent method, Newton's method, and the coordinate descent method for the *UniSAT7* problem.

Proposition 7 *[216] Suppose f has second partial derivatives which are continuous on E^m. Suppose further that at the local minimum point $\mathbf{y}^*$ the Hessian matrix of f, $\mathbf{H}(\mathbf{y}^*)$, is positive definite. If $\{\mathbf{y}_k\}$ is a sequence generated by the steepest descent method that converges to $\mathbf{y}^*$, then the sequence of objective values $\{f(\mathbf{y}_k)\}$ converges to $f(\mathbf{y}^*)$ linearly with a convergence ratio no greater than $[(A-a)/(A+a)]^2$, where $A \geq a > 0$ are the largest and smallest eigenvalues of the Hessian matrix $\mathbf{H}(\mathbf{y}^*)$, respectively.*

Proposition 8 *[216] Suppose f has third partial derivatives which are continuous on E^m. Suppose further that at the local minimum point $\mathbf{y}^*$ the Hessian matrix of f, $\mathbf{H}(\mathbf{y}^*)$, is positive definite. Then if started sufficiently close to $\mathbf{y}^*$, the points generated by Newton's method converges to $\mathbf{y}^*$. The order of convergence is two.*

Lemma 9 *Suppose f has second partial derivatives which are continuous on E^m. Suppose further that at the local minimum point $\mathbf{y}^*$ the Hessian matrix of f, $\mathbf{H}(\mathbf{y}^*)$, is positive definite. If $\{\mathbf{y}_k\}$ is a sequence generated by the coordinate descent method where at each stage the coordinate corresponding to the largest (in absolute value) component of the gradient vector is selected (the Gauss-Southwell Method [216]) that converges to $\mathbf{y}^*$, then the sequence of objective values $\{f(\mathbf{y}_k)\}$ converges to $f(\mathbf{y}^*)$ linearly with a convergence ratio no greater than $1 - \frac{a}{A(m-1)}$, where $A \geq a > 0$ are the largest and smallest eigenvalues of the Hessian matrix $\mathbf{H}(\mathbf{y}^*)$, respectively.*

Proof: See [137].

Theorem 10 *Let f be the function defined in the UniSAT7 problem model. If $\{\mathbf{y}_k\}$ is a sequence of vectors generated by the steepest descent method that converges to* *a solution* $\mathbf{y}^*$ *of* f*, then the sequence of the objective values* $\{f(\mathbf{y}_k)\}$ *converges to* $f(\mathbf{y}^*)$ *linearly with a convergence ratio* $[(A-a)/(A+a)]^2 < 1$*, where* $A \geq a > 0$ *are the largest and smallest eigenvalues of the Hessian matrix* $\mathbf{H}(\mathbf{y}^*)$ *of* f*, respectively.*

Proof: Clearly, f has second partial derivatives which are continuous on E^m. Therefore, the theorem follows from Theorem 6 and Proposition 7. □

Theorem 11 *Let* f *be the function defined in the UniSAT7 problem model. If started sufficiently close to a solution point* $\mathbf{y}^*$*, the sequence* $\{\mathbf{y}_k\}$ *generated by Newton's method converge to* $\mathbf{y}^*$*. The order of convergence is two.*

Proof: The theorem follows from Theorem 6 and Proposition 8. □

Theorem 12 *Let* f *be the function defined in the UniSAT7 problem model. If* $\{\mathbf{y}_k\}$ *is a sequence generated by the coordinate descent method where at each stage the coordinate corresponding to the largest (in absolute value) component of the gradient vector is selected (the Gauss-Southwell Method [216]) that converges to a solution* $\mathbf{y}^*$ *of* f*, then the sequence of the objective values* $\{f(\mathbf{y}_k)\}$ *converges to* $f(\mathbf{y}^*)$ *linearly with a convergence ratio* $\left(1 - \frac{a}{A(m-1)}\right) < 1$*, where* $A \geq a > 0$ *are the largest and smallest eigenvalues of the Hessian matrix* $\mathbf{H}(\mathbf{y})$*, respectively.*

Proof: The theorem follows from Theorem 6 and Lemma 9. □

7.3.2. *Efficiency Estimation*

From the convergence properties given above, we can roughly estimate the efficiency of the steepest descent method and the coordinate descent method for solving the *UniSAT7* problem.

Let $\{\mathbf{y}_k\}$ be a sequence of vectors generated by the steepest descent method that convergences to a solution point $\mathbf{y}^*$ and let the initial value of the vector $\mathbf{y}_0$ to be $(0, \ldots, 0)$. Then $f(\mathbf{y}_0) = n + m$. From Theorem 10, we have that the convergence ratio of the steepest descent method for f is

$$\left(\frac{A-a}{A+a}\right)^2,$$

where $A \geq a > 0$ are the largest and smallest *eigenvalues* of the *Hessian* matrix $\mathbf{H}(\mathbf{y}^*)$ of f, respectively. From this, we have

$$f(\mathbf{y}_{k+1}) \leq \left(\frac{A-a}{A+a}\right)^2 f(\mathbf{y}_k), \tag{42}$$

for sufficiently large k. From $A \geq a > 0$, clearly there exists a constant $\beta < 1$ such that

$$\left(\frac{A-a}{A+a}\right)^2 \leq \beta.$$

Therefore, if (42) holds for every $k \geq 1$, then for $k > -\log(m+n)/\log\beta$, we have

$$\begin{aligned} f(\mathbf{y}_k) &\leq \beta^k(n+m) \\ &< \frac{1}{n+m}(n+m) \\ &= 1. \end{aligned}$$

Thus, from Theorem 4, the *UniSAT7* problem can be solved in $O(\log(n+m))$ iterations by the steepest descent method on the assumption that (42) holds for every $k \geq 1$.

Let $\{\mathbf{y}_k\}$ be a sequence of vectors generated by the coordinate descent method where at each stage the coordinate corresponding to the largest (in absolute value) component of the gradient vector is selected (the Gauss-Southwell Method [216]) that converges to a solution point $\mathbf{y}^*$. Then by Theorem 12, we have

$$f(\mathbf{y}_{k+1}) \leq \left(1 - \frac{a}{A(m-1)}\right) f(\mathbf{y}_k), \tag{43}$$

for sufficiently large k. Since $A \geq a > 0$, clearly for some constant $\beta < 1$,

$$\left(1 - \frac{a}{A(m-1)}\right)^m \leq \beta.$$

Therefore, if (43) holds for all $k \geq 1$, then initially choosing $\mathbf{y}_0 = (0, \ldots, 0)$ and for $k > -m\log(m+n)/\log\beta$, we have

$$\begin{aligned} f(\mathbf{y}_k) &\leq \left(1 - \frac{a}{A(m-1)}\right)^k f(\mathbf{y}_0) \\ &< \left(1 - \frac{a}{A(m-1)}\right)^{m(-\log(m+n)/\log\beta)} (m+n) \\ &\leq \beta^{-\log(m+n)/\log\beta}(m+n) \\ &= \frac{1}{m+n}(m+n) = 1. \end{aligned}$$

From this and Theorem 4, we have that the *UniSAT7* problem can be solved in $O(m\log(n+m))$ iterations by the coordinate descent method on the assumption that (43) holds for all $k \geq 1$.

7.4. Average Time Complexity

We have made some preliminary analysis of the average time complexity of some global optimization SAT algorithms [136]. It shows that, the $SAT14.5$ algorithm, with probability at least $1-e^{-m}$, finds a solution within $k = O(m(\log m)^2)$ iterations of the *while* loop for a randomly generated satisfiable CNF formula with $l \geq 3$ and $n/m \leq \alpha 2^l/l$, where $\alpha < l$ is a constant. From this and the fact that the run time of procedure *enumerate*() is $O(2^m)$, the average time complexity of the $SAT14.5$ algorithm is

$$(1 - e^{-m})O(m(\log m)^2(lmn)) + e^{-m}O(2^m) = O(ln(m\log m)^2).$$

Clearly, the $SAT14.5$ algorithm can give an answer to an unsatisfiable CNF formula in $O(2^m)$ time.

7.5. Experimental Results

In this section, we give experimental results of a simplified $SAT14.5$ algorithm, a $SAT14.15$ algorithm which is a backtracking version of the $SAT14.5$ algorithm, a simplified $SAT14.6$ algorithm, and a $SAT14.16$ algorithm which is a backtracking version of the $SAT14.6$ algorithm [119, 122, 124, 126, 131]. We will test these algorithms with various SAT problem instance models including Pehoushek's problem instances [250] and will compare their performance with those of the Davis-Putnam algorithm [63, 62], the $GSAT$ algorithm [287], and an interior point zero-one integer programming algorithm [186].

Due to limited memory space on a workstation computer, in order to experiment with the large-scale SAT problem instances, these global optimization algorithms were implemented in C in a space-efficient manner.

To facilitate performance comparisons among other optimization algorithms developed for the SAT problem, we use the same problem instances and table documents as those used in Section 6.5 and [119, 121, 122, 126, 130, 131].

7.5.1. *Performance of the $SAT14.5$ Algorithm*

Real Execution Time of the $SAT14.5$ Algorithm. The initial problem instances were generated based on the exact l-SAT model. The real performance of the $SAT14.5$ algorithm running on a SUN workstation is illustrated in Table XI. The number of clauses (n), the number of variables (m), and the number of literals per clause (l), are give in the first three columns. Column 4 (*Global*) indicates the number of times that the algorithm succeeds in finding a solution, i.e., a global minimum point. Column 5 (*Local*) indicates the number of times that the algorithm was stuck at local minimum points.

From these results we can observe that, in terms of global convergence and local convergent rate, the $SAT14.5$ algorithm exhibits good convergent properties and fast computing speed. Because $SAT14.5$ algorithm is based on $SAT4$ and $UniSAT5$ models (Sections 6.4 and 7.1.2), it is much faster than most sequential local search algorithms and has comparable performance to parallel local search algorithms, such as the $SAT1.6$ algorithms.

As discussed in Section 7.4 and [136], beyond a certain range of hardness, for example, for $n = 8500, m = 1000$, and $l = 4$, the computing time of the $SAT14.5$ algorithm starts to increase.

The Probabilistic Behavior of the $SAT14.5$ Algorithm. The $SAT14.5$ algorithm uses a probabilistic local handler. Table XI also shows its probabilistic behavior, the number of iterations required, and its success in finding a solution. The table indicates that the algorithm succeeds for most SAT problem instances. For many practical problem instances we have encountered, $SAT14.5$'s performance seems

TABLE XI
Real execution performance of a $SAT14.5$ algorithm on a SUN SPARC 2 workstation (Time Units: seconds)

Problems (n,m,l)			Ten Trials		Execution Time			Iterations (k)		
Clauses	Variables	Literals	Global	Local	Min	Mean	Max	Min	Mean	Max
100	100	3	10	0	0.000	0.001	0.010	1	1.900	7
200	100	3	10	0	0.000	0.005	0.010	2	2.500	4
300	100	3	10	0	0.000	0.012	0.020	2	5.900	12
400	100	3	10	0	0.010	0.035	0.080	6	28.40	76
1000	1000	3	10	0	0.030	0.037	0.070	2	2.600	4
1500	1000	3	10	0	0.050	0.068	0.100	3	3.600	5
2000	1000	3	10	0	0.080	0.100	0.130	4	4.800	6
2500	1000	3	10	0	0.110	0.149	0.230	5	6.500	10
3000	1000	3	10	0	0.140	0.213	0.290	6	9.700	14
3500	1000	3	10	0	0.230	0.756	1.640	12	41.50	91
1000	1000	4	10	0	0.010	0.029	0.060	1	1.700	3
2000	1000	4	10	0	0.060	0.072	0.090	2	2.100	3
3000	1000	4	10	0	0.090	0.117	0.170	2	3.000	4
4000	1000	4	10	0	0.130	0.168	0.220	3	3.900	5
5000	1000	4	10	0	0.190	0.261	0.310	4	5.800	7
6000	1000	4	10	0	0.270	0.447	0.570	6	10.60	14
7000	1000	4	10	0	0.670	1.034	1.330	19	29.10	37
8000	1000	4	10	0	1.000	2.192	6.860	36	84.30	274
8500	1000	4	10	0	2.210	12.67	26.56	86	557.0	1232
10000	1000	5	10	0	0.410	0.558	0.990	4	6.300	12
11000	1000	5	10	0	0.400	0.552	0.700	5	6.700	9
12000	1000	5	10	0	0.500	0.680	0.800	6	9.100	11
13000	1000	5	10	0	0.630	0.921	1.480	8	12.90	22
14000	1000	5	10	0	0.870	1.339	2.080	12	21.40	38
15000	1000	5	10	0	0.950	1.957	3.410	15	37.50	72
10000	400	6	10	0	0.360	0.622	1.170	7	13.30	26
10000	500	6	10	0	0.240	0.473	0.660	3	7.200	11
10000	600	6	10	0	0.290	0.410	0.620	3	4.500	7
10000	700	6	10	0	0.260	0.361	0.530	3	3.600	5
10000	800	6	10	0	0.300	0.362	0.520	3	3.200	4
10000	900	6	10	0	0.270	0.372	0.570	2	2.600	4
10000	1000	6	10	0	0.270	0.343	0.440	2	2.300	3
20000	1000	7	10	0	0.430	0.653	1.080	2	2.400	4
30000	2000	7	10	0	0.870	1.130	1.420	2	2.300	3
40000	3000	7	10	0	1.600	1.798	2.390	2	2.200	3
50000	4000	7	10	0	1.920	2.131	2.430	2	2.000	2
60000	5000	7	10	0	2.340	2.721	3.340	2	2.300	3
10000	1000	10	10	0	0.040	0.076	0.130	1	1.000	1
20000	2000	10	10	0	0.130	0.185	0.260	1	1.000	1
30000	3000	10	10	0	0.280	0.370	0.500	1	1.000	1
40000	4000	10	10	0	0.310	0.459	0.680	1	1.000	1
50000	5000	10	10	0	0.480	0.652	0.880	1	1.000	1

fairly robust.

A notable feature is that, since $SAT14.5$ has the similar algorithm structure to $SAT1.7$, the iteration times of both algorithms are shown to be identical.

TABLE XII

Real performance comparison between the Davis-Putnam algorithm and a $SAT14.5$ algorithm on a SUN SPARC 2 workstation (Average of ten algorithm executions. Time Units: seconds. Symbol "S/F" stands for algorithm's *success/failure* of giving a solution within a time limit of $120 \times n/m$ seconds.)

Problems Clause,Variable,Literal			Execution Time Davis-Putnam Algorithm				Execution Time $SAT14.5$ Algorithm			
n	m	l	S/F	Min	Mean	Max	S/F	Min	Mean	Max
500	500	3	10/0	2.066	2.159	2.249	10/0	0.010	0.015	0.020
750	500	3	10/0	2.699	2.916	3.183	10/0	0.020	0.021	0.030
1000	500	3	10/0	3.149	3.657	6.316	10/0	0.020	0.042	0.060
1250	500	3	9/1	3.966	5.797	10.02	10/0	0.040	0.059	0.090
1500	500	3	6/4	7.499	9.147	11.60	10/0	0.060	0.122	0.250
1000	500	4	10/0	4.533	4.684	4.933	10/0	0.030	0.031	0.040
1500	500	4	10/0	6.766	7.960	16.90	10/0	0.040	0.051	0.090
2000	500	4	8/2	8.999	10.27	14.25	10/0	0.060	0.089	0.130
2500	500	4	2/8	13.63	15.96	18.28	10/0	0.090	0.103	0.120
3000	500	4	1/9	46.33	46.33	46.33	10/0	0.110	0.162	0.220
3000	500	5	10/0	16.23	16.90	18.82	10/0	0.090	0.117	0.160
4000	500	5	5/5	21.72	28.39	44.66	10/0	0.110	0.151	0.180
5000	500	5	0/10	> 1200	> 1200	> 1200	10/0	0.150	0.231	0.360
6000	500	5	0/10	> 1440	> 1440	> 1440	10/0	0.260	0.386	0.570
7000	500	5	0/10	> 1680	> 1680	> 1680	10/0	0.450	0.694	1.150
10000	1000	10	10/0	99.71	101.8	103.1	10/0	0.050	0.077	0.130
12000	1000	10	10/0	122.1	124.3	126.7	10/0	0.050	0.117	0.160
14000	1000	10	10/0	140.7	145.2	148.7	10/0	0.050	0.118	0.220
16000	1000	10	10/0	165.1	167.1	168.7	10/0	0.080	0.152	0.220
18000	1000	10	10/0	185.5	188.6	189.9	10/0	0.090	0.208	0.320

TABLE XIII

Performance comparison between a $SAT14.5$ algorithm running on a SUN SPARC 2 workstation and the $GSAT$ algorithm running on a MIPS computer with comparable computing power (Time Units: seconds)

Problems			Execution Time	
n	m	l	GSAT	SAT14.5
215	50	3	0.400	0.080
301	70	3	0.900	0.030
430	100	3	6.000	0.560
516	120	3	14.00	1.660
602	140	3	14.00	0.990
645	150	3	45.00	0.180
860	200	3	168.0	0.100
1062	250	3	246.0	0.870
1275	300	3	720.0	15.28

TABLE XIV
Execution performance of a $SAT14.5$ algorithm with Pehoushek's Benchmarks on a SUN SPARC 2 workstation (Time Units: seconds)

Problems				Ten Trials		Execution Time			Iterations (k)		
Names	n	m	l	Global	Local	Min	Mean	Max	Min	Mean	Max
Prob1	600	180	3	10	0	0.510	4.570	12.56	334	3221.9	8880
Prob2	600	180	3	10	0	0.200	0.629	1.240	116	418.90	879
Prob3	600	180	3	10	0	0.070	0.885	2.560	32	596.70	1745
Prob4	600	180	3	10	0	0.280	1.749	3.510	185	1181.9	2285
Prob5	600	180	3	10	0	0.240	0.887	4.010	154	572.90	2560
Prob6	600	180	3	10	0	0.080	0.878	1.990	40	567.20	1309
Prob7	600	180	3	10	0	0.230	1.260	4.700	135	866.10	3321
Prob8	600	180	3	10	0	0.030	1.021	2.320	16	675.90	1616
Prob9	600	180	3	10	0	0.120	1.416	3.350	76	881.30	2150
Prob10	600	180	3	10	0	0.080	0.674	3.270	51	437.60	2059

TABLE XV
Performance comparison between a $SAT14.5$ algorithm running on a SUN SPARC 2 workstation and an interior point zero-one integer programming algorithm running on a *KORBX(R)* parallel/vector computer (Time Units: seconds)

Problems Clause,Variable,Literal			Execution Time Interior Point Algorithm				Execution Time $SAT14.5$ Algorithm			
n	m	l	S/F	Min	Mean	Max	S/F	Min	Mean	Max
100	50	5.0000	52/0	0.5	0.7	7.7	10/0	0.000	0.002	0.010
200	100	4.9999	70/0	1.0	1.1	1.7	10/0	0.000	0.005	0.010
400	200	7.0010	69/0	2.4	3.5	6.3	10/0	0.000	0.007	0.010
800	400	9.9980	31/0	5.1	5.6	7.7	10/0	0.000	0.010	0.020
800	400	7.0020	20/0	4.7	7.8	16.1	10/0	0.010	0.019	0.040
1000	500	10.000	49/0	6.5	7.4	9.8	10/0	0.010	0.013	0.020
2000	1000	9.9960	10/0	14.9	18.5	20.5	10/0	0.020	0.032	0.050
2000	1000	7.0010	50/0	18.1	21.5	27.3	10/0	0.040	0.102	0.140
2000	1000	3.0030	49/1	8.7	420.7	9595.5	10/0	0.290	1.614	9.890
4000	1000	4.0010	1/1	8314.3	8314.3	8314.3	10/0	1.330	13.14	32.99
4000	1000	10.003	10/0	24.0	25.1	25.8	10/0	0.060	0.115	0.250
8000	1000	10.001	10/0	37.5	38.0	38.8	10/0	0.140	0.293	0.500
16000	1000	9.9980	10/0	46.2	66.4	92.1	10/0	0.720	0.905	1.210
32000	1000	10.004	10/0	80.1	232.4	311.3	10/0	1.500	2.506	4.040

Performance Comparison with the Davis-Putnam Algorithm. The execution results of the Davis-Putnam algorithm and the $SAT14.5$ algorithm for the exact l-SAT problem instances are given in Table XII. Ten algorithm executions were made for each algorithm. The minimum, maximum, and average execution times are reported. Because Davis-Putnam algorithm was slow for hard problem instances, a maximum execution time of $120 \times n/m$ seconds was set as the time limit of its execution. Symbol "S/F" in Column 4 stands for algorithm's *success/failure* of giving an answer within

TABLE XVI
Real execution performance of a *SAT*14.15 algorithm on a SUN SPARC 2 workstation (Time Units: seconds)

Problems (n,m,l)			Ten Trials		Execution Time			Iterations (k)		
Clauses	Variables	Literals	Global	Local	Min	Mean	Max	Min	Mean	Max
100	100	3	10	0	0.000	0.006	0.020	1	1.500	2
200	100	3	10	0	0.000	0.008	0.020	2	3.300	5
300	100	3	10	0	0.010	0.020	0.030	3	6.100	13
400	100	3	10	0	0.010	0.190	1.120	10	146.4	894
1000	1000	3	10	0	0.030	0.055	0.070	2	2.400	3
1500	1000	3	10	0	0.060	0.084	0.100	2	3.300	4
2000	1000	3	10	0	0.080	0.120	0.160	3	4.800	7
2500	1000	3	10	0	0.150	0.184	0.230	5	6.500	9
3000	1000	3	10	0	0.230	0.350	0.550	8	14.30	24
3500	1000	3	10	0	0.530	1.111	3.190	26	56.20	168
1000	1000	4	10	0	0.030	0.051	0.060	1	1.600	2
2000	1000	4	10	0	0.100	0.120	0.180	2	2.700	4
3000	1000	4	10	0	0.140	0.179	0.230	2	3.200	4
4000	1000	4	10	0	0.210	0.256	0.300	3	4.000	5
5000	1000	4	10	0	0.320	0.390	0.530	5	6.000	8
6000	1000	4	10	0	0.410	0.543	0.770	6	9.300	16
7000	1000	4	10	0	0.460	0.826	1.160	8	18.50	32
8000	1000	4	10	0	1.100	2.365	5.310	28	76.10	184
8500	1000	4	10	0	3.370	9.118	16.54	112	370.2	744
10000	1000	5	10	0	0.530	0.686	0.850	4	5.800	8
11000	1000	5	10	0	0.630	0.859	1.090	5	8.400	12
12000	1000	5	10	0	0.770	0.962	1.290	6	9.300	14
13000	1000	5	10	0	0.850	1.275	1.840	8	15.50	24
14000	1000	5	10	0	1.030	1.579	2.260	11	20.40	34
15000	1000	5	10	0	1.350	2.403	3.990	17	38.00	67
10000	400	6	10	0	0.580	0.719	0.890	8	10.90	13
10000	500	6	10	0	0.510	0.726	1.000	4	7.200	11
10000	600	6	10	0	0.430	0.553	0.640	3	4.000	5
10000	700	6	10	0	0.470	0.608	0.850	3	3.900	6
10000	800	6	10	0	0.440	0.529	0.680	2	2.900	4
10000	900	6	10	0	0.460	0.541	0.740	2	2.900	4
10000	1000	6	10	0	0.460	0.525	0.750	2	2.300	4
20000	1000	7	10	0	0.990	1.247	1.650	2	2.900	5
30000	2000	7	10	0	1.660	1.921	2.220	2	2.300	3
40000	3000	7	10	0	2.280	2.545	2.800	2	2.100	3
50000	4000	7	10	0	3.110	3.249	3.400	2	2.000	2
60000	5000	7	10	0	3.720	3.926	4.270	2	2.000	2
10000	1000	10	10	0	0.290	0.354	0.430	1	1.000	1
20000	2000	10	10	0	0.730	0.782	0.870	1	1.000	1
30000	3000	10	10	0	1.100	1.227	1.420	1	1.000	1
40000	4000	10	10	0	1.580	1.731	1.680	1	1.000	1
50000	5000	10	10	0	1.900	2.301	2.740	1	1.000	1

such a time limit. For the Davis-Putnam algorithm, the average execution time does not include the maximum execution time limit if some of the ten executions were successful; the average execution time was taken as the maximum execution time

TABLE XVII
Real performance comparison between the Davis-Putnam algorithm and a $SAT14.15$ algorithm on a SUN SPARC 2 workstation (Average of ten algorithm executions. Time Units: seconds. Symbol "S/F" stands for algorithm's *success/failure* of giving a solution within a time limit of $120 \times n/m$ seconds.)

Problems Clause,Variable,Literal			Execution Time Davis-Putnam Algorithm				Execution Time $SAT14.15$ Algorithm			
n	m	l	S/F	Min	Mean	Max	S/F	Min	Mean	Max
500	500	3	10/0	2.066	2.159	2.249	10/0	0.020	0.030	0.040
750	500	3	10/0	2.699	2.916	3.183	10/0	0.030	0.032	0.040
1000	500	3	10/0	3.149	3.657	6.316	10/0	0.040	0.058	0.070
1250	500	3	9/1	3.966	5.797	10.02	10/0	0.050	0.071	0.110
1500	500	3	6/4	7.499	9.147	11.60	10/0	0.090	0.140	0.240
1000	500	4	10/0	4.533	4.684	4.933	10/0	0.020	0.039	0.060
1500	500	4	10/0	6.766	7.960	16.90	10/0	0.070	0.085	0.110
2000	500	4	8/2	8.999	10.27	14.25	10/0	0.090	0.122	0.200
2500	500	4	2/8	13.63	15.96	18.28	10/0	0.110	0.157	0.190
3000	500	4	1/9	46.33	46.33	46.33	10/0	0.150	0.215	0.330
3000	500	5	10/0	16.23	16.90	18.82	10/0	0.120	0.162	0.240
4000	500	5	5/5	21.72	28.39	44.66	10/0	0.170	0.220	0.260
5000	500	5	0/10	> 1200	> 1200	> 1200	10/0	0.210	0.278	0.360
6000	500	5	0/10	> 1440	> 1440	> 1440	10/0	0.360	0.429	0.660
7000	500	5	0/10	> 1680	> 1680	> 1680	10/0	0.430	0.666	1.130
10000	1000	10	10/0	99.71	101.8	103.1	10/0	0.310	0.354	0.420
12000	1000	10	10/0	122.1	124.3	126.7	10/0	0.390	0.470	0.630
14000	1000	10	10/0	140.7	145.2	148.7	10/0	0.440	0.488	0.580
16000	1000	10	10/0	165.1	167.1	168.7	10/0	0.500	0.563	0.640
18000	1000	10	10/0	185.5	188.6	189.9	10/0	0.600	0.663	0.740

TABLE XVIII
Performance comparison between a $SAT14.15$ algorithm running on a SUN SPARC 2 workstation and the $GSAT$ algorithm running on a MIPS computer with comparable computing power (Time Units: seconds)

Problems			Execution Time	
n	m	l	GSAT	SAT14.15
215	50	3	0.400	0.150
301	70	3	0.900	0.010
430	100	3	6.000	0.040
516	120	3	14.00	1.040
602	140	3	14.00	0.030
645	150	3	45.00	0.260
860	200	3	168.0	1.180
1062	250	3	246.0	0.230
1275	300	3	720.0	14.47

TABLE XIX
Execution performance of a $SAT14.15$ algorithm with Pehoushek's Benchmarks on a SUN SPARC 2 workstation (Time Units: seconds)

Problems				Ten Trials		Execution Time			Iterations (k)		
Names	n	m	l	Global	Local	Min	Mean	Max	Min	Mean	Max
Prob1	600	180	3	10	0	0.800	5.568	16.95	528	3946.5	11936
Prob2	600	180	3	10	0	0.100	0.483	1.950	58	308.90	1289
Prob3	600	180	3	10	0	0.040	0.833	2.060	18	557.00	1441
Prob4	600	180	3	10	0	0.050	2.758	8.450	30	1869.0	5817
Prob5	600	180	3	10	0	0.080	0.783	2.420	43	508.80	1591
Prob6	600	180	3	10	0	0.050	0.598	2.030	29	380.10	1393
Prob7	600	180	3	10	0	0.280	7.407	47.36	182	1883.3	4920
Prob8	600	180	3	10	0	0.340	0.844	1.970	215	527.90	1249
Prob9	600	180	3	10	0	0.080	1.016	3.250	37	679.90	2324
Prob10	600	180	3	10	0	0.040	0.961	2.290	18	598.80	1497

TABLE XX
Performance comparison between a $SAT14.15$ algorithm running on a SUN SPARC 2 workstation and an interior point zero-one integer programming algorithm running on a $KORBX(R)$ parallel/vector computer (Time Units: seconds)

Problems Clause, Variable, Literal			Execution Time Interior Point Algorithm				Execution Time $SAT14.15$ Algorithm			
n	m	l	S/F	Min	Mean	Max	S/F	Min	Mean	Max
100	50	5.0020	52/0	0.5	0.7	7.7	10/0	0.000	0.004	0.020
200	100	4.9950	70/0	1.0	1.1	1.7	10/0	0.000	0.011	0.020
400	200	6.9430	69/0	2.4	3.5	6.3	10/0	0.010	0.017	0.030
800	400	9.9640	31/0	5.1	5.6	7.7	10/0	0.020	0.032	0.040
800	400	6.9770	20/0	4.7	7.8	16.1	10/0	0.020	0.043	0.070
1000	500	9.9970	49/0	6.5	7.4	9.8	10/0	0.030	0.042	0.050
2000	1000	10.008	10/0	14.9	18.5	20.5	10/0	0.080	0.095	0.110
2000	1000	7.0140	50/0	18.1	21.5	27.3	10/0	0.070	0.121	0.160
2000	1000	3.0000	49/1	8.7	420.7	9595.5	10/0	0.150	0.738	2.340
4000	1000	4.0060	1/1	8314.3	8314.3	8314.3	10/0	1.050	25.29	91.52
4000	1000	10.005	10/0	24.0	25.1	25.8	10/0	0.160	0.206	0.330
8000	1000	10.001	10/0	37.5	38.0	38.8	10/0	0.380	0.498	0.720
16000	1000	9.9990	10/0	46.2	66.4	92.1	10/0	1.150	1.226	1.330
32000	1000	10.007	10/0	80.1	232.4	311.3	10/0	2.530	3.281	3.820

limit only if all ten executions were failed.

From numerous algorithm executions, we have observed that, for random problem instances listed in the table, the Davis-Putnam algorithm was approximately hundreds to thousands times slower than the $SAT14.5$ algorithm. As the hardness of the problem instances increases, the number of failure, F, increases quickly. For some slightly hard problem instances, such as $n = 5000, m = 500$, and $l = 5$, all ten algorithm executions failed after a reasonably long time limit. Due to its $O(n^{O(m/l)})$

TABLE XXI
Real execution performance of a $SAT14.6$ algorithm on a SUN SPARC 2 workstation (Time Units: seconds)

Problems (n,m,l)			Ten Trials		Execution Time			Iterations (k)		
Clauses	Variables	Literals	Global	Local	Min	Mean	Max	Min	Mean	Max
100	100	3	10	0	0.000	0.002	0.010	1	1.500	3
200	100	3	10	0	0.000	0.005	0.010	2	3.000	5
300	100	3	10	0	0.000	0.014	0.030	3	8.800	19
400	100	3	10	0	0.010	0.040	0.170	10	37.30	166
1000	1000	3	10	0	0.020	0.033	0.050	2	2.900	4
1500	1000	3	10	0	0.030	0.058	0.100	2	3.500	6
2000	1000	3	10	0	0.070	0.093	0.140	4	5.100	8
2500	1000	3	10	0	0.090	0.133	0.180	4	7.000	10
3000	1000	3	10	0	0.140	0.241	0.350	6	13.70	21
3500	1000	3	10	0	0.290	0.919	2.750	18	65.80	208
1000	1000	4	10	0	0.010	0.029	0.040	1	1.900	3
2000	1000	4	10	0	0.050	0.057	0.070	2	2.200	3
3000	1000	4	10	0	0.080	0.109	0.160	2	3.200	5
4000	1000	4	10	0	0.110	0.162	0.220	3	4.300	6
5000	1000	4	10	0	0.170	0.267	0.420	4	7.000	12
6000	1000	4	10	0	0.260	0.388	0.510	7	10.60	14
7000	1000	4	10	0	0.360	0.756	0.980	11	26.40	37
8000	1000	4	10	0	1.210	2.595	4.750	50	126.3	240
8500	1000	4	10	0	1.280	12.79	32.89	60	755.0	1968
10000	1000	5	10	0	0.370	0.464	0.610	5	6.600	8
11000	1000	5	10	0	0.430	0.580	0.830	6	9.300	15
12000	1000	5	10	0	0.560	0.649	0.900	9	11.20	17
13000	1000	5	10	0	0.580	0.978	1.730	9	18.60	37
14000	1000	5	10	0	0.710	1.282	2.210	12	27.10	51
15000	1000	5	10	0	0.950	2.047	4.240	19	52.50	114
10000	400	6	10	0	0.350	0.514	0.890	8	14.90	33
10000	500	6	10	0	0.260	0.345	0.530	4	5.800	10
10000	600	6	10	0	0.190	0.331	0.440	3	4.600	6
10000	700	6	10	0	0.210	0.289	0.360	3	3.500	4
10000	800	6	10	0	0.250	0.287	0.380	2	3.200	5
10000	900	6	10	0	0.230	0.278	0.420	2	2.700	4
10000	1000	6	10	0	0.250	0.269	0.300	2	2.200	3
20000	1000	7	10	0	0.460	0.496	0.550	2	2.400	3
30000	2000	7	10	0	0.730	0.910	1.040	2	2.500	3
40000	3000	7	10	0	1.090	1.250	1.570	2	2.200	3
50000	4000	7	10	0	1.440	1.632	1.950	2	2.200	3
60000	5000	7	10	0	1.820	1.971	2.140	2	2.000	2
10000	1000	10	10	0	0.020	0.048	0.080	1	1.000	1
20000	2000	10	10	0	0.060	0.099	0.180	1	1.000	1
30000	3000	10	10	0	0.110	0.179	0.240	1	1.000	1
40000	4000	10	10	0	0.170	0.328	0.590	1	1.000	1
50000	5000	10	10	0	0.260	0.395	0.500	1	1.000	1

average run time complexity, even for some fairly easy problem instances, such as $n = 10000, m = 1000$, and $l = 10$, the Davis-Putnam algorithm took an excessive amount of time to find a solution. In comparison, the $SAT14.5$ algorithm was suc-

TABLE XXII

Real performance comparison between the Davis-Putnam algorithm and a $SAT14.6$ algorithm on a SUN SPARC 2 workstation (Average of ten algorithm executions. Time Units: seconds. Symbol "S/F" stands for algorithm's *success/failure* of giving a solution within a time limit of $120 \times n/m$ seconds.)

Problems Clause, Variable, Literal			Execution Time Davis-Putnam Algorithm				Execution Time $SAT14.6$ Algorithm			
n	m	l	S/F	Min	Mean	Max	S/F	Min	Mean	Max
500	500	3	10/0	2.066	2.159	2.249	10/0	0.010	0.013	0.020
750	500	3	10/0	2.699	2.916	3.183	10/0	0.020	0.021	0.030
1000	500	3	10/0	3.149	3.657	6.316	10/0	0.020	0.031	0.040
1250	500	3	9/1	3.966	5.797	10.02	10/0	0.040	0.067	0.150
1500	500	3	6/4	7.499	9.147	11.60	10/0	0.070	0.115	0.300
1000	500	4	10/0	4.533	4.684	4.933	10/0	0.020	0.024	0.030
1500	500	4	10/0	6.766	7.960	16.90	10/0	0.030	0.042	0.070
2000	500	4	8/2	8.999	10.27	14.25	10/0	0.040	0.069	0.100
2500	500	4	2/8	13.63	15.96	18.28	10/0	0.070	0.109	0.180
3000	500	4	1/9	46.33	46.33	46.33	10/0	0.110	0.153	0.210
3000	500	5	10/0	16.23	16.90	18.82	10/0	0.060	0.074	0.090
4000	500	5	5/5	21.72	28.39	44.66	10/0	0.080	0.144	0.250
5000	500	5	0/10	> 1200	> 1200	> 1200	10/0	0.130	0.196	0.280
6000	500	5	0/10	> 1440	> 1440	> 1440	10/0	0.230	0.313	0.410
7000	500	5	0/10	> 1680	> 1680	> 1680	10/0	0.370	0.604	0.910
10000	1000	10	10/0	99.71	101.8	103.1	10/0	0.020	0.049	0.080
12000	1000	10	10/0	122.1	124.3	126.7	10/0	0.020	0.052	0.080
14000	1000	10	10/0	140.7	145.2	148.7	10/0	0.030	0.063	0.110
16000	1000	10	10/0	165.1	167.1	168.7	10/0	0.040	0.078	0.130
18000	1000	10	10/0	185.5	188.6	189.9	10/0	0.060	0.095	0.220

TABLE XXIII

Performance comparison between a $SAT14.6$ algorithm running on a SUN SPARC 2 workstation and the $GSAT$ algorithm running on a MIPS computer with comparable computing power (Time Units: seconds)

Problems			Execution Time	
n	m	l	GSAT	SAT14.6
215	50	3	0.400	0.020
301	70	3	0.900	0.020
430	100	3	6.000	0.050
516	120	3	14.00	0.410
602	140	3	14.00	1.990
645	150	3	45.00	0.040
860	200	3	168.0	6.710
1062	250	3	246.0	12.43
1275	300	3	720.0	19.14

TABLE XXIV
Execution performance of a $SAT14.6$ algorithm with Pehoushek's Benchmarks on a SUN SPARC 2 workstation (Time Units: seconds)

Problems				Ten Trials		Execution Time			Iterations (k)		
Names	n	m	l	Global	Local	Min	Mean	Max	Min	Mean	Max
Prob1	600	180	3	10	0	0.490	12.83	31.49	428	10706	26154
Prob2	600	180	3	10	0	0.110	0.714	2.510	68	582.90	2022
Prob3	600	180	3	10	0	0.270	1.538	4.240	226	1225.0	3356
Prob4	600	180	3	10	0	0.240	3.037	11.74	221	2456.2	9534
Prob5	600	180	3	10	0	0.190	0.744	1.770	150	576.90	1323
Prob6	600	180	3	10	0	0.030	0.735	2.480	12	578.70	1921
Prob7	600	180	3	10	0	0.110	2.399	12.05	83	1968.5	9926
Prob8	600	180	3	10	0	0.080	1.392	5.060	54	1096.3	4156
Prob9	600	180	3	10	0	0.030	0.756	3.550	21	579.90	2740
Prob10	600	180	3	10	0	0.120	1.246	2.800	98	998.60	2225

TABLE XXV
Performance comparison between a $SAT14.6$ algorithm running on a SUN SPARC 2 workstation and an interior point zero-one integer programming algorithm running on a *KORBX(R)* parallel/vector computer (Time Units: seconds)

Problems Clause,Variable,Literal			Execution Time Interior Point Algorithm				Execution Time $SAT14.6$ Algorithm			
n	m	l	S/F	Min	Mean	Max	S/F	Min	Mean	Max
100	50	5.0000	52/0	0.5	0.7	7.7	10/0	0.000	0.001	0.010
200	100	4.9900	70/0	1.0	1.1	1.7	10/0	0.000	0.005	0.010
400	200	6.9900	69/0	2.4	3.5	6.3	10/0	0.000	0.007	0.020
800	400	9.9999	31/0	5.1	5.6	7.7	10/0	0.000	0.003	0.010
800	400	6.9930	20/0	4.7	7.8	16.1	10/0	0.000	0.009	0.020
1000	500	9.9910	49/0	6.5	7.4	9.8	10/0	0.000	0.006	0.010
2000	1000	9.9940	10/0	14.9	18.5	20.5	10/0	0.010	0.019	0.030
2000	1000	7.0040	50/0	18.1	21.5	27.3	10/0	0.020	0.055	0.090
2000	1000	3.0000	49/1	8.7	420.7	9595.5	10/0	0.090	3.917	33.56
4000	1000	4.0010	1/1	8314.3	8314.3	8314.3	10/0	1.900	6.826	25.54
4000	1000	10.003	10/0	24.0	25.1	25.8	10/0	0.030	0.044	0.060
8000	1000	9.9980	10/0	37.5	38.0	38.8	10/0	0.080	0.254	0.430
16000	1000	9.9999	10/0	46.2	66.4	92.1	10/0	0.530	0.625	0.820
32000	1000	10.004	10/0	80.1	232.4	311.3	10/0	0.970	1.611	2.530

cessful for all ten executions. It was able to find a solution to the given problem instances.

Since $SAT14.5$ algorithm is *incomplete*, it can neither find a solution for an unsatisfiable CNF formula nor give an answer if a CNF formula is not satisfiable.

Performance Comparison with the GSAT Algorithm. Table XIII gives real performance comparison between a $SAT14.5$ algorithm running on a SUN SPARC

TABLE XXVI

Real execution performance of a *SAT*14.16 algorithm on a SUN SPARC 2 workstation (Time Units: seconds)

Problems (n,m,l)			Ten Trials		Execution Time			Iterations (k)		
Clauses	Variables	Literals	Global	Local	Min	Mean	Max	Min	Mean	Max
100	100	3	10	0	0.000	0.003	0.010	1	1.700	2
200	100	3	10	0	0.000	0.008	0.010	1	2.800	5
300	100	3	10	0	0.000	0.012	0.020	4	5.300	8
400	100	3	10	0	0.020	0.215	0.640	14	189.8	539
1000	1000	3	10	0	0.040	0.051	0.080	2	2.800	4
1500	1000	3	10	0	0.060	0.081	0.100	2	3.600	5
2000	1000	3	10	0	0.080	0.115	0.170	3	5.000	9
2500	1000	3	10	0	0.140	0.180	0.240	5	7.600	11
3000	1000	3	10	0	0.230	0.359	0.550	11	19.30	32
3500	1000	3	10	0	0.530	1.357	4.350	36	98.80	348
1000	1000	4	10	0	0.020	0.040	0.050	1	1.600	2
2000	1000	4	10	0	0.080	0.092	0.110	2	2.200	3
3000	1000	4	10	0	0.130	0.166	0.200	2	2.500	4
4000	1000	4	10	0	0.210	0.234	0.300	4	4.700	7
5000	1000	4	10	0	0.250	0.330	0.450	4	6.500	10
6000	1000	4	10	0	0.380	0.478	0.640	7	10.60	16
7000	1000	4	10	0	0.510	0.840	1.950	13	26.20	73
8000	1000	4	10	0	1.000	2.641	4.370	34	120.2	214
8500	1000	4	10	0	5.250	12.12	26.75	256	672.0	1464
10000	1000	5	10	0	0.500	0.567	0.610	5	5.900	7
11000	1000	5	10	0	0.600	0.750	1.040	6	9.200	15
12000	1000	5	10	0	0.650	0.844	1.080	6	10.60	16
13000	1000	5	10	0	0.740	1.064	1.750	9	15.90	33
14000	1000	5	10	0	1.060	1.652	2.480	15	31.30	54
15000	1000	5	10	0	1.350	2.166	4.120	24	48.60	106
10000	400	6	10	0	0.530	0.771	1.550	9	17.20	48
10000	500	6	10	0	0.440	0.553	0.710	4	6.400	10
10000	600	6	10	0	0.480	0.534	0.740	4	5.000	8
10000	700	6	10	0	0.390	0.491	0.590	3	3.900	5
10000	800	6	10	0	0.400	0.452	0.560	2	2.900	4
10000	900	6	10	0	0.410	0.476	0.590	2	2.900	4
10000	1000	6	10	0	0.420	0.473	0.660	2	2.700	4
20000	1000	7	10	0	0.880	1.004	1.270	2	2.900	4
30000	2000	7	10	0	0.990	1.460	1.600	1	1.900	2
40000	3000	7	10	0	2.010	2.196	2.550	2	2.400	3
50000	4000	7	10	0	2.650	2.730	2.870	2	2.000	2
60000	5000	7	10	0	3.150	3.402	3.610	2	2.000	2
10000	1000	10	10	0	0.029	0.312	0.330	1	1.000	1
20000	2000	10	10	0	0.630	0.664	0.720	1	1.000	1
30000	3000	10	10	0	1.010	1.076	1.130	1	1.000	1
40000	4000	10	10	0	1.430	1.511	1.590	1	1.000	1
50000	5000	10	10	0	1.810	1.887	2.020	1	1.000	1

workstation and the *GSAT* algorithm [287] running on a MIPS computer with comparable computing power [286]. For the exact 3-SAT problem instances generated from the same problem model used in [231], *SAT*14.5 algorithm is approximately

TABLE XXVII
Real performance comparison between the Davis-Putnam algorithm and a $SAT14.16$ algorithm on a SUN SPARC 2 workstation (Average of ten algorithm executions. Time Units: seconds. Symbol "S/F" stands for algorithm's *success/failure* of giving a solution within a time limit of $120 \times n/m$ seconds.)

Problems Clause, Variable, Literal			Execution Time Davis-Putnam Algorithm				Execution Time $SAT14.16$ Algorithm			
n	m	l	S/F	Min	Mean	Max	S/F	Min	Mean	Max
500	500	3	10/0	2.066	2.159	2.249	10/0	0.020	0.027	0.030
750	500	3	10/0	2.699	2.916	3.183	10/0	0.030	0.033	0.040
1000	500	3	10/0	3.149	3.657	6.316	10/0	0.030	0.047	0.060
1250	500	3	9/1	3.966	5.797	10.02	10/0	0.050	0.072	0.090
1500	500	3	6/4	7.499	9.147	11.60	10/0	0.060	0.108	0.230
1000	500	4	10/0	4.533	4.684	4.933	10/0	0.030	0.038	0.040
1500	500	4	10/0	6.766	7.960	16.90	10/0	0.050	0.066	0.090
2000	500	4	8/2	8.999	10.27	14.25	10/0	0.070	0.088	0.110
2500	500	4	2/8	13.63	15.96	18.28	10/0	0.120	0.152	0.230
3000	500	4	1/9	46.33	46.33	46.33	10/0	0.170	0.234	0.360
3000	500	5	10/0	16.23	16.90	18.82	10/0	0.110	0.139	0.180
4000	500	5	5/5	21.72	28.39	44.66	10/0	0.150	0.188	0.210
5000	500	5	0/10	> 1200	> 1200	> 1200	10/0	0.240	0.288	0.340
6000	500	5	0/10	> 1440	> 1440	> 1440	10/0	0.320	0.471	0.760
7000	500	5	0/10	> 1680	> 1680	> 1680	10/0	0.490	0.623	0.950
10000	1000	10	10/0	99.71	101.8	103.1	10/0	0.290	0.315	0.340
12000	1000	10	10/0	122.1	124.3	126.7	10/0	0.350	0.382	0.440
14000	1000	10	10/0	140.7	145.2	148.7	10/0	0.400	0.430	0.470
16000	1000	10	10/0	165.1	167.1	168.7	10/0	0.480	0.517	0.590
18000	1000	10	10/0	185.5	188.6	189.9	10/0	0.550	0.583	0.630

tens to hundreds times faster than the $GSAT$ algorithm.

Performance Comparison with Pehoushek's Benchmarks. Pehoushek has recently provided an *i-j-regular 3-SAT* problem model [250]. In his model, one puts i positive and j negative copies of each variable in a set, pull out 3 at a time forming a clause. Table XIV shows real execution performance of the $SAT14.5$ algorithm for ten benchmark SAT problem instances provided by Pehoushek. According to Pehoushek's performance measure [250], $SAT14.5$ is *extremely* fast on his problem instances.

Performance Comparison with Interior Point Zero-One Integer Programming Algorithm Kamath et al. used an interior point zero-one integer programming algorithm to solve the SAT problem [186]. They implemented their algorithm in FORTRAN and C languages and ran the algorithm on a *KORBX(R)* parallel/vector computer with problem instances generated from the average 3-SAT problem model. The *KORBX(R)* parallel computer operates in scalar mode at approximately 1 MFlops and at 32 MFlops with full vector concurrent mode. Their execution results are given in Table XV.

TABLE XXVIII

Performance comparison between a $SAT14.16$ algorithm running on a SUN SPARC 2 workstation and the $GSAT$ algorithm running on a MIPS computer with comparable computing power (Time Units: seconds)

Problems			Execution Time	
n	m	l	GSAT	SAT14.16
215	50	3	0.400	0.030
301	70	3	0.900	0.020
430	100	3	6.000	0.370
516	120	3	14.00	0.040
602	140	3	14.00	0.170
645	150	3	45.00	0.870
860	200	3	168.0	0.490
1062	250	3	246.0	0.090
1275	300	3	720.0	4.510

TABLE XXIX

Execution performance of a $SAT14.16$ algorithm with Pehoushek's Benchmarks on a SUN SPARC 2 workstation (Time Units: seconds)

Problems				Ten Trials		Execution Time			Iterations (k)		
Names	n	m	l	Global	Local	Min	Mean	Max	Min	Mean	Max
Prob1	600	180	3	10	0	3.000	8.863	24.79	2428	7240.1	19990
Prob2	600	180	3	10	0	0.080	0.869	2.410	40	672.40	1905
Prob3	600	180	3	10	0	0.090	1.723	5.580	56	1345.4	4312
Prob4	600	180	3	10	0	0.260	1.374	4.390	201	1090.4	3527
Prob5	600	180	3	10	0	0.070	1.015	2.190	37	765.60	1641
Prob6	600	180	3	10	0	0.070	0.768	1.980	46	583.00	1468
Prob7	600	180	3	10	0	0.700	2.464	6.080	31	1988.7	4920
Prob8	600	180	3	10	0	0.290	0.899	2.170	222	699.60	1747
Prob9	600	180	3	10	0	0.170	1.054	3.450	112	803.70	2595
Prob10	600	180	3	10	0	0.120	1.097	4.180	74	854.50	3184

We ran the $SAT14.5$ algorithm for the same problem instances on a sequential SUN SPARC 2 workstation. The results are listed in the same table. Apparently, compared to the interior point zero-one integer programming algorithm running on a parallel computer, in addition to improved global convergence, $SAT1.5$ algorithm is much simpler and it achieves many orders of magnitude of performance improvements in terms of computing time.

TABLE XXX
Performance comparison between a $SAT14.16$ algorithm running on a SUN SPARC 2 workstation and an interior point zero-one integer programming algorithm running on a *KORBX(R)* parallel/vector computer (Time Units: seconds)

Problems Clause,Variable,Literal			Execution Time Interior Point Algorithm				Execution Time $SAT14.16$ Algorithm			
n	m	l	S/F	Min	Mean	Max	S/F	Min	Mean	Max
100	50	4.9940	52/0	0.5	0.7	7.7	10/0	0.000	0.004	0.010
200	100	5.0030	70/0	1.0	1.1	1.7	10/0	0.000	0.007	0.010
400	200	6.9980	69/0	2.4	3.5	6.3	10/0	0.010	0.018	0.030
800	400	10.002	31/0	5.1	5.6	7.7	10/0	0.030	0.030	0.030
800	400	7.0000	20/0	4.7	7.8	16.1	10/0	0.010	0.026	0.050
1000	500	9.9990	49/0	6.5	7.4	9.8	10/0	0.030	0.039	0.080
2000	1000	10.002	10/0	14.9	18.5	20.5	10/0	0.060	0.083	0.160
2000	1000	7.0000	50/0	18.1	21.5	27.3	10/0	0.020	0.055	0.090
2000	1000	3.0000	49/1	8.7	420.7	9595.5	10/0	21.85	27.19	34.81
4000	1000	4.0090	1/1	8314.3	8314.3	8314.3	10/0	0.500	9.555	39.68
4000	1000	10.005	10/0	24.0	25.1	25.8	10/0	0.140	0.163	0.290
8000	1000	10.006	10/0	37.5	38.0	38.8	10/0	0.290	0.353	0.560
16000	1000	10.000	10/0	46.2	66.4	92.1	10/0	1.000	1.052	1.140
32000	1000	10.004	10/0	80.1	232.4	311.3	10/0	2.060	2.434	2.800

7.5.2. *Performance of the $SAT14.15$ Algorithm*

Tables XVI to XX show the real experiment performance of the $SAT14.15$ algorithm which is a *complete* version of the $SAT14.5$ algorithm. Since $SAT14.15$ combines $SAT14.5$ algorithm with backtracking, it is able to verify satisfiability as well as unsatisfiability. For some problem instances, it takes slightly more execution time than the $SAT14.5$ to manage the "completeness" bookkeeping.

7.5.3. *Performance of the $SAT14.6$ Algorithm*

Tables XXI to XXV show the real experiment performance of the $SAT14.6$ algorithm. It has comparable and complementary performance to the $SAT14.5$ algorithm.

7.5.4. *Performance of the $SAT14.16$ Algorithm*

Tables XXVI to XXX show the real experiment performance of the $SAT14.16$ algorithm which is a *complete* version of the $SAT14.6$ algorithm. For most problem instances, it takes slightly more execution time than the $SAT14.6$ to manage the "completeness" bookkeeping.

7.5.5. *Experiments with the structured SAT models*

To assess the performance of the $SAT14$ algorithms with *non-binary* problem instances, we also tested SAT problem instances generated from regularly structured

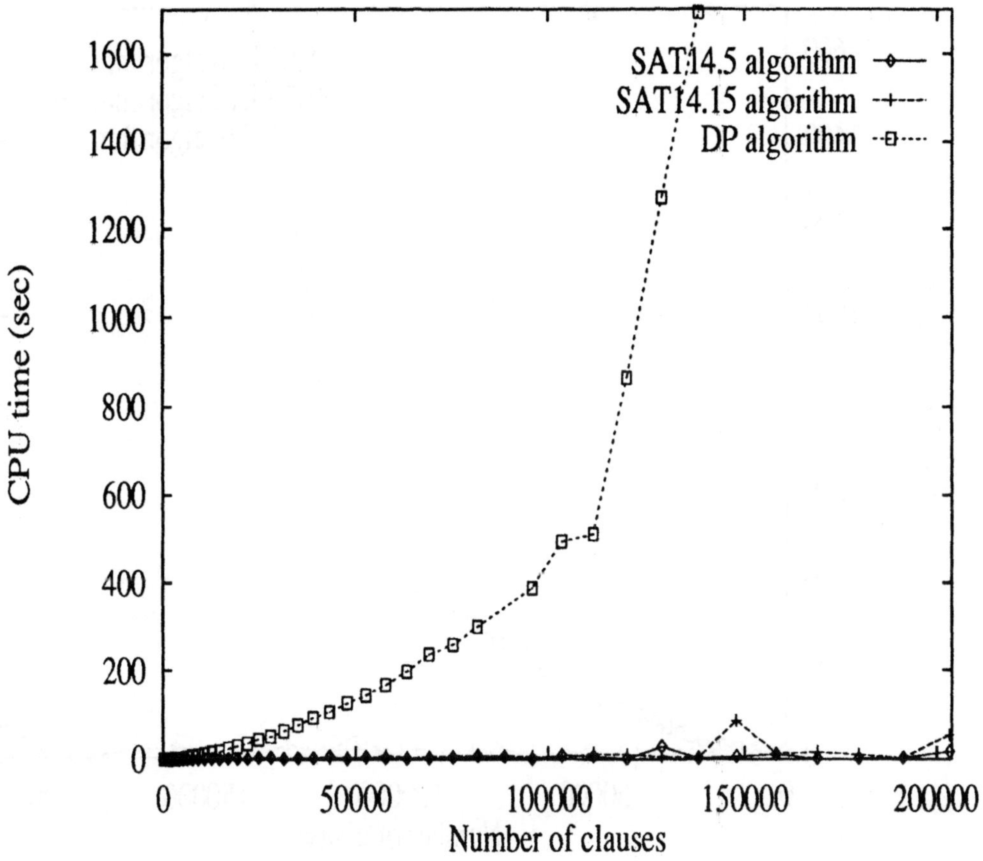

Fig. 19. The comparison of the Davis-Putnam algorithm with $SAT14.5$ and $SAT14.15$ algorithms for SAT problem instances generated from a CSP problem.

problems. Figures 19 and 20 show performance comparisons between the Davis-Putnam algorithm and a number of the $SAT14$ algorithms for SAT problem instances generated from a regular CSP problem.

8. Applications

There are three main purposes to discuss the applications of the SAT problem. First of all, useful applications, no matter practical or theoretical, are always the "driving forces" to SAT research. Second, application examples provide meaningful

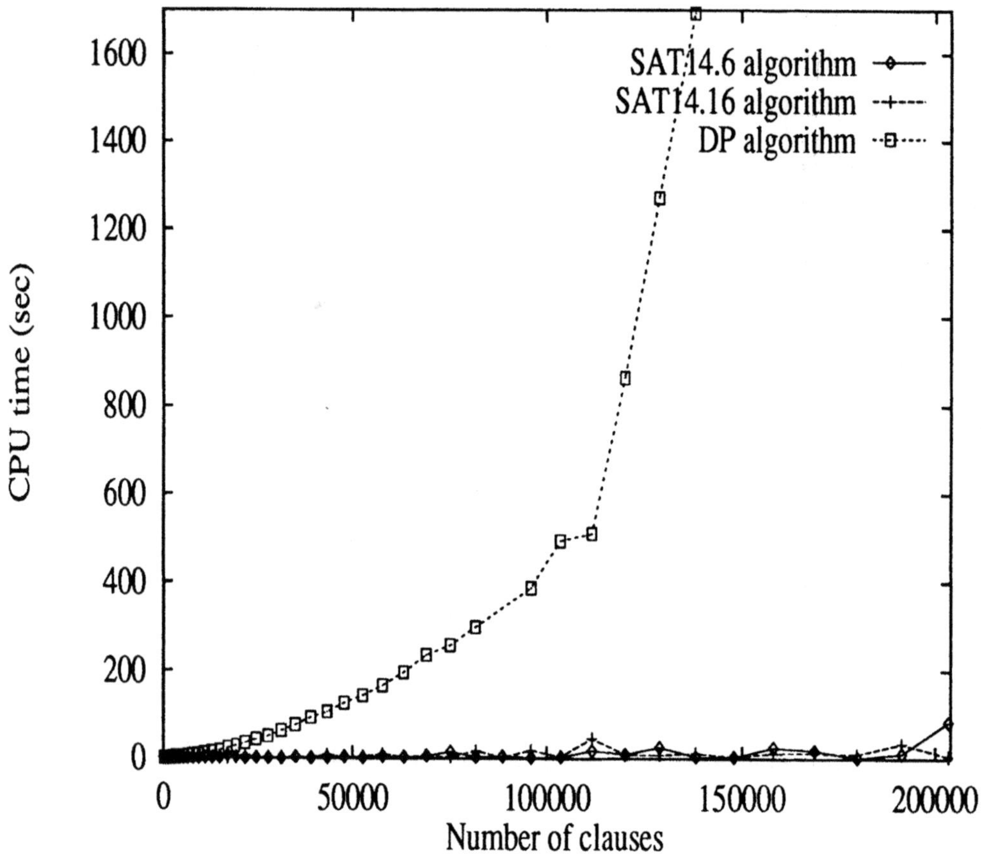

Fig. 20. The comparison of the Davis-Putnam algorithm with $SAT14.6$ and $SAT14.16$ algorithms for SAT problem instances generated from a CSP problem.

benchmarks to the development of new SAT algorithms. Third, techniques that solve SAT-related problems in different application areas may shed light on developing new SAT algorithms based on knowledge from these application areas. Finally, available techniques in one application area may shed light on solving problems in another area.

SAT problem has direct applications in mathematical logic, inference, machine learning, and VLSI engineering. It also has many *indirect* applications through the constraint satisfaction problem (see Section 2.3) and the constrained optimization problems [134]. Some of them are listed below:

- *Mathematics:* finding n-ary relations such as transitive closure [35], detecting graph and subgraph isomorphisms [55, 225, 227, 242, 268, 316, 329], the graph coloring problem [32, 146, 223, 227], mathematical cryptology [248, 271], the automata homomorphism problem [110], finding spanning trees and Euler tours in a graph [239], solving the traveling salesman problem [179, 180, 203, 245], and logical arithmetic [48].
- *Computer science and artificial intelligence:* the constraint satisfaction problem [7, 105, 118, 218, 275], the n-queen problem [105, 145, 295], extended inference [13], logical programming [50, 52, 77, 202], abductive inference for synthesizing composite hypotheses [183], semantic information processing [13, 90, 240], puzzles and cryptoarithmetic [104, 145, 221, 222, 237], truth maintenance [64, 65, 66, 76, 224], production system [172, 229, 230], the soma cube and instant insanity problem [105], theorem proving [164, 193, 255, 328], and neural network computing [7, 8, 67, 151, 212, 168].
- *Machine vision:* image matching problem [13, 45, 274, 327], line and edge labeling problems [39, 88, 313, 322, 331], stereopsis, scene analysis and semantics-based region growing [13, 39, 87, 88, 89, 313, 322], the shape and object matching problem [35, 60, 150], syntactic shape analysis [61, 149], shape from shading problem [6, 29, 96, 159, 160, 162, 161, 170, 219, 251], and image restoration [107].
- *Robotics:* for solving its vision problem [45, 169], packing problem [71], and trajectory and task planning problems [25, 84].
- *Computer-aided manufacturing:* task planning [236], design [234, 235], solid modeling, configuring task [97], design cellular manufacturing system, scheduling [91, 213], and 3-dimensional object recognition [140, 161].
- *Database systems:* operations on objects [315, 318], database consistency maintenance, query-answering and redundancy-checking, query optimization [41, 315], concurrency control [16, 85, 217], distributed database systems [100], truth and belief maintenance [64, 65, 66, 76, 224], the relational homomorphism problem [145, 315], and knowledge organization for recognition system [147].
- *Integrated circuit design automation:* circuit modeling [38, 311], logic minimization [152], state assignment [320, 321], state minimization [115, 267], I/O encoding for sequential machines [280], power dissipation estimation [73], logic partitioning [44, 81, 200, 241, 278], circuit layout and placement [5, 20, 26, 51, 58, 72, 143, 302], scheduling and high-level synthesis [27, 198, 249], pin assignment [24, 277], floorplanning [252, 308], interconnection analysis [79, 80], routing [1, 37, 69, 70, 142, 204, 247, 270, 272, 285], compaction [36, 78, 148, 188, 199, 209, 282, 290, 319], performance optimization [185, 194, 220, 276, 308], testing and test generation [74, 173, 201, 253], and verification [191, 309, 314] are essentially constrained search problems. Many of them demand efficient SAT algorithms.
- *Computer architecture design:* instruction set optimization [2, 59, 116, 176, 264, 266], computer controller optimization [15, 17, 189, 215, 263, 267], arithmetic logic circuit design [40], compiler system optimization [4, 208], scheduling [21, 75, 101, 102, 141, 207], fault-tolerant computing [14, 158], task partitioning and assignment [22, 23, 57, 171, 291], load balancing [210, 238, 330], real

time systems [153, 175, 195, 303, 304, 310], data flow consistency analysis [4], data module assignment in memory system [4], and parallel and distributed processing [269, 284].

- *Text processing:* optical character recognition [46, 233, 307], character constraint graph model [166], printed text recognition [12, 166], handwritten text recognition [293], automatic correction of errors in text [317].
- *Computer graphics:* construction of 2-dimensional pictures and 3-dimensional graphical objects from constraints, reasoning of the geometrical features of 3-dimensional objects [31, 98].

This list expands widely. In other areas such as industrial (chemical, transportation, construction, nuclear) engineering, management, medical research, social sciences, there are numerous applications of CSPs and SAT. A collection of such application problems is presented in [134].

9. Future Work

A number of research directions for satisfiability problem can be pursued in the future. We review a number of them briefly.

Algorithm integration. As is frequently observed, it is difficult to find a SAT algorithm that performs well for a wide range of SAT problem instances. A central principle in natural evolution is hybrid vigor. One of the future direction in SAT research is to integrate a number of SAT algorithms that each works well for certain range of problem instances. Another purpose for algorithm integration is to combine some incomplete algorithms with complete search framework so they can be made useful in practical application problems.

A simple integration solution is to put several SAT algorithms onto parallel processors, hardwire the solution output, and execute all of them simultaneously [125]. The execution may be terminated as long as one of the algorithms reports the results. Natural integration schemes would be directly merge different algorithm structures [120, 124, 125, 130]. We have implemented local search and global optimization algorithms with backtracking/resolution procedures. Previous experiments with these algorithms have shown some encouraging results [120, 124, 125, 121, 122, 130, 131].

Parallel hardware architectures. Implementing an algorithm on VLSI hardware architectures is now a conventional approach to speed up the algorithm execution. Not only it offers faster execution speed on hardware medium, certain sequential portions of the algorithm may be implemented in hardware architectures in parallel form.

For SAT problem *per se*, it contains certain granularity at the search tree level, clause level, and variable level which lend itself well to parallel processing. A number of parallel algorithms and architectures for the SAT problem were developed that perform at different levels of granularity. Two basic approaches were taken in this direction: implementing parallel SAT inference algorithms in special-purpose VLSI chips [117, 123, 132] or implementing parallel SAT algorithms on existing sequential computer machines [119, 130, 130, 301, 300] (also see Figure 10).

Algorithm space. The algorithm space described in Section 2.1 was instructive to develop continuous, unconstrained optimization algorithms for the SAT problem. Furthermore, a number of places in the algorithm space seems to be asymmetrical and irregular which can be investigated in the future. This may lead to some new SAT algorithms.

Theoretical issues. Recent research for the SAT problem has brought up some interesting theoretical problems, such as the average time complexity analysis, determining sat-and-unsat boundary, global convergence, local convergence rate, the structure and hardness of problem instance models, etc. Some of the problems, e.g., the average time complexity analysis, are extremely difficult [206, 258]. So far only some preliminary efforts based on simplified assumptions were given [135, 28].

Multispace search. The goal of traditional optimization algorithms is to find an assignment of values to variables such that all constraints are satisfied and the performance criteria are optimized. An optimization algorithm changes values and tries to find the "goal values." Traditional value search methods do not provide structural information to the search problem. It is difficult for them to handle difficult problems, e.g., local minimum points, in a hard search problem.

In multispace search [133, 134], any component related to the given search problem forms a new class of search space. For a given search problem, we define variable space, value space, constraint space, objective space, parameter space, and other search spaces. The totality of all the search spaces constitutes a *multispace.* During the search process, a multispace search algorithm not only changes values in the value space. It also scrambles across other spaces and dynamically *reconstructs* the problem structures that are related to the variables, constraints, objectives, parameters, and other components of the given search problem. Only at the *last moment* of the search, the "reconstructed" problem structure is replaced by the original problem structure, and thus the final value assignment represents the solution to the original search problem.

Presently some multispace search algorithms have been developed for the SAT problem [130, 133, 265].

10. Conclusions

The SAT problem is a *core* of the NP-complete problems. Traditional methods treat the SAT problem as a discrete, constrained decision problem. In this chapter, we show how to translate the SAT problem into an unconstrained optimization problem and use efficient local search and global optimization methods to solve the transformed optimization problem. The optimization method to the SAT problem gives significant performance improvements for certain classes of conjunctive normal form (CNF) formulas. It offers a complementary approach to the existing SAT algorithms.

The area of operations research and combinatorial optimization is a rich land of well-developed theory and methods. To combine optimization techniques with basic search framework seems an alternative way to handle the SAT problem. Not only the end results of such an endeavor have a major scientific impact, but in the process it will push optimization technology to its limit.

Acknowledgements

Ding-Zhu Du suggested this chapter be written. Ding-Zhu and Qianping Gu have made theoretical analyses to some optimization algorithms described in this chapter. Paul Purdom, Ding-Zhu Du, R.C.T. Lee, Vaughan Pratt, Qianping Gu, Xiaotie Deng, Ben Wah, Wei Wang, Rok Sosič, Ruchir Puri, Bin Du, Guojie Li, Wei Li, Zhiqiang Liu, Xiaofei Huang, and Richard Fujimoto have provided constructive comments for various early versions of this paper. Ding-Zhu Du, Zhiqiang Liu, Ben Wah, and Andy Sage have provided constant support to this project. Bin Du and Idan Shoham performed part of the experiments. Review comments from a well-known, anonymous referee are sincerely appreciated. Reviewers' comments helped to improve the presentation of this paper. Paul Purdom, R.C.T. Lee, Christos Papadimitriou, and Dennis Kibler have provided some recent papers and informed some recent work.

References

1. L. Abel. On the order of connections for automatic wire routing. *IEEE Trans. on Computers*, pages 1227–1233, Nov. 1972.
2. T. Agerwala. Microprogram Optimization : A Survey. *IEEE Trans. on Computers*, C-25:962–973, Oct. 1976.
3. A.V. Aho, J.E. Hopcroft, and J.D. Ullman. *The Design and Analysis of Computer Algorithms*. Addison-Wesley, Reading, 1974.
4. A.V. Aho, R. Sethi, and J.D. Ullman. *Compilers*. Addison-Wesley, Reading, 1986.
5. S.B. Akers. On the use of the linear assignment algorithm in module placement. In *Proc. of the 18th ACM/IEEE DAC*, pages 137–144, 1981.
6. J. Aloimonos. Visual shape computation. *Proceedings of the IEEE*, 76:899–916, Aug. 1988.
7. J. A. Anderson and G. E. Hinton. *Models of Information Processing in the Brain. In* G. E. Hinton and J. A. Anderson, editors, *Parallel Models of Associative Memory*, chapter 1, pages 9–48. Lawrence Erlbaum Associates, Publishers, Hillsdale, New Jersey, 1981.
8. J.A. Anderson and E. Rosenfeld, editors. *Neurocomputing: Foundations of Research*. MIT Press, Cambridge, 1988.
9. S. Arora, C. Lund, R. Motwani, M. Sudan, and M. Szegedy. Proof verification and hardness of approximation problems. In *Proceedings 33rd IEEE Symposium on the Foundations of Computer Science*, pages 14–23, 1992.
10. P. Ashar, A. Ghosh, and S. Devadas. Boolean satisfiability and equivalence checking using general Binary Decision Diagrams. *Integration, the VLSI journal*, 13:1–16, 1992.
11. B. Aspvall, M.F. Plass, and R.E. Tarjan. A linear-time algorithm for testing the truth of certain quantified Boolean formulas. *Information Processing Letters*, 8(3):121–132, Mar. 1979.
12. H. Baird. Anatomy of a versatile page reader. *Proceedings of the IEEE*, 80(7):1059–1065, Jul. 1992.
13. D. H. Ballard and C. M. Brown. *Computer Vision*. Prentice-Hall, Englewood Cliffs, New Jersey, 1982.
14. A.E. Barbour. Solutions to the minimization problem of fault-tolerant logic circuits. *IEEE Trans. on Computers*, 41(4):429–443, Apr. 1992.
15. R. G. Bennetts. An Improved Method for Prime C-Class Derivation in the State Reduction of Sequential Networks. *IEEE Trans. on Computers*, C-20:229–231, Feb. 1971.
16. P.A. Berstein, V. Hadzilacos, and N. Goodman. *Concurrency Control and Recovery in Database Systems*. Addison-Wesley Publication Co., 1987.
17. N. N. Biswas. State Minimization of Incompletely Specified Sequential Machines. *IEEE Trans. on Computers*, C-23:80–84, Jan. 1974.
18. J. R. Bitner and E. M. Reingold. Backtrack programming techniques. *Comm. of ACM*, 18(11):651–656, Nov. 1975.
19. C.E. Blair, R.G. Jeroslow, and J.K. Lowe. Some results and experiments in programming techniques for propositional logic. *Computers and Operations Research*, 5:633–645, 1986.
20. J.P. Blanks. Near-optimal placement using a quadratic objective function. In *Proc. of the 22th ACM/IEEE DAC*, pages 609–615, 1985.
21. J. Blazewicz. Scheduling dependent tasks with different arrival times to meet deadlines. In *Proc. of the International Workshop on Modelling and Performance Evaluation of Computer Systems*, pages 57–65, 1976.
22. S. H. Bokhari. Partitioning problems in parallel, pipelined, and distributed computing. *IEEE Trans. on Computers*, 37(1):48–57, Jan 1988.
23. F. Bonomi. On job assignment for a parallel system of processor sharing queues. *IEEE Trans. on Computers*, 39(7):858–869, July 1990.
24. H.N. Brady. An approach to topological pin assignment. *IEEE Trans. on Computer-Aided Design*, CAD-3:250–255, July 1984.
25. M. Brady, J.M. Hollerbach, T.L. Johnson, T. Lozano-Perez, and M.T. Mason, editors. *Robot Motion: Planning and Control*. The MIT Press, Cambridge, 1982.
26. M.A. Breuer. Min-cut placement. *J. of Design Automation and Fault Tolerant Computing*, 1:343–362, 1976.
27. F. Brewer and D. Gajski. Chippe: A system for constraint driven behavioral synthesis. *IEEE Trans. on Computer-Aided Design*, 9(7):681–695, July 1990.

28. A.Z. Broder, A.M. Frieze, and E. Upfal. On the satisfiability and maximum satisfiability of random 3-CNF formulas. In *Proceedings of the Fourth Annual ACM-SIAM Symposium on Discrete Algorithms*, pages 322–330, 1993.
29. M.J. Brooks and B.K.P. Horn. Shape and source from shading. In *Proc. of IJCAI'85*, pages 932–936, Aug. 1985.
30. C.A. Brown and P.W. Purdom. An average time analysis of backtracking. *SIAM J. Comput.*, 10(3):583–593, Aug. 1981.
31. B. Bruderlin. Constructing three dimensional geometric objects defined by constraints. In *Proceedings of 1986 ACM SIGGRAPH Workshop in Interactive Graphics*, 1986.
32. M. Bruynooghe. Graph coloring and constraint satisfaction. Technical Report CW 44, Dept. of Computer Science, Katholieke Universiteit Leuven, Dec. 1985.
33. R.E. Bryant. Graph-based algorithms for boolean function manipulation. *IEEE Trans. on Computers*, C-35(8):677–691, Aug. 1986.
34. K. Bugrara and P. Purdom. Clause order backtracking. Technical Report 311, 1990.
35. A. Bundy, editor. *Catalogue of Artificial Intelligence Tools*. Springer-Verlag, Berlin, 1984.
36. J.L. Burns and A.R. Newton. Efficient constraint generation for hierarchical compaction. In *Proc. Intl. Conf. on Computer Design*, pages 197–200. IEEE Computer Society, Oct. 1987.
37. M. Burstein and R. Pelavin. Hierarchical channel router. In *Proc. of the 20th ACM/IEEE DAC*, pages 591–597, Jun. 1983.
38. S. Chakradhar, V. Agrawal, and M. Bushnell. Neural net and boolean satisfiability model of logic circuits. *IEEE Design & Test of Computers*, pages 54–57, Oct. 1990.
39. I. Chakravarty. A generalized line and junction labelling scheme with applications to scene analysis. *IEEE Trans. on Pattern Analysis and Machine Intelligence*, PAMI-1(2):202–205, Apr. 1979.
40. P.K. Chan, M.D.F. Schlag, C.D. Thomborson, and V.G. Oklobdzija. Delay optimization of carry-skip adders and block carry-lookahead adders using multidimensional dynamic programming. *IEEE Trans. on Computers*, 41(8):920–930, Aug. 1992.
41. A.K. Chandra and P.M. Merlin. Optimal implementation of conjunctive queries in relational databases. In *Proceedings of the 9th ACM Symposium on Theory of Computing*, pages 77–90, 1977.
42. M.T. Chao and J. Franco. Probabilistic analysis of two heuristics for the 3-satisfiability problem. *SIAM J. on Computing*, 15:1106–1118, 1986.
43. M.T. Chao and J. Franco. Probabilistic analysis of a generalization of the unit-clause literal selection heuristics for the k satisfiable problem. *Information Science*, 51:289–314, 1990.
44. H.R. Charney and D.L. Plato. Efficient partitioning of components. In *Proc. of the 5th Annual Design Automation Workshop*, pages 16–21, 1968.
45. R. T. Chin and C. R. Dyer. Model-based recognition in robot vision. *ACM Computing Surveys*, 18(1):67–108, Mar. 1986.
46. Y. Chu, editor. *Special Section on Chinese/Kanji Text and Data Processing. IEEE Computer*, volume 18, No. 1. Jan. 1985.
47. V. Chvátal and B. Reed. Mick gets some (the odds are on his side). In *Proceedings on the Foundations of Computer Science*, 1992.
48. J. Cleary. Logical arithmetic. *Future Computing Systems*, 2(2), 1987.
49. M.B. Clowes. On seeing things. *Artificial Intelligence*, 2:79–116, 1971.
50. J. Cohen, editor. *Special Section on Logic Programming. Comm. of the ACM*, volume 35. 1992.
51. J.P. Cohoon and W.D. Paris. Genetic placement. In *Digest of the Intl. Conf. on Computer-Aided Design*, pages 422–425, 1986.
52. A. Colmerauer. Opening the Prolog III universe. *BYTE Magazine*, pages 177–182, Aug. 1987.
53. S.A. Cook. The complexity of theorem-proving procedures. In *Proceedings of the Third ACM Symposium on Theory of Computing*, pages 151–158, 1971.
54. T.H. Cormen, C.E. Leiserson, and R.L. Rivest. *Introduction to Algorithms*. MIT Press, Cambridge, 1990.
55. D.G. Corneil and D.G. Kirkpatrick. A theoretical analysis of various heuristics for the graph isomorphism problem. *SIAM J. on Computing*, 9(2):281–297, 1980.
56. V. Černy. A thermodynamical approach to the travelling salesman problem: An efficient simulation algorithm. Technical report, Institute of Physics and Biophysics, Comenius Uni-

versity, Bratislava, 1982.

57. Z. Cvetanovic. The effects of problem partitioning, allocation, and granularity on the performance of multiple-processor systems. *IEEE Trans. on Computers*, C-36(4):421–432, Apr 1987.
58. W. Dai and E.S. Kuh. Hierarchical floorplanning for building block layout. In *Digest of Intl. Conf. on Computer-Aided Design*, pages 454–457, 1986.
59. S. R. Das, D. K. Banerji, and A. Chattopadhyay. On Control Memory Minimization in Microprogrammed Digital Computers. *IEEE Trans. on Computers*, C-22:845–848, Sept. 1973.
60. L. S. Davis. Shape matching using relaxation techniques. *IEEE Trans. on Pattern Analysis and Machine Intelligence*, PAMI-1(1):60–72, Jan. 1979.
61. L. S. Davis and T. C. Henderson. Hierarchical constraint processes for shape analysis. *IEEE Trans. on Pattern Analysis and Machine Intelligence*, PAMI-3(3):265–277, May 1981.
62. M. Davis, G. Logemann, and D. Loveland. A machine program for theorem proving. *Communications of the ACM*, 5:394–397, 1962.
63. M. Davis and H. Putnam. A computing procedure for quantification theory. *J. of ACM*, pages 201–215, 1960.
64. J. de Kleer. Exploiting locality in a TMS. In *Proceedings of AAAI'90*, pages 264–271, 1990.
65. R. Dechter. A constraint-network approach to truth maintenance. Technical Report R-870009, Computer Science Dept., UCLA, Los Angeles, 1987.
66. R. Dechter and A. Dechter. Belief maintenance in dynamic constraint networks. In *Proceedings of AAAI'88*, 1988.
67. J.S. Denker, editor. *Neural Networks for Computing*, volume 151 of *AIP (Snowbird, Utah) Conference Proceedings*. American Institute of Physics, New York, 1986.
68. N. Dershowitz, J. Hsiang, N. Josephson, and D. Plaisted. Associative-commutative rewriting. In *Proceedings of IJCAI*, pages 940–944, 1983.
69. D.N. Deutsch. A 'dogleg' channel router. In *Proc. of the 13th ACM/IEEE DAC*, pages 425–433, Jun. 1976.
70. D.N. Deutsch. Compacted channel routing. In *Digest Intl. Conf. on Computer-Aided Design*, pages 223–225, Nov. 1985.
71. J. P. A. Deutsch. A short cut for certain combinational problems. In *British Joint Comput. Conference*, 1966.
72. S. Devadas. Optimal layout via boolean satisfiability. In *Proceedings of ICCAD'89*, pages 294–297, Nov. 1989.
73. S. Devadas, K. Keutzer, and J. White. Estimation of power dissipation in cmos combinational circuits using boolean function manipulation. *IEEE Transactions on CAD*, 11(3):373–383, Mar. 1992.
74. S. Devadas, H.T. Ma, A.R. Newton, and A.L. Sangiovanni-Vincentelli. A synthesis and optimization procedure for fully and easily testable sequential machines. *IEEE Trans. on Computer-Aided Design*, 8(10):1100–1107, Oct. 1989.
75. S.K. Dhall and C.L. Liu. On a real-time scheduling problem. *Operations Research*, 26(1):127–140, Feb. 1978.
76. V. Dhar and A. Crocker. A problem-solver/TMS architecture for general constraint satisfaction problems. Technical report, Dept. of Information Systems, New York University, 1989.
77. M. Dincbas, H. Simonis, and P.V. Hentenryck. Solving a cutting-stock problem in constraint logic programming. In *Proceedings of the 5th International Conference on Logic Programming*, 1988.
78. J. Doenhardt and T. Lengauer. Algorithm aspects of one-dimensional layout. *IEEE Trans. on Computer-Aided Design*, CAD-6(5):863–878, 1987.
79. W.E. Donath. Placement and average interconnection lengths of computer logic. *IEEE Trans. on Circuits and Systems*, CAS-26(4):272–277, Apr. 1979.
80. W.E. Donath. Wire length distribution for placements of computer logic. *IBM J. of Research and Development*, 25(3):152–155, May 1981.
81. W.E. Donath and A.J. Hoffman. Algorithms for partitioning of graphs and computer logic based on eigenvectors of connection matrices. *IBM Technical Disclosure Bulletin 15*, pages 938–944, 1972.
82. O. Dubois. Counting the number of solutions for instances of satisfiability. *Theoretical*

Computer Science, 81:49–64, 1991.

83. O. Dubois and J. Carlier. Probabilistic approach to the satisfiability problem. *Theoretical Computer Science*, 81:65–75, 1991.
84. C. Eastman. Preliminary report on a system for general space planning. *Comm. of ACM*, 15(2):76–87, Feb. 1972.
85. K.P. Eswaran, J.N. Gray, R.A. Lorie, and I.L. Traiger. The notion of consistency and predicate look in a database system. *Comm. of ACM*, 19(11), Nov. 1976.
86. S. Even, A. Itai, and A. Shamir. On the complexity of timetable and multi-commodity flow problems. *SIAM J. on Computing*, 5(4):691–703, 1976.
87. O. Faugeras and K. Price. Semantic description of aerial images using stochastic labeling. *IEEE Trans. on Pattern Analysis and Machine Intelligence*, PAMI-3(6):633–642, Nov. 1981.
88. O. D. Faugeras, editor. *Fundamentals in Computer Vision*. Cambridge University Press, London, 1983.
89. J. A. Feldman and Y. Yakimovsky. Decision theory and artificial intelligence: I. a semantics-based region analyser. *Artificial Intelligence*, 5:349–371, 1974.
90. J. D. Findler. *Associative Networks: Representation and Use of Knowledge by Computers*. Academic Press, New York, 1979.
91. M.S. Fox. *Constraint-Directed Search: A Case Study of Job-Shop Scheduling*. Pitman, London, 1987.
92. J. Franco Probabilistic analysis of the pure literal heuristic for the satisfiability problem. *Annals of Operations Research*, 1:273–289, 1984.
93. J. Franco. On the probabilistic performance of algorithms for the satisfiability problem. *Information Processing Letters*, 23:103–106, 1986.
94. J. Franco and Y.C. Ho. Probabilistic performance of heuristic for the satisfiability problem. *Discrete Applied Mathematics*, 22:35–51, 1988/89.
95. J. Franco and M. Paull. Probabilistic analysis of the davis-putnam procedure for solving the satisfiability problem. *Discrete Applied Mathematics*, 5:77–87, 1983.
96. R.T. Frankot and R. Chellappa. A method for enforcing integrability in shape from shading algorithms. *IEEE Trans. on Pattern Analysis and Machine Intelligence*, PAMI-10(4):439–451, Jul. 1988.
97. F. Frayman and S. Mittal. *Cossack: A Constraints-based Expert System for Configuration Tasks*. Computational Mechanics Publications, Nadel, 1987.
98. T. W. Fuqua. Constraint kernels: Constraints and dependencies in a geometric modeling system. Master's thesis, Dept. of Computer Science, Univ. of Utah, Aug. 1987.
99. Z. Galil. On the complexity of regular resolution and the Davis-Putnam procedure. *Theoretical Computer Science*, pages 23–46, 1977.
100. H. Garcia-Molina. Using semantic knowledge for transaction processing in a distributed database. *ACM Trans. on Database Systems*, 8(2), Jun. 1983.
101. M.R. Garey and D.S. Johnson. Complexity results for multiprocessor scheduling under resource constraints. *SIAM J. Comput.*, 4:397–411, 1975.
102. M.R. Garey and D.S. Johnson. Two-processors scheduling with start-times and deadlines. *SIAM J. on Computing*, 6:416–426, 1977.
103. M.R. Garey and D.S. Johnson. *Computers and Intractability: A Guide to the Theory of NP-Completeness*. W.H. Freeman and Company, San Francisco, 1979.
104. J. Gaschnig. A constraint satisfaction method for inference making. In *Proceedings of 12th Annual Allerton Conf. Circuit System Theory*, 1974.
105. J. Gaschnig. *Performance Measurements and Analysis of Certain Search Algorithms*. PhD thesis, Carnegie-Mellon University, Dept. of Computer Science, May 1979.
106. A.V. Gelder. A satisfiability tester for non-clausal propositional calculus. *Information and Computation*, 79(1):1–21, Oct. 1988.
107. S. Geman and D. Geman. Stochastic relaxation, gibbs distributions, and the bayesian restoration of images. *IEEE Trans. on Pattern Analysis and Machine Intelligence*, PAMI-6(6):721–741, Nov. 1984.
108. M.R. Genesereth and N. J. Nilsson. *Logical Foundations of Artificial Intelligence*. Morgan Kaufmann Publishers, Los Altos, California, 1987.
109. P.C. Gilmore. A proof method for quantification theory. *IBM J. Res. Develop.*, 4:28–35, 1960.
110. A. Ginzberg. *Algebraic Theory of Automata*. Academic Press, New York, 1968.

111. A. Goldberg. Average case complexity of the satisfiability problem. In *Proc. Fourth Workshop on Automated Deduction*, pages 1–6, 1979.
112. A. Goldberg. On the complexity of the satisfiability problem. Technical Report Courant Computer Science No. 16, New York University, 1979.
113. A. Goldberg, P.W. Purdom, and C.A. Brown. Average time analysis of simplified davis-putnam procedures. *Information Processing Letters*, 15(2):72–75, 6 Sept. 1982.
114. D.E. Goldberg. *Genetic Algorithms in Search, Optimization, and Machine Learning*. Addison-Wesley, Reading, 1989.
115. A. Grasselli and F. Luccio. A Method for Minimizing the Number of Internal States in Incompletely Specified Sequential Networks. *IEEE Trans. on Computers*, EC-14:350–359, June 1965.
116. A. Grasselli and U. Montanari. On the Minimization of READ-ONLY Memories in Microprogrammed Digital Computers. *IEEE Trans. on Computers*, C-19:1111–1114, Nov. 1970.
117. J. Gu, W. Wang, and T. C. Henderson. A parallel architecture for discrete relaxation algorithm. *IEEE Trans. on Pattern Analysis and Machine Intelligence*, PAMI-9(6):816–831, Nov. 1987.
118. J. Gu. Parallel algorithms and architectures for very fast search. Technical Report UUCS-TR-88-005, Jul. 1988.
119. J. Gu. How to solve Very Large-Scale Satisfiability (VLSS) problems. Technical Report. 1988 (Present in part in, J. Gu, Benchmarking SAT Algorithms, Technical Report UCECE-TR-90-002, 1990).
120. J. Gu. Local search with backtracking: A complete algorithm for SAT problem. 1989.
121. J. Gu. Local search for satisfiability (SAT) problem. Submitted for publication, 1989.
122. J. Gu. Global optimization for satisfiability (SAT) problem. Submitted for publication, 1989.
123. J. Gu and W. Wang. *VLSI Architectures for Discrete Relaxation Algorithm. In* Discrete Relaxation Techniques, chapter 9, pages 111–136. Oxford University Press, New York, 1990.
124. J. Gu. Optimization algorithms with backtracking search. 1990.
125. J. Gu. Local search, global optimization, and resolution: A legal marriage of three. Submitted for publication, 1990.
126. J. Gu. Efficient local search for very large-scale satisfiability problem. *SIGART Bulletin*, 3(1):8–12, Jan. 1992, ACM Press.
127. J. Gu. *On Optimizing a Search Problem. In* Advanced Series on Artificial Intelligence, Vol. 1, chapter 2, pages 63–105. World Scientific, New Jersey, Jan. 1992.
128. J. Gu. *Design Efficient Local Search Algorithms. In* F. Belli and F.J. Radermacher, editors, *Lecture Notes in Computer Science, Vol. 604: IEA/AIE*, pages 651–654. Springer-Verlag, Berlin, Jun. 1992.
129. J. Gu. The *UniSAT* problem models (appendix). *IEEE Trans. on Pattern Analysis and Machine Intelligence*, 14(8):865, Aug. 1992.
130. J. Gu. Local search for satisfiability (SAT) problem. *IEEE Trans. on Systems, Man, and Cybernetics*, 23(4):1108-1129, Jul./Aug. 1993.
131. J. Gu. Global optimization for satisfiability (SAT) problem. *IEEE Trans. on Knowledge and Data Engineering*, 6(2), Apr. 1994, in press.
132. J. Gu and W. Wang. A novel discrete relaxation architecture. *IEEE Trans. on Pattern Analysis and Machine Intelligence*, 14(8):857–865, Aug. 1992.
133. J. Gu. *Multispace Search: A New Optimization Approach. In* New Advances in Optimization and Approximation. *Ding-Zhu Du (ed)*. Kluwer Academic Publishers, Boston, MA, Jan. 1994.
134. J. Gu. *Constraint-Based Search*. Cambridge University Press, New York, 1994.
135. J. Gu and Q.P. Gu. Average time complexities of several local search algorithms for the satisfiability problem. Technical Report UCECE-TR-91-004, submitted for publication. 1991.
136. J. Gu and Q.P. Gu. Average time complexity of SAT14.5 algorithm. Submitted for publication, 1992.
137. J. Gu, Q.P. Gu, and D.-Z. Du. Convergence properties of some optimization algorithms for the satisfiability problem. Submitted for publication, 1992.
138. J. Gu and X. Huang. Implementation and performance of the SAT14.7 algorithm. Submitted for publication. Feb. 1991.
139. J. Gu and X. Huang. Efficient local search with search space smoothing. *IEEE Trans. on Systems, Man, and Cybernetics*, 24(4), Apr. 1994, in press.

140. A. Guzman. *Computer Recognition of Three-Dimensional Objects in a Visual Scene.* PhD thesis, MIT, 1968.
141. S. Ha and E.A. Lee. Compile-time scheduling and assignment of data-flow program graphs with data-dependent iteration. *IEEE Trans. on Computers*, 40(11):1225–1238, Nov. 1991.
142. G.T. Hamachi and J.K. Ousterhout. A switchbox router with obstacle avoidance. In *Proc. of the 21th ACM/IEEE DAC*, pages 173–179, Jun. 1984.
143. M. Hanan, P.K. Wolff, and B.J. Agule. Some experimental results on placement techniques. *J. of Design Automation and Fault-Tolerant Computing*, 2:145–168, May 1978.
144. P. Hansen and B. Jaumard. Algorithms for the maximum satisfiability problem. *Computing*, 44:279–303, 1990.
145. R. M. Haralick and L. G. Shapiro. The consistent labeling problem: Part 1. *IEEE Trans. on Pattern Analysis and Machine Intelligence*, PAMI-1(2):173–184, Apr. 1979.
146. F. Harary. *Graph Theory.* Addison-Wesley, Reading, 1969.
147. W. S. Havens. A theory of schema labelling. *Computational Intelligence*, 1(3 & 4):127–139, 1985.
148. T. Hedges, W. Dawson, and Y.E. Cho. Bitmap graph build algorithm for compaction. In *Digest Intl. Conf. on Computer-Aided Design*, pages 340–342, Sep. 1985.
149. T. C. Henderson and L. S. Davis. Hierarchical models and analysis of shape. *Pattern Recognition*, 14(1-6):197–204, 1981.
150. T. C. Henderson and A. Samal. Multiconstraint shape analysis. *Image and Vision Computing*, 4(2):84–96, May 1986.
151. G.E. Hinton and T.J. Sejnowski. *Learning and Relearning in Boltzmann Machine. In* D. E. Rumelhart and J. L. McClelland, editors, *Parallel Distributed Processing: Explorations in the Microstructure of Cognition. Vol. 1: Foundations*, volume 1, chapter 7, pages 282–317. The MIT Press, Cambridge, 1986.
152. S.J. Hong and S. Muroga. Absolute minimization of completely specified switching functions. *IEEE Trans. on Computers*, 40(1):53–65, Jan. 1991.
153. P. Hood and Grover. Designing real time systems in Ada. Technical Report 1123-1, SofTech, Inc., Waltham, Jan. 1986.
154. J.N. Hooker. Generalized resolution and cutting planes. *Annals of Operations Research*, 12:217–239, 1988.
155. J.N. Hooker. A quantitative approach to logical inference. *Decision Support Systems*, 4:45–69, 1988.
156. J.N. Hooker. Resolution vs. cutting plane solution of inference problems: Some computational experience. *Operations Research Letter*, 7(1):1–7, 1988.
157. J.N. Hooker and C. Fedjki. Branch-and-cut solution of inference problems in propositional logic. Technical Report 77-88-89, GSIA, Carnegie Mellon University, Aug. 1989.
158. A.L. Hopkins and et. al. FTMP - a highly reliable fault-tolerant multiprocessor for aircraft. In *Proceedings of the IEEE*, pages 1221–1239, Oct. 1978.
159. B.K.P. Horn. *Obtaining Shape from Shading Information, in* The Psychology of Computer Vision, *P.H. Winston, editor*, pages 115–155. McGraw-Hill, New York, 1975.
160. B.K.P. Horn. Understanding image intensity. *Artificial Intelligence*, 8:301–231, 1977.
161. B.K.P. Horn and M. J. Brooks, editors. *Shape from Shading.* The MIT Press, Cambridge, 1989.
162. B.K.P. Horn and M.J. Brooks. The variational approach to shape from shading. *Computer Vision, Graphics & Image Processing*, 33(2):174–208, 1986.
163. J. Hsiang. Refutational theorem proving using term-rewriting systems. *Artificial Intelligence*, pages 255–300, 1985.
164. T.H. Hu, C.Y. Tang, and R.C.T. Lee. An average case analysis of a resolution principle algorithm in mechanical theorem proving. *Annals of Mathematics and Artificial Intelligence*, 6:235–252, 1992.
165. X. Huang and J. Gu. A quantitative solution for constraint satisfaction. Submitted for publication. Mar. 1991.
166. X. Huang, J. Gu, and Y. Wu. A constrained approach to multifont character recognition. *IEEE Transactions on Pattern Analysis And Machine Intelligence*, 15(8):838–843, Aug. 1993.
167. D.A. Huffman. *Impossible Objects as Nonsense Sentences. In* B. Meltzer and D. Michie, Eds., *Machine Intelligence*, pages 295–323. Edinburgh University Press, Edinburgh, Scotland,

1971.

168. R. A. Hummel and S. W. Zucker. On the foundations of relaxation labeling processes. *IEEE Trans. on Pattern Analysis and Machine Intelligence*, PAMI-5(3):267–287, May 1983.

169. K. Ikeuchi. Model-based interpretation of range imagery. In *Proc. of the DRAPA Image Understanding Workshop*, pages 321–339. DRAPA, DoD, USA, DARPA, DoD, Feb. 23-25 1987.

170. K. Ikeuchi and B.K.P. Horn. Numerical shape from shading and occluding boundaries. *Artificial Intelligence*, 17(1-3):141–184, 1981.

171. B. Indurkhya, H. S. Stone, and L. Xi-Cheng. Optimal partitioning of randomly generated distributed programs. *IEEE Trans. on Software Engineering*, SE-12:483–495, Mar 1986.

172. T. Ishida. Parallel rule firing in production systems. *IEEE Trans. on Knowledge and Data Engineering*, 3(1):11–17, Mar. 1991.

173. N. Itazaki and K. Kinoshita. Test pattern generation for circuits with tri-state modules by z-algorithm. *IEEE Trans. on Computer-Aided Design*, 8(12):1327–1334, Dec. 1989.

174. K. Iwama. CNF satisfiability test by counting and polynomial average time. *SIAM J. on Computing*, pages 385–391, 1989.

175. F. Jahanian and A.K. Mok. Safety analysis of timing properties in real-time systems. *IEEE Trans. on Software Engineering*, SE-12(9):890–904, Sept. 1986.

176. T. Jayasri and D. Basu. An Approach to Organizing Microinstructions Which Minimizes the Width of Control Store Words. *IEEE Trans. on Computers*, C-25:514–521, May 1976.

177. R.E. Jeroslow and J. Wang. Solving propositional satisfiability problems. *Annals of Mathematics and AI*, 1:167–187, 1990.

178. R.G. Jeroslow. Computation-oriented reductions of predicate to propositional logic. *Decision Support Systems*, 4:183–197, 1988.

179. D.S. Johnson. More approaches to the traveling salesman guide. *Nature*, 330:525, 1987.

180. D.S. Johnson. *Local Optimization and the Traveling Salesman Problem. In* M.S. Paterson, editor, *Lecture Notes in Computer Science, Vol. 443: Automata, Languages and Programming*, pages 446–461. Springer-Verlag, Berlin, 1990.

181. J.L. Johnson. A neural network approach to the 3-satisfiability problem. *J. of Parallel and Distributed Computing*, 6:435–449, 1989.

182. W. Lewis Johnson. Letter from the editor. *SIGART Bulletin*, 2(2):1, April 1991, ACM Press.

183. J. R. Josephson, B. Chandrasekaran, J. W. Smith Jr., and M. C. Tanner. A mechanism for forming composite explanatory hypotheses. *IEEE Trans. on Systems, Man, and Cybernetics*, SMC-17(3):445–454, May/June 1987.

184. P.C. Jackson Jr. Heuristic search algorithms for the satisfiability problem. Submitted to the third IEEE TAI Conference, Jul. 1991.

185. Y.C. Ju, B. Rao, and R.A. Saleh. Consistency checking and optimization of macromodels. *IEEE Trans. on Computer-Aided Design*, 10(8):957–967, Aug. 1991.

186. A.P. Kamath, N.K. Karmarkar, K.G. Ramakrishnan, and M.G.C. Resende. Computational experience with an interior point algorithm on the satisfiability problem. *Working paper. Mathematical Sciences Research Center, AT&T Bell Laboratories*, Oct. 1989.

187. R.M. Karp. *The Probabilistic Analysis of Some Combinatorial Search Algorithms. In* Algorithms and Complexity: New Directions and Recent Results, pages 1–19. Academic Press, New York, 1976.

188. G. Kedem and H. Watanabe. Graph optimization techniques for IC layout and compaction. *IEEE Trans. on Computer-Aided Design*, CAD-3:12–20, 1984.

189. J. Kella. State Minimization of Incompletely Specified Sequential Machines. *IEEE Trans. on Computers*, C-19:342–348, April 1970.

190. S. Kirkpatrick, C.D. Gelat, and M.P. Vecchi. Optimization by simulated annealing. *Science*, 220:671–680, 1983.

191. K.L. Kodandapani and E.J. McGrath. A wirelist compare program for verifying VLSI layouts. *IEEE Design and Test of Computers*, 3(3):46–51, 1986.

192. E. Koutsoupias and C.H. Papadimitriou. On the greedy algorithm for satisfiability. *Information Processing Letters*, 43:53–55, 10 August 1992.

193. R. Kowalski. A proof procedure using connection graphs. *J. ACM*, 22(4):572–595, Oct. 1975.

194. M.R. Kramer and J. van Leeuwen. *The Complexity of Wire Routing and Finding Minimum Area Layouts for Arbitrary VLSI Circuits*, volume 2, chapter VLSI Theory, pages 129–146.

Jai Press Inc., Greenwich, CT, 1984.

195. C.M. Krishna, K.G. Shin, and I.S. Bhandari. Processor tradeoffs in distributed real-time systems. *IEEE Trans. on Computers*, C-36(9):1030–1040, Sept. 1987.

196. V. Kumar. Algorithms for constraint satisfaction problems: A survey. Technical Report TR-91-28, Dept. of Computer Science, Univ. of Minnesota, 1991.

197. V. Kumar. Algorithms for constraint satisfaction problems: A survey. *The AI Magazine*, 13(1):32–44, 1992.

198. E.D. Lagnese and D.E. Thomas. Architectural partitioning for system level synthesis of integrated circuits. *IEEE Trans. on Computer-Aided Design*, 10(7):847–860, July 1991.

199. G. Lakhani and R. Varadarajan. A wire-length minimization algorithm for circuit layout compaction. In *Proc. of ISCAS'87*, pages 276–279, May 1987.

200. B.S. Landman and R.L. Russo. On a pin versus block relationship for partitions of logic graphs. *IEEE Trans. on Computers*, C-20:1469–1479, Dec. 1971.

201. T. Larrabee. Test pattern generation using boolean satisfiability. *IEEE Trans. on Computer-Aided Design*, 11(1):4–15, Jan. 1992.

202. C. Lassez. Constraint logic programming. *BYTE Magazine*, pages 171–176, Aug. 1987.

203. E.L. Lawler, J.K. Lenstra, A.H.G. Rinnooy Kan, and D.B. Shmoys, editors. *The Traveling Salesman Problem*. John Wiley & Sons, New York, 1985.

204. C. Lee. An algorithm for path connections and its applications. *IEEE Trans. on Electronic Computers*, VEC-10:346–365, Sept. 1961.

205. E.A. Lee. Consistency in dataflow graphs. *IEEE Trans. on Parallel and Distributed Systems*, 2(2):223–235, Apr. 1991.

206. R.C.T. Lee. Private Communications, 1992-1993.

207. J.P. Lehoczky and L. Sha. Performance of real-time bus scheduling algorithms. *ACM Performance Evaluation Review*, 14(1), May 1986. Special Issue.

208. J. Li and M. Chen. Compiling communication-efficient programs for massively parallel machines. *IEEE Trans. on Parallel and Distributed Systems*, 2(3):361–376, July 1991.

209. Y.-Z. Liao and C.K. Wong. An algorithm to compact a VLSI symbolic layout with mixed constraints. *IEEE Trans. on Computer-Aided Design*, CAD-2(2):62–69, 1983.

210. F.C.H. Lin and R.M. Keller. The gradient model load balancing method. *IEEE Trans. on Software Engineering*, SE-13(1):32–38, Jan. 1987.

211. S. Lin. Computer solutions of the traveling salesman problem. *Bell Sys. Tech. Journal*, 44(10):2245–2269, Dec. 1965.

212. R.P. Lippmann. An introduction to computing with neural net. *IEEE ASSP Magazine*, 4(2):4–22, April 1987.

213. C.D. Locke. *Best-Effort Decision Making for Real-Time Scheduling*. PhD thesis, Carnegie-Mellon University, May 1985.

214. D.W. Loveland. *Automated Theorem Proving: A Logical Basis*. North-Holland, Amsterdam, 1978.

215. F. Luccio. Extending the Definition of Prime Compatibility Classes of States in Incompletely Specified Sequential Machine Reduction. *IEEE Trans. on Computers*, C-18:537–540, Jun. 1969.

216. D.G. Luenberger. *Linear and Nonlinear Programming*. Addison-Wesley, Reading, 1984.

217. N.A. Lynch. Multi-level atomicity - a new correctness criterion for database concurrency control. *ACM Trans. on Database Systems*, 8(4), Dec. 1983.

218. A. K. Mackworth. Consistency in networks of relations. *Artificial Intelligence*, 8:99–119, 1977.

219. J. Malik and D. Maydan. Recovering three-dimensional shape from a single image of curved objects. *IEEE Trans. on Pattern Analysis and Machine Intelligence*, PAMI-11(6):555–566, Jun. 1989.

220. D. Marple and A.E. Gamal. Area-delay optimization of programmable logic arrays. Technical report, Stanford University, Sept. 1986.

221. T.A. Marsland and M. Campbell. Parallel search of strongly ordered game trees. *Computing Surveys*, 14(4):533–551, Dec. 1982.

222. T.A. Marsland and J. Schaeffer. *Computers, Chess, and Cognition*. Springer-Verlag, New York, 1990.

223. D.W. Matula, W.G. Marble, and J.D.Isaacson. *Graph coloring algorithms. In* Graph Theory and Computing. *R.C. Read, editor*, pages 109–122. Academic Press, New York, 1972.

224. D. McAllester. Truth maintenance. In *Proceedings of AAAI'90*, pages 1109–1116, 1990.
225. J.J. McGregor. Relational consistency algorithms and their application in finding subgraph and graph isomorphisms. *Information Sciences*, 19:229–250, 1979.
226. C.R. McLean and C.R. Dyer. An analog relaxation processor. In *Proceedings of the 5th International Conference on Pattern Recognition*, pages 58–60, 1980.
227. K. Mehlhorn. *Data Structures and Algorithms: Graph Algorithms and NP-Completeness.* Springer-Verlag, Berlin, 1984.
228. S. Minton, M.D. Johnston, A.B. Philips, and P. Laird. Solving large-scale constraint satisfaction and scheduling problems using a heuristic repair method. In *Proceedings of AAAI'90*, pages 17–24, Aug. 1990.
229. D.P. Miranker. *TREAT: A New and Efficient Match Algorithm for AI Production Systems.* Pitman, London, 1990.
230. D.P. Miranker and B.J. Lofaso. The organization and performance of a treat-based production system compiler. *IEEE Trans. on Knowledge and Data Engineering*, 3(1):3–10, Mar. 1991.
231. D. Mitchell, B. Selman, and H. Levesque. Hard and easy distributions of SAT problems. In *Proceedings of AAAI'92*, pages 459–465, Jul. 1992.
232. R. Mohr and T. C. Henderson. Arc and path consistency revisited. *Artificial Intelligence*, 28:225–233, 1986.
233. S. Mori, C.Y. Suen, and K. Yamamoto. Historical review of ocr research and development. *Proceedings of the IEEE*, 80(7):1029–1058, Jul. 1992.
234. B.A. Nadel and J. Lin. Automobile transmission design as a constraint satisfaction problem: A collaborative research project with ford motor co. Technical report, Wayne State University, 1990.
235. D. Navinchandra. *Exploration and Innovation in Design.* Springer Verlag, Sadeh, 1990.
236. D. Navinchandra and D.H. Marks. Layout planning as a consistent labeling optimization problem. In *Proceedings of 4th International Symposium on Robotics and AI in Construction*, 1987.
237. A. Newell and H. A. Simon. *Human Problem Solving.* Prentice-Hall, Englewood Cliffs, New Jersey, 1972.
238. L.M. Ni, C. Xu, and T.B. Gendreau. A distributed drafting algorithm for load balancing. *IEEE Trans. on Software Engineering*, SE-11(10):1153–1161, Oct. 1985.
239. A. Nijenhuis and H. S. Wilf. *Combinatorial Algorithms.* Academic Press, New York, 1975.
240. N.J. Nilsson. *Principles of Artificial Intelligence.* Tioga Publishing Company, Palo Alto, California, 1980.
241. M. Nishizawa. Partitioning of logic units. *Fujitsu Scientific and Technical Journal*, 7(2):1–13, Jun. 1971.
242. D. Pager. On the efficient algorithm for graph isomorphism. *J. ACM*, 17(4):708–714, Jan. 1970.
243. C.H. Papadimitriou. On selecting a satisfying truth assignment. In *Proceedings of the 32nd Annual Symposium of the Foundations of Computer Science*, pages 163–169, 1991.
244. C.H. Papadimitriou. Private Communications, Mar. 1992.
245. C.H. Papadimitriou and K. Steiglitz. On the complexity of local search for the traveling salesman problem. *SIAM J. on Computing*, 6(1):76–83, 1977.
246. C.H. Papadimitriou and K. Steiglitz. *Combinatorial Optimization: Algorithms and Complexity.* Prentice Hall, Englewood Cliffs, 1982.
247. A.M. Patel, N.L. Soong, and R.K. Korn. Hierarchical VLSI routing - - an approximate routing procedure. *IEEE Trans. on Computer-Aided Design*, CAD-4(2):121–126, Apr. 1985.
248. W. Patterson. *Mathematical Cryptology.* Rowman & Littlefield, New Jersey, 1987.
249. P.G. Paulin and J.P. Knight. Force-directed scheduling for the behavioral synthesis of asic's. *IEEE Trans. on Computer-Aided Design*, 8(6):661–679, June 1989.
250. D. Pehoushek. SAT problem instances and algorithms. Private Communications, Stanford University, 1992.
251. M.A. Penna. A shape form shading analysis for a single perspective image of a polyhedron. *IEEE Trans. on Pattern Analysis and Machine Intelligence*, PAMI-11(6):545–554, Jun. 1989.
252. D.P. La Potin and S.W. Director. Mason: a global floorplanning approach for VLSI design. *IEEE Trans. on Computer-Aided Design*, CAD-5(4):477–489, Oct. 1986.
253. D. K. Pradhan, editor. *Fault-Tolerant Computing: Theory and Practice.* Prentice-Hall,

Englewood Cliffs, 1986.
254. D. Prawitz. An improved proof procedure. *Theoria*, 26(2):102–139, 1960.
255. P. Purdom. A survey of average time analyses of satisfiability algorithms. *J. of Information Processing*, 13(4):449–455, 1990.
256. P.W. Purdom. Tree size by partial backtracking. *SIAM J. Comput.*, 7(4):481–491, Nov. 1978.
257. P.W. Purdom. Search rearrangement backtracking and polynomial average time. *Artificial Intelligence*, 21:117–133, 1983.
258. P.W. Purdom. Private Communications, 1992-1993.
259. P.W. Purdom and C.A. Brown. An analysis of backtracking with search rearrangement. *SIAM J. Comput.*, 12(4):717–733, Nov. 1983.
260. P.W. Purdom and C.A. Brown. The pure literal rule and polynomial average time. *SIAM J. Comput.*, 14:943–953, 1985.
261. P.W. Purdom and C.A. Brown. Polynomial-average-time satisfiability problems. *Information Science*, 41:23–42, 1987.
262. P.W. Purdom, C.A. Brown, and E.L. Robertson. Backtracking with multi-level dynamic search rearrangement. *Acta Informatica*, 15:99–113, 1981.
263. R. Puri and J. Gu. An efficient algorithm to search for minimal closed covers in sequential machines. *IEEE Transactions on CAD*, 12(6):737–745, Jun. 1993.
264. R. Puri and J. Gu. An efficient algorithm for microword length minimization. *IEEE Transactions on CAD*, 12(10):1449–1457, Oct. 1993.
265. R. Puri and J. Gu. A modular partitioning approach for asynchronous circuit synthesis. *IEEE Transactions on CAD*, to appear.
266. C. D. V. P. Rao and N. N. Biswas. On the Minimization of Wordwidth in the Control Memory of a Microprogrammed Digital Computer. *IEEE Trans. on Computers*, C-32(9):863–868, Sept. 1983.
267. C. V. S. Rao and N. N. Biswas. Minimization of Incompletely Specified Sequential Machines. *IEEE Trans. on Computers*, C-24:1089–1100, Nov. 1975.
268. R.C. Read and D.G. Corneil. The graph isomorphism disease. *J. of Graph Theory*, 1:339–363, 1977.
269. G.M. Reed and A.W. Roscoe. A timed model for communicating sequential processes. In *Proc. of ICALP'86*, pages 314–323. Springer LNCS 226, 1986.
270. J. Reed, A. Sangiovanni-Vincentelli, and M. Santomauro. A new symbolic channel router: YACR2. *IEEE Trans. on Computer-Aided Design*, CAD-4(3):208–219, July 1985.
271. R.L. Rivest. *Cryptography. In* Handbook of Theoretical Computer Science. *J.V. Leeuwen, editor*, chapter 13, pages 719–756. The MIT Press, Cambridge, 1990.
272. R.L. Rivest and C.M. Fiduccia. A 'greedy' channel router. In *Proc. of the 19th ACM/IEEE DAC*, pages 418–424, Jun. 1982.
273. J.A. Robinson. A machine-oriented logic based on the resolution principle. *Journal of the ACM*, pages 23–41, 1965.
274. A. Rosenfeld. Computer vision: Basic principles. *Proceedings of the IEEE*, 76(8):863–868, Aug. 1988.
275. A. Rosenfeld, R. A. Hummel, and S. W. Zucker. Scene labeling by relaxation operations. *IEEE Trans. on Systems, Man, and Cybernetics*, SMC-6(6):420–433, June 1976.
276. A.E. Ruehli, P.K. Wolff, and G. Goertzel. Analytical power/timing optimization technique for digital system. In *Proc. of the 14th ACM/IEEE DAC*, pages 142–146, Jun. 1977.
277. R.L. Russo. On the tradeoff between logic performance and circuit-to-pin ratio for LSI. *IEEE Trans. on Computers*, C-21:147–153, 1972.
278. R.L. Russo, P.H. Oden, and P.K. Wolff. A heuristic procedure for the partitioning and mapping of computer logic graphs. *IEEE Trans. on Computers*, C-20:1455–1462, Dec. 1971.
279. Y.G. Saab and V. B. Rao. Combinatorial optimization by stochastic evolution. *IEEE Transactions on CAD*, CAD-10(4):525–535, Apr. 1991.
280. A. Saldanha, T. Villa, R.K. Brayton, and A.L. Sangiovanni-Vincentelli. A Framework for Satisfying Input and Output Encoding Constraints. In *Proceedings of ACM/IEEE Design Automation Conference*, pages 170–175, 1991.
281. T.J. Schaefer. The complexity of satisfiability problems. In *Proceedings of the 10th ACM Symposium on Theory of Computing*, pages 216–226, 1978.
282. W.L. Schiele. Improved compaction by minimized length of wires. In *Proc. of the 20th*

ACM/IEEE DAC, pages 121–127, 1983.

283. B.M. Schwartzschild. Statistical mechanics algorithm for monte carlo optimization. *Physics Today*, 35:17–19, 1982.

284. P. Schwarz and A. Spector. Synchronizing shared abstract types. *ACM Trans. on Computer Systems*, Aug. 1984.

285. C. Sechen and A.L. Sangiovanni-Vincentelli. The timberwolf placement and routing package. In *Proc. of the 1984 Custom Integrated Circuit Conf.*, May 1984.

286. B. Selman. Private Communications, Aug. 1992.

287. B. Selman, H. Levesque, and D. Mitchell. A new method for solving hard satisfiability problems. In *Proceedings of AAAI'92*, pages 440–446, Jul. 1992.

288. M.J. Shensa. A computational structure for the propositional calculus. In *Proceedings of IJCAI*, pages 384–388, 1989.

289. R.C.-J. Shi. An improvement on Karmarkar's algorithm for integer programming. Jun. 1992.

290. H. Shin, A.L. Sangiovanni-Vincentelli, and C.H. Sequin. Two-dimensional compaction by 'zero refining'. In *Proc. of the 23th ACM/IEEE DAC*, pages 115–122, Jun. 1986.

291. K.G. Shin and M.-S. Chen. On the number of acceptable task assignments in distributed computing systems. *IEEE Trans. on Computers*, 39(1):99–110, Jan. 1990.

292. P. Siegel. *Representation et Utilization de la Connaissances en Calcul Propositionel.* PhD thesis, University Aix-Marseille II, 1987.

293. J.C. Simon. Off-line cursive word recognition. *Proceedings of the IEEE*, 80(7):1150–1161, Jul. 1992.

294. R. Sosič and J. Gu. Quick n-queen search on VAX and Bobcat machines. CS 547 AI Class Project, Winter Quarter Feb. 1988.

295. R. Sosič and J. Gu. How to search for million queens. Technical Report UUCS-TR-88-008, Dept. of Computer Science, Univ. of Utah, Feb. 1988.

296. R. Sosič and J. Gu. A polynomial time algorithm for the n-queens problem. *SIGART Bulletin*, 1(3):7–11, Oct. 1990, ACM Press.

297. R. Sosič and J. Gu. Fast search algorithms for the n-queens problem. *IEEE Trans. on Systems, Man, and Cybernetics*, SMC-21(6):1572–1576, Nov./Dec. 1991.

298. R. Sosič and J. Gu. 3,000,000 queens in less than one minute. *SIGART Bulletin*, 2(2):22–24, Apr. 1991, ACM Press.

299. R. Sosič and J. Gu. Efficient local search with conflict minimization. *IEEE Trans. on Knowledge and Data Engineering*, Vol. 6, 1994.

300. R. Sosič, J. Gu, and R. Johnson. The Unison algorithm: Fast evaluation of Boolean expressions. *Communications of ACM*, accepted for publication in 1992.

301. R. Sosič, J. Gu, and R. Johnson. A universal Boolean evaluator. to appear.

302. J. Soukup. Circuit layout. In *Proceedings of the IEEE*, pages 1281–1304, Oct. 1981.

303. J. Stankovic, K. Ramamritham, and S. Cheng. Evaluation of a bidding algorithm for hard real-time distributed systems. *IEEE Trans. on Computers*, C-34(12):1130–1143, Dec. 1985.

304. J.A. Stankovic. Real-time computing systems: The next generation. Manuscript. Feb. 18, 1988.

305. H.S. Stone and J.M. Stone. Efficient search techniques – an empirical study of the n-queens problem. *IBM J. Res. Develop.*, 31(4):464–474, July 1987.

306. G. Strang. *Linear Algebra and Its Applications.* Academic Press, New York, 1976.

307. C.Y. Suen, M. Berthod, and S. Mori. Automatic recognition of handprinted characters – the state of the art. *Proceedings of the IEEE*, 68(4):469–487, Apr. 1980.

308. S. Sutanthavibul, E. Shragowitz, and J.B. Rosen. An analytical approach to floorplan design and optimization. *IEEE Trans. on Computer-Aided Design*, 10(6):761–769, June 1991.

309. M. Takashima, T. Mitsuhashi, T.Chiba, and K. Yoshida. Programs for verifying circuit connectivity of MOS/VLSI artwork. In *Proc. of the 19th ACM/IEEE DAC*, pages 544–550, 1982.

310. H. Tokuda, J. Wendorf, and H. Wang. Implementation of a time driven scheduler for real-time operating systems. In *Proc. of the Real-Time Systems Symposium*, Dec. 1987.

311. P. P. Trabado, A. Lloris-Ruiz, and J. Ortega-Lopera. Solution of Switching Equations Based on a Tabular Algebra. *IEEE Trans. on Computers*, C-42:591–596, May 1993.

312. G.S. Tseitin. *On the Complexity of Derivations in Propositional Calculus. In* Structures in Constructive Mathematics and Mathematical Logic, Part II, *A.O. Slisenko, ed.*, pages 115–125. 1968.

313. K. J. Turner. *Computer Perception of Curved Objects Using a Television Camera.* PhD thesis, Univ. Edinburgh, 1974.
314. J.D. Tygar and R. Ellickson. Efficient netlist comparison using hierarchy and randomization. In *Proc. of the 22th ACM/IEEE DAC*, pages 702–708, 1985.
315. J.D. Ullman. *Principles of Database Systems.* Computer Science Press, Rockville, 1982.
316. J.R. Ullman. An algorithm for subgraph isomorphism. *J. ACM*, 23(1):31–42, Jan. 1976.
317. J.R. Ullman. A binary *n*-gram technique for automatic correction of substitution, deletion, insertion and reversal errors in words. *The Computer Journal*, 20(2):141–147, Feb. 1977.
318. S.D. Urban and L.M.L. Delcambre. Constraint analysis: A design process for specifying operations on objects. *IEEE Trans. on Knowledge and Data Engineering*, 2(4):391–400, Dec. 1990.
319. M. van der Woude and X. Timermans. Compaction of hierarchical cells with minimum and maximum compaction constraints. In *Proc. Intl. Symposium on Circuits and Systems*, pages 1018–1021, 1983.
320. P. Vanbekbergen, B. Lin, G. Goossens, and H. De Man. A Generalized State Assignment Theory for Transformations on Signal Transition Graphs. In *Proceedings of ICCAD'92*, pages 112–117, 1992.
321. P. Vanbekbergen, B. Lin, G. Goossens, and H. De Man. A Generalized State Assignment Theory for Transformations on Signal Transition Graphs. *J. of VLSI Signal processing*, 1993. In Press.
322. D. Waltz. Generating semantic descriptions from drawings of scenes with shadows. Technical Report AI271, MIT, Nov. 1972.
323. D. Waltz. *Understanding Line Drawings of Scenes with Shadows. In* P. H. Winston, *The Psychology of Computer Vision*, chapter 2, pages 19–92. McGraw-Hill Book Company, New York, 1975.
324. H.P. Williams. Linear and integer programming applied to the propositional calculus. *Systems Research and Information Sciences*, 2:81–100, 1987.
325. P.H. Winston. *Artificial Intelligence.* Addison-Wesley, Reading, 1984.
326. L.C. Wu and C.Y. Tang. Solving the satisfiability problem by using randomized approach. *Information Processing Letters*, 41:187–190, 1992.
327. T. Y. Young and K. S. Fu, editors. *Handbook of Pattern Recognition and Image Processing.* Academic Press, Orlando, 1986.
328. R. Zabih and D. McAllester. A rearrangement search strategy for determining propositional satisfiability. In *Proceedings of AAAI'88*, pages 155–160, 1988.
329. V.N. Zemlyachenko, N.M. Korneeko, and R.I. Tyshkevich. Graph isomorphism problem. *J. of Soviet Mathematics*, 29:1426–1481, 1985.
330. S. Zhou. A trace-driven simulation study of dynamic load balancing. *IEEE Trans. on Software Engineering*, SE-14(9):1327–1341, Sept. 1988.
331. S. W. Zucker, R. A. Hummel, and A. Rosenfeld. An application of relaxation labeling to line and curve enhancement. *IEEE Trans. on Computers*, C-26:394–403,922–929, 1977.

ERGODIC CONVERGENCE IN PROXIMAL POINT ALGORITHMS WITH BREGMAN FUNCTIONS

OSMAN GÜLER*
Department of Mathematics and Statistics
University of Maryland Baltimore County
Baltimore, Maryland 21228, USA
e-mail: guler@math.umbc.edu.

Abstract. *We prove ergodic convergence results for the proximal point algorithm with Bregman kernels. In the function minimization case, the iterates converge to a minimizer of the function in the ergodic sense. We give global convergence rate estimate for the residual* $f(z^k) - \min f(x)$. *In the general case of finding a zero of a maximal monotone operator* T, *our main result states that, under very general conditions, all limit points of the ergodic solution sequence converge to a zero of* T . *Under slightly less general conditions, we show that either the original solution sequence convergence to a zero of* T, *or it stays bounded away from the zero set of* T.

1. Introduction

The goal of this paper is to establish ergodic convergence results for the nonlinear proximal point algorithms using Bregman kernels. These methods have been introduced recently [9], [12], [10], [24], etc. and generalize the usual proximal point algorithm [22], [6], [13]. In this section, we start by recalling the proximal point algorithm and some elementary concepts concerning maximal monotone operators. We then state the definition of the Bregman kernels. Finally, we discuss the concept of ergodic convergence and its role in convex optimization.

We first recall the usual proximal point algorithm with the quadratic kernel. Consider the convex minimization problem

$$\min_{x \in \mathbb{R}^n} f(x), \tag{1}$$

where $f : \mathbb{R}^n \to \mathbb{R} \cup \{\infty\}$ is a closed, proper convex function, using the terminology established in Aubin and Ekeland [1] or Rockafellar [21].

The classical proximal point algorithm is introduced into optimization literature by Martinet [15], based on an earlier work of Moreau [17]. It solves (1) by the recursion

$$x^{k+1} = \arg \min_{x \in \mathbb{R}^n} \{f(x) + \frac{1}{2\lambda_{k+1}} ||x - x^k||^2\}, \tag{2}$$

where $\{\lambda_k\}$ is a sequence of positive numbers.

* Research partially supported by the National Science Foundation under grant DMS-9306318.

Ding-Zhu Du and Jie Sun (eds.), Advances in Optimization and Approximation, 155–165.

The minimization problem (1) is general enough to include a generic convex programming problem of the form

$$\begin{array}{ll} \min\limits_{x \in C} & f_0(x) \\ s.t. & f_i(x) \leq 0, \quad i = 1, \ldots, m, \end{array}$$

where C is a closed convex subset of $\mathbb{R}^n$ and $f_i : \mathbb{R}^n \to \mathbb{R}$, $i = 0, 1, \ldots, m$ are convex functions.

The proximal point algorithm has been popularized by Rockafellar [22, 23] who shows that the application of the proximal point algorithm to the dual of the above convex program is equivalent to the augmented Lagrangian method applied to the original problem. Augmented Lagrangian methods have many advantages over penalty methods, see for example Bertsekas [3] and Rockafellar [23].

The relation (2) can be rewritten as the inclusion $0 \in \partial f(x^{k+1}) + (x^{k+1} - x^k)/\lambda_{k+1}$, or

$$x^{k+1} = (I + \lambda_{k+1}\partial f)^{-1} x^k, \tag{3}$$

see for example Aubin and Ekeland [1], section 6.7. This form of the proximal point method suggests that it be extended to an arbitrary maximal monotone operator T by substituting T for ∂f above to obtain

$$x^{k+1} = (I + \lambda_{k+1}T)^{-1} x^k. \tag{4}$$

This is indeed the case and follows from a celebrated result of Minty [16]. We refer the reader to the books Brézis [5] and Aubin and Ekeland [1] for more information about monotone operators, and to the papers Brézis and Lions [6] and Rockafellar [22] for basic results about the proximal point method.

We now recall some elementary concepts pertaining to maximal monotone operators. A set–valued mapping $T : \mathbb{R}^n \to \mathbb{R}^n$ is said to be a *monotone operator* if $y' \in A(x')$ and $y \in A(x)$ imply $\langle y' - y, x' - x \rangle \geq 0$. A monotone operator T is said to be *maximal monotone* if its graph $G(T) = \{(y, x) \in \mathbb{R}^n \times \mathbb{R}^n : y \in T(x)\}$ is not properly contained in the graph of any other monotone operator $T' : \mathbb{R}^n \to \mathbb{R}^n$. A *solution* to T is a point $x^* \in \mathbb{R}^n$ such that $0 \in T(x^*)$. The point x^* is also called a *zero* of T. The *domain* of T is the set $D(T) = \{x \in \mathbb{R}^n : T(x) \neq \emptyset\}$, and its *range* is the set $R(T) = \cup\{T(x) : x \in D(T)\}$.

Many problems which involve convexity can be formulated as finding the solution of a maximal monotone operator. For example, convex minimization, concave–convex saddle point problems, and solutions of games can be formulated in this way. In particular, the subdifferential $T = \partial f$ is a maximal monotone operator, and a point $x^* \in \mathbb{R}^n$ minimizes f if and only if $0 \in \partial f(x^*)$.

Recently, efforts have been made to replace the quadratic kernel $||x - y||^2/2$ in (2) with more general functions $D(x, y)$ satisfying certain conditions [9], [12], [24], [25]. One motivation for these efforts come from the possibility of choosing the kernel to suit the structure of the problem at hand. Another motivation is to obtain more differentiability in the corresponding augmented Lagrangian method than the quadratic kernel can provide [25].

One possible choice for the kernel $D_\psi(x,y)$ is provided by the Bregman kernel introduced in [4]. (The term "Bregman kernel" is coined later by Censor and Lent [8]). Given a differentiable function ψ, a measure of "distance" based on Bregman's distance [4] is defined by

$$D_\psi(x,y) = \psi(x) - \psi(y) - \langle \nabla\psi(y), x - y\rangle$$

where $\langle\cdot,\cdot\rangle$ denotes the inner product in $\mathbb{R}^n$ and $\nabla\psi$ is the gradient of ψ. The function ψ is called a *Bregman function* or "D–function", if it satisfies the following conditions in the definition below.

Definition 1.1 *Let S be an open convex set in $\mathbb{R}^n$. A function $\psi : \overline{S} \to \mathbb{R}$ is called a Bregman function with zone S if*
(i) ψ is continuously differentiable on S.
(ii) ψ is strictly convex on $\overline{S}$.
(iii) For every $\alpha \in \mathbb{R}$, the partial level sets $L_1(y,\alpha) = \{x \in \overline{S} : D_\psi(x,y) \le \alpha\}$ and $L_2(x,\alpha) = \{y \in S : D_\psi(x,y) \le \alpha\}$ are bounded for every $y \in S$ and $x \in \overline{S}$.
(iv) If $\{y^k\} \subseteq S$ converges to y^*, then $D_\psi(y^*, y^k) \to 0$.
(v) If $\{x^k\}$ and $\{y^k\}$ are sequences such that $y^k \to y^* \in \overline{S}$, $\{x^k\}$ is bounded, and $D_\psi(x^k, y^k) \to 0$, then $x^k \to y^*$.

Note that $D_\psi(x,x) = 0$, and the convexity of ψ implies $D_\psi(x,y) \ge 0$ for $x \in \overline{S}$ and $y \in S$.

Examples of Bregman distance function are given in Eckstein [12] and Chen and Teboulle [10]. For example, the Kullback–Liebler relative entropy distance

$$D_\psi(x,y) = \sum_{i=1}^{n} x_i \ln\frac{x_i}{y_i} + y_i - x_i$$

is a Bregman distance function on the set $\{(x,y) \in \mathbb{R}^n \times \mathbb{R}^n : x \ge 0, y > 0\}$ for the entropy function $\psi(x) = \sum_{i=1}^{n} x_i \ln x_i - x_i$ defined for $x \ge 0$ (with the convention that $0 \ln 0 = 0$). The corresponding augmented Lagrangian method is the *exponential multiplier method* (see for example [12]) and extensively studied in [3, 25].

The concept of ergodic convergence is quite useful in convex minimization and in finding fixed points of nonexpansive maps. If $\{x^k\}$ is a sequence of points, one forms the sequence $\{z^n\}$ of weighted averages given by

$$z^n = \sum_{i=1}^{n} \frac{\lambda_k}{\sigma_n} x^k \quad \text{and} \quad \sigma_n = \sum_{i=1}^{n} \lambda_k,$$

where $\lambda_k > 0$. Quite often, the sequence $\{z^n\}$ has better properties than the original sequence $\{x^k\}$. In a sense, taking averages smooths out the sequence $\{x^k\}$. We refer the reader the survey article of Bruck [7] for more information on ergodic convergence and for an extensive list of references.

The rest of the paper is organized as follows. In section 2 we state the proximal point algorithm with a Bregman kernel for the convex minimization problem (1) and present our ergodic convergence results. In section 3, we present the general version of the algorithm for an arbitrary maximal monotone operator T. Theorem 3.1 is the main result of the paper and gives an ergodic convergence result under very general conditions. Finally, we investigate when ergodic convergence can be improved to ordinary convergence. Theorem 3.2 shows that under the restriction $D(T) \subseteq S$, either the sequence $\{x^k\}$ converges to a point in $T^{-1}(0)$, or it stays bounded away from the zero set $T^{-1}(0)$. Although the latter possibility does not seem plausible, we cannot remove it at present.

2. Convergence for Function Minimization

In this section we present ergodic convergence results for the proximal point method with a Bregman kernel $D_\psi(x, y)$ applied to the convex minimization problem (1). We shall refer to this algorithm as the proximal minimization method with D–functions (PMD), following the terminology in Chen and Teboulle [10]. It is obtained by substituting $D_\psi(x, x^k)$ for $||x - x^k||^2/2$ in (2), that is,

$$x^{k+1} = \arg \min_{x \in \mathbb{R}^n} \{f(x) + \frac{1}{\lambda_{k+1}} D_\psi(x, x^k)\},$$

where $\{\lambda_k\}$ is a sequence of positive numbers. Since x^{k+1} is characterized by the condition $0 \in \partial f(x^{k+1}) + (\nabla\psi(x^{k+1}) - \nabla\psi(x^k))/\lambda_{k+1}$, or

$$x^{k+1} = (\nabla\psi + \lambda_{k+1}\partial f)^{-1}\nabla\psi(x^k). \tag{5}$$

The following two simple lemmas are proved in Chen and Teboulle [10] and will be needed later.

Lemma 2.1 *Let ψ be a Bregman function. Then for any three points $a, b \in S$ and $c \in \overline{S}$, the following identity holds true*

$$D_\psi(c, a) + D_\psi(a, b) - D_\psi(c, b) = \langle \nabla\psi(b) - \nabla\psi(a), c - a\rangle.$$

Lemma 2.2 *Let f be a closed, proper convex function on $\mathbb{R}^n$ and let $\{\lambda_k\}$ be an arbitrary sequence of positive numbers. Suppose that $ri(dom(f)) \subseteq S$. Then PMD is well–defined, and for any $u \in \overline{S}$,*

$$\begin{aligned}\lambda_k(f(x^k) - f(u)) &\le \langle \nabla\psi(x^{k-1}) - \nabla\psi(x^k), x^k - u\rangle \\ &= D_\psi(u, x^{k-1}) - D_\psi(u, x^k) - D_\psi(x^k, x^{k-1}).\end{aligned}$$

Thus, if ψ satisfies $ri(dom(f)) \subseteq S$, then PMD is well–defined, see also [12]. Eckstein [12] shows that the sequence $\{x^k\}$ converges to a minimizer of f under the condition that $\{\lambda_k\}$ is bounded away from zero, and later Chen and Teboulle [10] prove the same result under the more general condition that $\sum_{k=1}^{\infty} \lambda_k = \infty$. Generalizing the result of Güler [13] to the present situation, Chen and Teboulle give for any $u \in \overline{S}$ the following global convergence rate

$$f(x^n) - f(u) \leq \frac{D_\psi(u, x^0)}{\sigma_n} = O(\frac{1}{\sigma_n}).$$

Since convexity of f implies $f(z^n) \leq f(x^n)$, we also have the global convergence rate estimate

$$f(z^n) - f(u) \leq \frac{D_\psi(u, x^0)}{\sigma_n} = O(\frac{1}{\sigma_n}).$$

Also, the sequences $\{x^n\}$ and $\{z^n\}$ converge to the same point, by the virtue of the Silverman–Toeplitz theorem [11], pp. 75.

In the case $\nabla\psi$ is Lipschitz continuous, that is, there exists $L > 0$ such that for any $x, y \in \mathbb{R}^n$

$$||\nabla\psi(x) - \nabla\psi(y)|| \leq L||x - y||,$$

then it is possible to improve the convergence rate above. The proof below is similar to the proof of Theorem 3.1 in Güler [13]. It is essentially based on Polyak's proof of the convergence rate of the steepest descent method for a convex function with Lipschitz continuous derivative [19].

Theorem 2.1 *Suppose $\nabla\psi$ is Lipschitz continuous with Lipschitz constant L. Then the convergence rate of the PMD can be improved to*

$$f(z^n) - \min f(x) = o(1/\sigma_n),$$

that is, $\sigma_n(f(z^n) - \min f(x)) \to 0$.

Proof. Let x^* be a minimizer of f. For brevity, we denote $W_k = f(x^k) - f(x^*)$. Setting $u = x^{k-1}$ in Lemma 2.2, we obtain

$$\langle \nabla\psi(x^{k-1}) - \nabla\psi(x^k), x^{k-1} - x^k \rangle \leq \lambda_k(W_{k-1} - W_k). \tag{6}$$

Since $\nabla\psi$ is Lipschitz continuous, a result of Baillon and Haddad [2] (Corollary 10, pp. 150) implies that

$$\langle \nabla\psi(x^{k-1}) - \nabla\psi(x^k), x^{k-1} - x^k \rangle \geq L^{-1}||\nabla\psi(x^{k-1}) - \nabla\psi(x^k)||^2. \tag{7}$$

(This result is proved independently by Gol'shtein and Treti'akov [19, 20].) From (6) and (7), we get

$$L^{-1}||\nabla\psi(x^{k-1}) - \nabla\psi(x^k)||^2 \leq \lambda_k(W_{k-1} - W_k). \tag{8}$$

Also, setting $u = x^*$ in Lemma 2.2 gives

$$\begin{aligned} \lambda_k W_k &\leq \langle \nabla\psi(x^{k-1}) - \nabla\psi(x^k), x^k - x^* \rangle \\ &= \langle \nabla\psi(x^{k-1}) - \nabla\psi(x^k), x^k - x^{k-1} \rangle + \langle \nabla\psi(x^{k-1}) - \nabla\psi(x^k), x^{k-1} - x^* \rangle \\ &\leq \langle \nabla\psi(x^{k-1}) - \nabla\psi(x^k), x^{k-1} - x^* \rangle \\ &\leq ||\nabla\psi(x^{k-1}) - \nabla\psi(x^k)|| \cdot ||x^{k-1} - x^*||, \end{aligned} \tag{9}$$

where the second inequality follows from the monotonicity of $\nabla\psi$. From (9) and (8) we obtain

$$\begin{aligned}\lambda_k^2 W_k^2 &\le \|\nabla\psi(x^{k-1}) - \nabla\psi(x^k)\|^2 \cdot \|x^{k-1} - x^*\|^2 \\ &\le L\lambda_k(W_{k-1} - W_k)\|x^{k-1} - x^*\|^2,\end{aligned}$$

which simplifies to

$$W_k\left(1 + \frac{\lambda_k W_k}{L\|x^{k-1} - x^*\|^2}\right) \le W_{k-1},$$

or

$$W_{k-1}^{-1} \le W_k^{-1}\left(1 + \frac{\lambda_k W_k}{L\|x^{k-1} - x^*\|^2}\right)^{-1}.$$

Choosing $u = x^*$ in Lemma 2.2, we obtain

$$\begin{aligned}\lambda_k W_k &\le D_\psi(x^*, x^{k-1}) \\ &= \psi(x^*) - \psi(x^{k-1}) - \langle\nabla\psi(x^{k-1}), x^* - x^{k-1}\rangle \\ &\le \frac{L}{2}\|x^{k-1} - x^*\|^2,\end{aligned}$$

where the last inequality is well known (see for example Ortega and Rheinboldt [18], pp. 73). This inequality implies that $\lambda_k W_k/(L\|x^{k-1} - x^*\|^2) \le 1/2$. Since $(1+t)^{-1} \le 1 - 2t/3$ in $[0, 1/2]$, we have

$$W_{k-1}^{-1} \le W_k^{-1}\left(1 - \frac{2\lambda_k W_k}{3L\|x^{k-1} - x^*\|^2}\right) = W_k^{-1} - \frac{2\lambda_k}{3L\|x^{k-1} - x^*\|^2}.$$

Summing this inequality for $k = 1, \ldots, n$, we get

$$0 \le W_0^{-1} \le W_n^{-1} - \frac{2}{3L}\sum_{k=1}^{n} \lambda_k\|x^{k-1} - x^*\|^{-2}.$$

Since x^* is an arbitrary minimizer of f, we have

$$\sigma_n W_n \le \frac{3L}{2\sum_{k=1}^{n}(\lambda_k/\sigma_n)\rho(x^{k-1}, X^*)^{-2}},$$

where X^* is the set of minimizers of f. Since $\rho(x^{k-1}, X^*) \to 0$, $\rho(x^{k-1}, X^*)^{-2} \to \infty$. By the Silverman–Toeplitz theorem, the denominator in the above inequality converges to infinity. Therefore, $\sigma_n W_n = \sigma_n(f(x^n) - \min f(x)) \to 0$. Since $f(z^n) \le f(x^n)$, we also have $\sigma_n(f(z^n) - \min f(x)) \to 0$. □

It is shown in Güler [14] that a suitable conjugate gradient type modification of the usual proximal point method applied to (1) enjoys a better global convergence rate than the ones obtained above. It is unknown at present whether a similar acceleration is possible for PMD.

3. Convergence for Arbitrary Maximal Monotone Operators

In this section we prove convergence results for the generalization of PMD to an arbitrary maximal monotone operator T. This algorithm is obtained by substituting T for ∂f in the recursion (5), that is,

$$x^{k+1} = (\nabla\psi + \lambda_{k+1}T)^{-1}\nabla\psi(x^k). \tag{10}$$

At present, there are no known conditions which are both necessary and sufficient for the existence of the sequence $\{x^k\}$; this seems to be a hard question. Sufficient conditions are given in Eckstein [12] on ψ for the existence of the sequence $\{x^k\}$. In this section, we will assume the equation (10) is well–defined.

The following is the main result of this paper.

Theorem 3.1 *Suppose T is a maximal monotone operator and $\sigma_n \to \infty$. Then $T^{-1}(0) \neq \emptyset$ if and only if the sequence $\{x^k\}$ is bounded. If $T^{-1}(0) \neq \emptyset$, then every limit point of the sequence $\{z^n\}$ is a zero of the maximal monotone operator T.*

Proof. Let $(u, v) \in G(T)$. Using $c = u$, $a = x^k$, and $b = x^{k-1}$ in Lemma 2.1, we obtain

$$\begin{aligned} \lambda_k\langle y^k, u - x^k\rangle &= \langle \nabla\psi(x^{k-1}) - \nabla\psi(x^k), u - x^k\rangle \\ &= D_\psi(u, x^k) + D_\psi(x^k, x^{k-1}) - D_\psi(u, x^{k-1}) \end{aligned} \tag{11}$$

where

$$y^k := \frac{\nabla\psi(x^{k-1}) - \nabla\psi(x^k)}{\lambda_k} \in T(x^k).$$

Since $(x^k, y^k) \in G(T)$, and T is monotone,

$$\langle y^k - v, x^k - u\rangle \geq 0,$$

so that from (11) we obtain

$$\lambda_k\langle v, u - x^k\rangle \geq D_\psi(u, x^k) + D_\psi(x^k, x^{k-1}) - D_\psi(u, x^{k-1}). \tag{12}$$

If $T^{-1}(0) \neq \emptyset$, then choosing $(u, v) = (x^*, 0) \in G(T)$ in (12), we see that

$$D_\psi(x^*, x^k) \leq D_\psi(x^*, x^{k-1}), \tag{13}$$

so that $\{D_\psi(x^*, x^k)\}$ is a decreasing sequence. From condition (iii) in Definition 1.1, we see that the sequence $\{x^k\}$ is bounded.

To establish the converse, we sum (12) for $k = 1, \ldots, n$,

$$\sum_{k=1}^{n} \lambda_k\langle v, x^k - u\rangle \leq D_\psi(u, x^0) - D_\psi(u, x^n) - \sum_{k=1}^{n} D_\psi(x^k, x^{k-1}).$$

Dividing both sides of this inequality by σ_n, we obtain

$$\langle v, z^n - u\rangle \le \frac{D_\psi(u,x^0) - D_\psi(u,x^n) - \sum_{k=1}^n D_\psi(x^k,x^{k-1})}{\sigma_n} \le \frac{D_\psi(u,x^0)}{\sigma_n}. \tag{14}$$

Since $\{x^n\}$ is a bounded sequence, so is the sequence $\{z^n\}$. Suppose $z^{n(k)} \to z^\infty$ for some subsequence $\{n(k)\}$. Since $\sigma_n \to \infty$, the last term in (14) converges to zero. Taking the limit of both sides of (14) as $k \to \infty$, we obtain

$$\langle v, z^\infty - u\rangle \le 0.$$

Since $(u,v) \in G(T)$ is arbitrary and T is maximal monotone, we have $0 \in T(z^\infty)$. □

Obviously, the convergence of the iterates $\{z^k\}$ is not as strong as the convergence of the original iterates $\{x^k\}$. Thus, it is of some interest to investigate when one can replace ergodic convergence with ordinary convergence. As we saw in Section 2, this question is completely settled for the case $T = \partial f$. In the general case, the current knowledge is less satisfactory. Eckstein [12] shows that the iterates $\{x^k\}$ again converge to a point $x^* \in T^{-1}(0)$, but under the very restrictive assumption that $\overline{D(T)} \subseteq S$. He also assumes the nonrestrictive condition that $\lambda_k > c$ for some positive constant c.

We improve on Eckstein's results below. In the remainder of this section, we shall assume that $D(T) \subseteq S$. We begin with the following technical result.

Lemma 3.1 *Let $\{x^n\}$ be the sequence generated by the PMD in Theorem 3.1 under the condition that $T^{-1}(0) \neq \emptyset$. Then the projection sequence $\{Px^k\}$ converges, where*

$$Px^k = \arg\min_{u \in T^{-1}(0)} D_\psi(u, x^k).$$

Proof. Setting $x^* = Px^{k-1}$ in (13), we get

$$D_\psi(Px^k, x^k) \le D_\psi(Px^{k-1}, x^k) \le D_\psi(Px^{k-1}, x^{k-1}),$$

where the first inequality follows from the definition of Px^k. Thus $\{D_\psi(Px^k, x^k)\}$ is a decreasing sequence.

Also, setting $a = Px^l$, $b = x^l$, and $c = x^* \in T^{-1}(0)$ in Lemma 2.1 gives

$$D_\psi(x^*, Px^l) + D_\psi(Px^l, x^l) - D_\psi(x^*, x^l) = \langle \nabla\psi(x^l) - \nabla\psi(Px^l), x^* - Px^l\rangle.$$

Since Px^l minimizes $D_\psi(u, x^l)$ over $u \in T^{-1}(0)$, we have from the variational inequality

$$\langle \nabla\psi(Px^l) - \nabla\psi(x^l), x^* - Px^l\rangle \ge 0.$$

This means that

$$D_\psi(x^*, Px^l) + D_\psi(Px^l, x^l) \le D_\psi(x^*, x^l) \le D_\psi(x^*, x^1), \tag{15}$$

where the second inequality follows from (13). Consequently, $D_\psi(x^*, Px^l)$ is bounded and condition (iii) of Definition 2.1 implies that the sequence $\{Px^l\}$ is bounded.

Choose $k \leq l$ and set $x^* = Px^k$ in (15). We get

$$D_\psi(Px^k, Px^l) + D_\psi(Px^l, x^l) \leq D_\psi(Px^k, x^l) \leq D_\psi(Px^k, x^k),$$

where the last inequality follows from (13). Since $\{D_\psi(Px^k, x^k)\}$ is a decreasing sequence, we have

$$D_\psi(Px^k, Px^l) \leq D_\psi(Px^k, x^k) - D_\psi(Px^l, x^l) \to 0,$$

as $k, l \to \infty$, that is,

$$\lim_{k,l\to\infty} D_\psi(Px^k, Px^l) = 0. \tag{16}$$

Now let $\{Px^{k_1(i)}\}$ and $\{Px^{k_2(i)}\}$ be two convergent subsequences of $\{Px^k\}$ converging to Px_1^∞ and Px_2^∞, respectively. Suppose that $Px_1^\infty \neq Px_2^\infty$. Without losing generality, we may assume that there exists a subsequence i_j such that $k_1(i_j) < k_2(i_j)$. Then relation (16) implies that $\lim_{j\to\infty} D_\psi(Px^{k_1(i_j)}, Px^{k_2(i_j)}) = 0$. It follows from condition (v) of Definition 2.1 that $Px_1^\infty = Px_2^\infty$, contradicting our assumption that $Px_1^\infty \neq Px_2^\infty$. Thus the whole sequence $\{Px^k\}$ converges. □

Theorem 3.2 *Suppose T is a maximal monotone operator with $T^{-1}(0) \neq \emptyset$ and satisfies the condition $D(T) \subseteq S$. Let $\sigma_n \to \infty$. Then either the sequence $\{x^k\}$ generated by PMD converges to a point $T^{-1}(0) \neq \emptyset$, or it is bounded away from the set $T^{-1}(0)$.*

Proof. Suppose that the sequence $\{x^k\}$ is not bounded away from the zero set $T^{-1}(0)$. Then there exists a subsequence $\{x^{k_1(i)}\}$ converging to a point $x^\infty \in T^{-1}(0)$. It follows from (15) that

$$D_\psi(x^\infty, Px^{k_1(i)}) + D_\psi(Px^{k_1(i)}, x^{k_1(i)}) \leq D_\psi(x^\infty, x^{k_1(i)}) \to 0, \tag{17}$$

which implies $D_\psi(x^\infty, Px^{k_1(i)}) \to 0$. Condition (iii) of Definition (2.1) implies that $Px^{k_1(i)}) \to x^\infty$, and Lemma 3.1 further implies $Px^k \to x^\infty$.
Note that (17) also implies that $\lim_{i\to\infty} D_\psi(Px^{k_1(i)}, x^{k_1(i)}) = 0$. Since $D_\psi(Px^k, x^k)$ is a decreasing sequence, this means that the whole sequence $\{D_\psi(Px^k, x^k)\}$ converges to zero. Now let $\{x^{k_2(i)}\}$ be another subsequence converging to a point x^0. Since $\lim_{i\to\infty} D_\psi(Px^{k_2(i)}, x^{k_2(i)}) = 0$, condition (v) of Definition 2.1 implies that $x^0 = x^\infty$. Thus, the whole sequence $\{x^k\}$ converges to $x^\infty \in T^{-1}(0)$. □

References

1. Aubin, J. P. and Ekeland I. (1984) *Applied Nonlinear Analysis*. Interscience Publications, John Wiley and Sons, New York.
2. Baillon, J. B. and Haddad, G. (1977). Quelques propriétés des opérateurs angle-bornés et n-cycliquement monotones. *Israel Journal of Mathematics* **26** 137–150.
3. Bertsekas, D. P. (1982) *Constrained Optimization and Lagrange Multiplier Methods*. Academic Press, New York.

4. Bregman, L. (1967) The relaxation method of finding the common points of convex sets and its applications to the solution of problems in convex programming. *USSR Computational Mathematics and Mathematical Physics* **7** 200–217.

5. Brézis, H. (1973) *Opérateurs Maximaux Monotones.* Mathematics Studies No. 5, North–Holland, Amsterdam.

6. Brézis, H. and Lions, P. L. (1978) Produits infinis de résolvantes. *Israel Journal of Mathematics* **29** 329–345.

7. Bruck, R. E. (1983) Asymptotic behavior of nonexpansive mappings. In *Fixed Points and Nonexpansive Mappings*, (R. Sine, ed.) Contemporary Mathematics vol. 18, American Mathematical Society, Providence, Rhode Island, 1–47.

8. Censor, Y. and Lent, A. (1981) An interval row action method for interval convex programming. *Journal of Optimization Theory and Applications* **34** 321–353.

9. Censor, Y. and Zenios, S. (1992) The proximal minimization algorithm with D–functions. *Journal of Optimization Theory and Applications* **73** 451–464.

10. Chen, G. and Teboulle, M. (1993) Convergence analysis of a proximal–like minimization algorithm using Bregman functions. *SIAM Journal on Optimization* **3** 538–543.

11. Dunford, N. and Schwartz, J. T. (1988) *Linear Operators, Part I: General Theory.* Interscience Publications, John Wiley and Sons, New York.

12. Eckstein, J. (1993) Nonlinear proximal point algorithms using Bregman functions, with applications to convex programming. *Mathematics of Operations Research* **18** 202–226.

13. Güler, O. (1991) On the convergence of the proximal point algorithm for convex minimization. *SIAM Journal on Control and Optimization* **29** 403–419.

14. Güler, O. (1992) New proximal point algorithms for convex minimization. *SIAM Journal on Optimization* **2** 649–664.

15. Martinet, B. (1970) Regularisation, d'inéquations variationelles par approximations succesives. *Revue Française d'Informatique et de Recherche Operationelle* **4** 154–159.

16. Minty, G. (1962) Monotone (nonlinear) operators in a Hilbert space. *Duke Mathematics Journal* **29** 341–348.

17. Moreau, J. J. (1965) Proximité et dualité dans un espace Hilbertien. *Bull. Soc. Math., Fr.* **93** 273–299.

18. Ortega, J. M. and Rheinboldt, W. C. (1970) *Iterative Solutions of Nonlinear Equations in Several Variables.* Academic Press, New York.

19. Polyak, B. T. (1987) *Introduction to Optimization.* Optimization Software, New York.

20. Polyak, B. T. (1989) Personal communication.

21. Rockefellar, R. T. (1970) *Convex Analysis.* Princeton Univ. Press, Princeton, New Jersey.

22. Rockafellar, R. T. (1976) Monotone operators and the proximal point algorithm. *SIAM Journal on Control and Optimization* **14** 877–898.

23. Rockafellar, R. T. (1976) Augmented Lagrangians and applications of the proximal point algorithm in convex programming. *Mathematics of Operations Research* 1 97–116.

24. Teboulle, M. (1992) Entropic proximal mappings with applications to nonlinear programming. *Mathematics of Operations Research* 17 670–690.

25. Tseng, P. and Bertsekas, D. P. (1993) On the convergence of the exponential multiplier method for convex programming. *Mathematical Programming* 60 1–19.

ADDING AND DELETING CONSTRAINTS IN THE LOGARITHMIC BARRIER METHOD FOR LP *

D. DEN HERTOG, C. ROOS and T. TERLAKY†
Delft University of Technology
Faculty of Technical Mathematics and Computer Science
P.O. Box 5031
2600 GA Delft

Abstract. We analyze the effect of adding and deleting a constraint of the linear program on the position of the central point, the "distance" to the path, and the change in the barrier function value. Based on these results we propose a column generation and deletion variant of the logarithmic barrier method for linear programming. The algorithm starts with a (small) subset of the dual constraints, and follows the corresponding central path until the iterate is close to or violates one of the constraints, which is in turn added to the current system. It also has the possibility of deleting constraints which are likely to be nonbinding in the optimal solutions. A complexity analysis for this algorithm will be given.

Key words: interior point method, linear programming, logarithmic barrier function, polynomial algorithm, column generation and deletion.

1. Introduction

Karmarkar [9] pioneered the rapidly developing field of interior point methods for linear programming. These methods not only have nice theoretical properties, but are very efficient from the practical point of view, especially for large problems. One drawback to all interior point methods is the great computational effort required in each iteration. In each iteration the search direction p is obtained by solving a linear system with normal matrix $AD^{-2}A^T$, where A is the constraint matrix $(m \times n)$ and D a positive diagonal matrix depending on the current iterate. Therefore, working with a subset of the dual constraints rather than the full system, would save a great deal of computation, especially if $n \gg m$.

The first attempt to save computations is the so–called "build–down" or "column–deletion" method, proposed by Ye [14, 15]. In his approach, a criterion for detecting (non)binding constraints is derived on the basis of an ellipsoid which circumscribes the optimal facet. If a constraint is detected to be non–binding in the optimal set, it is removed from the system. Consequently, the system becomes increasingly smaller, which reduces the computational effort for computing the normal matrix $AD^{-2}A^T$. However, the speed of the detection process is crucial. If the non–binding constraints are only detected during the last stage of the algorithm, the reduction in computation is negligible. To the best of our knowledge, there are no computational results to be found in the literature concerning this build–down process.

* This work is completed with the support of a research grant from SHELL.
† On leave from the Eötvös University, Budapest, and partially supported by OTKA No. 2116.

Ding-Zhu Du and Jie Sun (eds.), Advances in Optimization and Approximation, 166–185.

The second attempt to save computations is the "build–up" or "column generation" method. Papers on column generation techniques within interior point methods were first written by Mitchell [10] and Goffin and Vial [7] for the projective method. In [2] computational results for Goffin and Vial's method applied to convex programming are reported. However, these papers provide no theoretical analysis for the effect on the potential function and/or the number of iterations after the addition of a column/row.

Ye [17] proposed a (non–interior) potential reduction method for the linear feasibility problem which allows column generation. In each iteration an inequality violated at the current center is added to the system (in a shifted position), until a feasible point has been found. He proved that such a point can be found in $O(\sqrt{n}L)$ iterations, where L is the input length of the problem. Although each linear programming problem can be formulated as a linear feasibility problem, this is an inefficient way of solving linear programming problems.

Dantzig and Ye [3] proposed a build–up scheme for the dual affine scaling algorithm. This method differs from the "standard" affine scaling method in that the ellipsoid chosen to generate the search direction p is constructed from a set of m "promising" dual constraints. If the next iterate $y + p$ violates one of the other constraints, this constraint is added to the current system and a new ellipsoid and search direction (using the new set of constraints) are calculated. After making the step, a new set of m promising dual constraints is selected.

Tone [12] proposed an active–set strategy for Ye's [16] dual potential reduction method. In this strategy the search direction is also constructed from a subset of constraints which have small dual slacks in the current iterate. More constraints are added if no sufficient potential reduction is obtained. After making the step a new set of dual constraints, with small slack values, is selected. This algorithm converges to an optimal solution in $O(\sqrt{n}L)$ iterations. In [8] some computational results for this active–set strategy are reported.

Vaidya [13] proposed a cutting plane algorithm for the problem of finding a point in a convex set. This method is based on the same ideas as the method of Goffin and Vial [7], but instead of using the analytic center he used the so–called "volumetric center". In this algorithm constraints are added and removed with the nice property that at each stage the number of constraints is bounded by $O(m)$. Atkinson and Vaidya [1] came up with a similar algorithm based on analytic centers. For linear programming the complexity result of this method is much worse then the standard interior point methods.

In [4] the present authors proposed a build–up strategy for the long–step logarithmic barrier method. This strategy starts with a (small) subset of the constraints, and follows the corresponding central path until the iterate is close to (or violates) one of the other constraints. To be more precise: if some slack value s_i satisfies $s_i < 2^{-t_a}$, for some "proximity" parameter t_a, then the i–th constraint, if not already in the current system, is added to it. This process is repeated until the iterate is close to the optimum. It was proved that this Build–Up Algorithm ends up with a 2^{-2L} solution within $O(q^* L)$ Newton iterations, where q^* is the number of constraints in the final system. Such a build–up method has two advantages. Firstly, it has the before mentioned property of using only subsets of the set of all constraints, and hence decreases the computational effort per iteration. Secondly, since it is likely that $q^* < n$, the (theoretical) complexity is better than that for the standard logarithmic barrier method.

This paper is an extension to our work in [4]. In the Build–Up and Down Algorithm proposed here, it is also allowed to delete "unpromising" dual constraints. The results for the Buid–Up part obtained in this paper are sharper than those obtained in [4], since we do not use global upper bounds for the slack values ($s_i \leq 2^L$).

The algorithm starts with a (small) subset of the constraints. If the current iterate is close to (or violates) a constraint, then this constraint is added. A constraint is deleted if the corresponding slack value of the current approximately centered iterate is "large". An essential tool in our approach is the analysis of the effect of shifting, adding and deleting a constraint on the position of the central point, the "distance" to the path, and the change in the barrier function. These basic properties are interesting in itself, but it also enables us to analyze the Build–Up and Down Algorithm.

The paper is built up as follows. In Section 2 we recall some results for the "standard" logarithmic barrier method, obtained in [11, 5], which we need here. In Section 3 we analyze the effect of shifting, adding and deleting a constraint on the position of the central point, the distance to the path and the barrier function value. In Section 4 we state the Build–Up and Down Algorithm. Its complexity is analyzed in Section 5.

As far as notations are concerned, e denotes the vector of all ones, e_i the i–th unit vector and I the identity matrix. Given an n–dimensional vector s we denote by S the $n \times n$ diagonal matrix whose diagonal entries are the coordinates s_j of s; s^T is the transpose of the vector s and the same notation holds for matrices. Finally $||s||$ denotes the l_2 norm.

2. The Logarithmic Barrier Method

In this section we present the problem, its central path and a measure for the distance to the central path. Some well–known results for logarithmic barrier methods, which are needed in the sequel, are recalled from [11, 5].

Consider the well–known standard linear programming problem:

$$(P) \qquad \min\left\{c^T x \;:\; Ax = b,\; x \geq 0\right\}.$$

Here A is an $m \times n$ matrix, b and c are $m-$ and $n-$ dimensional vectors, respectively. In this paper we deal with the dual of (P):

$$(D) \qquad \max\left\{b^T y \;:\; A^T y + s = c,\; s \geq 0\right\},$$

where y and s, $m-$ and n–dimensional vectors respectively, are the variables in which the minimization is carried out. Without loss of generality, we assume that all the coefficients are integer. L denotes the length of the binary encoded input data of (D).

We make the standard assumptions that the interior of the feasible region of (D) is non–empty and that the optimal set is bounded. In order to simplify the analysis, we also assume that A has full rank, though this assumption is not essential.

The dual logarithmic barrier function is given by

$$f(y,\mu) := \frac{b^T y}{\mu} + \sum_{j=1}^{n} \ln s_j, \tag{1}$$

where μ is a positive barrier parameter. Under the above assumptions, this function achieves a maximum value at a unique interior point. The necessary and sufficient first order optimality conditions for this point are:

$$\begin{aligned} A^T y + s &= c, \quad s \geq 0, \\ Ax &= b, \quad x \geq 0, \\ Sx &= \mu e. \end{aligned} \tag{2}$$

Let us denote the unique solution of this system by $(x(\mu), y(\mu), s(\mu))$. The primal and dual central path is defined as the solution set $x(\mu)$ and $y(\mu)$ respectively, for $\mu \geq 0$.

Roos and Vial [11] introduced the following measure for the distance of an interior feasible point y to the central point $y(\mu)$:

$$\delta(y,\mu) := \min_x \left\{ \|\frac{Sx}{\mu} - e\| : Ax = b \right\}. \tag{3}$$

The unique solution of the minimization problem in the definition of $\delta(y,\mu)$ is denoted by $x(y,\mu)$. Let us denote the Newton direction for (1) by p. A closed formula for p is:

$$p = (AS^{-2}A^T)^{-1} \left(\frac{b}{\mu} - AS^{-1}e \right). \tag{4}$$

It can be verified (see [5]) that $\delta(y,\mu) = \|S^{-1}A^T p\|$. Moreover, if y is feasible then

$$y = y(\mu) \Longleftrightarrow \delta(y,\mu) = 0.$$

The following lemmas and theorem are now well–known. Proofs can be found in [11, 5]. Whenever this is convenient and gives no confusion we write δ instead of $\delta(y,\mu)$.

Lemma 1 *If $\delta := \delta(y,\mu) \leq 1$ then $x := x(y,\mu)$ is primal feasible. Moreover,*

$$\mu(n - \delta\sqrt{n}) \leq c^T x - b^T y \leq \mu(n + \delta\sqrt{n}).$$

Lemma 2 *If $\delta(y,\mu) < 1$ then $y^* = y+p$ is a strictly feasible point for (D). Moreover,*

$$\delta(y^*,\mu) \le \delta(y,\mu)^2.$$

Lemma 3 *If $\delta := \delta(y,\mu) < 1$ then*

$$f(y(\mu),\mu) - f(y,\mu) \le \frac{\delta^2}{1-\delta^2}.$$

Lemma 4 *Let $\tilde{\alpha} := (1+\delta)^{-1}$. Then $y + \tilde{\alpha}p$ is strictly feasible and*

$$\Delta f := f(y+\tilde{\alpha}p,\mu) - f(y,\mu) \ge \delta - \ln(1+\delta).$$

Lemma 5 *If $\overline{\mu} := (1-\theta)\mu$, where $0 < \theta < 1$, and $\delta := \delta(y,\mu) \le \frac{1}{2}$ then*

$$f(y(\overline{\mu}),\overline{\mu}) - f(y,\overline{\mu}) \le \frac{\theta}{1-\theta}\,(\theta n + 3\sqrt{n}) + \frac{1}{3}.$$

The logarithmic barrier method approximately follows the central path, which ends in an optimal solution of the problem. This method is therefore called a path–following method. More precisely: line searches along Newton directions are carried out until the iterate is in the vicinity of the current central point; then the barrier parameter is reduced by a factor $1-\theta$, $0 < \theta < 1$, whereafter Newton iterations are carried out to reach the vicinity of the new central point. Suppose the logarithmic barrier method, as defined in [5], starts with barrier parameter μ_0 and initial iterate y^0 such that $\delta(y^0,\mu_0) \le \frac{1}{2}$. Then the following theorem gives an upper bound for the total number of iterations (see [5]).

Theorem 1 *After at most*

$$O\left(\frac{1}{\theta}\ln\frac{n\mu_0}{\epsilon}\right)$$

reductions of the barrier parameter, the logarithmic barrier algorithm ends up with a primal and a dual solution such that $x^Ts \le \epsilon$. Each reduction of the barrier parameter requires at most

$$O\left(\frac{\theta^2 n}{1-\theta}+1\right)$$

Newton iterations to reach the vicinity of the new central point.

3. The Effects of Shifting, Adding and Deleting Constraints

In the algorithm we will add and delete constraints. In the analysis we also need to consider the effect of shifting a constraint. That is why we start with some lemmas dealing with shifting a constraint. We note that if we take $b = 0$ then some of Ye's [17] results follow from some lemmas in this section.

In the sequel of the paper we will use the following notation. For a given feasible pair (y, s), $s > 0$, the matrix $(AS^{-2}A^T)_Q$ denotes $AS^{-2}A^T$ restricted to the columns of A in the index set Q, i.e.

$$(AS^{-2}A^T)_Q = \sum_{i\in Q} \frac{a_i a_i^T}{s_i^2}.$$

Moreover, we define

$$\|z\|_Q := \sqrt{z^T(AS^{-2}A^T)_Q^{-1} z},$$

assuming that $AS^{-2}A^T$ has full rank. The full index set $\{1, \cdots, n\}$ is denoted by N, and

$$\delta_i := \frac{s_i}{\|a_i\|_N},$$

for $i = 1, \cdots, n$. Note that δ_i is the distance to the i–th constraint measured in a certain metric.

3.1. Shifting a Constraint

Suppose the first constraint is shifted by a fraction β of the current slack s_1. So we replace the constraint

$$a_1^T y \leq c_1$$

by

$$a_1^T y \leq c_1 - \beta s_1, \ 0 \leq \beta \leq 1.$$

Let the asterisk $*$ refer to this new situation; so $s_1^* = (1-\beta)s_1$ and $s_i^* = s_i$ for $i = 2, \cdots, n$. The following lemma shows the effect on the δ–measure.

Lemma 6

$$\delta^*(y, \mu) \leq \delta(y, \mu) + \beta(1 + \delta(y, \mu)).$$

Proof. By definition

$$\delta^*(y, \mu) = \min_{x^*} \left(\left\| \frac{Sx^* - \beta s_1 x_1^* e_1}{\mu} - e \right\| \ : \ Ax^* = b \right).$$

For simplicity we denote $x = x(y, \mu)$, where $x(y, \mu)$ solves the minimization problem in (3). Taking $x^* = x$, it follows that

$$\delta^*(y, \mu) \leq \left\| \frac{Sx}{\mu} - e - \frac{\beta s_1 x_1 e_1}{\mu} \right\|$$

$$\leq \delta(y,\mu) + \frac{\beta s_1 x_1}{\mu}\|e_1\|$$
$$= \delta(y,\mu) + \frac{\beta s_1 x_1}{\mu}.$$

Now using that $|\frac{x_1 s_1}{\mu} - 1| \leq \delta(y,\mu)$, and hence $\frac{s_1 x_1}{\mu} \leq 1 + \delta(y,\mu)$, we obtain

$$\delta^*(y,\mu) \leq \delta(y,\mu) + \beta(1+\delta(y,\mu)),$$

which proves the lemma. □

The following lemma shows that after shifting the first constraint by an amount $\beta s_1(\mu)$, the slack variable $s_1^*(\mu)$ in the new central point $y^*(\mu)$ is smaller than $s_1(\mu)$, but the difference is at most the amount of shifting.

Lemma 7

$$(1-\beta)s_1(\mu) \leq s_1^*(\mu) \leq s_1(\mu)$$

Proof. We analyze the effect of shifting the first constraint by an amount η. For the moment, the central point for this new situation (with respect to μ) is denoted by $y(\mu,\eta)$. So $y^*(\mu) = y(\mu, \beta s_1(\mu))$, and $y(\mu) = y(\mu, 0)$. First note that $y(\mu,\eta)$ is the unique maximum of

$$f^*(y,\mu) = \frac{b^T y}{\mu} + \sum_{i=2}^{n} \ln(c_i - a_i^T y) + \ln(c_1 - a_1^T y - \eta),$$

and hence satisfies the following Kuhn–Tucker conditions:

$$\begin{aligned} Ax &= b \\ A^T y + \eta e_1 + s &= c \\ Sx &= \mu e. \end{aligned} \tag{5}$$

Taking the derivative with respect to η, denoted by a prime, we get

$$\begin{aligned} Ax' &= 0 \\ A^T y' + e_1 + s' &= 0 \\ Xs' + Sx' &= 0. \end{aligned} \tag{6}$$

From (5) and (6) we derive that

$$0 = AX^2 s' = -AX^2 A^T y' - AX^2 e_1,$$

and hence $y' = -(AX^2A^T)^{-1}AX^2 e_1$. This means that $s' = A^T(AX^2A^T)^{-1}AX^2e_1 - e_1$ and $Xs' = -P_{AX}(Xe_1) = -x_1 P_{AX}(e_1)$. Consequently,

$$s_1' = -e_1^T P_{AX}(e_1).$$

Since the eigenvalues of the projection matrix P_{AX} are zero or one, the result is that $-1 \le s_1' \le 0$, from which the lemma follows. □

The following lemma shows that the barrier function value in the central point decreases by at least β after shifting a constraint by a fraction β of the corresponding slack value.

Lemma 8

$$f^*(y^*(\mu), \mu) \le f(y(\mu), \mu) - \beta$$

Proof. Using the notation introduced in the proof of Lemma 7, we have

$$\frac{d\, f^*(y(\mu, \eta), \mu)}{d\, \eta} = \frac{b^T y'}{\mu} + \sum_{i=1}^{n} \frac{s_i'}{s_i}.$$

Since, by using (5) and (6),

$$b^T y' = x^T A^T y' = \mu e^T S^{-1} A^T y' = -\mu e^T S^{-1}(s' + e_1),$$

this results into

$$\frac{d\, f^*(y(\mu, \eta), \mu)}{d\, \eta} = -e^T S^{-1}(s' + e_1) + \sum_{i=1}^{n} \frac{s_i'}{s_i} = -e^T S^{-1} e_1 = -\frac{1}{s_1(\mu, \eta)}.$$

Now using the mean value theorem we obtain

$$\begin{aligned} f^*(y^*(\mu), \mu) &= f^*(y(\mu, \beta s_1(\mu)), \mu) \\ &= f^*(y(\mu, 0), \mu) + \beta s_1(\mu) \left. \frac{d\, f^*(y(\mu, \eta), \mu)}{d\, \eta} \right|_{0<\eta<\beta s_1(\mu)} \\ &= f(y(\mu), \mu) - \beta s_1(\mu) \left. \frac{1}{s_1(\mu, \eta)} \right|_{0<\eta<\beta s_1(\mu)} \\ &\le f(y(\mu), \mu) - \beta. \end{aligned}$$

The last inequality follows from $s_1(\mu, \eta) \le s_1(\mu, 0) = s_1(\mu)$, for $\eta \ge 0$, according to Lemma 7. This proves the lemma. □

3.2. Adding a Constraint

Suppose we add the constraint $a_0^T y \le c_0$. Let $s_0 > 0$ be the corresponding current slack variable. The next lemma states what the effect is on δ. An asterisk $*$ refers to the situation after adding the constraint.

Lemma 9 *Let $\delta := \delta(y, \mu)$. Then*

$$\delta^*(y, \mu) \le \begin{cases} \dfrac{1 + \delta\delta_0}{\sqrt{1 + \delta_0^2}} & \text{if } \delta_0 \ge \delta \\ \sqrt{1 + \delta^2} & \text{if } \delta_0 < \delta. \end{cases}$$

Proof. By definition

$$\begin{aligned}
\delta^*(y,\mu)^2 &= \min_{x^*,\xi}\left(\left\| \frac{1}{\mu}\begin{pmatrix} Sx^* \\ s_0\xi \end{pmatrix} - \begin{pmatrix} e \\ 1 \end{pmatrix} \right\|^2 \; : \; Ax^* + \xi a_0 = b \right) \\
&= \min_{x^*,\xi}\left(\left\| \frac{Sx^*}{\mu} - e \right\|^2 + \left(\frac{s_0\xi}{\mu} - 1\right)^2 \; : \; Ax^* + \xi a_0 = b \right).
\end{aligned}$$

Let $\Delta x := x^* - x(y,\mu)$, where $x(y,\mu)$ solves the minimization problem in (3). Then

$$\begin{aligned}
\delta^*(y,\mu)^2 &= \min_{\Delta x,\xi}\left(\left\| \frac{Sx(y,\mu)}{\mu} - e + \frac{S\Delta x}{\mu} \right\|^2 + \left(\frac{s_0\xi}{\mu} - 1\right)^2 \; : \; A\Delta x = -\xi a_0 \right) \\
&\leq \min_{\Delta x,\xi}\left(\left(\delta + \left\| \frac{S\Delta x}{\mu} \right\| \right)^2 + \left(\frac{s_0\xi}{\mu} - 1\right)^2 \; : \; A\Delta x = -\xi a_0 \right) \\
&= \min_{\Delta x,\xi}\left((\delta + \|S\Delta x\|)^2 + (s_0\xi - 1)^2 \; : \; A\Delta x = -\xi a_0 \right) \\
&= \min_{\xi}\left((\delta + |\xi| \|a_0\|_N)^2 + (s_0\xi - 1)^2 \right).
\end{aligned}$$

It is an easy task to prove that if $s_0 \geq \delta(y,\mu)\|a_0\|_N$, then the right–hand side is minimal for

$$\xi = \frac{s_0 - \delta\|a_0\|_N}{\|a_0\|_N^2 + s_0^2} = \frac{\delta_0 - \delta}{\|a_0\|_N(1+\delta_0^2)},$$

else $\xi = 0$ is optimal. Substituting these values gives the lemma. □

Note that according to Lemma 9, adding a constraint hardly influences the δ–measure if δ_0 is large, i.e. if the constraint which is added is far away from the current iterate. The next lemma states that if we add a constraint, then the corresponding slack value in the new central point is larger than in the old central point.

Lemma 10

$$s_0^*(\mu) \geq s_0(\mu)$$

Proof. First note that $y^*(\mu)$ is feasible for the old situation. Moreover, if $s_0(\mu) \leq 0$, then the lemma is obviously true, since $s_0^*(\mu) > 0$. So, we may assume that $s_0(\mu) > 0$, i.e. $y(\mu)$ is also feasible for the new situation. Since $y(\mu)$ maximizes $f(y,\mu)$ we have $f(y(\mu),\mu) \geq f(y^*(\mu),\mu)$. Since $y^*(\mu)$ maximizes $f^*(y,\mu)$ we have $f^*(y^*(\mu),\mu) \geq f^*(y(\mu),\mu)$. Now it is clear that

$$\begin{aligned}
\ln s_0(\mu) &= f^*(y(\mu),\mu) - f(y(\mu),\mu) \\
&\leq f^*(y^*(\mu),\mu) - f(y(\mu),\mu) \\
&\leq f^*(y^*(\mu),\mu) - f(y^*(\mu),\mu) \\
&= \ln s_0^*(\mu),
\end{aligned}$$

which means that $s_0^*(\mu) \geq s_0(\mu)$. □

The next lemma gives an upper bound for the barrier function value in the new central point after adding the constraint.

Lemma 11 *If* $\delta := \delta(y,\mu) \leq \frac{1}{4}$, *then*

$$f^*(y^*(\mu),\mu) - f^*(y,\mu) \leq \frac{1}{3} + \max\left(0, \ln\frac{4}{\delta_0}\right).$$

Proof. Suppose that $\delta_0 \geq 4$. Then according to Lemma 9

$$\delta^*(y,\mu) \leq \frac{1+\delta\delta_0}{\sqrt{1+\delta_0^2}} \leq \delta + \frac{1}{\sqrt{1+\delta_0^2}} \leq \frac{1}{4} + \frac{1}{\sqrt{1+16}} < \frac{1}{2}.$$

Consequently, we obtain, due to Lemma 3,

$$f^*(y^*(\mu),\mu) - f^*(y,\mu) \leq \frac{1}{3}. \tag{7}$$

Now suppose that $\delta_0 < 4$. Then we add the constraint in a shifted position such that δ_0 is exactly 4. Let us refer to this situation by using the superscript 0. Now we may write

$$\begin{aligned} f^*(y^*(\mu),\mu) - f^*(y,\mu) &= f^*(y^*(\mu),\mu) - f^0(y^0(\mu),\mu) \\ &+ f^0(y^0(\mu),\mu) - f^0(y,\mu) \\ &+ f^0(y,\mu) - f^*(y,\mu). \end{aligned} \tag{8}$$

Now we deal with the three pairs of terms in the right–hand side of (8) separately.

The first term $f^*(y^*(\mu),\mu) - f^0(y^0(\mu),\mu)$ is smaller than or equal to 0, since according to Lemma 8 the barrier function value in the exact central point decreases after shifting.

Note that according to Lemma 9

$$\delta^0(y,\mu) \leq \frac{1}{4} + \frac{1}{\sqrt{1+16}} < \frac{1}{2}.$$

From Lemma 3 it now follows that

$$f^0(y^0(\mu),\mu) - f^0(y,\mu) \leq \frac{1}{3}.$$

For the third term we simply have

$$f^0(y,\mu) - f^*(y,\mu) = \ln s_0^0 - \ln s_0 = \ln\frac{4\|a_0\|_N}{s_0} = \ln\frac{4}{\delta_0}.$$

Substituting all this into (8) yields

$$f^*(y^*(\mu),\mu) - f^*(y,\mu) \leq \frac{1}{3} + \ln\frac{4}{\delta_0}. \tag{9}$$

Combining (7) and (9) gives the lemma. □

3.3. Deleting a Constraint

In this section we consider the case that the constraint $a_j^T y \leq c_j$ is removed from the given problem, while assuming that the remaining constraint matrix has still full rank. As before, let the asterisk $*$ refer to the new situation. Now, letting $N^* := N \setminus \{j\}$, we have the following lemma, which will be needed in Lemma 13.

Lemma 12 *Let* $\delta_j := \frac{s_j}{\|a_j\|_N}$ *and* $\delta_j^* := \frac{s_j}{\|a_j\|_{N^*}}$. *Then* $\delta_j > 1$ *and*

$$\delta_j^* = \sqrt{\delta_j^2 - 1}.$$

Proof. To simplify the notation we will assume that $s = e$. This amounts to rescaling the columns of A by the slack values, and gives no loss of generality. Let A^* denote the matrix obtained from A by removing the j–th column. Then

$$A^*(A^*)^T = AA^T - a_j a_j^T.$$

Since $A^*(A^*)^T$ is invertible, Sherman–Morrison's formula states that

$$(A^*(A^*)^T)^{-1} = (AA^T)^{-1} + \frac{(AA^T)^{-1} a_j a_j^T (AA^T)^{-1}}{1 - a_j^T (AA^T)^{-1} a_j}.$$

Multiplying this from the left with a_j^T and from the right with a_j we obtain

$$\|a_j\|_{N^*}^2 = \|a_j\|_N^2 + \frac{\|a_j\|_N^4}{1 - \|a_j\|_N^2} = \frac{\|a_j\|_N^2}{1 - \|a_j\|_N^2}.$$

This can be rewritten as

$$\delta_j^{*2} = \delta_j^2 \left(1 - \frac{1}{\delta_j^2}\right) = \delta_j^2 - 1.$$

Note that certainly $\delta_j \geq 1$. To prove that $\delta_j > 1$, note that $\delta_j = 1$ if and only if $a_j^T (AA^T)^{-1} a_j = 1$. This can be written as $e_j^T A^T (AA^T)^{-1} A e_j = 1$. Since $A^T (AA^T)^{-1} A e_j$ equals the orthogonal projection of e_j onto the row space of A, we conclude that $\delta_j = 1$ holds if and only if e_j is in the row space of A. This is equivalent to having $x_j = 0$ whenever $Ax = 0$. In other words, no dependence relation between the columns of A contains the j–th column with nonzero coefficient. From this we conclude that $\delta_j = 1$ if and only if removing the j–th column from A decreases the rank. This completes the proof of the lemma, because we assumed that the remaining constraint matrix has still full rank. □

Lemma 13 *Let* $\delta := \delta(y, \mu)$. *Then*

$$\delta^*(y, \mu) \leq \delta + \frac{1 + \delta}{\sqrt{\delta_j^2 - 1}}.$$

Proof. Let A^* and S^* be defined as in the previous lemma. Then by definition

$$\delta^*(y,\mu) = \min_{x^*}\left(\left\|\frac{S^*x^*}{\mu} - e\right\| \;:\; A^*x^* = b\right).$$

Let $\Delta x := x^* - x^*(y,\mu)$, where $x(y,\mu)^T = (\xi, x^*(y,\mu)^T)$ solves the minimization problem for $\delta = \delta(y,\mu)$. Then

$$\delta^2 = \|\frac{S^*x^*(y,\mu)}{\mu} - e\|^2 + (\frac{s_j\xi}{\mu} - 1)^2.$$

Hence

$$|\frac{\xi}{\mu}| \le \frac{1+\delta}{s_j}. \tag{10}$$

So, we may write

$$\begin{aligned}
\delta^*(y,\mu) &= \min_{\Delta x}\left(\left\|\frac{S^*x^*(y,\mu)}{\mu} - e + \frac{S^*\Delta x}{\mu}\right\| \;:\; A^*\Delta x = \xi a_j\right) \\
&\le \delta + \min_{\Delta x}\left(\frac{1}{\mu}\|S^*\Delta x\| \;:\; A^*\Delta x = \xi a_j\right) \\
&= \delta + \frac{\xi}{\mu}\|a_j\|_{N^*} \\
&\le \delta + \frac{1+\delta}{s_j}\|a_j\|_{N^*},
\end{aligned}$$

where the last inequality follows from (10). Using the definition of δ_j^* we conclude that

$$\delta^*(y,\mu) \le \delta + \frac{1+\delta}{\delta_j^*}.$$

By the previous lemma, this completes the proof. □

4. The Build–Up and Down Algorithm

To keep the formulas simple we will assume in this section (as in Ye [17]) that the constraints $-e \le y \le e$ are included in the dual constraints $A^Ty \le c$. Let $J \subseteq N$ be the index set corresponding to the constraints $-e \le y \le e$. We will also assume that the columns of A have length 1. Note that each bounded problem can be formulated in this way by rescaling. At the end of this section we will show that these assumptions are not essential.

Under the above assumptions it can easily be proved (see Ye [17]), that if $Q \supseteq J$ then

$$\|a_i\|_Q^2 = a_i^T(AS^{-2}A^T)_Q^{-1}a_i \le \frac{1}{2}. \tag{11}$$

Based on the analysis in the previous section we will give strategies for working with subsets of the constraints instead of the full constraint matrix A. Both a build–up strategy (adding constraints) and a build–down strategy (deleting constraints) will be given.

In our approach we start with a certain (small) subset Q of the index set N, such that $Q \supseteq J$. Then we start the logarithmic barrier method, with respect to the constraints in Q. This means that we work with the problem

$$(D_Q) \qquad \max\left\{b^T y \ : \ a_i^T y \leq c_i, \ i \in Q\right\},$$

instead of problem (D). If the current iterate is close to (or violates) a constraint which is not in Q, it is added to the current system, and we go back to the previous iterate. To be more precise: we check in each iteration if there is an index $i \notin Q$ such that

$$s_i < 2^{-t_a} \ \text{ or } \ s_i < 2^{-t_a}\hat{s}_i, \tag{12}$$

where t_a is some "adding" parameter, and $\hat{s}$ is the slack vector of the dual iterate which was almost centered with respect to the previous value of the barrier parameter. If there is such a constraint, we add it to our system, go back to the previous iterate (for which $s_i \geq 2^{-t_a}$ and $s_i \geq 2^{-t_a}\hat{s}_i$) and continue the process. Consequently, all iterates are feasible for the whole problem. This is the build–up part.

If the slack value of a constraint in the current iterate is sufficiently large (say $s_i \geq t_d$, where t_d is a "deleting" parameter), we will remove it from our current system, since it is then likely that this constraint will be nonbinding in an optimal solution. After removing constraints, we recenter when necessary. Note that it is not sure that this constraint is nonbinding in an optimal solution, but, since we have a strategy for adding constraints, this causes no problem: constraints which are incorrectly removed, will be added later on. To avoid "cycling", i.e. the case that a constraint is deleted and added many (in fact infinitely many) times, we will only delete constraints if we are close to the central path.

Before describing the algorithm we introduce some notation. Let $\delta_Q(y,\mu)$, $f_Q(y,\mu)$ and p_Q denote the δ–measure, the barrier function and the Newton direction, respectively, with respect to the subsystem Q. Recall that computing $(AS^{-2}A^T)_Q$ requires less computational time than computing $AS^{-2}A^T$, especially when $|Q|$ is relatively small.

The algorithm goes as follows.

Build–Up and Down Algorithm

Input:
$\mu = \mu_0$ is the barrier parameter value;
ϵ is the desired accuracy of the final solution;
θ is the reduction parameter, $0 < \theta < 1$;
Q is the initial subset of constraints, $Q \supseteq J$;
$y = \hat{y}$ is a given interior feasible point for (D) such that $\delta_Q(\hat{y},\mu) \leq \frac{1}{4}$;

begin

```
while s^T x(y, μ) > ε do
    begin
      Delete–Constraints;
      μ := (1 − θ)μ;
      Center–and–Add–Constraints
    end
end.
```

The procedures **Delete–Constraints** and **Center–and–Add–Constraints** are defined as follows.

Procedure Delete–Constraints

Input:
$t_d \geq 4$ is a "deleting" parameter;

```
begin
  for i := 1 to n do
    if i ∈ Q \ J and s_i ≥ t_d then
    begin
      Q := Q \ {i};
      if δ_Q(y, μ) > 1/4 then Center–and–Add–Constraints
    end
end.
```

Procedure Center–and–Add–Constraints

Input:
t_a is an "adding" parameter;

```
begin
  while δ_Q(y, μ) > 1/4 do
    begin
      α̃ := arg max_{α>0} { f_Q(y + α p_Q, μ) : s_i − α a_i^T p_Q > 0, ∀i ∈ Q };
      if ∃i ∉ Q : s_i − α̃ a_i^T p_Q < 2^{−t_a} max(1, ŝ_i) then
        Q := Q ∪ {i : s_i − α̃ a_i^T p_Q < 2^{−t_a} max(1, ŝ_i), i ∉ Q}
      else y := y + α̃ p_Q
    end
  ŷ := y
end.
```

In [4] it is shown how to obtain a starting point y and set Q such that y is feasible for the whole problem and $\delta_Q(y,\mu) \leq \frac{1}{4}$.

5. Complexity Analysis

First we analyze the procedure Center–and–Add–Constraints. Let q be the cardinality of the current subset Q. Lemmas 4 and 5 give an upper bound for the number of Newton iterations required by the standard logarithmic barrier method between two reductions of the barrier parameter. Lemma 15 below gives a similar upper bound if also constraints are added between the two reductions. The following lemma will be needed in the proof of Lemma 15. It generalizes Lemma 5 and Lemma 11.

Lemma 14 *If $\delta = \delta(y,\mu) \leq \frac{1}{4}$, and $\overline{\mu} := (1-\theta)\mu$, then*

$$f^r(y^r(\overline{\mu}),\overline{\mu}) - f^r(y,\overline{\mu}) \leq \frac{1}{3} + \frac{\theta}{1-\theta}\left(\theta(q+r) + 3\sqrt{q+r}\right) + \sum_{k=1}^{r} \max\left(0, \ln\frac{2r\sqrt{2}}{s_{i_k}}\right),$$

where the superscript r refers to the situation that constraints $i_1, \cdots, i_r$ are added.

Proof. The proof generalizes the proof of Lemma 11. We construct situation 0 by adding constraint $i_1, \cdots, i_r$ in a shifted position such that

$$s^0_{i_k} = \max(s_{i_k}, 2r\sqrt{2}),$$

which means that $\delta^0_{i_k} \geq 4r$.

Note that

$$\begin{aligned} f^r(y^r(\overline{\mu}),\overline{\mu}) - f^r(y,\overline{\mu}) &= f^r(y^r(\overline{\mu}),\overline{\mu}) - f^0(y^0(\overline{\mu}),\overline{\mu}) \\ &+ f^0(y^0(\overline{\mu}),\overline{\mu}) - f^0(y,\overline{\mu}) \\ &+ f^0(y,\overline{\mu}) - f^r(y,\overline{\mu}). \end{aligned} \tag{13}$$

Now we deal with the three pairs of terms in the right–hand side of (13) separately.

The first term $f^r(y^r(\overline{\mu}),\overline{\mu}) - f^0(y^0(\overline{\mu}),\overline{\mu})$ is smaller than or equal to 0, since according to Lemma 8 the barrier function value in the exact central point decreases after shifting.

Repeatedly applying Lemma 9 and using $\delta^0_{i_k} \geq 4r$, we have that

$$\delta^0(y,\mu) \leq \delta + \sum_{k=1}^{r} \frac{1}{\sqrt{1+(\delta^0_{i_k})^2}} \leq \frac{1}{4} + r\frac{1}{\sqrt{1+16r^2}} < \frac{1}{2}.$$

From Lemma 5 it now follows that

$$f^0(y^0(\overline{\mu}),\overline{\mu}) - f^0(y,\overline{\mu}) \leq \frac{1}{3} + \frac{\theta}{1-\theta}\left(\theta(q+r) + 3\sqrt{q+r}\right).$$

For the third term we simply have

$$f^0(y,\overline{\mu}) - f^r(y,\overline{\mu}) = \sum_{k=1}^{r}(\ln s^0_{i_k} - \ln s_{i_k}) = \sum_{k=1}^{r} \max\left(0, \ln\frac{2r\sqrt{2}}{s_{i_k}}\right).$$

Substituting all this into (13) gives the lemma. □

Lemma 15 *Between two reductions of the barrier parameter, the Build-Up and Down Algorithm requires at most*

$$O\left(1+\frac{\theta}{1-\theta}(\theta(q+r)+3\sqrt{q+r})+rt_a+r\ln r\right)$$

Newton iterations if r constraints were added.

Proof. Let $\overline{\mu}$ be the current value and μ the previous value of the barrier parameter (i.e. $\overline{\mu}:=(1-\theta)\mu$). Let $i_1,\cdots,i_r$ denote the indices of the r constraints which are added to Q while the barrier parameter has value $\overline{\mu}$, and let y^k, for $k=1,\cdots,r$, denote the iterate just after the i_k-th constraint has been added, whereas $y^k(\overline{\mu})$ and $f^k(y^k,\overline{\mu})$ denote the corresponding central point and barrier function value, respectively. Finally, let y^0 denote the iterate at the moment that μ is reduced to $\overline{\mu}$ (i.e. y^0 is in the vicinity of $y^0(\mu)$), and $f^0(y^0,\mu)$ the corresponding barrier function value.

Since in each iteration the barrier function value increases by a constant (Lemma 4), the number of Newton iterations needed to go from y^0 to the vicinity of $y^r(\overline{\mu})$ is proportional to at most

$$P:=f^0(y^1,\overline{\mu})-f^0(y^0,\overline{\mu})+\sum_{k=1}^{r-1}(f^k(y^{k+1},\overline{\mu})-f^k(y^k,\overline{\mu}))+f^r(y^r(\overline{\mu}),\overline{\mu})-f^r(y^r,\overline{\mu}). \tag{14}$$

Since $f^k(y^k,\overline{\mu})=f^{k-1}(y^k,\overline{\mu})+\ln s^k_{i_k}$,

$$\begin{aligned}\sum_{k=1}^{r-1}(f^k(y^{k+1},\overline{\mu})-f^k(y^k,\overline{\mu})) &= \sum_{k=1}^{r-1}(f^k(y^{k+1},\overline{\mu})-f^{k-1}(y^k,\overline{\mu})-\ln s^k_{i_k})\\ &= f^{r-1}(y^r,\overline{\mu})-f^0(y^1,\overline{\mu})-\sum_{k=1}^{r-1}\ln s^k_{i_k}.\end{aligned}$$

Substituting this into (14), while using $f^r(y^r,\overline{\mu})=f^{r-1}(y^r,\overline{\mu})+\ln s^r_{i_r}$, we obtain

$$\begin{aligned}P &= f^r(y^r(\overline{\mu}),\overline{\mu})-f^0(y^0,\overline{\mu})-\sum_{k=1}^{r}\ln s^k_{i_k}\\ &= f^r(y^r(\overline{\mu}),\overline{\mu})-f^r(y^0,\overline{\mu})+\sum_{k=1}^{r}\ln\frac{s^0_{i_k}}{s^k_{i_k}}.\end{aligned} \tag{15}$$

From Lemma 14 we have

$$f^r(y^r(\overline{\mu}),\overline{\mu})-f^r(y^0,\overline{\mu})\le\frac{1}{3}+\frac{\theta}{1-\theta}(\theta(q+r)+3\sqrt{q+r})+\sum_{k=1}^{r}\max\left(0,\ln\frac{2r\sqrt{2}}{s^0_{i_k}}\right).$$

Consequently,

$$P\le\frac{1}{3}+\frac{\theta}{1-\theta}(\theta(q+r)+3\sqrt{q+r})+\sum_{k=1}^{r}\ln\frac{\max(s^0_{i_k},2r\sqrt{2})}{s^k_{i_k}}.$$

The lemma follows because

$$s^k_{i_k} \geq 2^{-t_a} \max(s^0_{i_k}, 1).$$

□

Let us now analyze the procedure Delete–Constraints. Lemma 13 gives that if the i–th constraint, for which $s_i \geq t_d \geq 4$, is deleted in an iterate near the path, then

$$\delta^*(y, \mu) \leq \frac{1}{4} + \frac{1 + \frac{1}{4}}{4} \frac{1}{\sqrt{2}} < \frac{1}{2}.$$

From this point we have to recenter. As a consequence of Lemma 15 we obtain that recentering costs at most $O(1 + rt_a + r \ln r)$ Newton iterations, if r constraints have to be added.

Note that it is not excluded in the algorithm that a constraint is removed at each outer iteration, and added during each inner loop. To avoid this undesirable behavior, several strategies can be added to the algorithm; e.g. a constraint may be deleted only once (or a fixed number of times). In the sequel K will denote the maximal number of times a constraint may be deleted. Note that K is certainly not larger than the number of updates of μ, so using Theorem 1

$$K \leq \frac{1}{\theta} \ln \frac{q^* \mu_0}{\epsilon} := K_{\max},$$

where q^* denotes the maximum number of constraints in the subsystem during the whole process.

Theorem 2 *After at most*

$$O\left((K+1)(q^* \ln q^* + q^* t_a) + \left(\frac{\theta q^*}{1-\theta} + \frac{1}{\theta}\right) \ln \frac{q^* \mu_0}{\epsilon}\right)$$

Newton iterations an ϵ–optimal solution has been found for (D), where q^* denotes the maximal number of constraints in the subsystem during the whole process.

Proof. From Theorem 1 it follows that the standard logarithmic barrier algorithm needs

$$O\left(\left(\frac{\theta q^*}{1-\theta} + \frac{1}{\theta}\right) \ln \frac{q^* \mu_0}{\epsilon}\right)$$

Newton iterations to obtain an ϵ–optimal solution for (D_{Q^*}). Since this solution is also feasible for (D) and the subsystem forms a relaxation of (D), it is also an ϵ–optimal solution for (D).

During the process of the algorithm at most Kq^* constraints were deleted, which costs $O(Kq^*)$ additional Newton steps. Moreover, at most $(K+1)q^*$ constraints were added, which costs, according to Lemma 15, at most $O\left((K+1)(q^* t_a + q^* \ln q^*)\right)$ additional Newton iterations. Summing all Newton iterations proves the theorem. □

Suppose we want a 2^{-2L}–optimal solution to obtain an optimal solution, then we have to take $\epsilon = 2^{-2L}$. Now, according to Theorem 2, we have the following corollary:

Corollary: To obtain a 2^{-2L}–optimal solution the algorithm requires

- $O(\sqrt{q^*}L)$ Newton iterations if we take $K = O(1)$, $t_a = O\left(\frac{L}{\sqrt{n}}\right)$, and $\theta = \frac{\nu}{\sqrt{n}}$, where q is the current number of constraints in our subsystem, $\nu > 0$ and independent of q, n and L;
- $O(q^* L)$ Newton iterations if we take $K = O(1)$, $t_a = O(L)$ and $0 \leq \theta \leq 1$;
- $O(q^* L \ln q^*)$ Newton iterations if we take $K = K_{\max}$ and $t_a = O(\ln q)$, where q is the current number of constraints in our subsystem, and $0 \leq \theta \leq 1$.

□

Note that, since it is likely that $q^* < n$, the iteration bounds in the first two cases are better than that for the standard logarithmic barrier method (Theorem 1).

We conclude with several remarks.

Remark 1. The same results can be obtained without assuming that the box constraints $-e \leq y \leq e$ are included and that the columns of A have unit length. For this purpose we have to change the criterion (12) for adding constraints into

$$s_i < 2^{-t_a} \max(\hat{s}_i, \|a_i\|_Q),$$

and the criterion for deleting a constraint into

$$s_i > t_d \|a_i\|_Q.$$

Remark 2. At the end of Ye's [17] paper, it is said to be a further research topic how deleting inequalities will affect the potential function (the potential function he used is $f(y,\mu)$, with $b = 0$). The answer can be obtained very easily from Lemmas 13 and 3. Suppose that the current point is centered, i.e. $\delta(y,\mu) \leq \frac{1}{4}$, and that an inequality is removed if the corresponding slack value is larger than 4. Then according to Lemma 13 we have $\delta^*(y,\mu) \leq \frac{1}{2}$. Now using Lemma 3 we have

$$\begin{aligned} f^*(y^*(\mu),\mu) &\leq f^*(y,\mu) + \frac{1}{3} \\ &\leq f(y,\mu) - \ln 4 + \frac{1}{3} \\ &\leq f(y(\mu),\mu) - \ln 4 + \frac{1}{3} \\ &< f(y(\mu),\mu) - 1. \end{aligned}$$

This means that removing such a constraint reduces the potential function value by at least one. Removing a constraint which is close to the current iterate will, of course, increase the potential function value. It can easily be verified that a decrease can still be guaranteed if the corresponding slack value of the removed inequality is larger than 2.16.

Remark 3. It will be a major improvement if we could prove that $q^* = \Theta(m)$ for (a variant of) the Build–Up and Down Algorithm, i.e. that at most a multiple of m constraints are in the working set. This is a subject of further research. The following observation is important in this respect[1]. Suppose that Q is the current working set, then

$$\sum_{i\in Q}\frac{1}{\delta_i^2} = \sum_{i\in Q} e_i^T S_Q^{-1} A_Q^T (AS^{-2}A^T)_Q^{-1} A_Q S_Q^{-1} e_i = \text{trace}(I - P_{A_Q S_Q^{-1}}) = m,$$

where the last equality follows since the projection matrix $I - P_{A_Q S_Q^{-1}}$ has m eigenvalues 1, and the trace of a matrix equals its sum of the eigenvalues. Now it easily follows that, if there are $q = \gamma m$ constraints in the current working set, then for at least one of the constraints we must have

$$\delta_i \geq \sqrt{\frac{q}{m}} = \sqrt{\gamma},$$

which means that if γ is not too small then this constraint is "far" from the current iterate and can therefore be deleted.

Another subject of further research is to generalize the approach given in this section to (non–differentiable) convex programming. Starting with an LP–relaxation of the problem, we can apply the standard logarithmic barrier method. If the iterate is close to the boundary (or out) of the feasible region, a new linear constraint can be added to the LP–relaxation[2].

References

1. Atkinson, D.S., and Vaidya, P.M.: 1992, 'A Cutting Plane Algorithm that Uses Analytic Centers', Working Paper, Department of Mathematics, University of Illinois at Urbana–Champaign.
2. Bahn, O., Goffin, J.L., Vial, J.–Ph., and Du Merle, O.: 1991, 'Implementation and Behavior of an Interior Point Cutting Plane Algorithm for Convex Programming: an Application to Geometric Programming', Working Paper, Université de Genève, Genève, Switzerland. To Appear in *Annals of Discrete Mathematics*.
3. Dantzig, G.B., and Ye, Y.: 1990, 'A Build–Up Interior Method for Linear Programming: Affine Scaling Form', Technical Report SOL 90–4, Systems Optimzation Laboratory, Department of Operations Research, Stanford University, Stanford, California.
4. Den Hertog, D., Roos, C., and Terlaky, T.: 1992, 'A Build–Up Variant of the Path–Following Method for LP', *Operations Research Letters* Vol. no. **12**, pp. 181–186.
5. Den Hertog, D., Roos, C., and Vial, J.–Ph.: 1992, 'A Complexity Reduction for the Long–Step Path–Following Algorithm for Linear Programming', *SIAM Journal on Optimization* Vol. no. **2**, pp. 71–87.
6. Den Hertog, D., Kaliski, J.A., Roos, C., and Terlaky, T.: 1992, 'A Logarithmic Barrier Cutting Plane Method for Convex Programming', Report No. 93–43, Faculty of Mathematics and Informatics, Delft University of Technology, Delft, Holland.
7. Goffin, J.L., and Vial, J.–Ph.: 1990, 'Cutting Planes and Column Generation Techniques with the Projective Algorithm', *Journal of Optimization Theory and Applications* Vol. no. **65**, pp. 409–429.

---

[1] This observation was made by Kurt Anstreicher. See also [13].

[2] During his stay at Delft, February–August 1992, John Kaliski developed such an implementation and obtained very promising results; see [6].

8. Kaliski, J.A., and Ye, Y.: 1991, 'A Decomposition Variant of the Potential Reduction Algorithm for Linear Programming', Working Paper 91–11, Department of Management Sciences, The University of Iowa, Iowa City, Iowa.
9. Karmarkar, N.: 1984, 'A New Polynomial–Time Algorithm for Linear Programming', *Combinatorica* Vol. no. **4**, pp. 373–395.
10. Mitchell, J.E.: 1988, 'Karmarkar's Algorithm and Combinatorial Optimization Problems', Ph.D. Thesis, Cornell University, Ithaca.
11. Roos, C. and Vial, J.-Ph.: 1992, 'A Polynomial Method of Approximate Centers for Linear Programming', *Mathematical Programming* Vol. no. **54**, pp. 295–305.
12. Tone, K.: 1991, 'An Active–Set Strategy in Interior Point Method for Linear Programming', Working Paper, Graduate School of Policy Science, Saitama University, Urawa, Saitama 338, Japan.
13. Vaidya, P.M.: 1989, 'A New Algorithm for Minimizing Convex Functions over Convex Sets', *Proceedings of 30th Annual IEEE Symposium on Foundations of Computer Science*, IEEE Computer Society Press, Los Alamitos, California, 338–343. To Appear in *Mathematical Programming*.
14. Ye, Y.: 1989, 'Eliminating Columns and Rows in Potential Reduction and Path–Following Algorithms for Linear Programming', Working Paper Series No. 89–7, Department of Management Sciences, The University of Iowa, Iowa City, Iowa.
15. Ye, Y.: 1990, 'The "Build–Down" Scheme for Linear Programming', *Mathematical Programming* Vol. no. **46**, pp. 61–72.
16. Ye, Y.: 1991, 'An $O(n^3L)$ Potential Reduction Algorithm for Linear Programming', *Mathematical Programming* Vol. no. **50**, pp. 239–258.
17. Ye, Y.: 1992, 'A Potential Reduction Algorithm Allowing Column Generation', *SIAM Journal on Optimization* Vol. no. **2**, pp. 7–20.

A PROJECTION METHOD FOR SOLVING INFINITE SYSTEMS OF LINEAR INEQUALITIES

H. HU
Department of Mathematical Sciences
Northern Illinois University
DeKalb, IL 60115-2888

Abstract. This paper presents a projection method for solving an infinite system of linear inequalities. The solution set of the system may be bounded or unbounded, and the interior of the solution set may be empty or nonempty. The method is extended to an infinite system of convex inequalities.

1. Introduction

In many theoretical and practical problems one needs to find a solution of an infinite system of linear inequalities:

$$\text{(I)} \qquad a(u)^T x \leq b(u) \quad \text{for all } u \in U,$$

where $x \in R^n$, $U \subseteq R^m$ is an infinite parameter set, $a : U \to R^n$, and $b : U \to R^1$. For example, finding a positive diagonal solution to the Lyapunov matrix equation is equivalent to solving a certain infinite system of linear inequalities. The solution set is always unbounded and the interior may be empty [4]. For a second example, many methods for semi-infinite programming need a feasible solution as a starting point [2,3]. This paper presents a projection method for solving a general infinite system of linear inequalities. The basic idea of the method is: Let x^0 be an arbitrary point. If the current iterate x^k is not a solution of the system to be solved, then let x^{k+1} be the orthogonal projection of x^k on the boundary of a linear inequality near the most violated inequality by x^k. If the system is consistent, then the sequence $\{x^k : k = 1, 2, \ldots\}$ generated by the method converges to a solution of the system. The convergence rate is linear if certain regularity conditions are satisfied. The method can be considered as the adaptation to an infinite system of inequalities of the method by Agmon [1]. The convergence proofs, which would apply if the system were finite, are quite different from that in [1]. The advantages of the method include: It allows the solution set to be unbounded and to have empty interior. It does not require finding the most violated constraint at each iteration. It does not need solving a sequence of linear programs whose dimensions tend to infinity. The method is extended to an infinite systems of convex inequalities.

2. The Projection Method

Throughout this paper we assume that $a(u) \neq 0$ for all $u \in U$, $\sup\{\|a(u)\| : u \in U\} < \infty$, and $\inf\{b(u) : u \in U\} > -\infty$.

Ding-Zhu Du and Jie Sun (eds.), Advances in Optimization and Approximation, 186–194.

Let $S \equiv \{x \in R^n : a(u)^T x \leq b(u) \text{ for all } u \in U\}$ be the solution set of (I).

For any real number a, we define

$$a^+ = \begin{cases} a, & \text{if } a \geq 0; \\ 0, & \text{if } a < 0. \end{cases}$$

Let $d(x) = \min\{\, \|x - y\| : y \in S\}$ be the Euclidean distance from x to S.

Let $H(x) = \sup\{\, (a(u)^T x - b(u))^+ : u \in U\}$ be the "biggest violation" by x of (I). It is obvious that $0 \leq H(x) < \infty$ for all $x \in R^n$, $H(x) = 0$ if and only if $x \in S$, and $H(x)$ is a continuous convex function on R^n. For a fixed x, if there exists a $\bar{u} \in U$ such that $a(\bar{u})^T x - b(\bar{u}) > 0$, then $H(x) = \sup\{a(u)^T x - b(u) : u \in U\}$.

Let $\{\epsilon_k : k = 0,\ 1,\ \ldots\}$ be a sequence of positive numbers such that $\lim_{k \to \infty} \epsilon_k = 0$.

Method 1.

Step 0.

Let $k := 0$;

let x^0 be an arbitrary point in R^n.

Step 1.

Find a $\bar{u} \in U$ such that $a(\bar{u})^T x^k - b(\bar{u}) \geq \sup\{a(u)^T x^k - b(u) : u \in U\} - \epsilon_k$.

Step 2.

If $a(\bar{u})^T x^k - b(\bar{u}) \leq -\epsilon_k$, then x^k is a solution of (I), stop.

If $-\epsilon_k < a(\bar{u})^T x^k - b(\bar{u}) \leq 0$, then $\epsilon_k := \epsilon_k / 2$, go to Step 1.

If $a(\bar{u})^T x^k - b(\bar{u}) > 0$, then $u^k := \bar{u}$, go to Step 3.

Step 3.

Let $x^{k+1} := x^k - (a(u^k)^T x^k - b(u^k)) a(u^k) / \|a(u^k)\|^2$

be the orthogonal projection of x^k on $a(u^k)^T x = b(u^k)$;

$k := k + 1$, go to Step 1.

Comments on Method 1.

(1) Generally speaking, $\sup\{a(u)^T x^k - b(u) : u \in U\}$ is a nonlinear program. We assume that, given x^k and ϵ_k, there are nonlinear programming algorithms that can find an approximate solution $\bar{u}$ as described in Step 1. While many methods for semi-infinite programming require finding an exact optimal solution of this nonlinear programming problem at every iteration [2,3].

(2) If $-\epsilon_k < a(\bar{u})^T x^k - b(\bar{u}) \leq 0$, then $\sup\{a(u)^T x^k - b(u) : u \in U\} \leq \epsilon_k$. The x^k satisfying this inequality is called an ϵ_k-solution of (I). The distance from an ϵ_k-solution to the solution set S is dominated by a constant multiple of ϵ_k if certain regularity conditions are satisfied [5]. Thus one may want to stop the method in Step 2 if ϵ_k is sufficiently small and $-\epsilon_k < a(\bar{u})^T x^k - b(\bar{u}) \leq 0$.

(3) If at a certain iteration k the method repeats Steps 1 and 2 infinitely many times before going to Step 3, then ϵ_k becomes arbitrarily small and hence x^k is an exact solution of (I).

(4) ϵ_k may be specified as the method goes on.

(5) If, for every x^k, there exists a $u^k \in U$ such that

$$a(u^k)^T x^k - b(u^k) = \max\{a(u)^T x^k - b(u) : u \in U\}$$

and if there is a method that can find such a u^k, then Steps 1 and 2 of Method 1 can be simplified as follows. Find a $u^k \in U$ such that $a(u^k)^T x^k - b(u^k) = \max\{a(u)^T x^k - b(u) : u \in U\}$. If $a(u^k)^T x^k - b(u^k) \leq 0$, then x^k is a solution of (I), stop.

Without loss of generality, we may assume that the method generates an infinite sequence $\{x^k : k = 0, 1, \ldots\}$, i.e., at any iteration k the method repeats Steps 1 and 2 finitely many times before going to Step 3.

Theorem 1. If (I) is consistent and Method 1 does not stop after finitely many iterations, then the sequence $\{x^k : k = 0, 1, \ldots\}$ generated by Method 1 converges to a solution of (I).

Proof. For any $y \in S$ and any $k \geq 0$, since $a(u^k)^T x^k - b(u^k) > 0$, $a(u^k)^T y - b(u^k) \leq 0$, and x^{k+1} is the orthogonal projection of x^k on $a(u^k)^T x = b(u^k)$,

$$\|x^{k+1} - y\|^2 \leq \|x^k - y\|^2 - \|x^{k+1} - x^k\|^2. \tag{1}$$

For all $k = 0, 1, \ldots$, let y^k be the point in S nearest to x^k. Using (1) repeatedly, we have that for any fixed $k > 0$,

$$\|x^{n+k} - y^k\|^2 \leq \|x^k - y^k\|^2 = d(x^k)^2 \quad \text{for all } n = 0,\ 1,\ \ldots. \tag{2}$$

That is, $\{x^{n+k} : n = 0, 1, \ldots\}$ is contained in the closed ball $B(y^k, d(x^k))$ with center y^k and radius $d(x^k)$. It follows that the sequence $\{x^k : k = 0, 1, \ldots\}$ is bounded. By (1) and the fact that y^{k+1} is the point in S nearest to x^{k+1},

$$d(x^{k+1}) = \|x^{k+1} - y^{k+1}\| \leq \|x^{k+1} - y^k\| < \|x^k - y^k\| = d(x^k). \tag{3}$$

That is, x^k approaches S steadily. Moreover,

$$\begin{aligned}\|x^{k+1} - y^{k+1}\|^2 &\leq \|x^{k+1} - y^k\|^2 \\ &\leq \|x^k - y^k\|^2 - \|x^{k+1} - x^k\|^2 \\ &= \|x^k - y^k\|^2 \left(1 - \|x^{k+1} - x^k\|^2 / \|x^k - y^k\|^2\right).\end{aligned}$$

Now let $\tau_k \equiv \left(1 - \|x^{k+1} - x^k\|^2 / \|x^k - y^k\|^2\right)^{\frac{1}{2}}$ for all $k = 0, 1, \ldots$, then $0 < \tau_k < 1$ and

$$\|x^{k+1} - y^{k+1}\| \leq \tau_k \|x^k - y^k\|. \tag{4}$$

It follows that

$$d(x^{k+1}) \leq (\Pi_{i=0}^{k} \tau_i) d(x^0). \tag{5}$$

In order to show that $\{x^k : k = 0, 1, \ldots\}$ is a convergent sequence, it suffices to show that $d(x^k) \to 0$ for this implies that $\cap_{k=0}^{\infty} B(y^k, d(x^k))$ is a singleton. As $\{\Pi_{i=0}^{k} \tau_i \geq 0 : k = 1, 2, \ldots\}$ is a decreasing sequence, we discuss two cases.

Case 1. $\Pi_{i=0}^{\infty} \tau_i = 0$. By (5), $d(x^k) \to 0$.

Case 2. $\Pi_{i=0}^{\infty} \tau_i > 0$. This implies that $\tau_i \to 1$. By the definition of τ_k and (3),

$$\|x^{k+1} - x^k\|^2 = (1 - \tau_k^2) d(x^k)^2 \leq (1 - \tau_k^2) d(x^0)^2.$$

Hence,

$$\|x^{k+1} - x^k\| \to 0. \tag{6}$$

As the method does not stop at iteration k and $a(u^k)^T x^k - b(u^k) > 0$,

$$\begin{aligned} 0 \le H(x^k) &= \sup\{\,(a(u)^T x^k - b(u))^+ : u \in U\} \\ &= \sup\{\,a(u)^T x^k - b(u) : u \in U\} \\ &\le a(u^k)^T x^k - b(u^k) + \epsilon_k \\ &= \|x^{k+1} - x^k\|\,\|a(u^k)\| + \epsilon_k \\ &\le \sup\{\,\|a(u)\| : u \in U\}\,\|x^{k+1} - x^k\| + \epsilon_k. \end{aligned} \tag{7}$$

It follows from (6) and (7) that $\lim_{k\to\infty} H(x^k) = 0$. Since $\{x^k : k = 1,\ 2,\ \ldots\}$ is bounded and $H(x)$ is a continuous function on R^n, for a convergent subsequence $x^{k_j} \to x^*$ we have

$$\lim_{j\to\infty} H(x^{k_j}) = H(x^*) = 0 \quad \text{and} \quad x^* \in S. \tag{8}$$

Recall that $d(x^k)$ is a decreasing sequence, (8) implies $\lim_{k\to\infty} d(x^k) = 0$, and thus $\cap_{k=0}^{\infty} B(y^k, d(x^k))$ is a singleton. ∎

Corollary 1. (I) is consistent if and only if $\{x^k : k = 0,\ 1,\ \ldots\}$ is bounded and $\lim_{k\to\infty} \|x^{k+1} - x^k\| = 0$, if and only if $\{x^k : k = 0,\ 1,\ \ldots\}$ converges to a solution of (I). ∎

3. Convergence Rate

In this section we show that the convergence rate of Method 1 is linear if S contains an interior point and $\inf\{\,\|a(u)\| : u \in U\} > 0$, and if we slightly modify (in a way we will explain later) Step 2 of Method 1. The following lemma plays an important role in the proof of linear convergence rate. It is based on a theorem in [5].

Lemma 1. Let the sequence $\{x^k : k = 0,\ 1,\ \ldots\}$ be generated by Method 1. If S contains an interior point and $\inf\{\,\|a(u)\| : u \in U\} > 0$, then there exists a constant $0 < \gamma < 1$ such that $H(x^k) \ge \gamma d(x^k)$ for all $k = 0, 1, \ldots$.

Proof. We assume, temporarily, that the origin is in the interior of S. Therefore, there exists $\alpha > 0$ such that $B(0, \alpha)$ is contained in S, which implies that the distance from the origin to any $a(u)^T x = b(u)$ is at least α, i.e., $b(u)/\|a(u)\| \ge \alpha$ for all $u \in U$. Hence, there exists $\delta > 0$ satisfying

$$b(u) \ge \delta \ \text{ for all } u \in U. \tag{9}$$

For any *fixed* $x^k \in \{x^j : j = 0,\ 1,\ \ldots\}$, let y^k be the point in S nearest to x^k. It is well known that the halfspace $(x^k - y^k)^T z \le (x^k - y^k)^T y^k$ contains S. By an infinite version of Farkas's Lemma (see, e.g., [6] pp.159-160), there exist vectors $u^{ij} \in U$ and coefficients $\lambda_{ij} \geq 0$, $i = 1, 2, \ldots, j = 0, \ldots, n$ such that

$$\lim_{i\to\infty} \sum_{j=1}^{n} \lambda_{ij} a(u^{ij}) = x^k - y^k \tag{10}$$

and

$$\lim_{i\to\infty} (\lambda_{i0} + \sum_{j=1}^{n} \lambda_{ij} b(u^{ij})) = (x^k - y^k)^T y^k. \tag{11}$$

Therefore,

$$\lim_{i\to\infty} (\lambda_{i0} + \sum_{j=1}^{n} \lambda_{ij} (b(u^{ij}) - a(u^{ij})^T y^k)) = 0. \tag{12}$$

As $\lambda_{ij} \geq 0$ and $b(u^{ij}) - a(u^{ij})^T y^k \geq 0$ for all $i = 1, 2, \ldots, j = 0, \ldots, n$,

$$\lim_{i\to\infty} \lambda_{i0} = 0 \tag{13}$$

and

$$\lim_{i\to\infty} \sum_{j=1}^{n} \lambda_{ij} (b(u^{ij}) - a(u^{ij})^T y^k) = 0. \tag{14}$$

Let $M = \sup\{ \|y^k\| : k = 0, 1, \ldots\}$. $M < +\infty$ because $x^k \to x^*$ and $\|y^k - x^k\| = d(x^k) \to 0$. It follows from (9), (11) and (13) that

$$\limsup_{i\to\infty} \sum_{j=1}^{n} \lambda_{ij} \leq \delta^{-1} M \|x^k - y^k\|. \tag{15}$$

By (10), (14), and (15),

$$\begin{aligned}
\|x^k - y^k\|^2 &= \lim_{i\to\infty} \sum_{j=1}^{n} \lambda_{ij} a(u^{ij})^T (x^k - y^k) \\
&= \lim_{i\to\infty} (\sum_{j=1}^{n} \lambda_{ij} (a(u^{ij})^T x^k - b(u^{ij})) + \sum_{j=1}^{n} \lambda_{ij} (b(u^{ij}) - a(u^{ij})^T y^k)) \\
&\leq (\limsup_{i\to\infty} \sum_{j=1}^{n} \lambda_{ij}) H(x^k) \\
&\leq \delta^{-1} M \|x^k - y^k\| H(x^k).
\end{aligned}$$

Letting $\gamma = 1/(1 + \delta^{-1} M)$ we have proved $\gamma d(x^k) \leq H(x^k)$ for all $k = 0, 1, \ldots$ provided the origin is in the interior of S. Now suppose that $\hat{x}$ is in the interior of S. Define $\hat{S} = \{z : a(u)^T z \leq b(u) - a(u)^T \hat{x} \text{ for all } u \in U\}$, $\hat{d}(x) = \min\{ \|x - z\| : z \in \hat{S}\}$, and $\hat{H}(x) = \sup\{ (a(u)^T x - b(u) + a(u)^T \hat{x})^+ : u \in U\}$. As the origin is in the interior of $\hat{S}$, we can prove that $\hat{H}(x^k - \hat{x}) \geq \gamma \hat{d}(x^k - \hat{x})$ for all $k = 1, 2, \ldots$. Lemma 1 then follows from the fact $H(x^k) = \hat{H}(x^k - \hat{x})$ and $d(x^k) = \hat{d}(x^k - \hat{x})$. ∎

Theorem 2. If S contains an interior point and $\inf\{\,\|a(u)\| : u \in U\} > 0$, then there exist $0 < \theta < 1$ and $x^* \in S$ such that $x^* = \lim_{k\to\infty} x^k$ and $\|x^k - x^*\| \le 2\theta^k \|x^0 - x^*\|$ for all $k = 1, 2, \ldots$.

Proof. Let $K = \max\{1, \sup\{\,\|a(u)\| : u \in U\}\}$ and $0 < \alpha < 1$. We suppose, temporarily, that ϵ_k is chosen in such a way that $0 < \epsilon_k \le \alpha H(x^k)$ and $\epsilon_k \to 0$. Then,

$$\|x^{k+1} - x^k\| = \frac{a(u^k)^T x^k - b(u^k)}{\|a(u^k)\|} \ge \frac{H(x^k) - \epsilon_k}{\|a(u^k)\|} \ge \frac{(1-\alpha)H(x^k)}{K}. \tag{16}$$

It follows from (1), (16), and Lemma 1 that

$$\begin{aligned}\|x^{k+1} - y^{k+1}\|^2 &\le \|x^k - y^k\|^2 - \|x^{k+1} - x^k\|^2 \\ &\le \|x^k - y^k\|^2 - (1-\alpha)^2 K^{-2} H(x^k)^2 \\ &\le \|x^k - y^k\|^2 - (1-\alpha)^2\gamma^2 K^{-2}\|x^k - y^k\|^2 \\ &= \big(1 - (1-\alpha)^2\gamma^2 K^{-2}\big)\|x^k - y^k\|^2. \end{aligned} \tag{17}$$

Letting $0 < \theta = \big(1 - (1-\alpha)^2\gamma^2 K^{-2}\big)^{\frac{1}{2}} < 1$ and applying (17) repeatedly, we have

$$\|x^{k+1} - y^{k+1}\| \le \theta^{k+1}\|x^0 - y^0\|.$$

Since x^* and x^k are in the ball $B(y^k, \|x^k - y^k\|)$, Theorem 3 follows provided $\epsilon_k \le \alpha H(x^k)$. To achieve this, the third line of Step 2 should be replaced by: If $a(\bar u)^T x^k - b(\bar u) > 0$ and $\epsilon_k \le \alpha(a(\bar u)^T x^k - b(\bar u))$, then $u^k := \bar u$, go to Step 3. Otherwise, let $\epsilon_k := \alpha(a(\bar u)^T x^k - b(\bar u))$; find a $\bar u \in U$ such that $a(\bar u)^T x^k - b(\bar u) \ge \sup\{a(u)^T x^k - b(u) : u \in U\} - \epsilon_k$. If $a(\bar u)^T x^k - b(\bar u) \le -\epsilon_k$, then x^k is a solution of (I), stop. If $-\epsilon_k < a(\bar u)^T x^k - b(\bar u) \le 0$, then $\epsilon_k := \epsilon_k/2$, go to Step 1. If $a(\bar u)^T x^k - b(\bar u) > 0$, then $u^k := \bar u$, go to Step 3. ∎

4. Infinite Systems of Convex Inequalities

Consider a general infinite system of convex inequalities:

$$\text{(II)} \qquad g_u(x) \le 0 \quad \text{for all } u \in U,$$

where $g_u : R^n \to R^1$ is a real-valued convex function for any $u \in U$ and U is the parameter set.

Let $\tilde S \equiv \{x \in R^n : g_u(x) \le 0 \text{ for all } u \in U\}$ be the solution set of (II) and $\tilde d(x) = \min\{\,\|x - y\| : y \in \tilde S\}$ be the Euclidean distance from x to $\tilde S$.

Let $\tilde H(x) \equiv \sup\{g_u(x)^+ : u \in U\}$ be the "biggest violation" by x of (II). $\tilde H(x)$ is a nonnegative convex function, and $\tilde H(x) = 0$ if and only if $x \in \tilde S$. For a fixed x, if there exists a $\bar u \in U$ such that $g_{\bar u}(x) > 0$, then $\tilde H(x) = \sup\{g_u(x) : u \in U\}$.

Let $\partial g_u(x)$ be the subdifferential of $g_u(x)$. It is well known that a real-valued convex function on R^n is continuous and subdifferentiable on R^n.

The projection method can be extended to the infinite system of convex inequalities (II). Here we project the current iterate x^k on a hyperplane that separates x^k and a convex inequality near the most violated convex inequality by x^k. We show, under some regularity conditions, that if the system is consistent, then the new iterate x^{k+1} is closer to the solution set than x^k, the sequence $\{x^k : k = 0, 1, \ldots\}$ converges to a solution of the system, and the convergence rate is linear. Let $\{\epsilon_k : k = 0, 1, \ldots\}$ be a sequence of positive numbers such that $\lim_{k\to\infty} \epsilon_k = 0$.

Method 2.

Step 0.
  Let $k := 0$;
  let x^0 be an arbitrary point in R^n.

Step 1.
  Find a $\bar{u} \in U$ such that $g_{\bar{u}}(x^k) \geq \sup\{g_u(x^k) : u \in U\} - \epsilon_k$.

Step 2.
  If $g_{\bar{u}}(x^k) \leq -\epsilon_k$, then x^k is a solution of (II), stop.
  If $-\epsilon_k < g_{\bar{u}}(x^k) \leq 0$, then $\epsilon_k := \epsilon_k/2$, go to Step 1.
  If $g_{\bar{u}}(x^k) > 0$, then $u^k := \bar{u}$;
  find a $w^k \in \partial g_{u^k}(x^k)$;
  if $w^k = 0$, then (II) has no solution, stop.

Step 3.
  Let $x^{k+1} := x^k - g_{u^k}(x^k)w^k/\|w^k\|^2$ be the orthogonal projection of x^k on the separating hyperplane $g_{u^k}(x^k) + {w^k}^T(x - x^k) = 0$;
  $k := k + 1$, go to Step 1.

Comments on Method 2.

(1) If $w^k = 0$, then $0 < g_{u^k}(x^k) = \min\{g_{u^k}(x) : x \in R^n\}$, which implies that $g_{u^k}(x) \leq 0$ has no solution and thus (II) has no solution.

(2) All comments on Method 1 still hold here with $a(u)^T x - b(u)$ replaced by $g_u(x)$.

Regularity Conditions.

(C1) $\sup\{ g_u(x) : u \in U\} < +\infty$ for all x in $B(y^0, \tilde{d}(x^0) + \delta)$, where δ is a small positive number.

(C2) $\bigcup_{u\in U} \bigcup_{x\in B(y^0,\tilde{d}(x^0))} \{w : w \in \partial g_u(x)\}$ is bounded.

(C3) $\tilde{S}$ contains an interior point.

(C4) $\inf\{ \|\xi_{\hat{u}}(\hat{x})\| : g_{\hat{u}}(\hat{x}) = 0,\ \hat{u} \in U,\ \xi_{\hat{u}}(\hat{x}) \in \partial g_{\hat{u}}(\hat{x})\} > 0$, where $\xi_{\hat{u}}(\hat{x}) = \operatorname{argmax}\{ \|w\| : w \in \partial g_{\hat{u}}(\hat{x})\}$.

Theorem 3. Suppose that (II) is consistent, and conditions (C1) and (C2) are satisfied. If Method 2 does not stop after finitely many iterations, then the sequence $\{x^k : k = 0, 1, \ldots\}$ converges to a solution of (II).

Proof. Let y^k be the point in $\tilde{S}$ closest to x^k. By (C1) $\tilde{H}(x)$ is well defined and continuous on $B(y^0, \tilde{d}(x^0))$. As in the proof of Theorem 1, $\{x^{n+k} : n = 0, 1, \ldots\}$ is contained in the closed ball $B(y^k, \tilde{d}(x^k))$ and $\tilde{d}(x^{k+1}) \leq (\Pi_{i=0}^k \tau_i)\tilde{d}(x^0)$, where $0 < \tau_k \equiv (1 - \|x^{k+1} - x^k\|^2/\|x^k - y^k\|^2)^{\frac{1}{2}} < 1$. If $\Pi_{i=0}^\infty \tau_i = 0$, then $\tilde{d}(x^k) \to 0$ and

$\cap_{k=0}^{\infty} B(y^k, \tilde{d}(x^k))$ is a singleton. If $\Pi_{i=0}^{\infty}\tau_i > 0$, then $\tau_i \to 1$ and thus

$$0 \le \|x^{k+1} - x^k\| \le (1 - \tau_k^2)\tilde{d}(x^0) \to 0. \tag{18}$$

By Steps 1 and 3 of Method 2,

$$\|x^{k+1} - x^k\| = g_{u^k}(x^k)/\|w^k\| \ge (\tilde{H}(x^k) - \epsilon_k)/\|w^k\|. \tag{19}$$

It follows from (C2), (18), and (19) that $\tilde{H}(x^k) \to 0$. As $\tilde{H}(x)$ is continuous on $B(y^0, \tilde{d}(x^0))$, for a convergent subsequence $x^{k_j} \to x^*$ we have

$$\lim_{j\to\infty} \tilde{H}(x^{k_j}) = \tilde{H}(x^*) = 0 \quad \text{and} \quad x^* \in \tilde{S}. \tag{20}$$

Since $\tilde{d}(x^k)$ is a decreasing sequence, (20) implies that $\lim_{k\to\infty} \tilde{d}(x^k) = 0$, and therefore $\cap_{k=0}^{\infty} B(y^k, \tilde{d}(x^k))$ is a singleton. ∎

Corollary 2. Suppose that conditions (C1) and (C2) are satisfied. Then (II) is consistent if and only if $\{x^k : k = 0, 1, \ldots\}$ is bounded and $\lim_{k\to\infty} \|x^{k+1} - x^k\| = 0$, if and only if $\{x^k : k = 0, 1, \ldots\}$ converges to a solution of (II). ∎

Lemma 2. If all regularity conditions are satisfied, then there exists a constant $0 < \gamma < 1$ such that $\tilde{H}(x^k) > \gamma\tilde{d}(x^k)$ for all $k = 0, 1, \ldots$.

Proof. Since $\tilde{S}$ has an interior point, we can represent it as (see, e.g., [6] p.169)

$$\tilde{S} = \{x : \xi_{\hat{u}}(\hat{x})^T(x - \hat{x}) \le 0 \text{ for all } \hat{x} \text{ such that } g_{\hat{u}}(\hat{x}) = 0 \text{ for some } \hat{u} \in U\}.$$

By Lemma 1 there exists a constant $0 < \gamma < 1$ such that for all $k = 0, 1, \ldots$

$$\gamma\tilde{d}(x^k) \le \sup\{ (\xi_{\hat{u}}(\hat{x})^T(x^k - \hat{x}))^+ : \text{ for all } \hat{x} \text{ such that } g_{\hat{u}}(\hat{x}) = 0 \text{ for some } \hat{u} \in U\}.$$

Lemma 2 then follows from the above inequality and the fact $(\xi_{\hat{u}}(\hat{x})^T(x^k - \hat{x}))^+ \le g_{\hat{u}}(x^k)^+ \le \tilde{H}(x^k)$. ∎

Theorem 4. If all regularity conditions are satisfied, then there exist $0 < \theta < 1$ and $x^* \in \tilde{S}$ such that $x^* = \lim_{k\to\infty} x^k$ and $\|x^k - x^*\| \le 2\theta^k\|x^0 - x^*\|$ for all $k = 1, 2, \ldots$. ∎

The proof of Theorem 4 is similar to that of Theorem 3 and is omitted.

5. Application

We use Method 1 to solve the following infinite system of linear inequalities:

$$-(D(u)Au)^T x + 1 \le 0 \text{ for all } u \in \{u \in R^n : u^T u = 1\},$$

where A is a given $n \times n$ real matrix and $D(u)$ is a diagonal matrix with diagonal entries $u_1, \ldots, u_n$. It is known that any solution x^* to this system is a scaling parameter of the given matrix A, i.e., $D(x^*)A$ is positive definite [4]. Here the nonlinear program $\sup\{-(D(u)Au)^T x^k + 1 : u^T u = 1\}$ to be solved in Step 1 is equivalent to $\min\{u^T(D(x^k)A + A^T D(x^k))u : u^T u = 1\}$ and the latter can be solved by finding the smallest eigenvalue and a corresponding eigenvector of the real symmetric matrix $D(x^k)A + A^T D(x^k)$. Method 1 is coded in FORTRAN and the subroutine RS of EISPACK is used to calculate eigenvalues and eigenvectors [4,7]. The input matrix A is randomly generated.

Table 1

Problem dimension	5×5	6×6	7×7	8×8	9×9
# of iterations	27	15	32	50	30

The computational result shows that the projection method provides a practical and efficient way for solving infinite systems of linear inequalities.

References

1. S. Agmon, The relaxation method for linear inequalities, *Canadian Journal of Mathematics* 6:382–392 (1954).
2. A. V. Fiacco and K. O. Kortanek ed., *Semi-infinite programming and applications*, Springer-Verlag, New York, 1983.
3. R. Hettich ed., *Semi-infinite programming*, Springer-Verlag, New York, 1979.
4. H. Hu, A projective method for rescaling a diagonally stable matrix to be positive definite, *SIAM J. Matrix Analysis and Applications*, 13:1255–1263 (1992).
5. H. Hu and Q. Wang, On approximate solutions of infinite systems of linear inequalities, *Linear Algebra and its Applications* 114/115:429-438 (1989).
6. R. T. Rockafellar, *Convex analysis*, Princeton University Press, Princeton, New Jersey, 1970.
7. B.T. Smith, J.M. Boyle, Y. Ikebe, V.C. Klema, and C.B. Moler, *Matrix eigensystem routines guide*, Springer, Berlin, 1970.

OPTIMIZATION PROBLEMS IN MOLECULAR BIOLOGY

TAO JIANG*
McMaster University

and

MING LI†
University of Waterloo

Abstract. Molecular biology has raised many interesting and deep mathematical and computational questions. For example, given the DNA or protein sequences of several organisms, can we know how much they are related to each other by computing an optimal alignment or, in particular, a longest common subsequence, of the sequences? Is it possible to reconstruct the evolutionary process for a set of extant species from their DNA sequences? Given many small overlapping fragments of a DNA molecule, how do we recover the DNA sequence? Will the shortest common superstring of these fragments give a good estimate? How many fragments suffice to guarantee that the reconstructed sequence is within 99% of the true DNA sequence? An organism can evolve by chromosome inversions, and this raises the question of how to transform one sequence into another with the smallest number of reversals.

Rather than an extensive literature survey, the purpose of this article is to introduce in depth several prominent optimization problems arising in molecular biology. We will emphasize recent developments and provide proof sketches for the results whenever possible.

Key words: multiple sequence alignment, shortest common supersequence, longest common subsequence, shortest common superstring, computational complexity, approximation algorithm, average-case analysis.

1. Introduction and Preliminaries.

In some sense, computer science and molecular biology are similar: they both have enjoyed explosive growth since 1950's and they both deal with *sequences*. While a sequence in computer science consists of elements such as **if**, **then**, **else**, **goto**, and so on, and encodes possibly a program, a DNA sequence in molecular biology is made of characters from the set $\{A, C, G, T\}$, each stands for a nucleotide, and encodes possibly a human being.

The DNA sequences range from several thousand characters long for a simple virus to 3×10^9 characters long for a human. The massive lengths of such sequences give rise to many interesting combinatorial optimization questions. Most of these questions are computationally hard and their practical solutions are still open as yet of today. For example, given the DNA or protein sequences of several organisms,

* Supported in part by NSERC Operating Grant OGP0046613. Address: Department of Computer Science and Systems, McMaster University, Hamilton, Ont. L8S 4K1, Canada. E-mail: jiang@maccs.mcmaster.ca

† Supported in part by the NSERC Operating Grant OGP0046506. Address: Department of Computer Science, University of Waterloo, Waterloo, Ont. N3L 3G1, Canada. E-mail: mli@math.uwaterloo.ca

Ding-Zhu Du and Jie Sun (eds.), Advances in Optimization and Approximation, 195–216.

molecular biologists would like to find how much they are related to each other by computing an optimal alignment or, sometimes, a longest common subsequence. The problem is NP-hard in general. It is very important to find efficient algorithms for these problems that give *approximately good* solutions. Another example, given the DNA sequences of a set of related extant species, geneticists would like to recover their evolutionary process by the means of building a minimum-cost phylogenetic tree. A phylogenetic tree is, in some sense, a Steiner tree on networks, where the vertices are all possible DNA sequences, the regular points are the given sequences, and the cost of an edge is the *edit distance* between the two involved sequences. Again the problem is NP-hard.

Consider yet one more example. Given many small overlapping fragments of a DNA molecule, biochemists usually (as in Sanger's method) reconstructs the DNA sequence by computing the shortest common superstring of these fragments. Since computing the shortest common superstring is NP-hard, the biochemists use a simple greedy algorithm routinely without knowing how well the algorithm performs. It has been conjectured for years that the greedy algorithm linearly approximates the shortest common superstring. This problem was only solved recently in [7]. In the last example, consider the evolution of single-chromosome organisms. It is known that such organisms can evolve by chromosome inversions, *i.e.*, a segment of the chromosome gets reversed in an evolutionary step. This raises the question of computing the smallest number of reversals to transform one sequence into another. Although some nice approximation algorithms have been obtained, presently we do not even know if the problem is NP-hard.

This article will treat, in depth, these and some other combinatorial optimization problems that molecular biologists face. Our main interest is in the computational complexity and approximation algorithms for these problems. In section 2, we consider two most important versions of multiple sequence alignment: multiple alignment with SP-score and multiple tree alignment. Multiple sequence alignment is known as one of the most important and difficult problems in computational biology. Longest common subsequences and shortest common supersequences are studied in section 3, and shortest common superstrings are treated in section 4. Note that these problems can all be formulated as multiple sequence alignment problems under a proper score scheme.

We assume the reader has the basic knowledge of algorithms and computational complexity (such as NP and P). Consult, *e.g.*, [15] otherwise. Below we introduce the concept of linear reduction and MAX SNP, which is not in [15].

In 1988, Papadimitriou and Yannakakis introduced a class of natural optimization problems, called *MAX SNP*, which includes the Vertex Cover and Independent Set problems on bounded-degree graphs, Max Cut, various version of Maximum Satisfiability, etc., and an approximability-preserving reduction, called *linear reduction* [43]. It is known that every problem in MAX SNP can be approximated within *some* constant factor and MAX SNP-hard problems do not have a polynomial-time approximation scheme (PTAS) unless P = NP [2]. (A problem has a PTAS if for every fixed $\epsilon > 0$, the problem can be approximate within factor $1+\epsilon$ in polynomial time and a problem is MAX SNP-hard if every problem in MAX SNP can be linearly reduced to it.) The following is the definition of a linear reduction:

Definition 1 Suppose that Π and Π' are two optimization (minimization, in particular) problems. (The definition for maximization problems is analogous.) We say that Π ***L-reduces*** (for ***linearly reduces***) to Π' if there are two polynomial time algorithms f and g and constants $\alpha, \beta > 0$ such that, for any instance I of Π,

1. $OPT(f(I)) \leq \alpha \cdot OPT(I)$
2. Given any solution of $f(I)$ with weight w', algorithm g produces in polynomial time a solution of I with weight w satisfying $|w - OPT(I)| \leq \beta |w' - OPT(f(I))|$.

The following two facts hold for L-reductions. First, the composition of two L-reductions is also an L-reduction. Second, if problem Π L-reduces to problem Π' and Π' can be approximated in polynomial time within a factor of $1+\epsilon$, then Π can be approximated within factor $1+\alpha\beta\epsilon$. In particular, if Π' has a PTAS, so does Π.

2. Multiple Sequence Alignment.

Multiple sequence alignment has been identified as one of the most important and challenging problems in computational biology [33, 35]. It plays an essential role in two related areas of molecular biology: finding highly conserved subregions among a set of biological sequences, and inferring the evolutionary history of some species from their associated sequences. A huge number of papers have been written on effective and efficient methods for computing an optimal multiple sequence alignment. For a comprehensive survey, see [9, 62]. In this section we will focus on two popular version of multiple alignment: *multiple alignment with SP-score* and *multiple tree alignment*. We will survey recent results on the complexity of these problems and efficient approximation algorithms.

2.1. Some Basic Concepts and Terminology.

A *sequence* is a string over some alphabet Σ. For DNA sequences, the alphabet Σ contains four letters A, C, G, and T representing four distinct nucleotides, and for protein sequences, Σ contains 20 letters, each represents a unique amino acid. An *alignment* of two sequences s_1 and s_2 is obtained by inserting spaces into or at either end of s_1 and s_2 such that the two resulting sequences s_1' and s_2' are of the same length. That is, every letter in s_1' is opposite to a unique letter in s_2'. A space is viewed as a new letter and is denoted Δ. Two opposing identical letters form a *match* and two opposing nonidentical letters form a *mismatch*, which can also be viewed as a replacement. A space in one sequence opposite to a letter x in the other can be viewed as a deletion of x from the second sequence, or an insertion of x into the first sequence.

Suppose that l is the length of the sequences s_1' and s_2'. The value of the alignment is defined as $\sum_{i=1}^{l} s(s_1'(i), s_2'(i))$, where $s_1'(i)$ and $s_2'(i)$ denote the two letters at the i-th column of the alignment, and $s(s_1'(i), s_2'(i))$ denotes the *score* of the two opposing letters under some given *score scheme* s. There are several popular score schemes for amino acids and for nucleotides [30, 50]. A standard assumption about a score scheme s is that it satisfies *triangle inequality*, *i.e.*, for any three letters x, y, and z, $s(x,z) \leq s(x,y) + s(y,z)$. An *optimal alignment* of two sequences is an alignment that *minimizes* the value over all possible alignments. The *edit distance* between

```
ACGTACG TACGTAC
 CG ACGG  CGTTGA
```

Fig. 1. An alignment of ACGTACGTACGTAC and CGACGGCGTTGA.

two sequences s and s', denoted $d(s, s')$, is defined as the minimum alignment value of s and s'.

Figure 1 gives an example alignment of sequences ACGTACGTACGTAC and CGACGGCGTTGA. Its value is 10 if a match costs 0, an insertion/deletion costs 1, and a mismatch costs 2.

The concept of an alignment can be easily extended to more than 2 sequences. A *multiple alignment* $\mathcal{A}$ of $k \geq 2$ sequences is obtained as follows: spaces are inserted into each sequence so that the resulting sequences have the same length l, and the sequences are arrayed in k rows of l columns each. Again, a score value is defined on each column under some score scheme and the value of $\mathcal{A}$ is simply the sum of the scores of all columns. Many score schemes have been suggested. Among them, *SP-score* seems to be very sensible and has received a lot of attention [3, 8, 49, 17]. (Here, SP stands for *sum of all pairs.*) The SP scheme defines the score value of a column as the sum of the scores of all (unordered) pairs of the letters in the column. That is, the value of the alignment $\mathcal{A}$ is the sum of the values of pairwise alignments induced by $\mathcal{A}$.

In the analysis of genetic evolution, we are given a set X of k sequences, each stands for an extant species. Let Y be a set of hypothetical sequences, where $Y \cap X = \emptyset$. (Usually each sequence in Y could represent an extinct species.) An *evolutionary tree* $T_{X,Y}$ for X is a *weighted* (sometimes *rooted*) tree of $|X \cup Y|$ nodes, where each node is associated with a unique sequence in $X \cup Y$ [17]. The *cost* of an edge is the edit distance between the two sequences associated with the ends of the edge. The cost $c(T_{X,Y})$ of the tree $T_{X,Y}$ is the total cost of all edges in $T_{X,Y}$. Given sequences X, the *optimal evolutionary tree* or *multiple tree alignment* or, simply, *tree alignment* problem is to find a set of sequences Y as well as an evolutionary tree $T_{X,Y}$ for X which minimizes $c(T_{X,Y})$ over possible sets Y and trees $T_{X,Y}$.

An important variant of tree alignment is that we are not only given the sequences of some species, but also the phylogenetic structure (*i.e.*, the tree structure). More precisely, we are given a set X of k sequences and a tree structure with k leaves, each of them is associated with a unique sequence in X. Then we would like to find the hypothetical sequences Y and assign them to the internal nodes of the given tree so that the total cost is minimized [45]. We will refer to this problem as *tree alignment with a given phylogeny*.

2.2. Multiple Sequence Alignment with SP-score.

The best algorithm to compute an optimal alignment under SP measure is based on dynamic programming and requires a running time which is in the order of the product of the lengths of input sequences [1]. It has recently been shown that computing an optimal multiple alignment with SP-score (actually, a decision version

TABLE I
Score scheme I.

S	0	1	a	b	Δ
0	2	2	1	2	1
1	2	2	2	1	1
a	1	2	0	2	1
b	2	1	2	0	1
Δ	1	1	1	1	0

of it) is NP-complete [60].

Theorem 1 *Multiple alignment with SP-score is NP-complete.*

The proof is a reduction from the shortest common supersequence problem on the binary alphabet $\{0, 1\}$, which is shown NP-complete in [41]. The alphabet $\Sigma = \{0, 1, a, b\}$ and the score scheme is defined in Table 1. Clearly the scheme satisfies triangle inequality.

Gusfield first proposed a polynomial-time approximation algorithm for this problem that achieves ratio $2 - \frac{1}{k}$ on k input sequences [16, 17]. His construction uses a so called *center star method*, which is illustrated below.

Algorithm Center Star

1. Let S be the set of input sequences.
2. Find a sequence $s \in S$ such that $\sum_{s' \neq s} d(s, s')$ is minimized.
3. Align each sequence $s' \neq s$ optimally against s. This induces a multiple alignment $\mathcal{A}$ of the strings in S.
4. Output $\mathcal{A}$.

Pevzner improved Gusfield's algorithm to obtain a ratio of $2 - \frac{2}{k}$ [44]. Besides a minimal center star, he also finds a minimal matching between the sequences other than s. Recently Bafna and Pevzner pushed the ratio to $2 - \frac{l}{k}$ [4] for any fixed l. It remains an interesting question if the problem has a PTAS.

2.3. Multiple Tree Alignment.

The construction of an optimal evolutionary tree from a given set of sequences can also be viewed as a type of multiple sequence alignment, called multiple tree alignment or, simply, tree alignment. Foulds and Graham proved that a variant of tree alignment, where the distance between two sequences is defined as Hamming distance, is NP-complete [11]. Recently, Sweedyk and Warnow proved that tree alignment is NP-complete [54]. Several approximate methods have been proposed in the literature [19, 20, 46, 47]. Gusfield showed that a minimum-cost spanning tree of the input sequences has a cost that is at most twice the cost of an optimal evolutionary tree [16, 17]. An interesting question is whether one can find efficient algorithms with approximation ratio better that 2.

Tree alignment can be naturally formulated as a Steiner tree problem on networks [23]. Let S be a set of sequences. Define an infinite weighted graph $G = (V, E)$

TABLE II
Score scheme II.

	0	1	Δ
0	0	1	2
1	1	0	2
Δ	2	2	0

as follows. V contains all possible sequences and each edge in E represents an optimal pairwise alignment and thus has the corresponding edit distance as its weight. The set of *regular* points (*i.e.*, the nodes that have to be included in the Steiner tree) is S. Then the Steiner minimal tree on G gives an optimal evolutionary tree for S. Of course this does not yield an effective algorithm for tree alignment as G is infinite. Nonetheless, it provides a way of approximating the problem.

Modify the graph G as follows. For any three sequences $x, y, z \in S$, let $s_{x,y,z}$ be a sequence such that

$$d(s_{x,y,z}, x) + d(s_{x,y,z}, y) + d(s_{x,y,z}, z)$$

is minimized. The sequence can be found in polynomial time by dynamic programming. Let $V = S \cup \{s_{x,y,z} | x, y, z \in S\}$. The edge set E is modified accordingly. Zelikovsky has given a polynomial-time approximation algorithm for Steiner minimal tree on networks that achieves performance ratio 11/6. The algorithm only considers components involving three regular points and is thus applicable to graph G. Running this algorithm on G produces an evolutionary tree whose cost is at most 11/6 times the optimum. Recently, Berman and Ramaiyer have improved Zelikovsky's ratio to 1.747 [6].

But can we make the approximation ratio arbitrarily close to 1? The answer is negative since it has been shown that tree alignment is MAX SNP-hard [60].

Theorem 2 *Tree alignment is MAX SNP-hard.*

The reduction goes as follows. We first L-reduce Vertex Cover-B [43] to a restricted version of tree alignment:

Restricted Tree Alignment: Given two sets of sequences X and Y, find a subset $Y' \subseteq Y$ and an evolutionary tree $T_{X,Y'}$ with the smallest cost.

The alphabet Σ used in the reduction is $\{0, 1\}$ and the score scheme is given in Table 3, which satisfies triangle inequality.

Then this restricted problem is L-reduced to tree alignment.

2.4. TREE ALIGNMENT WITH A GIVEN PHYLOGENY.

An important variant of tree alignment is that we are not only given the sequences of some species, but also the phylogenetic structure (*i.e.*, the tree structure). Although the problem seems easier, the algorithms proposed in [1, 20, 47] all run in

TABLE III
Score scheme III.

	0	1	a	b	Δ
0	0	1	1	1	1
1	1	0	1	1	1
a	1	1	0	2	2
b	1	1	2	0	2
Δ	1	1	2	2	0

exponential time in the worst case. It is worth mentioning that, in practice, the heuristic algorithm of Sankoff and Cedergren seems to generate a good result within a feasible number of steps [47].

Among the many possible phylogenetic structures, binary tree and star are the most common ones. We can prove that tree alignment with a given phylogeny is NP-complete even when the phylogeny is a binary tree [60]. Furthermore, the problem is MAX SNP-hard if the phylogeny is a star. In contrast, when the given phylogeny is a binary tree, tree alignment is *not* MAX SNP-hard, for the problem has a PTAS (see below).

Theorem 3 *It is NP-complete to construct an optimal evolutionary tree even when the given phylogeny is a binary tree.*

The reduction is again from the shortest common supersequence problem. In the proof, the alphabet $\Sigma = \{0, 1, a, b\}$ and the score scheme is as in Table 3, which satisfies triangle inequality.

It has also been shown that tree alignment with a given binary phylogeny has a PTAS under *any* score scheme satisfying triangle inequality [61]. The result in fact holds for trees with bounded degree.

Theorem 4 *Consider tree alignment when the phylogeny is given and is a binary tree. For any $\epsilon > 0$, there exists an algorithm that constructs an evolutionary tree with cost at most $1 + \epsilon$ times the optimum in time polynomial in input size and $\frac{1}{\epsilon}$.*

On the other hand, approximating an optimal evolutionary tree is much harder when the given phylogenetic tree has unbounded degree [60].

Theorem 5 *It is MAX SNP-hard to construct an optimal evolutionary tree when the given phylogeny is a star.*

The reduction is from a bounded version of Max Cut, Max Cut-k, where the constant k upper bounds the degree of the graph. The problem is shown MAX SNP-hard in [60]. In the proof, the alphabet $\Sigma = \{0, 1, a, b, *, \#, \$\}$. Unfortunately, the score scheme, given in Table 4, does *not* satisfy inequality.

It remains an interesting open question if the score scheme can be made to satisfy triangle inequality. If so, then Theorem 4 implies that the degree of the tree really makes a difference in the approximability of the problem.

TABLE IV
score scheme

	#	$	0	1	*	Δ	a	b
#	0	2k	4k	4k	4k	4k	4k	4k
$	2k	0	4k	4k	4k	4k	4k	4k
0	4k	4k	0	1	0	1	1	1
1	4k	4k	1	0	0	1	1	1
*	4k	4k	0	0	0	0	0	2
Δ	4k	4k	1	1	0	0	0	2
a	4k	4k	1	1	0	0	0	2
b	4k	4k	1	1	2	2	2	2

3. Longest Common Subsequence and Shortest Common Supersequence.

For two sequences $s = s_1 \cdots s_m$ and $t = t_1 \cdots t_n$, we say that s is a *subsequence* of t and equivalently, t is a *supersequence* of s, if for some $i_1 < \ldots < i_m$, $s_j = t_{i_j}$. Given a finite set of sequences, S, a *shortest common supersequence* (SCS) of S is a shortest possible sequence s such that each sequence in S is a subsequence of s. A *longest common subsequence* (LCS) of S is a longest possible sequence s such that each sequence in S is a supersequence of s. Computing an LCS or an SCS can be viewed as special cases of the multiple sequence alignment problem [44]. A longest common subsequence (of some DNA sequences) is commonly used as a measure of similarity in the comparison of biological sequences [10]. Enormous amount of papers in molecular biology have been written on this issue. We refer the reader to [10, 12, 52].

These problems also arise naturally in many practical situations other than biology. In artificial intelligence (specifically, planning), the actions of a robot (and human for that matter) need to be planned. A robot usually has many goals to achieve. To achieve each goal the robot needs to perform a (linearly ordered) sequence of operations where identical operations in different sequences may be factored out and only performed once. Optimally merging these sequences of actions of the robot implies efficiency. Practical examples include robotic assembly lines [13] and metal-cutting in the domain of automated manufacturing [18, 32]. The essence of solutions to such problems are good approximation algorithms for the SCS problem. The SCS problem also has applications in text editing [48], data compression [53], and query optimization in database systems [51]. Other applications of the LCS problem include data compression and syntactic pattern recognition [39].

3.1. The Hardness of Approximating LCS and SCS.

For a long time, it is well-known that the SCS (even with alphabet size 5) and LCS (even with alphabet size 2) problems are all NP-hard [15, 40]. For k sequences of length n, it is known that, by dynamic programming techniques, both SCS and LCS problems can be solved in $O(n^k)$ time, independently due to many authors in computer science (e.g., [21, 59]) and biology. However, since the parameter n is usually extremely large in practice (*e.g.*, in computer text editing and DNA/RNA sequence comparison), the time requirement $O(n^k)$ is intolerable even for small to moderate k. There have been attempts to speed up the dynamic programming solution for LCS [22, 24]. The improved algorithms still run in $O(n^k)$ time in the worst case.

In the very first paper [40] that the NP-hardness of the SCS and LCS problems was proved, Maier asked for approximation algorithms. For the past many years, various groups of people, in the diversified fields of artificial intelligence, theoretical computer science, and biology have looked for heuristic algorithms to approximate SCS and LCS problems. No polynomial time algorithms with guaranteed approximation bound have been found. Recently we [25] have proved that a good approximation is impossible, unless P = NP.

Theorem 6 *There exists a constant $\delta > 0$ such that, if the LCS problem has a polynomial time approximation algorithm with performance ratio n^δ, where n is the number of input sequences, then $P = NP$.*

Proof. We reduce the Largest Clique problem to LCS. Let $G = (V, E)$ be a graph and $V = \{v_1, ..., v_n\}$ the vertex set. Our alphabet Σ is chosen to be V.

Consider a vertex v_i and suppose that v_i is adjacent to vertices $v_{i_1}, \ldots, v_{i_q}$, where $i_1 < \ldots < i_q$. For convenience, let $i_0 = 0$ and $i_{q+1} = n + 1$. Let p be the unique index such that $0 \le p \le q$ and $i_p < i < i_{p+1}$. We include two sequences:

$$x_i = v_{i_1} \cdots v_{i_p} v_i v_1 \cdots v_{i-1} v_{i+1} \cdots v_n$$

$$x_i' = v_1 \cdots v_{i-1} v_{i+1} \cdots v_n v_i v_{i_{p+1}} \cdots v_{i_q}$$

Let $S = \{x_i, x_i' | 1 \le i \le n\}$.

Lemma 7 *The graph G has a clique of size k if and only if the set S has a common subsequence of length k, for any k.*

The proof is completed by recalling the result of Arora *et al.* that, unless P = NP, the Largest Clique problem does not have a polynomial-time approximation algorithm with performance ratio n^δ on graphs with n vertices, for some constant $\delta > 0$ [2]. ■

The above result shows that, for a set S of n sequences, it is impossible to find a common subsequence of length at least $OPT(S)/n^\delta$ in polynomial time for some constant $\delta > 0$. But it still remains open whether one can find a common subsequence of length at least $OPT(S)/n^\delta$ in polynomial time for any constant $\delta > 0$. We conjecture the answer to be in the negative.

Remark. It is also natural to measure the performance of an approximation algorithm for LCS in terms of the size of alphabet or the maximum length of input sequences. Clearly the statement of Theorem 6 holds when n is replaced by these parameters.

We now consider the approximation of LCS on a fixed alphabet. Suppose that Σ is an alphabet of k letters: $a_1, ..., a_k$. It is easy to show that LCS on Σ can be approximated with ratio k.

Theorem 8 *For any set S of sequences from Σ, the following algorithm finds a common subsequence for S of length at least $OPT(S)/k$.*

Algorithm Long-Run.
Find maximum m such that a_i^m is a common subsequence of all input sequences, for some $a_i \in \Sigma$. Output a_i^m.

It is an interesting question whether LCS on a bounded alphabet is MAX SNP-hard. Although we believe that the answer is "yes", even when the alphabet is binary, we have not been able to establish an L-reduction from any known MAX SNP-hard problem. Maier's original proof of the NP-hardness for LCS on bounded alphabets is not useful here, since the reduction is not linear.

Conjecture. LCS on a binary alphabet is MAX SNP-hard.

We now switch our attention to the SCS problem. Maier proved the NP-hardness of SCS by reducing the Vertex Cover problem to SCS [40]. It is easy to see that his reduction is actually linear if the graph is of bounded degree. Since the Vertex Cover problem on bounded degree graphs is MAX SNP-hard, SCS is also MAX SNP-hard. Thus [2] implies that SCS problem does not have a PTAS. But we [25] will prove a much stronger result in the following.

Theorem 9 *(i) There does not exist a polynomial-time linear approximation algorithm for SCS, unless $P = NP$. (ii) There exists a constant $\delta > 0$ such that, if SCS has a polynomial-time approximation algorithm with ratio $\log^\delta n$, where n is the number of input sequences, then NP is contained in* **DTIME**$(2^{\text{polylog } n})$.

The proof consists of three steps. We first show that a restricted version of SCS in which each input sequence is of length 2 and each letter of the alphabet appears at most 3 times in the input sequences is MAX SNP-hard. Thus this restricted version of SCS does not have a PTAS, assuming that $P \neq NP$. Then we define the product of sets of sequences and relate the SCS of such a product to the SCS's of the components constituting the product. Finally, we demonstrate that a polynomial-time constant ratio approximation algorithm for SCS would imply a PTAS for the restricted SCS, by blowing up instances using the product of sets of sequences.

• A Restricted Version of SCS and MAX SNP-hardness.

Maier proved the NP-hardness of SCS by reducing the Vertex Cover problem to SCS [40]. It is easy to see that his reduction is actually linear if the graph is of bounded degree. Since Vertex Cover on bounded degree graphs is MAX SNP-hard [43], SCS is MAX SNP-hard.

Let SCS(l, r) denote the restricted version of SCS in which each input sequence is of length l and each letter appears at most r times totally in all sequences. Such restricted problems have been recently studied by Timkovskii [57]. We will need the version SCS$(2, 3)$ later on to prove the above theorem. It is known that SCS$(2, 2)$ can be solved in polynomial time and SCS$(2, 3)$ is NP-hard [57]. Obviously, SCS(l, r) can be approximated with ratio r. The following lemma is proven in [25].

$$(a,1)(b,1)(c,1)(a,2)(b,2)(c,2)$$
$$(a,1)(b,1)(c,1)(d,2)(e,2)(f,2)$$
$$(d,1)(e,1)(f,1)(a,2)(b,2)(c,2)$$
$$(d,1)(e,1)(f,1)(d,2)(e,2)(f,2)$$

$$(a,3)(b,3)(c,3)(a,4)(b,4)(c,4)$$
$$(a,3)(b,3)(c,3)(d,4)(e,4)(f,4)$$
$$(d,3)(e,3)(f,3)(a,4)(b,4)(c,4)$$
$$(d,3)(e,3)(f,3)(d,4)(e,4)(f,4)$$

Fig. 2. The product of $\{aabb, abab\}$ and $\{121, 212\}$.

Lemma 10 *SCS(2,3) does not have a PTAS, unless P = NP.*

• The Product of Sets of Sequences.

First, we extend the operation "concatenation" to sets of sequences. Let X, Y be sets of sequences. The *concatenation* of X and Y, denoted $X \cdot Y$, is the set $\{x \cdot y | x \in X, y \in Y\}$. For example, if $X = \{abab, aabb\}$ and $Y = \{123, 231, 312\}$, then $X \cdot Y = \{abab123, aabb123, abab231, aabb231, abab312, aabb312\}$. The following lemma is useful in our construction. We can prove

Lemma 11 *Let $X = X_1 \cdot X_2 \cdots X_n$. Suppose that y is a supersequence for X. Then there exist $y_1, y_2, \ldots, y_n$ such that $y = y_1 \cdot y_2 \cdots y_n$ and each y_i is a supersequence for X_i, $1 \le i \le n$.*

We are now ready to define the product of two sets of sequences. The symbol $\times$ will be used to denote the product operation. We start by defining the product of single letters. Let Σ and Σ' be two alphabets, and $a \in \Sigma$, $b \in \Sigma'$ two letters. The product of a and b is just the composite letter $(a,b) \in \Sigma \times \Sigma'$. The product of a sequence $x = a_1 \cdot a_2 \cdots a_n$ and a letter b is $a_1 \times b \cdot a_2 \times b \cdots a_n \times b$. The product of a set $X = \{x_1, x_2, \ldots, x_n\}$ of sequences and a letter a is the set $\{x_1 \times a, x_2 \times a, \ldots x_n \times a\}$. The product of a set X of sequences and a sequence $y = a_1 \cdot a_2 \cdots a_n$ is the set $X \times a_1 \cdot X \times a_2 \cdots X \times a_n$. Finally, let X and $Y = \{y_1, y_2, \ldots, y_n\}$ be two sets of sequences. The product of X and Y is the set $\bigcup_{i=1}^{n} X \times y_i$. For example, if $X = \{abc, def\}$ and $Y = \{12, 34\}$, then $X \times Y$ contains the 8 sequences in Figure 2.

Note that if each sequence in X has length l_1 and each sequence in Y has length l_2, then $X \times Y$ contains $|Y| \cdot |X|^{l_2}$ sequences of length $l_1 \cdot l_2$.

We will only be interested in products in which the second operand has a special property. Let Y be a set of sequences with the following property: each sequence is of even length and every letter at an even position is *unique* in Y, *i.e.*, the letter appears only once (totally) in all sequences in Y. We will refer to such unique letters as *delimiters*. The following lemma relates the SCS of a product to the SCS's of its operands and is crucial to the proof of the theorem. We omit the proof.

Lemma 12 *Let X and Y be sets of sequences. Suppose that Y has the above special property. Then $OPT(X \times Y) = OPT(X) \cdot OPT(Y)$. Moreover, given a supersequence for $X \times Y$ of length l, we can find in polynomial time supersequences for X and Y of lengths l_1 and l_2, respectively, such that $l_1 \cdot l_2 \leq l$.*

• Proving the Theorem.

The basic idea is to use the product operation to blow up a given instance of SCS. However, the product of sets of sequences cannot be performed in polynomial time, unless the the second operand contain sequences of bounded length. Thus we consider the restricted version SCS(2, 3). A nice thing about SCS(2, 3) is that for any instance S of SCS(2, 3), the total length of the sequences in S is at most $3 \cdot OPT(S)$. Hence we can insert unique delimiters into the sequences as required in the previous subsection without affecting the MAX SNP-hardness. So let SCS$(2,3)'$ denote the version whose instances are obtained from instances of SCS(2, 3) by inserting unique delimiters. It is easy to see that SCS$(2,3)'$ is also MAX SNP-hard.

For any set S, S^k denotes the product of k S's. The next lemma follows from Lemma 12.

Lemma 13 *Let $k \geq 1$ be a fixed integer. For any instance S of SCS$(2,3)'$, $OPT(S^k) = OPT(S)^k$. Moreover, given a supersequence for S^k of length l, we can find in polynomial time a supersequence for S of length $l^{1/k}$.*

Observe that, if $|S| = n$, then $|S^k| = n^{O(4^k)}$, since each sequence in S has length 4.

We only prove (i) of Theorem 9. The idea is to show that if SCS has a polynomial time linear approximation algorithm, then SCS$(2,3)'$ has a PTAS. Suppose that SCS has a polynomial-time approximation algorithm with performance ratio α. For any given $\epsilon > 0$, let $k = \lceil \log_{1+\epsilon} \alpha \rceil$. Then by Lemma 13, we have an approximation algorithm for SCS$(2,3)'$ with ratio $\alpha^{1/k} \leq 1 + \epsilon$. The algorithm runs in time $n^{O(4^k)}$, thus polynomial in n. This implies a PTAS for SCS$(2,3)'$. ■

3.2. Algorithms That Work Well on the Average.

We have seen that the LCS and SCS problems are not only NP-hard to solve exactly, but also NP-hard to approximate. The approximation of these problems restricted to fixed alphabets also seem to be hard. Does this really imply that the LCS and SCS problems are *practically* intractable? The answer is, however, no. We [25] will show that for random inputs, the LCS and SCS problems can be approximated up to a small additive error. In the following, let $\Sigma = \{a_1, \ldots, a_k\}$ be a fixed alphabet.

Let's assume, for convenience, that the input is always n sequences over the alphabet Σ, each of length n, although our results actually hold when the number of input sequences is polynomial in n. We will show that for random LCS and SCS instances, some remarkably simple greedy algorithms can do very well. Kolmogorov complexity is a key tool for proving the following results. The readers who do not know Kolmogorov complexity can either skip the proof or see [37] for definitions.

In the proof of Theorem 8, algorithm Long-Run was used to produce a trivial common subsequence. Here we show that this algorithm in fact performs extremely well on the average.

Theorem 14 *For any set S containing n sequences of length n over alphabet Σ, Long-Run achieves $OPT(S) + OPT(S)^{\frac{1}{2}+\epsilon}$ for any $\epsilon > 0$, on the average.*

Proof. Fix a long Kolmogorov random string x of length n^2 over Σ. Thus

$$K(x) \geq n^2 - 2\log n. \tag{1}$$

Cut x into n equal length pieces $x_1, \ldots, x_n$, and consider the set $S = \{x_1, \ldots, x_n\}$ as our input. The following lemma is a standard fact of Kolmogorov complexity [37].

Lemma 15 *If any letter $a \in \Sigma$ appears in x_i, for any i,*

$$n^{\frac{1}{2}+\epsilon} \tag{2}$$

less (more) than $\frac{n}{k}$ times, for some $\epsilon > 0$, then for some constant $\delta > 0$,

$$K(x_i) \leq n - \delta n^{2\epsilon}.$$

It follows from Lemma 15 and the fact that x is random that, for any $a \in \Sigma$, a appears in each of $x_1, \ldots, x_n$ at least $\frac{n}{k} - O(n^{1/2+\epsilon})$ times, for any small $\epsilon > 0$. Thus

$$a^{n/k - O(n^{1/2+\epsilon})} \tag{3}$$

is a common subsequence of sequences $x_1, \ldots, x_n$. In the next lemma, we show that the true LCS cannot really be much longer than Formula (3). Thus the theorem is true for this particular input $x_1, \ldots, x_n$. And since most inputs are Kolmogorov random, by simple averaging, the theorem will follow.

Lemma 16 *If s is a common subsequence of $x_1, \ldots, x_n$, then*

$$|s| < \frac{n}{k} + n^{\frac{1}{2}+\epsilon}. \tag{4}$$

Proof. For the purpose of a contradiction, suppose that the lemma is wrong. That is,

$$|s| \geq \frac{n}{k} + n^{\frac{1}{2}+\epsilon}.$$

We will try to compress x using s. The idea is to save n^δ digits for some $\delta > 0$ on each x_i using s.

Fix an x_i. We will do another encoding for x_i. Let $s = s_1 s_2 \ldots s_p$, $p = |s|$. We align s with the corresponding letters in x_i as to the left as possible, and rewrite x_i as follows:

$$\alpha_1 s_1 \alpha_2 s_2 \ldots \alpha_p s_p x_i'. \tag{5}$$

Here α_1 is the longest prefix of x_i containing no s_1; α_2 is the longest substring of x_i starting from s_1 containing no s_2, and so on. x_i' is the remaining part of x_i after s_n. Thus α_j does not contain letter s_j, for $j = 1, \ldots, p$. That is, each α_j contains at most $k-1$ letters in Σ. In other words, $\alpha_j \in (\Sigma - s_j)^*$.

We now show that x_i can be compressed, by at least n^δ digits for some $\delta > 0$, with the help of s. In fact, it is sufficient to prove that the prefix

$$y = \alpha_1 s_1 \alpha_2 s_2 \ldots \alpha_p s_p, \tag{6}$$

can be compressed by this amount.

Using s, we know which $k-1$ letters in Σ appear in α_i, for each i. Thus we can recode y as follows. For each i, if $s_i = a_k$, then do nothing. Otherwise, replace each s_i by a_k, the last letter in Σ; in each α_i, replace every a_k by s_i, which is also a letter in Σ. We can convert this transformed string y' back to y, using s, by reversing above process.

Now this transformed string y' is also a string over Σ. But in y', letter a_k appears $n^{\frac{1}{2}+\epsilon}$ times, since $|s|$ is least this long. By Lemma 15, for some constant $\delta > 0$, we have

$$K(y') \le |y'| - \delta n^{2\epsilon}.$$

But since from y' and s we can obtain y and hence, together with a self-delimiting form of x_i', obtain x_i. We conclude that

$$K(x_i|s) \le n - \delta n^{2\epsilon} + O(\log n),$$

where the term $O(\log n)$ takes care of the extra digits required for the self-delimiting representation of x_i' and the description of above procedure.

We repeat this for every x_i. In total, we save more than $\Omega(n^{1+2\epsilon})$ digits to encode x. Thus,

$$K(x) \le n^2 - \Omega(n^{1+2\epsilon}) + |s| + O(n \log n) < |x| - \log |x|.$$

Therefore, x is not random, and a contradiction! ■

Now, we consider all possible inputs of n sequences of length n. For each such input, we concatenate the n sequences together to obtain one string. Only about $O(1/n)$ of them are not random. That is, only $O(1/n)$ of them do not satisfy $K(x) \ge |x| - \log |x|$. For all others, Lemma 16 applies. Observe that the worst case performance of algorithm Long-Run is $OPT(S)/k$. Thus on the average, algorithm Long-Run outputs a common subsequence of length at least $OPT(S) - O(OPT(S)^{\frac{1}{2}+\epsilon})$ for any fixed $\epsilon > 0$. ■

In [13], the average performance of the following algorithm for the SCS problem is analyzed.

Algorithm Majority-Merge

1. Input: n sequences, each of length n.
2. Set supersequence $s := \epsilon$;
3. Let a be the majority among the leftmost letters of the remaining sequences. Set $s := sa$ and delete the front a from these sequences. Repeat this step until no sequences are left.

4. Output s.

It is shown in [13] that Majority-Merge produces a common supersequence of length $(k+1)n/2 + O(\sqrt{n})$ on the alphabet Σ, on the average. We have shown that this is indeed near optimal [25].

Theorem 17 *For any set S containing n sequences over the alphabet Σ of length n, the algorithm Majority-Merge produces a common supersequence of length $OPT(S) + O(OPT(S)^{\delta})$, on the average, where $\delta = \frac{\sqrt{2}}{2} \approx 0.707$.*

Both Theorems 14 and 17 actually hold for inputs consisting of $p(n)$ sequences of length n, where $p()$ is some fixed polynomial.

4. Shortest Common Superstring.

Given a finite set of strings $S = \{s_1, \ldots, s_m\}$, the *shortest common superstring* of S is the shortest string s such that each s_i appears as a substring (a consecutive block) of s.

Example. A shortest common superstring of all words in the following sentence: "Alf ate half lethal alpha alfalfa" is "lethalphalfalfate".

Given a DNA molecule, determining the corresponding sequence or, *sequencing* the molecule, is a crucial step towards understanding the biological functions of molecule. In fact, today no problem in biochemistry can be studied in isolation from its genetic background. However with current laboratory methods, such as Sanger procedure, it is by far impossible to sequence a long molecule (of 3×10^9 base pairs for human) directly as a whole. Each time, a rather randomly chosen fragment of less than 500 base pairs can be sequenced. In general, biochemists "cut", using different restriction enzymes, long DNA sequences into pieces of 200-500 nucleotides (characters). A biochemist "samples" the fragments and Sanger procedure is applied to sequence the sampled fragment. From hundreds of these random fragments, a biochemist has to *assemble* the superstring representing the whole molecule.

An algorithm, that tries to find a shortest common superstring of a given set of strings (fragments) is routinely used in the laboratories, by computers or by hands. The algorithm, called Greedy, works as follows: repeatedly merge a pair of strings with maximum overlap until only one string is left. Clearly, Greedy does not always find the shortest superstring. In fact it can be two times off as the following example shows.

Example. Let $S = \{c(ab)^k, (ba)^k, (ab)^k c\}$. Then Greedy will merge $c(ab)^k$ and $(ab)^k c$ first, and thus it outputs a string almost twice as long as the optimal one, $c(ab)^{k+1}c$.

This raises the following questions:

- Can we compute the shortest common superstring precisely in polynomial time?
- If not, how well can we approximate it?
- How well does Greedy do? The conjecture for years has been that Greedy achieves linear approximation. More exactly, it is conjectured that Greedy achieves twice the optimal length in the worst case.

In this section, we try to answer the above questions, and also leave some interesting open question.

4.1. Some Hardness Results.

The first question was answered by Gallant, Maier, and Storer [14] in 1980 when the authors were studying some data compression problems. A reduction from the directed Hamiltonian path problem gives the following theorem.

Theorem 18 *Given $S = \{s_1, \ldots, s_m\}$, it is NP-hard to find a shortest superstring of S.*

The problem remains NP-hard even if

- each string in S is of length at most 3 with unbounded Σ; or
- $|\Sigma| = 2$. But in this case, some strings in S need to have length about $O(\log \Sigma_{i=1}^{m} |s_i|)$.

The following stronger result was shown in [7]:

Theorem 19 *Finding a shortest superstring is MAX SNP-hard.*

The reduction is from a special case of the TSP (Traveling Salesperson Problem) on directed graphs. Let TSP(1, 2)-B be the TSP restricted to instances where the graph has a bounded degree B and all the distances are either 1 or 2. It has been shown that TSP(1, 2)-B is MAX SNP-hard. It is possible to convert the construction in [14] for proving the NP-completeness of the superstring decision problem into an L-reduction from TSP(1, 2)-B. The details are omitted here. By this theorem and [2], we immediately have:

Corollary 20 *If the shortest superstring problem has a PTAS, then $P = NP$.*

4.2. Linear Approximation Algorithms.

The follow is the best we can do in polynomial time [14].

Theorem 21 *Let $S = \{s_1, \ldots, s_m\}$. If $|s_i| \leq 2$ for all i, then a shortest superstring of S can be found in polynomial time.*

Despite of the hardness results, many people believed the Greedy-style algorithms work well in practice. It has been conjectured that superstrings produced by Greedy have length at most $2n$. Blum, Jiang, Li, Tromp, and Yannakakis [7] have proved that indeed the shortest superstring can be approximated to within a constant factor, and in fact Greedy achieves $4n$ approximation. Here, due to space constraint, we will prove that a simple variant of Greedy achieves $4n$. Some definitions are needed.

For two strings s and t, not necessarily distinct, let v be the longest string such that $s = uv$ and $t = vw$ for some *non-empty* strings u and w. We call $|v|$ the (amount of) *overlap* between s and t, and denote it as $ov(s,t)$. Furthermore, u is called the *prefix* of s with respect to t, and is denoted $pref(s,t)$. Finally, we call $|pref(s,t)| = |u|$ the *distance* from s to t, and denote it as $d(s,t)$. So, the string $uvw = pref(s,t)t$ is obtained by maximally merging s to the left of t and is called the *merge* of s and t. For $s_i, s_j \in S$, we will abbreviate $pref(s_i, s_j)$ to simply $pref(i,j)$. As an example of *self-overlap*, we have for the string s = alfalfa an overlap of $ov(s,s) = 4$. Also, $pref(s,s)$ = alf and $d(s,s) = 3$.

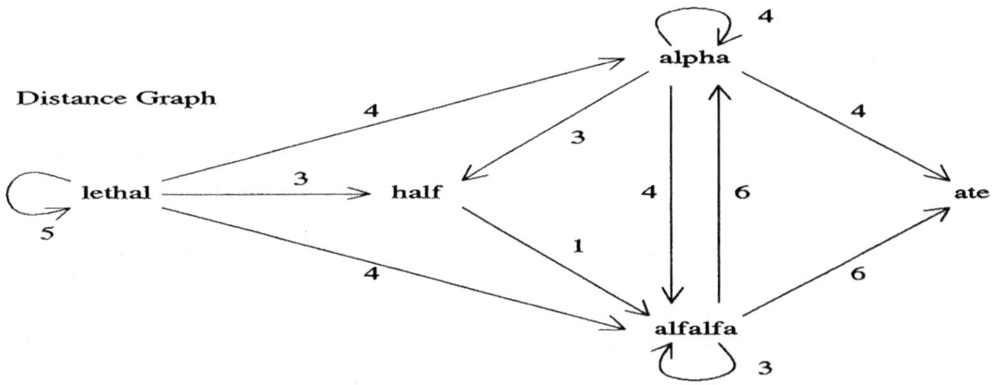

Fig. 3. A distance graph.

Given a list of strings $s_{i_1}, s_{i_2}, \ldots, s_{i_r}$, we define the superstring $s = \langle s_{i_1}, \ldots, s_{i_r} \rangle$ to be the string $pref(i_1, i_2) \cdots pref(i_{r-1}, i_r) s_{i_r}$. That is, s is the shortest string such that $s_{i_1}, s_{i_2}, \ldots, s_{i_r}$ appear *in order* in that string. For a superstring of a substring-free set, this order is well-defined, since substrings cannot 'start' or 'end' at the same position, and if substring s_j starts before s_k, then s_j must also end before s_k. Note that there exists a permutation π on the set $\{1, \ldots, m\}$, such that $S_\pi = \langle s_{\pi(1)}, \ldots, s_{\pi(m)} \rangle$ is a shortest superstring for S and every shortest superstring for S equals S_π for some permutation π.

We will consider TSP on a weighted directed complete graph G_S derived from S and show that one can achieve a factor of 4 approximation for TSP on that graph, yielding a factor of 4 approximation for the shortest superstring problem. Graph $G_S = (V, E, d)$ has m vertices $V = \{1, \ldots, m\}$, and m^2 edges $E = \{(i, j) : 1 \leq i, j \leq m\}$. Here we take as weight function the distance $d(,)$: edge (i, j) has weight $w(i, j) = d(s_i, s_j)$, to obtain the *distance graph*. We will call s_i the string *associated* with vertex i, and let $pref(i, j) = pref(s_i, s_j)$ be the string associated to edge (i, j). As an example, the distance graph for the set { ate, half, lethal, alpha, alfalfa } is given in Figure 3. All edges not shown have overlap 0.

For any weighted graph G, a *cycle cover* of G is a set of vertex-disjoint cycles, covering all vertices of G. The cycle cover is said to be *optimal* if it has the smallest weight. Let $TSP(G_S)$ and $CYC(G_S)$ denote the weight of an optimal Hamiltonian cycle and the weight of an optimal cycle cover of G_S, respectively. It is easy to see that

$$CYC(G_S) \leq TSP(G_S) \leq OPT(S)$$

Also observe that, although finding an optimal Hamiltonian cycle in a weighted graph is in general a hard problem, one can use the well-known Hungarian algorithm to find an optimal cycle cover in polynomial-time.

We now define some notation for dealing with (directed) cycles in G_S. If $c = i_0, \ldots, i_{r-1}, i_0$ is a cycle in G_S with r vertices, we define $strings(c)$ to be the equivalence class $[pref(i_0, i_1) \cdots pref(i_{r-1}, i_0)]$ and $strings(c, i_k)$ the rotation starting with i_k, *i.e.*, the string $pref(i_k, i_{k+1}) \cdots pref(i_{k-1}, i_k)$, where subscript arithmetic is modulo r. Denote the weight of a cycle c as $w(c)$. The following simple fact about cycles in G_S is easy to prove.

Claim 22 *Let $c = i_0, \cdots, i_{r-1}, i_0$ be a cycle. The superstring $\langle s_{i_0}, \cdots, s_{i_{r-1}} \rangle$ is a substring of strings$(c, i_0)s_{i_0}$.*

We present an algorithm Mgreedy which finds a superstring of length at most $4n$. This result is due to [7]. The construction proceeds in two stages. We first show that an algorithm that finds an optimal cycle cover on the distance graph G_S, then opens each cycle into a single string, and finally concatenates all such strings together has a performance at most $4n$. We then show that in fact, for distance graphs, a greedy strategy can be used to find optimal cycle covers.

Consider the following algorithm for finding a superstring of the strings in S.

Algorithm Concat-Cycles

1. On input S, create graph G_S and find an optimal cycle cover $C = \{c_1, \ldots, c_p\}$ on G_S.
2. For each cycle $c_i = i_1, \cdots, i_r, i_1$, let $\tilde{s}_i = \langle s_{i_1}, \ldots, s_{i_r} \rangle$ be the string obtained by opening c_i, where i_1 is arbitrarily chosen. The string $\tilde{s}_i$ has length at most $w(c_i) + |s_{i_1}|$ by Claim 22.
3. Concatenate together the strings $\tilde{s}_i$ and produce the resulting string $\tilde{s}$ as output.

Theorem 23 *Algorithm Concat-Cycles produces a string of length at most $4 \cdot OPT(S)$.*

Before proving Theorem 23, we need a lemma giving an upper bound on the overlap between strings in different cycles of C. A proof of the lemma can be found in [7].

Lemma 24 *Let c and c' be two cycles in C with $s \in c$ and $s' \in c'$. Then, $ov(s, s') < w(c) + w(c')$.*

Proof of Theorem 23. Since $C = \{c_1, \ldots, c_p\}$ is an optimal cycle cover, $CYC(G_S) = \sum_{i=1}^{p} w(c_i) \leq OPT(S)$. A second lower bound on $OPT(S)$ can be determined as follows: For each cycle c_i, let $w_i = w(c_i)$ and l_i denote the length of the longest string in c_i. By Lemma 24, if we consider the longest string in each cycle and merge them together optimally, the total amount of overlap will be at most $2\sum_{i=1}^{p} w_i$. So the resulting string will have length at least $\sum_{i=1}^{p} l_i - 2w_i$. Thus $OPT(S) \geq \max(\sum_{i=1}^{p} w_i, \sum_{i=1}^{p} l_i - 2w_i)$.

The output string $\tilde{s}$ of algorithm Concat-Cycles has length at most $\sum_{i=1}^{p} l_i + w_i$ (Claim 22). So,

$$
\begin{aligned}
|\tilde{s}| &\leq \sum_{i=1}^{p} l_i + w_i \\
&= \sum_{i=1}^{p} l_i - 2w_i \; + \; \sum_{i=1}^{p} 3w_i \\
&\leq OPT(S) \; + \; 3 \cdot OPT(S) \\
&= 4 \cdot OPT(S). \quad \blacksquare
\end{aligned}
$$

Now let's modify the algorithm Greedy slightly:

Algorithm Mgreedy

1. Let S be the input set of strings and T be empty.
2. While S is non-empty, do the following: Choose $s, t \in S$ (not necessarily distinct) such that $ov(s,t)$ is maximized, breaking ties arbitrarily. If $s \neq t$, remove s and t from S and replace them with the merged string $\langle s, t \rangle$. If $s = t$, just remove s from S and add it to T.
3. When S is empty, output the concatenation of the strings in T.

It is shown in [7] that Mgreedy in fact mimics algorithm Concat-Cycles and creates optimal cycle cover and hence also achieves $4n$ approximation. [7] also introduces an algorithm Tgreedy which operates in the same way as Mgreedy except that, in the last step, it merges the strings in T by running Greedy on them. It is shown that Tgreedy achieves a better bound of $3n$. The bound of $3n$ has been improved to $2.89n$ by Teng and Yao [56] recently.

One might expect that an analysis similar to that of Concat-Cycles or Mgreedy would also work for the original algorithm Greedy. This turns out not to be the case. The analysis of Greedy is severely complicated by the fact that it continues processing the "self-overlapping" strings. Concat-Cycles and Mgreedy was especially designed to avoid these complications, by separating such strings. With a more complicated analysis we can nevertheless prove the following theorem [7].

Theorem 25 *Greedy produces a superstring of length at most* $4n$.

The proof of this theorem is quite difficult and omitted here. Note that the $4n$ upper bound is still a factor 2 away from the conjectured $2n$ bound. Other variations of the superstring problem were also studied. For example, it was proved [29] that one can still achieve linear approximation if the substrings are allowed to be flipped.

5. Some Unsolved Questions in Computational Biology.

In this section, we list some important open questions arising as optimization problems in molecular biology.

1. With bounded alphabet, is the longest common subsequence problem still MAX SNP-hard? We conjecture that the answer is yes.
2. With bounded alphabet, is the shortest common supersequence problem still MAX SNP-hard?
3. Devise non-trivial approximation algorithms for shortest common supersequences and longest common subsequences. In particular, can one find a common subsequence of length at least $OPT(S)/n^{\delta}$ in polynomial time for some constant $\delta < 1$?
4. Prove or disprove that algorithm Greedy produces a superstring of length at most $2 \cdot OPT$. Obtain better approximation bounds for the shortest common superstring problem than the bounds in [7] and [56].
5. Consider the following generalized version of the superstring problem: Given sets of strings P (the *positive* examples) and N (the *negative* examples), find a shortest string that contains every positive example but no negative example [26,

27, 28]. The best approximation ratio for this problem is $\log n$ [26]. It will be of interest to know if the problem has a polynomial-time linear approximation algorithm.

6. Improve the approximation ratio 2 for multiple alignment with SP-score. In particular, does the problem have a PTAS?
7. Devise better approximation algorithms for tree alignment. Can we approximate the problem without using Steiner trees? (The Steiner minimal trees tend to generate localized component, which is very unusual in an evolutionary process.)
8. The edit distance between two strings can be extended to more complicated operations. When comparing the chromosomes of some organisms, we often need consider the operation that reverses a substring of arbitrary length in some chromosome. Define the *reversal distance* between s and t as the minimum number of reversal required to transform s into t. Currently it is open if the reversal distance can be computed in polynomial time. Some approximation algorithms with ratios 2 and 7/4 have been found in [34] and [5].

References

1. S. Altschul and D. Lipman, Trees, stars, and multiple sequence alignment, *SIAM Journal on Applied Math.* 49, pp. 197-209, 1989
2. S. Arora, C. Lund, R. Motwani, M. Sudan and M. Szegedy, Proof verification and hardness of approximation problems, *Proc. IEEE 32nd FOCS*, pp. 14-23, 1992.
3. D. Bacon and W. Anderson, Multiple sequence alignment, *Journal of Molecular Biology* 191, pp. 153-161, 1986.
4. V. Bafna and P. Pevzner, Approximate methods for multiple sequence alignment, *Manuscript*, 1993.
5. V. Bafna and P. Pevzner, Genome rearrangements and sorting by reversals, to be presented at *34th IEEE FOCS*, Oct. 1993.
6. P. Berman and V. Ramaiyer, Improved approximations for the Steiner tree problem, *Manuscript*, 1993.
7. A. Blum, T. Jiang, M. Li, J. Tromp, and M. Yannakakis. Linear approximation of shortest superstrings. *Proc. 23rd ACM Symp. on Theory of Computing*, 1991, 328-336; also to appear in *J.ACM*.
8. H. Carrillo and D. Lipman, The multiple sequence alignment problem in biology, *SIAM Journal on Applied Math.* 48, pp. 1073-1082, 1988.
9. S. C. Chan, A. K. C. Wong and D. K. T. Chiu, A survey of multiple sequence comparison methods, *Bulletin of Mathematical Biology* 54(4), pp. 563-598, 1992.
10. M.O. Dayhoff. Computer analysis of protein evolution. *Scientific American* 221:1(July, 1969), 86-95.
11. L. R. Foulds and R.L. Graham, The Steiner problem in phylogeny is NP-complete, *Advances in Applied Mathematics* 3, pp. 43-49, 1982.
12. D.E. Foulser. On random strings and sequence comparisons. *Ph.D. Thesis*, Stanford University, 1986.
13. D.E. Foulser, M. Li, and Q. Yang. Theory and algorithms for plan merging. *Artificial Intelligence Journal*, 57(1992), 143-181.
14. J. Gallant, D. Maier, J. Storer. On finding minimal length superstring. *Journal of Computer and System Sciences*, 20(1980), 50-58.
15. M. Garey and D. Johnson. *Computers and Intractability.* Freeman, New York, 1979.
16. D. Gusfield, Efficient methods for multiple sequence alignment with guaranteed error bounds, *Tech. Report, CSE-91-4, UC Davis*, 1991.
17. D. Gusfield, Efficient methods for multiple sequence alignment with guaranteed error bounds, *Bulletin of Mathematical Biology* 55, pp. 141-154, 1993.

18. C.C. Hayes. A model of planning for plan efficiency: Taking advantage of operator overlap. *Proceedings of the 11th International Joint Conference of Artificial Intelligence*, Detroit, Michigan. (1989), 949-953.
19. J. J. Hein, A tree reconstruction method that is economical in the number of pairwise comparisons used, *Mol. Biol. Evol.* 6(6), pp. 669-684, 1989.
20. J. J. Hein, A new method that simultaneously aligns and reconstructs ancestral sequences for any number of homologous sequences, when the phylogeny is given, *Mol. Biol. Evol.* 6(6), pp. 649-668, 1989.
21. D.S. Hirschberg. The longest common subsequence problem. *Ph.D. Thesis*, Princeton University, 1975.
22. W.J. Hsu and M.W. Du. Computing a longest common subsequence for a set of strings. *BIT* 24, 1984, 45-59.
23. F. K. Hwang and D. S. Richards, Steiner tree problems, *Networks* 22, pp. 55-89, 1992.
24. R.W. Irving and C.B. Fraser. Two algorithms for the longest common subsequence of three (or more) strings. *Proc. 2nd Symp. Combinatorial Pattern Matching*, 1992.
25. T. Jiang and M. Li. On the approximation of shortest common supersequences and longest common subsequences. submitted to *SIAM J. Computing*, 1992.
26. T. Jiang and M. Li. Towards a DNA sequencing theory (revised version). Submitted for publication, 1991.
27. T. Jiang and M. Li. On the complexity of learning strings and sequences. *Proc. 4th Workshop on Computational Learning*, 1991; also to appear in *Theoret. Comp. Sci.*
28. T. Jiang and M. Li. Approximating shortest superstrings with constraints. *Proc. 3rd Workshop on Algorithms and Data Structures*, 1993, pp. 385-396; also to appear in *Theoret. Comp. Sci.*
29. T. Jiang, M. Li, and D-Z. Du, A note on shortest superstrings with flipping, *Inform. Process. Lett.*, 44:4(1992), 195-199.
30. T.H. Jukes and C.R. Cantor, Evolution of protein molecules, in H.N. Munro, ed., *Mammalian Protein Metabolism*, Academic Press, pp. 21-132, 1969.
31. D. Karger, R. Motwani, and G.D.S. Ramkumar. On approximating the longest path in a graph. *Proc. 3rd WADS*, 1993.
32. R. Karinthi, D.S. Nau, and Q. Yang. Handling feature interactions in process planning. Department of Computer Science, University of Maryland, College Park, MD. (1990).
33. R. M. Karp, Mapping the genome: some combinatorial problems arising in molecular biology, *Proc. 25th ACM STOC*, pp. 278-285, 1993.
34. J. Kececioglu and D. Sankoff, Exact and approximation algorithms for the inversion distance between two chromosomes, to appear in *Algorithmica*.
35. E.S. Lander, R. Langridge and D.M. Saccocio, Mapping and interpreting biological information, *Communications of the ACM* 34(11), pp. 33-39, 1991.
36. A. Lesk (Edited). *Computational Molecular Biology, Sources and Methods for Sequence Analysis.* Oxford University Press, 1988.
37. M. Li and P.M.B. Vitányi. *An Introduction to Kolmogorov Complexity and Its Applications.* Springer-Verlag, 1993.
38. M. Li and P.M.B. Vitányi. Combinatorial properties of finite sequences with high Kolmogorov complexity. To appear in *Math. Syst. Theory.*
39. S.Y. Lu and K.S. Fu. A sentence-to-sentence clustering procedure for pattern analysis. IEEE Trans. Syst., Man, Cybern. Vol. SMC-8(5), 1978, 381-389.
40. D. Maier. The complexity of some problems on subsequences and supersequences. *J. ACM*, 25:2(1978), 322-336.
41. M. Middendorf, More on the complexity of common superstring and supersequence problems, to appear in *Theoret. Comp. Sci.*
42. C. Papadimitriou and K. Steiglitz. *Combinatorial Optimization: Algorithms and Complexity.* Prentice-Hall, 1982.
43. C.H. Papadimitriou and M. Yannakakis. Optimization, approximation, and complexity classes. Extended abstract in *Proc. 20th ACM Symp. on Theory of Computing.* 1988, 229-234; full version in *Journal of Computer and System Sciences* 43, 1991, 425-440.
44. P. Pevzner, Multiple alignment, communication cost, and graph matching, *SIAM J. Applied Math.* 56(6), pp. 1763-1779, 1992.
45. D. Sankoff, Minimal mutation trees of sequences, *SIAM J. Applied Math.* 28(1), pp. 35-42, 1975.

46. D. Sankoff, R. J. Cedergren and G. Lapalme, Frequency of insertion-deletion, transversion, and transition in the evolution of 5S ribosomal RNA, *J. Mol. Evol.* 7, pp. 133-149, 1976.
47. D. Sankoff and R. Cedergren, Simultaneous comparisons of three or more sequences related by a tree, In D. Sankoff and J. Kruskal, editors, *Time warps, string edits, and macromolecules: the theory and practice of sequence comparison,* pp. 253-264, Addison Wesley, 1983.
48. D. Sankoff and J. Kruskal (Eds.) *Time Warps, String Edits, and Macromolecules: The Theory and Practice of Sequence Comparison.* Addison-Wesley, Reading, MA., 1983.
49. G.D. Schuler, S.F. Altschul, and D.J. Lipman. A workbench for multiple alignment construction and analysis, in *Proteins: Structure, function and Genetics,* in press.
50. R.Schwarz and M. Dayhoff, Matrices for detecting distant relationships in M. Dayhoff, ed., *Atlas of protein sequences,* National Biomedical Research Foundation, 1979, pp.353-358.
51. T. Sellis. Multiple query optimization. *ACM Transactions on Database Systems,* 13:1(1988), 23-52
52. T.F. Smith and M.S. Waterman. Identification of common molecular subsequences. Journal of Molecular Biology, 147(1981), 195-197.
53. J. Storer. *Data compression: methods and theory.* Computer Science Press, 1988.
54. E. Sweedyk and T. Warnow, The tree alignment problem is NP-hard, *Manuscript,* 1992.
55. J. Tarhio and E. Ukkonen. A greedy approximation algorithm for constructing shortest common superstrings. *Theoretical Computer Science 57 131-145 1988*
56. S.H. Teng and F. Yao. Approximating shortest superstrings. *34th IEEE Symp. Foundat. Comput. Sci.,* 1993.
57. V.G. Timkovskii. Complexity of common subsequence and supersequence problems and related problems. *English Translation from Kibernetika,* 5(1989), 1-13.
58. J. Turner. Approximation algorithms for the shortest common superstring problem. *Information and Computation 83, 1989, 1-20.*
59. R.A. Wagner and M.J. Fischer. The string-to-string correction problem. J. ACM, 21:1(1974), 168-173.
60. L. Wang and T. Jiang, On the complexity of multiple sequence alignment, submitted to *Journal of Computational Biology,* 1993.
61. L. Wang and T. Jiang, Approximation algorithms for tree alignment with a given phylogeny, submitted to *Algorithmica,* 1993.
62. M.S. Waterman, Sequence alignments, in *Mathematical Methods for DNA Sequences,* M.S. Waterman (ed.), CRC, Boca Raton, FL, pp. 53-92, 1989.
63. A.Z. Zelikovsky, The 11/6 approximation algorithm for the Steiner problem on networks, to appear in *Information and Computation.*

A DUAL AFFINE SCALING BASED ALGORITHM FOR SOLVING LINEAR SEMI-INFINITE PROGRAMMING PROBLEMS

CHIH-JEN LIN
Department of Mathematics
National Taiwan University
Taipei, Taiwan, China.

SHU-CHERNG FANG*
Operations Research and Industrial Engineering
North Carolina State University
Raleigh, NC 27695-7913, U.S.A.

and

SOON-YI WU
Institute of Applied Mathematics
National Cheng-Kung University
Tainan, Taiwan, hina.

Abstract. In this paper we study a class of linear semi-infinite programming problems with finitely many variables and infinitely many constraints over a compact metric space. A dual affine scaling based algorithm which adds one constraint at a time is introduced. The convergence proof of the proposed algorithm together with some numerical experience are also included.

Key words: Semi-infinite programming, linear programming, dual affine scaling method.

1. Introduction

Consider the following linear semi-infinite programming problem which has n nonnegative variables and infinitely many inequality constraints :

$$\text{Minimize } \sum_{j=1}^{n} c_j x_j$$

(LSIP)

$$\text{subject to } \sum_{j=1}^{n} x_j a_j(t) \geq b(t), \quad \forall t \in T \tag{1.1}$$

$$x_j \geq 0 \;\; j = 1 \ldots n \tag{1.2}$$

where T is a compact metric space with an infinite cardinality, $a_j(t), j = 1, \ldots, n$, and $b(t)$ are real valued continuous functions defined on T.

Its dual problem can be formulated in the following form:

* This research work is partially supported by the NCSC-Cray Research Grant and the National Science Council Grant #NCS81-0415-E-007-10 in Taipei of China.

Ding-Zhu Du and Jie Sun (eds.), Advances in Optimization and Approximation, 217–234.

(DLSIP)

$$\text{Maximize} \quad \int_T g(t)d\nu(t)$$

$$\text{subject to} \quad \int_T f_j(t)d\nu(t) \le c_j, \quad j = 1, 2, \cdots, n \tag{1.3}$$

$$\nu \in M^+(T) \tag{1.4}$$

where $M^+(T)$ is the space of all non–negative bounded regular Borel measures on T. When (LSIP) has a nonempty feasible domain, under some regularity conditions, it was shown [2, 15] that there is no duality gap between (LSIP) and (DLSIP). Moreover, (LSIP) achieves its optimum at least at one "extreme point" of its feasible domain.

Some basic concepts and methods for solving (LSIP) can be found in [2, 10, 12, 13]. One direct way of solving (LSIP) is to discretize T into a finite number of grid points $\{t_1, t_2, ..., t_m\}$ and then an approximation can be obtained by solving a regular linear programming problem with n non-negative variables and m inequality constraints obtained by evaluating (1.1) at $\{t_1, t_2, ..., t_m\}$. Usually, a case with more grid points results in better approximation at the cost of solving a larger size linear program. Also, the choice of grid points has to take into consideration of the behavior of the functions $a_j(t)$, $j = 1, ..., n$, and $b(t)$.

To avoid having too many constraints at one time, an "adding constraint" (or "cutting plan") method was introduced, see references [10, 11, 12]. Basically, it constructs a sequence of finite linear programs whose solutions converge to an optimal solution of (LSIP). To describe this method, we first denote the following problem by (LP_k), since it is a linear programming problem with k explicit constraints.

(LP_k)

$$\text{Minimize} \quad \sum_{j=1}^{n} c_j x_j$$

$$\text{subject to} \quad \sum_{j=1}^{n} x_j a_j(t_i) \ge b(t_i), \qquad i = 1, 2, \cdots, k \tag{1.5}$$

$$x_j \ge 0, \qquad j = 1, 2, \cdots, n \tag{1.6}$$

Then the method works according to the following scheme:

Step 1: Set $k = 1$, choose any $t_1 \in T$, and set $T_1 = \{t_1\}$.
Step 2: Solve (LP_k) with an optimal solution $x(k)$.
Step 3: Find a minimizer t_{k+1} of $\phi_k(t)$ over T where

$$\phi_k(t_{k+1}) = \sum_{j=1}^{n} x_j(k) a_j(t_{k+1}) - b(t_{k+1}).$$

Step 4: If $\phi_k(t_{k+1}) \ge 0$, then stop. $x(k)$ is optimal for (LSIP).
Otherwise, set $T_{k+1} = T_k \cup \{t_{k+1}\}$, increment $k \leftarrow k + 1$, and go to Step 2.

In the algorithm, one constraint is added at a time and the computational bottleneck falls either in solving a linear program (LP_k) or in finding a global minimizer t_{k+1} of $\phi_k(t)$.

Recently, some research has been done to extend the interior-point approach to solving semi-infinite programming problems [8, 9, 19, 20]. Also the asymptotic behavior of interior-point methods was studied from the view of semi-infinite programming [21]. In this paper, instead of *directly extending* the interior point methods for solving (LSIP), we try to *incorporate* the interior point approach into the "adding constraint" method for experiment.

Notice that when the dimensionality of the compact metric space T is low, finding an optimizer of a continuous function $\phi_k(t)$ over T usually causes little computational problem. In this case, solving (LP_k) in Step 2 repeatly consumes most time in applying the "adding constraint" method. With this understanding, when an interior point method[14] is incorporated into the "adding constraint" method for solving (LP_k), we may want to focus on reducing the total number of Cholesky factorizations involved.

To reduce the computational effort in solving each (LP_k), one idea is to find an "inexact solution", instead of an "exact optimal solution" [6, 7]. Potential difficulties of this "inexact approach" include (1) replacing $x(k)$ by an inexact solution, for $k = 1, 2, ...$, may affect the the choice of t_{k+1} in Step 3, and hence the convergence of the algorithm; and (2) Unless the inexact solution $x(k)$ and/or related information obtained in the kth iteration can be somehow utilized for the $k+1^{th}$ iteration, solving (LP_{k+1}) from a "cold start" is not efficient. The objective of this paper is to design an *affine scaling based* algorithm [3, 4, 24] which takes care of the above mentioned difficulties. Observing that (LP_{k+1}) has one more constraint than (LP_k), we know the solution $x(k)$ of (LP_k) may become infeasible for (LP_{k+1}) and decide to consider the dual affine scaling method [1, 16, 17, 18].

In the rest of this paper, we propose a dual affine scaling based "adding constraint" algorithm in Section 2, provide a convergence proof in Section 3, and then report some computational experience in Section 4 before concluding remarks are made in Section 5.

2. A New Algorithm for (LSIP)

To work on the standard form, we first convert (LP_k) into a standard form problem (SLP_k).

$$
(SLP_k)\qquad \begin{aligned} &\text{Minimize } c(k)^T x \\ &\text{subject to } A(k)x = B(k) \qquad (2.1) \\ &x_j \geq 0, \quad j = 1, \cdots, n+k \end{aligned}
$$

where

$$A(k) = \begin{bmatrix} a_1(t_1) & \dots & a_n(t_1) & -1 & 0 & \cdots & 0 \\ a_1(t_2) & \dots & a_n(t_2) & 0 & -1 & \cdots & 0 \\ \vdots & & \vdots & \vdots & \vdots & \ddots & \vdots \\ a_1(t_k) & \cdots & a_n(t_k) & 0 & 0 & \cdots & -1 \end{bmatrix}$$

is a k by $n+k$ dimensional matrix, $B(k) = [b(t_1), \dots, b(t_k)]^T$ is a k dimensional column vector, and $c(k) = [c_1, \dots, c_n, 0, \dots, 0]^T$ and $x = [x_1, \dots, x_n, x_{n+1}, \dots, x_{n+k}]^T$ are $n+k$ dimensional column vectors.

Note that $x_{n+1}, \dots, x_{n+k}$ are slack variables, also $A(k)$ and $A(k+1)$ are related in the following way :

$$A(k+1) = \begin{bmatrix} & & 0 \\ & A(k) & \vdots \\ & & 0 \\ a_1(t_{k+1}), \cdots, a_n(t_{k+1}), 0, \cdots, 0, & & -1 \end{bmatrix}.$$

It is easy to see that $A(k)$ has full row rank, for all k, and the linear dual of (SLP_k) can be written as

$$\begin{aligned} (\text{DLP}_k) \qquad & \text{Maximize } B(k)^T w \\ & \text{subject to } A(k)^T w + s = c(k) \qquad (2.2) \\ & s_j \geq 0, \quad j = 1, 2, \cdots, n+k \end{aligned}$$

where $w = [w_1, \dots, w_k]^T$ is a k dimensional column vector and $s = [s_1, \dots, s_{n+k}]^T$ is an $n+k$ dimensional column vector.

From the special structure of $A(k)$ and $c(k)$, we have $-w_j + s_{j+n} = 0, \forall j = 1, \dots, k$. Since s_j is restricted to be nonnegative, we further know that $w_j \geq 0, \ \forall j$.

As discussed in the introduction section, when an interior point algorithm is incorporated into Step 2 of the "adding constraint" method, we expect it exhibits the following properties : (1) An initial interior feasible solution of (SLP_1) or (DLP_1) can be found. (2) The information obtained from solving (SLP_{k-1}) or (DLP_{k-1}) can be used to easily obtain an initial interior feasible solution of (SLP_k) or (DLP_k), for $k \geq 2$. (3) Although $x(k)$ is only an inexact solution of (SLP_k), it has to converge to an optimal solution of (LSIP).

In order to achieve these objectives, we make the following assumptions:

(A1) (LSIP) has a nonempty feasible domain.
(A2) (SLP_1) has a bounded feasible domain.
(A3) (DLP_1) has an interior feasible solution.
(A4) (SLP_k) and (DLP_k) are nondegenerate for all k.

Note that the first and the second assumptions imply that (LSIP) has a finite optimal solution. Assumption 3 allows us to start with a dual interior feasible solution, and Assumption 4 will be used to simplify our convergence proof. Assumption 2 also implies that, for each k, (DLP_k) is bounded above. Therefore, when the dual affine scaling algorithm is applied to (DLP_k), it iterates (from the ith solution to $i+1^{th}$ solution according to following procedure:

(Procedure : dual affine scaling iteration)
Input : $A(k), B(k), c(k), \bar{w}(k,i), \bar{s}(k,i)$, such that

$$A(k)^T \bar{w}(k,i) + \bar{s}(k,i) = c(k),$$

$$\bar{s}(k,i) > 0.$$

Steps : Let $S(k,i)$ be the diagonal matrix formed by the elements of $\bar{s}(k,i)$, compute

$$dw(k,i) = (A(k)S(k,i)^{-2}A(k)^T)^{-1}B(k),$$
$$ds(k,i) = -A(k)^T dw(k,i).$$

Since (DLP$_k$) is bounded above, if $ds(k,i) = 0$, we let

$$\bar{x}(k,i+1) = -S(k,i)^{-2}ds(k,i) = 0,$$

$$\bar{w}(k,i+1) = \bar{w}(k,i),$$

$$\bar{s}(k,i+1) = \bar{s}(k,i).$$

Otherwise, some components of $ds(k,i)$ must be negative and we let

$$\bar{x}(k,i+1) = -S(k,i)^{-2}ds(k,i),$$

$$\beta = \min{}_j\{\frac{\alpha\bar{s}_j(k,i)}{-ds_j(k,i)} \mid ds_j(k,i) < 0\}, \text{ where } 0 < \alpha < 1,$$

$$\bar{w}(k,i+1) = \bar{w}(k,i) + \beta dw(k,i),$$

$$\bar{s}(k,i+1) = c(k) - A(k)^T \bar{w}(k,i+1).$$

Output : $\bar{x}(k,i+1), \bar{w}(k,i+1), \bar{s}(k,i+1)$.

It is clear to see that the major computation involved in this procedure lies in the Cholesky factorization of $(A(k)S(k,i)^{-2}A(k)^T)$. Repeating this procedure either identifies that $\bar{x}(k,i+1) = 0$ is an optimal solution of (SLP$_k$), as $ds(k,i) = 0$ for some i, or produces a sequence of $\bar{x}(k,i)$ which converges to an optimal solution of (SLP$_k$), as $i \to \infty$. Also note that, if $ds(k,i) \neq 0$, we have

$$\begin{aligned}
& B(k)^T \bar{w}(k,i+1) - B(k)^T \bar{w}(k,i) \\
= \; & \beta B(k)^T dw(k,i) \\
= \; & \beta(A(k)S(k,i)^{-2}A(k)^T dw(k,i))^T dw(k,i) \\
= \; & \beta dw(k,i)^T A(k)S(k,i)^{-2}A(k)^T dw(k,i) \\
= \; & \beta ds(k,i)^T S(k,i)^{-2} ds(k,i) \\
= \; & \beta((\tfrac{ds_1(k,i)}{\bar{s}_1(k,i)})^2 + \cdots + (\tfrac{ds_{n+k}(k,i)}{\bar{s}_{n+k}(k,i)})^2) \\
> \; & 0.
\end{aligned} \tag{2.3}$$

This further indicates that, unless $ds(k,i) = 0$, the dual affine scaling procedure will not stop with an exact optimal solution of (DLP$_k$). Moreover, the procedure validates that, for each i,

$$B(k)^T \bar{w}(k,i+1) \geq B(k)^T \bar{w}(k,i), \tag{2.4}$$

no matter $ds(k,i) = 0$ or not. Also, by the way we define $\bar{x}(k,i+1)$, $ds(k,i)$, and $dw(k,i)$, $A(k)\bar{x}(k,i+1) = B(k)$ always holds, but $\bar{x}(k,i+1) \geq 0$ may not be true.

Several observations can be made before we outline our algorithm.

Observation 1: When $t_1 \in T$ is given, since there is only one constraint involved, it is not difficult to find an initial interior feasible solution of (DLP_1), namely, $\bar{s}(1,1) > 0$ and $\bar{w}(1,1)$ such that

$$A(1)^T\bar{w}(1,1) + \bar{s}(1,1) = c(1).$$

This can be done by applying the regular two-phase method [5]. Also note that $A(1) = [a_1(t_1), \ldots, a_n(t_1), -1]$ and $c(1) = [c_1, \ldots, c_n, 0]^T$, by checking the following inequalities

$$\begin{aligned}
\bar{s}_1(1,1) &= c_1 - a_1(t_1)\bar{w}(1,1) > 0,\\
\bar{s}_2(1,1) &= c_2 - a_2(t_1)\bar{w}(1,1) > 0,\\
&\vdots\\
\bar{s}_n(1,1) &= c_n - a_n(t_1)\bar{w}(1,1) > 0,\\
\bar{s}_{n+1}(1,1) &= \bar{w}(1,1),
\end{aligned}$$

under Assumption 3, an appropriate value of $\bar{w}(1,1)$ may be identified quickly and hence the vector $\bar{s}(1,1) > 0$ can be calculated.

Observation 2: For a given $T_k = \{t_1, \ldots, t_k \in T\}$, suppose that we have an initial interior feasible solution of (DLP_k), namely $\bar{s}(k,1) > 0$ and $\bar{w}(k,1)$ such that

$$A(k)^T\bar{w}(k,1) + \bar{s}(k,1) = c(k).$$

By repeating the dual affine scaling procedure, a sequence of triplets $\bar{x}(k,i)$, $\bar{w}(k,i)$, $\bar{s}(k,i)$ will be obtained such that (1) $\bar{w}(k,i)$, $\bar{s}(k,i)$ is interior feasible for (DLP_k), for each i; and (2) $\bar{x}(k,i)$ converges to an optimal solution $x(k)$ of (SLP_k). Therefore, for any small number $\epsilon > 0$, there is a large enough positive integer i^k such that $min_j\{\bar{x}_j(k,i^k) \mid \bar{x}_j(k,i^k) < 0, j = 1, .., n+k\} \geq -\epsilon$. In this case, $\bar{x}(k,i^k)$ can be viewed as an approximate of $x(k)$ and we can find an optimizer t_{k+1,i^k} of $\phi_{k,i^k}(t)$ over T with

$$\phi_{k,i^k}(t_{k+1,i^k}) = \sum_{j=1}^{n} \bar{x}_j(k,i^k)a_j(t_{k+1,i^k}) - b(t_{k+1,i^k}).$$

Observation 3: When $t_{k+1,i^k} \notin T_k$ is found, we may set $x(k) = \bar{x}(k,i^k)$, $w(k) = \bar{w}(k,i^k)$, $s(k) = \bar{s}(k,i^k)$, and $t_{k+1} = t_{k+1,i^k}$. Our objective is to find a "good" initial interior feasible solution of (DLP_{k+1}), namely, $\bar{s}(k+1,1) > 0$ and $\bar{w}(k+1,1)$ such that

$$A(k+1)^T\bar{w}(k+1,1) + \bar{s}(k+1,1) = c(k+1),$$

and

$$B(k+1)^T\bar{w}(k+1,1) > B(k)^T w(k).$$

This can be done by considering the following mechanism:

– Case 1: $b(t_{k+1}) > 0$.
First, we define a scalar

$$w = \begin{cases} 1 & \text{, if } a_j(t_{k+1}) \leq 0, \forall j = 1, \ldots, n. \\ min_j \{ \frac{s_j(k)}{\gamma a_j(t_{k+1})} \mid a_j(t_{k+1}) > 0, \gamma > 1 \} & \text{, otherwise.} \end{cases} \tag{2.5}$$

Then we define

$$\bar{w}(k+1, 1) = \begin{bmatrix} w(k) \\ w \end{bmatrix},$$

and

$$\bar{s}(k+1,1) = \begin{bmatrix} s_1(k) - a_1(t_{k+1})w \\ \vdots \\ s_n(k) - a_n(t_{k+1})w \\ s_{n+1}(k) - 0w \\ \vdots \\ s_{n+k}(k) - 0w \\ w \end{bmatrix} = \begin{bmatrix} s_1(k) - a_1(t_{k+1})w \\ \vdots \\ s_n(k) - a_n(t_{k+1})w \\ s_{n+1}(k) \\ \vdots \\ s_{n+k}(k) \\ w \end{bmatrix}.$$

Since $s(k) > 0$, we know that $w > 0$ and $\bar{s}(k+1, 1) > 0$. Also remember that $A(k)^T w(k) + s(k) = c(k)$, we have

$$\begin{aligned} & A(k+1)^T \bar{w}(k+1,1) + \bar{s}(k+1,1) \\ = & \begin{bmatrix} & a_1(t_{k+1}) \\ & \vdots \\ A(k)^T & a_n(t_{k+1}) \\ & 0 \\ & \vdots \\ & 0 \\ 0 \ldots 0 & -1 \end{bmatrix} \begin{bmatrix} w(k) \\ w \end{bmatrix} + \begin{bmatrix} s_1(k) - a_1(t_{k+1})w \\ \vdots \\ s_n(k) - a_n(t_{k+1})w \\ s_{n+1}(k) \\ \vdots \\ s_{n+k}(k) \\ w \end{bmatrix} \\ = & \begin{bmatrix} c(k) \\ 0 \end{bmatrix} = c(k+1). \end{aligned} \tag{2.6}$$

Moreover, we see that

$$B(k+1)^T \bar{w}(k+1,1) = B(k)^T w(k) + b(t_{k+1})w > B(k)w(k) \tag{2.7}$$

Hence a "warm" starting solution $(\bar{w}(k+1,1), \bar{s}(k+1,1))$ of (DLP$_{k+1}$) is found.

– Case 2: $b(t_{k+1}) \leq 0$.

Remember that $w(k) = \bar{w}(k, i^k)$ and $s(k) = \bar{s}(k, i^k)$, we may take $w(k)$ and $s(k)$ as the input and apply the dual affine scaling procedure *one* more time to produce $\bar{w}(k, i^k + 1)$, $\bar{s}(k, i^k + 1)$, and $\bar{x}(k, i^k + 1)$.

As discussed before, if $ds(k, i^k) = 0$, then $\bar{x}(k, i^k + 1) = -S(k, i^k)^{-2} ds(k, i^k) = 0$ is an optimum solution of (SLP$_k$). Now we find an optimizer t_{k+1, i^k+1} of $\phi_{k, i^k+1}(t)$

over T. If

$$\phi_{k,i^k+1}(t_{k+1,i^k+1}) = \sum_{j=1}^{n} \bar{x}_j(k, i^k+1)a_j(t_{k+1,i^k+1}) - b(t_{k+1,i^k+1}) \geq 0, \tag{2.8}$$

then we know $\bar{x}(k, i^k+1)_{1\ldots n}$ (the first n elements of $\bar{x}(k, i^k+1)$) is an optimal solution of (LSIP). Therefore, we can output $[0, \ldots, 0]$ as an optimal solution of (LSIP). Otherwise, we have

$$\phi_{k,i^k+1}(t_{k+1,i^k+1}) = \sum_{j=1}^{n} \bar{x}_j(k, i^k+1)a_j(t_{k+1,i^k+1}) - b(t_{k+1,i^k+1}) < 0. \tag{2.9}$$

Since $\bar{x}_j(k, i^k+1) = 0, \forall\, j = 1, \ldots, n$, it implies that $b(t_{k+1,i^k+1}) > 0$, we can reset $x(k) = \bar{x}(k, i^k+1), w(k) = \bar{w}(k, i^k+1), s(k) = \bar{s}(k, i^k+1)$, $t_{k+1} = t_{k+1,i^k+1}$, and i^k by i^k+1. Then following the mechanism described in Case 1, we can find a warm starting solution $(\bar{w}(k+1, 1), \bar{s}(k+1, 1))$ for (DLP_{K+1}).

Suppose that $ds(k, i^k) \neq 0$, by (2.3), we know that

$$\begin{aligned} & B(k)^T\bar{w}(k, i^k+1) - B(k)^T w(k) \\ = \ & B(k)^T\bar{w}(k, i^k+1) - B(k)^T\bar{w}(k, i^k) \\ > \ & 0. \end{aligned} \tag{2.10}$$

In this case, we can find a scalar w satisfying that $0 < w \leq min_j\{\frac{\bar{s}_j(k,i^k+1)}{\gamma a_j(t_{k+1})} \mid a_j(t_{k+1}) > 0, \gamma > 1\}$ and $0 \geq b(t_{k+1})w > B(k)^T w(k) - B(k)^T\bar{w}(k, i^k+1)$. Consequently, we can define

$$\bar{w}(k+1, 1) = \begin{bmatrix} \bar{w}(k, i^k+1) \\ w \end{bmatrix},$$

and

$$\bar{s}(k+1,1) = \begin{bmatrix} \bar{s}_1(k, i^k+1) - a_1(t_{k+1})w \\ \vdots \\ \bar{s}_n(k, i^k+1) - a_n(t_{k+1})w \\ \bar{s}_{n+1}(k, i^k+1) - 0w \\ \vdots \\ \bar{s}_{n+k}(k, i^k+1) - 0w \\ w \end{bmatrix} = \begin{bmatrix} \bar{s}_1(k, i^k+1) - a_1(t_{k+1})w \\ \vdots \\ \bar{s}_n(k, i^k+1) - a_n(t_{k+1})w \\ \bar{s}_{n+1}(k, i^k+1) \\ \vdots \\ \bar{s}_{n+k}(k, i^k+1) \\ w \end{bmatrix}$$

Similar to Case 1, we can easily show that $A(k+1)^T\bar{w}(k+1, 1) + \bar{s}(k+1, 1) = c(k+1)$ and $\bar{s}(k+1, 1) > 0$. Moreover,

$$\begin{aligned} & B(k+1)^T\bar{w}(k+1, 1) \\ = \ & B(k)^T\bar{w}(k, i^k+1) + b(t_{k+1})w \\ > \ & B(k)^T w(k) \\ = \ & B(k)^T w(k, i^k). \end{aligned} \tag{2.11}$$

With these observations, we now propose the following dual affine scaling based algorithm for solving (LSIP):

Algorithm 1

Step 1: Set $k = 1$, choose any $t_1 \in T$ and set $T_1 = \{t_1\}$. Select θ such that $0 < \theta < 1$ and $\epsilon_1 > 0$. Also select $\delta_1 > 0$, $\delta_2 > 0$, and $\delta_3 > 0$ to be sufficiently small.

Find $\bar{s}(1,1) > 0$ and $\bar{w}(1,1)$ such that

$$A(1)^T \bar{w}(1,1) + \bar{s}(1,1) = c(1)$$

and go to Step 3.

Step 2: Use the mechanism described in **Observation 3** either to find $\bar{s}(k,1) > 0$ and $\bar{w}(k,1)$ such that

$$A(k)^T \bar{w}(k,1) + \bar{s}(k,1) = c(k),$$

and

$$B(k)^T \bar{w}(k,1) > B(k-1)^T w(k-1),$$

or to identify that $[0,\ldots,0]$ is an optimal solution of (LSIP) and terminate the algorithm.

Step 3: Take $A(k), B(k), c(k), \bar{w}(k,1)$, and $\bar{s}(k,1)$ as input and repeat the dual affine scaling procedure (for i^k times) until

either $\bar{x}(k,i^k), \bar{w}(k,i^k), \bar{s}(k,i^k)$ satisfy the following two conditions :

(1) $\min_j \{\bar{x}_j(k,i^k) \mid \bar{x}_j(k,i^k) < 0, j = 1,..,n+k\} \geq -\epsilon_k$,

(2) A minimizer $t_{k+1,i^k} \notin T_k$ of $\phi_{k,i^k}(t)$ over T is found such that

$$\phi_{k,i^k}(t_{k+1,i^k}) = \sum_{j=1}^{n} \bar{x}_j(k,i^k) a_j(t_{k+1,i^k}) - b(t_{k+1,i^k}) < 0,$$

or

$\bar{x}_j(k,i^k) \geq -\delta_1$ for $j = 1,...,n+k$, $c(k)^T \bar{x}(k,i^k) - B(k)^T \bar{w}(k,i^k) \leq \delta_2$, and $\phi_k(t_{k+1,i^k}) \geq -\delta_3$.

Set $\phi_k(t) = \phi_{k,i^k}(t)$, $x(k) = \bar{x}(k,i^k)$, $w(k) = \bar{w}(k,i^k)$, $s(k) = \bar{s}(k,i^k)$, and $t_{k+1} = t_{k+1,i^k}$.

Step 4: If ($x_j(k) \geq -\delta_1$, for $j = 1,...,n+k$, $c(k)^T x(k) - B(k)^T w(k) \leq \delta_2$, and $\phi_k(t_{k+1}) \geq -\delta_3$), then Stop ! Output $x(k)_{1\ldots n} = [x_1(k),\ldots,x_n(k)]^T$ as an optimal solution of (LSIP).

Otherwise, set $\epsilon_{k+1} = (1-\theta)\epsilon_k$, $T_{k+1} = T_k \cup \{t_{k+1}\}$, and increment k by 1, then go to Step 2.

One more observation can be made here.

Observation 4: As in **Observation 2**, in theory, repeating the dual affine scaling procedure in Step 3 until it satisfies conditions (1), and (2) may become a loop and generate an infinite sequence of $\{\bar{x}(k,i)\}$. In this case, by a following theorem, we see that the sequence actually converges to an optimal solution of (SLP_k) and its first n elements solves (LSIP). The criteria of meeting $\bar{x}_j(k,i^k) \geq -\delta_1$, for $j = 1,...,n+k$, $c(k)^T \bar{x}(k,i^k) - B(k)^T \bar{w}(k,i^k) \leq \delta_2$, and $\phi_k(t_{k+1,i^k}) \geq -\delta_3$ simply intend to stop the loop once a limit point of the sequence $\{\bar{x}(k,i)\}$ is identified. If no such infinite sequence is generated, we show later that Algorithm 1 finds an optimal solution of (LSIP).

Theorem 2.1 *In Step 3, if repeating the affine scaling procedure until conditions (1) and (2) are satisfied indeed generates an infinite sequence of* $\{\bar{x}(k,i)\}, i = 1,2,...,$ *then* $\{\bar{x}(k,i)_{1...n}\}$ *converges to an optimal solution of (LSIP), as* $i \to \infty$.

Proof. Since $A(k)$ has full rank, under the Assumption 4, by the theory of affine scaling algorithm [22, 24], $\{\bar{x}(k,i)\}$ converges to an optimal solution of (SLP$_k$). We denote $\lim_{i\to\infty} \bar{x}(k,i) = x^*(k)$.

Remember conditions (1) and (2). Since $x^*(k) \geq 0$, there exists a positive integer N such that

$$\min_j\{\bar{x}_j(k,i) \mid \bar{x}_j(k,i) < 0\} \geq -\epsilon_k, \text{ for } i \geq N+1,$$

and the loop is caused by condition (2). Hence we know that either $\phi_{k,i}(t_{k+1,i}) \geq 0$ or $t_{k+1,i} \in T_k$, for $i \geq N+1$. Then at least one of the following two cases occurs.

Case 1: There is a subsequence $\{t_{k+1,i_h}\}$ of $\{t_{k+1,i}\}$ such that $\phi_{k,i_h}(t_{k+1,i_h}) \geq 0$, $\forall h = 1,\ldots,\infty$.

Case 2: There is a subsequence $\{t_{k+1,i_h}\}$ of $\{t_{k+1,i}\}$ such that $t_{k+1,i_h} \in T_k$, $\forall h = 1,\ldots,\infty$.

In the first case, define $\phi_{k,*}(t) = \sum_{j=1}^{n} a_j(t)x_j^*(k) - b(t)$ on T and assume $t_{k+1,*}$ minimizes $\phi_{k,*}(t)$ over T. From the definition of $\phi_{k,i_h}(t)$, we know

$$0 \leq \phi_{k,i_h}(t_{k+1,i_h}) \leq \phi_{k,i_h}(t_{k+1,*}). \tag{2.12}$$

Let $h \to \infty$, then $\phi_{k,i_h}(t) \to \phi_{k,*}(t)$. Consequently, we see

$$0 \leq \phi_{k,*}(t_{k+1,*}). \tag{2.13}$$

This means that $x^*(k)_{1...n}$ is a feasible solution of (LSIP). Since $x^*(k)_{1...n}$ is an optimal solution for (LP$_k$), which has larger feasible domain than (LSIP) has, $x^*(k)_{1...n}$ must be an optimal solution of (LSIP).

In the second case, since T is a compact metric space, there is a subsequence $\{t_{k+1,i_l}\}$ of $\{t_{k+1,i_h}\}$ which converges to a limit point $\bar{t}_{k+1} \in T$. Same as in Case 1, we define $\phi_{k,*}(t) = \sum_{j=1}^{n} a_j(t)x_j^*(k) - b(t)$ on T and assume that $t_{k+1,*}$ minimizes $\phi_{k,*}(t)$ over T.

Note that T_k has no more than k elements, $\bar{t}_{k+1}$ must be one of the k elements, namely $t_1,\ldots,t_k$. Since

$$\begin{bmatrix} a_1(t_1) & \cdots & a_n(t_1) \\ & \vdots & \\ a_1(t_k) & \cdots & a_n(t_k) \end{bmatrix} \begin{bmatrix} x_1^*(k) \\ \vdots \\ x_n^*(k) \end{bmatrix} \geq \begin{bmatrix} b(t_1) \\ \vdots \\ b(t_k) \end{bmatrix},$$

we know $\phi_{k,*}(\bar{t}_{k+1}) \geq 0$.

The definition of $\phi_{k,i_l}(t)$ implies that

$$\phi_{k,i_l}(t_{k+1,i_l}) \leq \phi_{k,i_l}(t_{k+1,*}). \tag{2.14}$$

When $l \to \infty$, we have $\lim_{l\to\infty} \phi_{k,i_l}(t) = \phi_{k,*}(t)$ and $\lim_{l\to\infty} t_{k+1,i_l} = \bar{t}_{k+1}$. Consequently, we see

$$0 \leq \phi_{k,*}(\bar{t}_{k+1}) \leq \phi_{k,*}(t_{k+1,*}). \tag{2.15}$$

Therefore, $x^*(k)_{1\ldots n}$ is a feasible solution of (LSIP) and, hence, an optimal solution of (LSIP).

□

3. Convergence Proof

In this section, we show that if Algorithm 1 does not terminate in Step 2, nor identifies a limit point in Step 3, then it generates an optimal solution of (LSIP), under Assumptions (A1) - (A4). We start with two simple results:

Lemma 3.1 *The sequence $\{B(k)^T w(k)\}$ generated by Algorithm 1 is strictly monotone increasing and bounded above.*

Proof. After Step 3, we know that

$$B(k)^T \bar{w}(k, i^k) \geq B(k)^T \bar{w}(k, 1) > B(k-1)^T w(k-1). \tag{3.1}$$

Remember that $w(k) = \bar{w}(k, i^k)$, hence

$$B(k)^T w(k) > B(k-1)^T w(k-1).$$

This shows the sequence $\{B(k)^T w(k)\}$ is strictly monotone increasing.

Moreover, if $x^*(k)$ is an optimal solution of (SLP_k), $w^*(k)$ an optimal solution of (DLP_k), and x^* an optimal solution of (LSIP), then

$$B(k)^T w(k) \leq B(k)^T w^*(k) = c(k)^T x^*(k) \leq c^T x^*. \tag{3.2}$$

Since (LSIP) has a finite optimum, $B(k)^T w(k)$ must be bounded above.

□

Lemma 3.2 *The sequence $\{c(k)^T x(k) - B(k)^T w(k)\}$ converges to zero, as k approaches infinity.*

Proof. Note that, after Step 3,

$$B(k)^T w(k) = B(k)^T \bar{w}(k, i^k) \geq B(k)^T \bar{w}(k, i^k - 1) > B(k-1)^T w(k-1). \tag{3.3}$$

Since $\{B(k)^T w(k)\}$ is strictly monotone increasing and bounded above, $\{B(k)^T \bar{w}(k, i^k - 1)\}$ converges to the same limit as $\{B(k)^T w(k)\}$ does. Thus all the three sequences $\{B(k)^T w(k)\}, \{B(k)^T \bar{w}(k, i^k)\}$ and $\{B(k)^T \bar{w}(k, i^k - 1)\}$ converge to the same limit. Now, as k approaches infinity,

$$
\begin{aligned}
& B(k)^T \bar{w}(k,i^k) - B(k)^T \bar{w}(k,i^k-1) \\
= \; & \beta B(k)^T dw(k,i^k-1) \\
= \; & \beta (A(k)S(k,i^k-1)^{-2}A(k)^T dw(k,i^k-1))^T dw(k,i^k-1) \\
= \; & \beta dw(k,i^k-1)^T A(k)S(k,i^k-1)^{-2}A(k)^T dw(k,i^k-1) \\
= \; & \beta ds(k,i^k-1)^T S(k,i^k-1)^{-2} ds(k,i^k-1) \\
= \; & \beta ((\tfrac{ds_1(k,i^k-1)}{\bar{s}_1(k,i^k-1)})^2 + \cdots + (\tfrac{ds_{n+k}(k,i^k-1)}{\bar{s}_{n+k}(k,i^k-1)})^2) \\
\rightarrow \; & 0.
\end{aligned}
\tag{3.4}
$$

Therefore,

$$
\frac{ds_j(k,i^k-1)}{\bar{s}_j(k,i^k-1)} \rightarrow 0, \forall j = 1,\dots,n+k, \text{ as } k \rightarrow \infty. \tag{3.5}
$$

However,

$$
\begin{aligned}
& c(k)^T \bar{x}(k,i^k) - B(k)^T \bar{w}(k,i^k-1) \\
= \; & (A(k)^T \bar{w}(k,i^k-1) + \bar{s}(k,i^k-1))^T \bar{x}(k,i^k) - B(k)^T \bar{w}(k,i^k-1) \\
= \; & (\bar{s}(k,i^k-1))^T \bar{x}(k,i^k) + (\bar{w}(k,i^k-1))^T A(k)S(k,i^k-1)^{-2}A(k)^T dw(k,i^k-1) \\
& -B(k)^T \bar{w}(k,i^k-1) \\
= \; & (\bar{s}(k,i^k-1))^T \bar{x}(k,i^k) + (\bar{w}(k,i^k-1))^T B(k) - B(k)^T \bar{w}(k,i^k-1) \\
= \; & -(\bar{s}(k,i^k-1))^T S(k,i^k-1)^{-2} ds(k,i^k-1) \\
= \; & -(\tfrac{ds_1(k,i^k-1)}{\bar{s}_1(k,i^k-1)} + \cdots + \tfrac{ds_{n+k}(k,i^k-1)}{\bar{s}_{n+k}(k,i^k-1)}).
\end{aligned}
\tag{3.6}
$$

Hence, $c(k)^T \bar{x}(k,i^k) - B(k)^T \bar{w}(k,i^k-1) \rightarrow 0$, as $k \rightarrow \infty$. Moreover, by knowing that $\lim_{k\rightarrow\infty} B(k)^T \bar{w}(k,i^k) = \lim_{k\rightarrow\infty} B(k)^T \bar{w}(k,i^k-1)$ and $x(k) = \bar{x}(k,i^k)$, we can conclude that

$$
\lim_{k\rightarrow\infty} (c(k)^T x(k) - B(k)^T w(k)) = 0. \tag{3.7}
$$

□

Recall that if Algorithm 1 does not terminate in Step 2, nor identifies a limit point in Step 3, then after each iteration, it generates a new point $t_{k+1} \in T$ and an $n+k$ vector $x(k)$. Let us denote its first n elements by an n vector $x(k)_{1,..,n} = [x_1(k), ..., x_n(k)]^T$. Then we have the following result:

Lemma 3.3 *If $\{x(k_h)\}$ is a subsequence of $\{x(k)\}$ and $\{x(k_h)_{1...n}\}$ converges to a point $x^*_{1...n}$, then $\sum_{j=1}^{n} a_j(t_i)x^*_j - b(t_i) \geq 0$, for each t_i obtained in Algorithm 1.*

Proof. By the way we define $x(k_h), A(k)$ and $B(k)$, it is clear that

$$
\sum_{j=1}^{n} a_j(t_i)x_j(k_h) - b(t_i) = x_{n+i}(k_h) \geq -\varepsilon_{k_h}, \forall k_h \geq i. \tag{3.8}
$$

Since $\lim_{h\rightarrow\infty} x(k_h)_{1...n} = x^*_{1...n}$ and $\lim_{h\rightarrow\infty} \varepsilon_{k_h} = 0$, by letting $h \rightarrow \infty$ we know that

$$\sum_{j=1}^{n} a_j(t_i)x_j^* - b(t_i) \geq 0. \tag{3.9}$$

□

Now we are ready to show the main convergence result.

Theorem 3.1 *Under assumptions (A1) - (A4), if no infinite loop is identified in Step 3 and the algorithm does not stop in Step 2, Algorithm 1 generates an infinite sequence $\{x(k)\}$ which has a convergent subsequence $\{x(k_h)_{1...n}\}$ such that $\{x(k_h)_{1...n}\}$ converges to an optimal solution of (LSIP), as $k \to \infty$.*

Proof. By Assumption 2, there exists a convergent subsequence $\{x(k_h)_{1...n}\}$ of $\{x(k)_{1...n}\}$ which converges to a limit point $x^*_{1...n}$. Since, $x_j(k_h) \geq -\varepsilon_{k_h}, j = 1, ..., n, \forall h$, by letting $h \to \infty$, we see $x_j^* \geq 0$, $\forall j = 1, \ldots, n$.

Note that T is a compact metric space, there is a subsequence $\{t_{k_l+1}\}$ of $\{t_{k_h+1}\}$, which converges to a limit point $\bar{t} \in T$. Define $\phi_*(t) = \sum_{j=1}^{n} a_j(t)x_j^* - b(t)$ on T. The previous lemma implies that

$$\phi_*(t_k) = \sum_{j=1}^{n} a_j(t_k)x_j^* - b(t_k) \geq 0 \ \ , \text{ for } k = 1, 2, \ldots. \tag{3.10}$$

By the way we choose t_k in Step 3 , if we let t_* be a minimizer of $\phi_*(t)$ over T, then

$$\phi_{k_l}(t_*) \geq \phi_{k_l}(t_{k_l+1}). \tag{3.11}$$

Since $\lim_{l\to\infty} \phi_{k_l}(t) = \phi_*(t)$ and $\lim_{l\to\infty} t_{k_l+1} = \bar{t}$, we know

$$\phi_*(t_*) \geq \phi_*(\bar{t}) \geq 0. \tag{3.12}$$

Hence $x^*_{1...n}$ is a feasible solution of (LSIP).

Suppose that $x^*_{1...n}$ is not an optimal solution of (LSIP), then there exists a feasible solution $\bar{x}$ of (LSIP) such that $c^T\bar{x} < c^T x^*_{1...n}$. However, Lemma 3.2 implies that, as k approaches infinity,

$$c(k)^T x(k) - B(k)^T w(k) \to 0. \tag{3.13}$$

This means

$$c^T x^*_{1...n} = \lim_{k\to\infty} B(k)^T w(k). \tag{3.14}$$

Therefore, there exists a sufficiently large positive integer K such that $B(K)^T w(K) > c^T\bar{x}$. However, $\bar{x}$ is a feasible solution of (LSIP). Hence a contradiction is caused by $B(K)^T w(K) \leq c^T\bar{x}$.

□

4. Numerical Examples

In this section, we report some computational experience on implementing Algorithm 1 for solving (LSIP). Four sets of commonly seen problems were tested:

Problem 1 :

$$\min \sum_{j=1}^{n} \frac{x_j}{j}$$

$$\text{s.t.} \sum_{j=1}^{n} t^{j-1} x_j \geq e^t, t \in [0,1]$$

$$x_j \geq 0$$

Problem 2 :

$$\min \sum_{j=1}^{n} \frac{x_j}{j}$$

$$\text{s.t.} \sum_{j=1}^{n} t^{j-1} x_j \geq \frac{1}{1+t^2}, t \in [0,1]$$

$$x_j \geq 0$$

Problem 3 :

$$\min \sum_{j=1}^{n} \frac{x_j}{j}$$

$$\text{s.t.} \sum_{j=1}^{n} t^{j-1} x_j \geq \frac{1}{2-t}, t \in [0,1]$$

$$x_j \geq 0$$

Problem 4 :

$$\min \sum_{j=1}^{n} \frac{x_j}{j}$$

$$\text{s.t.} \sum_{j=1}^{n} t^{j-1} x_j \geq sin(t), t \in [0,1]$$

$$x_j \geq 0$$

In our tests, we set $n = 5$ as a baseline case. For the purposes of illustration and comparison, we first implemented an "exact version" of the "adding constraint" method simply by using LOQO, an interior-point LP and QP package provided by Vanderbei[23], to solve (LP_k) in Step 2 of each iteration. All default parameters of LOQO were kept unchanged and the stopping rule of LOQO was set when the duality gap $\leq 10^{-8}$. The stopping rule in Step 4 was Set when $\phi_k(t_{k+1}) \geq 10^{-4}$. A starting point with $t_1 = 0.5$ was used.

Then we implemented Algorithm 1 with the same stating point $t_1 = 0.5$. We also set $\alpha = 0.9$ for the dual affine scaling procedure, $\gamma = 2.5$ for the mechanism described in previous observations. Similar to the practical implementation of dual affine scaling methods in [1, 16, 17, 18], we used the following conditions to stop the algorithm:

1. $\bar{x}_j(k,i) > -0.025\|\bar{x}(k,i)\|$, for each j.
2. $\frac{\|(B(k)^T\bar{w}(k,i) - B(k)^T\bar{w}(k,i-1)\|}{\|1+B(k)^T\bar{w}(k,i-1)\|} < 5 * 10^{-7}$.
3. $\phi_{k,i}(t_{k+1,i}) \geq -10^{-4}$.

Hence the first condition in Step 3 was changed to $x_j(k,i^k) > -\epsilon_k\|x(k)\|$ from $min_j\{x_j(k,i^k)\} > -\epsilon_k$, for each j. Moreover, we set $\theta = 0.5$ and $\epsilon_1 = 1.0$.

To avoid spending too much effort in keeping finding a minimizer $t_{k+1,i^k} \in T_k$ in Step 3, we set an upper limit of 5 times. Once $t_{k+1,i} \in T_k$ happened more than 5 times, we stopped the dual affine scaling procedure and entered Step 4 by randomly selecting a $t_{k+1} \in T - T_k$.

Numerical results of the testing problems are reported in following tables:

Problem 1

TABLE 4.1

k	Cholesky	Exact version - obj value	Cholesky	Algorithm 1 - obj value
1	9	1.648700	1	1.566285
2	9	1.648700	1	1.634169
3	10	1.714841	3	1.651097
4	9	1.717759	3	1.710037
5	9	1.718265	1	1.710099
6			11	1.717746
7			1	1.717772
8			1	1.717774
9			1	1.717774
10			3	1.718238
11			2	1.718289
Total Chol.	46		28	

From the above table, we know the exact version took 5 iterations (i.e., solved 5 LP problems) while Algorithm 1 took 11 iterations to reach an optimal solution. The optimal objective value is very close. Actually, the optimal solution obtained by the "exact version" is $x_1 = 1.000000, x_2 = 0.998827, x_3 = 0.509786, x_4 = 0.139746, x_5 = 0.069939$, while Algorithm 1 provides an optimal solution with $x_1 = 1.000003$, $x_2 = 0.999300$, $x_3 = 0.507976$, $x_4 = 0.142198$, $x_5 = 0.068804$. However, in each iteration, Algorithm 1 required much fewer Cholesky factorizations to reach an "inexact" solution. Consequently, in total, Algorithm 1 required 28 Chôlesky factorizations while the exact version required 46. Also note that LOQO uses a predictor-corrector interior point method to solve LP, hence it takes longer time to complete each inner iteration. This example shows the computational potential of applying Algorithm 1 to solve (LSIP) with less computational burden than an exact version.

Similar results are shown for Problems 2, 3, and 4, as illustrated by the following tables:

Problem 2

TABLE 4.2

k	Cholesky	Exact version - obj value	Cholesky	Algorithm 1 - obj value
1	9	0.800000	1	0.641743
2	8	1.000000	2	0.799783
3			1	0.799262
4			1	0.992754
5			1	0.999936
6			3	1.000000
Total Chol.	17		9	

The optimal solution obtained by the "exact version" is $x_1 = 1.000000, x_2 = 0.000000, x_3 = 0.000000, x_4 = 0.000000, x_5 = 0.000000$, while Algorithm 1 provides an optimal solution with $x_1 = 1.000000, x_2 = 0.000000, x_3 = 0.000000, x_4 = 0.000000, x_5 = 0.000000$.

Problem 3

TABLE 4.3

k	Cholesky	Exact version- obj value	Cholesky	Algorithm 1 - obj value
1	9	0.666670	1	0.534786
2	9	0.666670	1	0.667756
3	9	0.692859	6	0.693194
4	9	0.693446	1	0.692870
5	9	0.693489	1	0.692857
6			1	0.692857
7			1	0.692857
8			6	0.693515
9			1	0.693454
10			1	0.693451
11			2	0.693451
12			6	0.693451
Total Chol.	45		29	

The optimal solution obtained by the "exact version" is $x_1 = 0.500000, x_2 = 0.252752, x_3 = 0.132483, x_4 = 0.000000, x_5 = 0.114763$, while Algorithm 1 provides an optimal solution with $x_1 = 0.500000$, $x_2 = 0.256053$, $x_3 = 0.124764$, $x_4 = 0.000000$, $x_5 = 0.119182$.

Problem 4

TABLE 4.4

k	Cholesky	Exact version- obj value	Cholesky	Algorithm 1- obj value
1	9	0.479430	1	0.384585
2	9	0.479430	1	0.480201
3	8	0.479430	1	0.479420
4	9	0.479430	1	0.479415
5			1	0.479425
6			1	0.479426
7			1	0.479426
8			1	0.479426
9			1	0.479426
10			5	0.479426
11				
Total Chol.	35		14	

The optimal solution obtained by the "exact version" is $x_1 = 0.040632, x_2 = 0.877595, x_3 = 0.000000, x_4 = 0.000000, x_5 = 0.000000$, while Algorithm 1 provides an optimal solution with $x_1 = 0.048627$, $x_2 = 0.861598$, $x_3 = 0.000000$, $x_4 = 0.000000$, $x_5 = 0.000000$.

5. Concluding Remarks

In this paper, we have proposed a dual affine scaling based algorithm for solving (LSIP). It incorporates the dual affine scaling techniques into the "adding constraint" method. The proposed algorithm has a good "warm" start and relaxes the requirement of finding an exact optimal solution of (SLP_k) to an inexact solution, at each iteration. Some preliminary computational experience shows the potential advantages of this proposed algorithm. However, it by no means implies that the proposed algorithm is better than other known methods. We are currently exploring the possibility of weakening the assumptions and incorporating the primal-dual method into a similar scheme to solve (LSIP).

References

1. I. Adler, N. K. Karmarkar, M. G. C. Resende, and G. Veiga, "An implementation of Karmarkar's algorithm for linear programming," *Mathematical Programming*, 44 (1989), 297–335. Errata in *Mathematical Programming*, 50 (1991), 415.
2. E.J. Anderson and P. Nash, *Linear programming in infinite-dimensional spaces*, Wiley, Chichester, 1987.
3. E. R. Barnes, "A variation of Karmarkar's algorithm for solving linear programming problems," *Mathematical Programming*, 36 (1986), 174–182.
4. I. I. Dikin, "Iterative solution of problems of linear and quadratic programming," *Doklady Akademii Nauk SSSR*, 174 (1967), 747–748. Translation in : *Soviet Mathematics Doklady*, 8 (1967), 674–675.
5. S. C. Fang, and S. C. Puthenpura, *Linear optimization and extensions: theory and algorithms*, Prentice Hall, Englewood Cliffs, New York, 1993.
6. S. C. Fang, and S. Y. Wu, "An inexact approach to solving linear semi-infinite programming problems," OR Report NO. 265, North Carolina State University, Raleigh, NC, November 1992, to appear in Optimization.

7. S. C. Fang , C. J. Lin, and S. Y. Wu, "On solving convex quadratic semi-infinite programming problems," OR Report NO. 268, North Carolina State University, Raleigh, NC, March 1993.
8. M. C. Ferris and A. B. Philpott, "An interior point algorithm for semi–infinite linear programming," *Mathematical Programming,* 43 (1989), 257–276.
9. M. C. Ferris and A. B. Philpott, "On affine scaling and semi–infinite programming," *Mathematical Programming,* 52 (1992), 361–364.
10. K. Glashoff and S. Å. Gustafson, *Linear optimization and approximation,* Springer–Verlag, New York, 1982.
11. S. Å. Gustafson, "On the computational solution of a class of generalized moment problems," *SIAM J. Numerical Analysis,* 7 (1970), 343–357.
12. S. Å. Gustafson and K. Kortanek, "Numerical treatment of a class of semi–infinite programming problems", *Naval Research Logistics Quarterly,* 20 (1973), 473–504.
13. R. Hettich and K. Kortanek, "Semi-infinite programming : theory, method and applications," Technical Report, College of Business Administration, The University of Iowa, Iowa City, Iowa, 1991.
14. N. K. Karmarkar, "A new polynomial–time algorithm for linear programming," *Combinatorica,* 4 (1984), 373–395.
15. H. C. Lai and S. Y. Wu, "On linear semi–infinite programming problems, an algorithm,", *Numerical Functional Analysis and Optimization,* 13 (1992), 287–304.
16. R. E. Marsten, M. J. Saltzman, D. F. Shanno, J. F. Ballintijn, and G. S. Pierce, "Implementation of a dual affine interior point algorithm for linear programming," *ORSA Journal on Computing,* 1 (1989), 287–297.
17. S. Mehrotra, "Implementation of affine scaling methods : Approximate solutions of systems of linear equations using preconditioned conjugate gradient methods," *ORSA Journal on Computing,* 4 (1992), 103–118.
18. C. L. Monma and A. J. Morton, "Computational experience with the dual affine variant of Karmarkar's method for linear programming," *Operations Research Letters,* 6 (1987), 261–267.
19. M. J. D. Powell, "Karmarkar's algorithm : A view from nonlinear programming," *Bulletin of the Institute of Mathematics and Its Applications,* 26 (1990), 165–181.
20. M. J. Todd, "Interior–point algorithms for semi–infinite programming," Technical Report 978, School of Operations Research and Industrial Engineering, Cornell University, Ithaca, NY, 1991.
21. L. Tunçel and M. J. Todd, "Asymptotic behavior of interior–point methods : A view from semi–infinite programming," Technical Report 1031, School of Operations Research and Industrial Engineering, Cornell University, Ithaca, NY, September 1992.
22. R. J. Vanderbei, "Affine–scaling for linear programs with free variables." *Mathematical Programming,* 43(1989), 31–44.
23. R. J. Vanderbei, LOQO User's Manual. Technical Report SOL 92–05, Dept. of Civil Engineering and Operations Research, Princeton University, Princeton, NJ 08544, 1992.
24. R. J. Vanderbei, M. S. Meketon, and B. A. Freedman, "A modification of Karmarkar's linear programming algorithm," *Algorithmica,* 1(1986), 395–407.

A GENUINE QUADRATICALLY CONVERGENT POLYNOMIAL INTERIOR POINT ALGORITHM FOR LINEAR PROGRAMMING *

ZHI-QUAN LUO
Room 225, Communications Research Laboratory, McMaster University, Hamilton, Ontario, L8S 4K1, Canada.

and

YINYU YE
Department of Management Sciences, University of Iowa, Iowa City, IA 52242, U.S.A.

Abstract. It is well-known that the predictor-corrector method for linear programming has $O(\sqrt{n}L)$ iteration complexity and local quadratic convergence. However, at each iteration the predictor-corrector method requires solving two linear systems, thus making it a two-step quadratically convergent method. In this paper, we propose a variant of the $O(\sqrt{n}L)$-iteration interior point algorithm for linear programming. Unlike the predictor-corrector method, the variant has a genuine quadratic convergence rate since each iteration involves solving only one linear system. Other features of the algorithm include the gradual phasing out of the centering direction when approaching optimality. Our work is based on the recent results of Gonzaga and Tapia [1, 2] about the iterate convergence of the predictor-corrector method, and it does not require any nondegeneracy assumption.

Key words: Linear programming, primal-dual interior point algorithms, quadratic convergence

1. Introduction

Consider a linear program in the following standard form:

$$\begin{array}{ll} \text{minimize} & c^T x \\ \text{subject to} & Ax = b, \; x \geq 0, \end{array} \tag{1.1}$$

where $c \in \Re^n$, $A \in \Re^{m \times n}$ and $b \in \Re^m$ are given, $x \in \Re^n$, and the superscript T denotes transpose. The linear program (1.1) is said to be feasible if the constraints are consistent.

The dual of (1.1) can be written as

$$\begin{array}{ll} \text{maximize} & b^T y \\ \text{subject to} & A^T y \leq c, \end{array} \tag{1.2}$$

where $y \in \Re^m$. The components of $s = c - A^T y \in \Re^n$ are called dual slacks. Denote by $\mathcal{F}$ the set of all x and (y, s) that are feasible for the primal and dual, respectively.

* Research of the first author is supported by the Natural Sciences and Engineering Research Council of Canada, Grant No. OPG009031; Research of the second author is supported in part by NSF Grant DDM-8922636, and an Interdisciplinary Research Grant of the University of Iowa Center for Advanced Studies.

Ding-Zhu Du and Jie Sun (eds.), Advances in Optimization and Approximation, 235–246.

Denote by $\mathcal{F}^0$ the set of points in $\mathcal{F}$ with $(x, s) > 0$. The optimality conditions for the above primal-dual pair of linear programs are the following.

$$\begin{aligned} Xs = 0,\ x &\geq 0,\ s \geq 0, \\ Ax &= b, \\ A^T y + s - c &= 0, \end{aligned} \tag{1.3}$$

where we have used the standard notation that, for any vector x in $\Re^n$, the capitalized letter X denotes the $n \times n$ diagonal matrix with $X_{ii} = x_i$, $i = 1, 2, ..., n$.

The predictor-corrector method, proposed by Mizuno et al. [9], generates a sequence of iterates $\{(x^k, y^k, s^k)\}$ in $\mathcal{F}^0$ with $(x^k)^T s^k \to 0$. In particular, each iteration of the predictor-corrector method consists of two steps, predictor step and corrector step. At the predictor step, the duality is reduced by moving along the Newton direction of the above nonlinear system; at the corrector step, the iterates are pulled back to a suitable neighborhood of the central path. In particular, the duality gap is unchanged in the corrector step. (The modified predictor-corrector method of [6] does allow a further constant factor reduction of the duality gap in the corrector step.) It is well known [9] that the method can be terminated at an optimal primal-dual pair in $O(\sqrt{n}L)$ iterations, where L is the length of binary encoding of the data. It is also known that the duality gap $(x^k)^T s^k$ converges to zero quadratically if one counts predictor step and corrector step together as one iteration (see [12]). However, the predictor-corrector method is not a truly quadratically convergent algorithm since at each iteration one needs to solve two different linear systems.

Can we find a genuine quadratically convergent $O(\sqrt{n}L)$-iteration interior point algorithm for linear programming? To the best of our knowledge, there have been two attempts in finding such an algorithm. In [11], Ye showed that it is possible to attain a convergence rate arbitrarily close to quadratic by gradually decreasing the frequency of performing the corrector step. In the recent work [2], Gonzaga and Tapia proposed a simplified version of predictor-corrector method whereby the coefficient matrix from the predictor step is used again in the corrector step. In this way, the corrector step involves only a back-solve of a linear system, taking $O(n^2)$ rather than $O(n^3)$ arithmetic operations as would be required by a standard corrector step. However, this simplied predictor-corrector method still does not attain true quadratic convergence due to the extra $O(n^2)$ operations required at each corrector step. In particular, the work required for n back-solves is $O(n^3)$, so the arithmetic complexity of the simplied predictor-corrector method is the same as that of performing one standard corrector step after every n predictor steps. Thus, its convergence order should be $2^{n/(n+1)}$.

In this paper, we propose a genuine quadratically convergent $O(\sqrt{n}L)$-iteration interior point algorithm for linear programming. A feature of this algorithm is the gradual phaseout of the centering direction when the iterates approach optimality. Our work is based on the recent results of Gonzaga and Tapia [1, 2] about the iterate convergence of the predictor-corrector method, and it does not require any nondegeneracy assumptions.

Throughput this paper, we will adopt the standard notations used in the interior point algorithm literature (see [9, 12]). For example, $\mu = \left(\sum_{i=1}^n x_i s_i\right)/n$ will denote the average duality gap for the iterate (x, s).

2. The Predictor–Corrector Method

We start by first describing a generic iteration of primal-dual interior point algorithms. Given any $(x, y, s) \in \mathcal{F}^0$ satisfying

$$\delta(x,s) := \left\| \frac{Xs}{\mu} - e \right\| \le \alpha < 1, \tag{2.1}$$

where e denotes the n-dimensional vector of 1's. Solve the linear system in (d_x, d_y, d_s)

$$\begin{aligned} Xd_s + Sd_x &= \gamma\mu e - Xs \\ Ad_x &= 0 \\ A^T d_y + d_s &= 0, \end{aligned} \tag{2.2}$$

where $\gamma \ge 0$ is the centering stepsize. Let (d_x, d_y, d_s) be the solution. Set the new iterate as

$$x^+ = x + \theta d_x, \qquad y^+ = y + \theta d_y, \qquad s^+ = s + \theta d_s,$$

where $0 < \theta \le 1$ is the stepsize.

Like many primal-dual interior point algorithms, each iteration of the predictor-corrector method takes the above generic form.

Algorithm 1: The Predictor-Corrector Method

– **Initialization**

Set $k = 0$. Let $\alpha \in (0,1)$ and take a strictly feasible pair $(x^0, y^0, s^0) \in \mathcal{F}^0$ such that

$$\delta(x^0,s^0) = \left\| \frac{X^0 s^0}{\mu^0} - e \right\| \le \frac{\alpha}{2}, \quad \text{and} \quad \mu^0 \le 2^L. \tag{2.3}$$

– **Predictor Step**

For each even integer $2k$ ($k = 0, 1, ...$), we have $(x^{2k}, y^{2k}, s^{2k}) \in \mathcal{F}^0$ satisfying $\delta(x^{2k}, s^{2k}) \le \frac{\alpha}{2}$. Solve the linear system (2.2) with $\gamma = 0$ to obtain the predictor direction $(d_{xp}^{2k}, d_{yp}^{2k}, d_{sp}^{2k})$. Let

$$x(\theta) = x^{2k} + \theta d_{xp}^{2k}, \quad y(\theta) = y^{2k} + \theta d_{yp}^{2k}, \quad s(\theta) = s^{2k} + \theta d_{sp}^{2k}, \quad \mu(\theta) = \frac{x^T(\theta) y(\theta)}{n}.$$

Then calculate

$$\theta^{2k} = \max\left\{ \theta \ : \ (x(\theta), y(\theta), s(\theta)) \in \mathcal{F}^0 \text{ and } \left\| \frac{X(\theta)s(\theta)}{\mu(\theta)} - e \right\| \le \alpha \right\}. \tag{2.4}$$

and set

$$x^{2k+1} = x(\theta^{2k}), \quad y^{2k+1} = y(\theta^{2k}), \quad s^{2k+1} = s(\theta^{2k}). \tag{2.5}$$

– **Corrector Step**

For each odd integer $2k+1$, we have $(x^{2k+1}, y^{2k+1}, s^{2k+1}) \in \mathcal{F}^0$ satisfying $\delta(x^{2k+1}, s^{2k+1}) \le \alpha$. Solve the linear system (2.2) to obtain the direction $(d_{xc}^{2k+1}, d_{yc}^{2k+1}, d_{sc}^{2k+1})$ for the corrector step. Then set

$$x^{2k+2} = x^{2k+1} + d_{xc}^{2k+1}, \qquad y^{2k+2} = y^{2k+1} + d_{yc}^{2k+1}, \qquad s^{2k+2} = s^{2k+1} + d_{sc}^{2k+1}.$$

Note that the predictor-corrector method chooses the centering stepsize γ alternately between 0 and 1. It is easily seen that, with $\gamma = 0$, the predictor step coincides with the pure Newton step for solving the nonlinear equations (1.3). The corrector step, which sets $\gamma = 1$, is used to ensure that the iterate $(x^{2k+2}, y^{2k+2}, s^{2k+2})$ remains close to the central path.

Let (x^*, s^*) denote the analytic center of the primal-dual optimal face. It follows from a standard result in linear programming that the index sets $B = \{i \in \{1, ..., n\} : x_i^* > 0\}$ and $N = \{i \in \{1, ..., n\} : s_i^* > 0\}$ forms a partition of $\{1, ..., n\}$. The following convergence results are well known [9, 12, 13].

Proposition 2.1 *For any $k \geq 0$, let $(x^{2k+1}, y^{2k+1}, s^{2k+1})$, $(x^{2k+2}, y^{2k+2}, s^{2k+2})$, $(d_{xp}^{2k}, d_{yp}^{2k}, d_{sp}^{2k})$ and $(d_{xc}^{2k+1}, d_{yc}^{2k+1}, d_{sc}^{2k+1})$ be defined as above. Then the following hold:*
(1) $\|d_{xp}^{2k}\| \leq \rho\mu^{2k}$ *and* $\|d_{sp}^{2k}\| \leq \rho\mu^{2k}$.
(2) $\|(d_{xc}^{2k+1})_N\| \leq \rho\mu^{2k+1}$ *and* $\|(d_{sc}^{2k+1})_B\| \leq \rho\mu^{2k+1}$.
(3) $\|x_N^{2k+2}\| \leq \rho\mu^{2k+1}$ *and* $\|s_B^{2k+2}\| \leq \rho\mu^{2k+1}$.
(4) $\mu^{2k+2} = \mu^{2k+1} = O\left((\mu^{2k})^2\right)$.
Here, ρ is a constant that is independent of k, and the constant in the big "O" notation is also independent of k.

Recently, Gonzaga and Tapia [1, 2] established strong convergence results about the predictor-corrector method. We summarize their main results in the following proposition. These results will be used in Section 4 to analyze our algorithm.

Proposition 2.2 (1) *Consider a sequence of feasible interior points $\{(x^k, y^k, s^k)\}$ with the property that $\delta(x^k, s^k) \leq 0.1$, $\mu^k \to 0$ and $(x^k, s^k) \to (x^*, s^*)$. (These points are not necessarily generated by the predictor-corrector method.) Let $(x^{k^+}, y^{k^+}, s^{k^+})$ be the vector obtained by applying one predictor-corrector iteration to (x^k, y^k, s^k) and let $(d_{xc}^k, d_{yc}^k, d_{sc}^k)$ be the direction generated in the corresponding corrector step. Then, $d_{xc}^k \to 0$, $d_{sc}^k \to 0$ and $(x^{k^+}, s^{k^+}) \to (x^*, s^*)$.*
(2) *Let $\{(x^k, y^k, s^k)\}$ be a sequence of interior points in $\mathcal{F}$. Suppose that there are infinite number of k's such that $(x^{k+1}, y^{k+1}, s^{k+1})$ is obtained by applying the predictor-corrector step to (x^k, y^k, s^k). Let $\mathcal{K}$ denote the set of such indices. Assume that*

$$\sum_{k \notin \mathcal{K}} \|(x^{k+1}, s^{k+1}) - (x^k, s^k)\| < \infty.$$

Then, $(x^k, s^k) \to (x^*, s^*)$.

Part (1) of Prop. 2.2 is due to Lemma 6.1 of [1], while part (2) follows from Proposition 5.3 of [2]. Roughly speaking, part (1) of Prop. 2.2 says that the changes in the primal-dual variables during the corrector step converge to zero as the variables approach the analytic center. Part (2) of Prop. 2.2 concerns about the iterate convergence of the predictor-corrector method. In particular, when $\{(x^k, y^k, s^k)\}$ is generated by the predictor-corrector method, then we have $\mathcal{K} = \{1, 2, ...\}$, in which case part (2) of Prop. 2.2 implies that (x^k, s^k) converges to the analytic center (x^*, s^*). When $\mathcal{K}$ is infinite by not equal to $\{1, 2, ...\}$, the iterate convergence to the analytic center is retained as long as the total variation due to non-predictor-corrector steps is bounded.

3. A Variant of the Primal-Dual Interior Point Algorithm

Similar to the predictor-corrector method, each iteration of our algorithm also takes the generic form (2.2).

Algorithm 2.

– **Initialization**
Set $k = 0$. Let $\alpha \in (0, 0.1)$ and take a strictly feasible pair $(x^0, y^0, s^0) \in \mathcal{F}^0$ such that

$$\delta(x^0, s^0) = \left\| \frac{X^0 s^0}{\mu^0} - e \right\| \leq \frac{\alpha}{2}, \quad \text{and} \quad \mu^0 \leq 2^L. \tag{3.1}$$

– **The k-th interation**
For each $k \geq 0$, we are given a strictly feasible primal-dual pair (x^k, y^k, s^k). Compute the direction (d_x^k, d_y^k, d_s^k) by solving the linear system (2.2) with $(x, y, s) = (x^k, y^k, s^k)$, $\mu = \mu^k$ and $\gamma = c(\mu^k)^2$, where $c > 0$ is a constant (independent of k) whose size will be determined later. Let $x^+ = x^k + \theta^k d_x^k$, $y^+ = y^k + \theta^k d_y^k$ and $s^+ = s^k + \theta^k d_s^k$ where

$$\theta^k := \frac{2}{\sqrt{1 + 8\|D_x^k d_s^k\|/(\alpha\mu^k)} + 1}. \tag{3.2}$$

If

$$c\mu^k < 0.9 \quad \text{and} \quad \frac{\|D_x^k d_s^k\|}{\alpha} \leq \frac{c(\mu^k)^2}{\sqrt{1 + 8\|D_x^k d_s^k\|/(\alpha\mu^k)} + 1}, \tag{3.3}$$

then set $(x^{k+1}, y^{k+1}, s^{k+1}) = (x^+, y^+, s^+)$. Otherwise, $(x^{k+1}, y^{k+1}, s^{k+1})$ is obtained by applying the standard predictor-corrector iteration to (x^k, y^k, s^k).

Notice that the direction (d_x^k, d_y^k, d_s^k) obtained by solving (2.2) is in fact a convex combination of the predictor direction $(d_{xp}^k, d_{yp}^k, d_{sp}^k)$ and the corrector direction $(d_{xc}^k, d_{yc}^k, d_{sc}^k)$. Specifically, it can be easily verified that

$$(d_x^k, d_y^k, d_s^k) = (1 - c\mu^k)(d_{xp}^k, d_{yp}^k, d_{sp}^k) + c\mu^k(d_{xc}^k, d_{yc}^k, d_{sc}^k). \tag{3.4}$$

Thus, as μ^k approaches zero, the corrector direction $(d_{xc}^k, d_{yc}^k, d_{sc}^k)$ will phase out and (d_x^k, d_y^k, d_s^k) will tend to the pure predictor direction $(d_{xp}^k, d_{yp}^k, d_{sp}^k)$.

Another point worth noting is that Algorithm 2 has two types of iteration: one is to set $(x^{k+1}, y^{k+1}, s^{k+1})$ to $(x^+(\theta^k), y^+(\theta^k), s^+(\theta^k))$ by using the combined predictor-corrector direction (d_x^k, d_y^k, d_s^k), and the other is the standard predictor-corrector iteration which is essentially a safeguard mechanism. We shall call the former the combined predictor-corrector iteration (or CPC for short). Notice that the condition $c\mu^k < 0.9$ specified by (3.3) is used to ensure that μ decreases after each CPC iteration, while the other condition in (3.3) ensures the positivity of (x^+, s^+) because without it either x^+ or s^+ may become nonpositive.

4. Convergence Analysis

In this section, we analyze the convergence behavior of Algorithm 2. In particular, we will show that the iterates generated by the algorithm converges and the duality gap converges to zero quadratically. Our proof is based on showing that the safeguard of the new algorithm is activated at most finite number times, as long as the constant c is chosen large enough (independent of k). We start by showing that the iterates lie in a fixed neighborhood of the central path and that the duality gap decreases at least linearly.

Lemma 4.1 *Let $\{(x^k, s^k)\}$ be the iterates generated by Algorithm 2. Then $(x^k, s^k) > 0$ and*

$$\|X^k s^k/\mu^k - e\| \le \alpha/2, \qquad \forall k \ge 0. \tag{4.1}$$

Moreover, there holds

$$\frac{\|D_x^k d_s^k\|}{\mu^k} \le \sqrt{2}n/4, \qquad k \ge 0 \tag{4.2}$$

and

$$\mu^{k+1} \le \left(1 - \frac{1}{10\left(\sqrt{1+2\sqrt{2}n/\alpha}+1\right)}\right)\mu^k, \qquad k \ge 0. \tag{4.3}$$

Proof. We show (4.1) by induction on k. Clearly, $(x^0, s^0) > 0$ and (4.1) holds for $k = 0$. Consider an arbitrary iteration $k > 0$ of Algorithm 2. If the safeguard is activated, namely, $(x^{k+1}, y^{k+1}, s^{k+1})$ is obtained by applying the predictor-corrector iteration to (x^k, y^k, s^k), then $(x^{k+1}, s^{k+1}) > 0$ and (4.1) holds by virtue of the predictor-corrector method.

If, on the other hand, the stepsize (3.2) is accepted at the kth iteration, then (3.3) holds. Let $x^+(\theta) = x^k + \theta d_x^k$ and $s^+(\theta) = s^k + \theta d_s^k$, where (d_x^k, d_s^k) is defined by Algorithm 2. Then, we have from (2.2) that

$$X^k d_s^k + S^k d_x^k = c(\mu^k)^2 e - X^k s^k, \quad (d_x^k)^T d_s^k = 0. \tag{4.4}$$

This implies that

$$\begin{aligned} X^+(\theta)s^+(\theta) &= (X^k + \theta D_x^k)(s^k + \theta d_s^k) \\ &= X^k s^k + \theta(X^k d_s^k + S^k d_x^k) + \theta^2 D_x^k d_s^k \\ &= (1-\theta)X^k s^k + \theta c(\mu^k)^2 e + \theta^2 D_x^k d_s^k. \end{aligned} \tag{4.5}$$

Multiplying both sides of (4.5) with e/n and using (4.4) gives

$$\mu^+(\theta) := \frac{(x^+(\theta))^T s^+(\theta)}{n} = (1-\theta)\mu^k + \theta c(\mu^k)^2. \tag{4.6}$$

By the inductive hypothesis, there holds $\|X^k s^k/\mu^k - e\| \le \alpha/2$. Also, combining (4.5) and (4.6) gives

$$\|X^+(\theta)s^+(\theta) - \mu^+(\theta)e\| = \|(1-\theta)(X^k s^k - \mu^k e) + \theta^2 D_x^k d_s^k\|.$$

Now we can use the same argument as given by Ye *et. al.* [12] for the predictor-corrector method (see the proof therein) to obtain

$$\|X^+(\theta)s^+(\theta) - \mu^+(\theta)e\| \le \alpha(1-\theta)\mu^k, \qquad \text{for all } 0 \le \theta \le \theta^k, \tag{4.7}$$

where θ^k is given by (3.2). Since $(1-\theta)\mu^k \le \mu^+(\theta)$ (cf. (4.6)), it follows that

$$\|X^+(\theta)s^+(\theta) - \mu^+(\theta)e\| \le \alpha\mu^+(\theta), \qquad \text{for all } 0 \le \theta \le \theta^k.$$

Thus, by a simple continuity argument (see Mizuno *et. al.* [9]), we easily see that $(x^{k+1}, s^{k+1}) = (x^+(\theta^k), s^+(\theta^k)) > 0$. Finally, it follows from (3.3) and (4.7) that

$$\begin{aligned}
\|X^{k+1}s^{k+1} - \mu^{k+1}e\| &= \|X^+(\theta^k)s^+(\theta^k) - \mu^+(\theta^k)e\| \\
&\le \alpha(1-\theta^k)\mu^k \\
&= \frac{\alpha}{2}\left((1-\theta^k)\mu^k + (1-\theta^k)\mu^k\right) \\
&\le \frac{\alpha}{2}\left((1-\theta^k)\mu^k + \theta^k c(\mu^k)^2\right) \\
&= \frac{\alpha}{2}\mu^{k+1},
\end{aligned}$$

where the last step follows from (4.6). This shows that (4.1) still holds in this case.

It remains to show (4.2) and (4.3). If a predictor-corrector iteration is performed, then (4.2) follows directly from Lemmas 1, 2 and 4 of [9]; in the other case where a CPC iteration is performed, (4.2) remains to hold. This is because of the same argument in [9]:

$$\begin{aligned}
\|D_x^k d_s^k\| &\le (\sqrt{2}/4)\|(X^kS^k)^{-1/2}(c(\mu^k)^2e - X^ks^k)\|^2 \\
&\le (\sqrt{2}/4)\left(c\mu^k\|(X^kS^k)^{-1/2}(\mu^ke - X^ks^k)\| + (1-c\mu^k)\|(X^kS^k)^{1/2}e\|\right)^2 \\
&\le (\sqrt{2}/4)\left(c\mu^k\frac{\alpha}{2-\alpha}\sqrt{\mu^k} + (1-c\mu^k)\sqrt{n\mu^k}\right)^2 \\
&\le (\sqrt{2}/4)\left(c\mu^k\sqrt{n\mu^k} + (1-c\mu^k)\sqrt{n\mu^k}\right)^2 \\
&= (\sqrt{2}/4)n\mu^k.
\end{aligned}$$

To prove (4.3), we also need to consider two cases. For the predictor-corrector iterations, the inequality

$$\mu^{k+1} \le \left(1 - \frac{1}{10\left(\sqrt{1+2\sqrt{2}n/\alpha} + 1\right)}\right)\mu^k$$

is well known [9, Theorem 4.1]. For the CPC iterations, we have $c\mu^k < 0.9$. By $\mu^{k+1} = \mu^+(\theta^k)$ and (4.6), we obtain

$$\mu^{k+1} = (1-\theta^k)\mu^k + \theta^k c(\mu^k)^2. \tag{4.8}$$

This together with (3.2) imply

$$\begin{aligned}
\mu^{k+1} &\le (1-\theta^k)\mu^k + 0.9\theta^k\mu^k \\
&= (1-0.1\theta^k)\mu^k
\end{aligned}$$

$$= \left(1 - \frac{1}{10\left(\sqrt{1+8\|D_x^k d_s^k\|/(\alpha\mu^k)}+1\right)}\right)\mu^k$$

$$\leq \left(1 - \frac{1}{10\left(\sqrt{1+2\sqrt{2}n/\alpha}+1\right)}\right)\mu^k$$

where the last step follows from (4.2). The proof is complete. **Q.E.D.**

The following lemma gives some additional simple properties of the iterates; it will be very useful in our subsequent analysis.

Lemma 4.2 *Assume that $c\mu^k < 0.9$. Then*

$$1/\rho \leq x_i^k \leq \rho \quad \text{for } i \in B, \qquad 1/\rho \leq s_i^k \leq \rho \quad \text{for } i \in N \tag{4.9}$$

$$\|d_x^k\| \leq (1+4c)\rho\mu^k, \quad \|d_s^k\| \leq (1+4c)\rho\mu^k, \quad 1-\theta^k \leq \frac{2(1+4c)^2\rho^2\mu^k}{\alpha} \tag{4.10}$$

where ρ is some constant (independent of k and c).

Proof. By Lemma 4.1, the iterates $\{(x^k, s^k)\}$ stay within a fixed neighborhood of the central path. It follows from [4] that $\{(x^k, s^k)\}$ is bounded and every limit point must satisfy strict complementarity. Therefore, (4.9) holds for all sufficiently large ρ. Also, by Prop. 2.1 (1), we have $\|d_{xp}^k\| \leq \rho\mu^k$ and $\|d_{sp}^k\| \leq \rho\mu^k$ for all ρ large enough.

We now proceed to prove (4.10). By (3.4) and $c\mu^k < 0.9$, we obtain

$$\begin{aligned} \|d_x^k\| &= \|(1-c\mu^k)d_{xp}^k + c\mu^k d_{xc}^k\| \\ &= \|d_{xp}^k\| + c\mu^k\|d_{xc}^k\| \\ &\leq \rho\mu^k + c\mu^k 4\rho \\ &= (1+4c)\rho\mu^k, \end{aligned}$$

where the last step is due to (4.9) and $\|(X^k)^{-1}d_{xc}^k\| \leq \frac{4\alpha}{(2-\alpha)^2} \leq 4$, which can be derived from

$$X^k d_{sc}^k + S^k d_{xc}^k = \mu^k e - X^k s^k$$

and

$$\|\mu^k e - X^k s^k\| \leq \alpha/2.$$

Finally, we use (3.2) to obtain

$$\begin{aligned} 1-\theta^k &= \frac{\sqrt{1+8\|D_x^k d_s^k\|/(\alpha\mu^k)}-1}{\sqrt{1+8\|D_x^k d_s^k\|/(\alpha\mu^k)}+1} \\ &= \frac{8\|D_x^k d_s^k\|}{\alpha\mu^k\left(\sqrt{1+8\|D_x^k d_s^k\|/(\alpha\mu^k)}+1\right)^2} \end{aligned}$$

$$
\begin{aligned}
&\le \frac{2\|D_x^k d_s^k\|}{\alpha\mu^k} \\
&\le \frac{2\|d_x^k\|\|d_s^k\|}{\alpha\mu^k} \\
&= \frac{2(1+4c)^2\rho^2\mu^k}{\alpha},
\end{aligned}
$$

as desired. **Q.E.D.**

Next we use Lemmas 4.2 and 4.1 to show the quadratic convergence of the duality gap.

Theorem 4.1 *Let $\{(x^k, y^k, s^k)\}$ be the iterates generated by Algorithm 2. Then $(x^k)^T s^k \to 0$ quadratically (viewing the predictor-corrector step as a single iteration).*

Proof. Consider an arbitrary iteration k of Algorithm 2. If the safeguard is activated, then $(x^{k+1}, y^{k+1}, s^{k+1})$ is obtained by applying the predictor-corrector iteration to (x^k, y^k, s^k). By Lemma 4.1, (x^k, s^k) lies in a fixed neighborhood of the central path, that is, $\|X^k s^k/\mu^k - e\| \le \alpha/2$. Then it follows from the two-step quadratic convergence of the predictor-corrector method (see Prop. 2.1 (4)) that $\mu^{k+1} = O((\mu^k)^2)$. On the other hand, if the CPC iteration is performed, then the condition (3.3) holds and the stepsize (3.2) is accepted. By (4.8), we have

$$\mu^{k+1} = (1-\theta^k)\mu^k + \theta^k c(\mu^k)^2.$$

This and (4.10) imply $\mu^{k+1} = O((\mu^k)^2)$. Therefore, in both cases the duality gap decreases quadratically. The proof is complete. **Q.E.D.**

We are now ready to prove the main convergence result.

Theorem 4.2 *We have the following:*

(1) *The algorithm can be terminated in $O(\sqrt{n}L)$ iterations with an optimal primal-dual pair.*

(2) *After $O(\sqrt{n}\log(c\mu^0))$ iterations there holds $c\mu^k < 0.9$.*

(3) *If*

$$c > c^* := \sqrt{n}\rho^2\left(\sqrt{1+2\sqrt{2}n/\alpha} + 1\right)/\alpha, \tag{4.11}$$

then the safeguard in Algorithm 2 will be activated at most finite number of times. Furthermore, $\{(x^k, s^k)\}$ converges R-quadratically to some optimal primal-dual pair.

Proof. (1) It follows from (4.3) that in $O(\sqrt{n}L)$ iterations we should have $\mu^k < 2^{-L}$, thus giving the desired polynomial time bound.

(2) By the definition of Algorithm 2, the predictor-corrector iteration (the safeguard) is performed as long as $c\mu^k \ge 0.9$. Thus, in $O(\sqrt{n}\log(c\mu^0))$ iterations, we will have $c\mu^{k'} < 0.9$ for some k'. Since μ^k is monotonically decreasing (cf. Lemma 4.1), we should have $c\mu^k < 0.9$ for all $k \ge k'$.

(3) Let $\mathcal{K}$ denote the set of indices at which the safeguard is activated. Suppose that $|\mathcal{K}|$ is infinite, we will derive a contradiction. Since $c\mu^k < 0.9$ is satisfied after a

finite number of iterations, we may assume without loss of generality that $c\mu^k < 0.9$ for all $k \in \mathcal{K}$. Then, by the condition (3.3), we have

$$\frac{\|D_x^k d_s^k\|}{\alpha} > \frac{c(\mu^k)^2}{\sqrt{1+8\|D_x^k d_s^k\|/(\alpha\mu^k)}+1}, \quad \forall k \in \mathcal{K}.$$

Using (4.2) to bound the right hand side term yields

$$\frac{\|D_x^k d_s^k\|}{\alpha} > \frac{c(\mu^k)^2}{\sqrt{1+2\sqrt{2}n/\alpha}+1}, \quad \forall k \in \mathcal{K}. \tag{4.12}$$

For each $k \notin \mathcal{K}$, we have from Lemma 4.2 that

$$\|(x^{k+1}, s^{k+1}) - (x^k, s^k)\| \le \|d_x^k\| + \|d_s^k\| \le 2(1+4c)\rho\mu^k.$$

By Theorem 4.1, the duality gap μ^k converges to zero quadratically. Thus, the total variation in (x, s) due to the CPC iterations must be bounded. In other words, we have

$$\sum_{k \notin \mathcal{K}} \|(x^{k+1}, s^{k+1}) - (x^k, s^k)\| < \infty.$$

By Prop. 2.2 (2), this and the assumption $|\mathcal{K}| = \infty$ imply that (x^k, s^k) converges to the analytic center (x^*, s^*). In light of Prop. 2.2 (1), this further implies that $\lim_{k\in\mathcal{K}, k\to\infty} d_{xc}^k = 0$, $\lim_{k\in\mathcal{K}, k\to\infty} d_{sc}^k = 0$. Consequently, there exists some $\epsilon^k \to 0$ such that

$$\|d_{xc}^k\| \le \epsilon^k, \quad \|d_{sc}^k\| \le \epsilon^k, \quad \text{for all } k \in \mathcal{K}. \tag{4.13}$$

Now we consider the product $(d_x^k)_i(d_s^k)_i$ for any $1 \le i \le n$. Without loss of generality, we assume $i \in B$, so we have $|(d_{sc}^k)_i| \le \rho\mu^k$ (cf. Prop. 2.1). This, together with (3.4) and $c\mu^k < 0.9$, yields

$$\begin{aligned}
(d_x^k)_i(d_s^k)_i &= \left((1-c\mu^k)(d_{xp}^k)_i + c\mu^k(d_{xc}^k)_i\right) \times \left((1-c\mu^k)(d_{sp}^k)_i + c\mu^k(d_{sc}^k)_i\right) \\
&\le \left(|(d_{xp}^k)_i| + c\mu^k|(d_{xc}^k)_i|\right) \times \left((1-c\mu^k)|(d_{sp}^k)_i| + c\mu^k(\rho\mu^k)\right) \\
&\le \left(\rho\mu^k + c\mu^k\epsilon^k\right)\rho\mu^k \\
&= \left(\rho + c\epsilon^k\right)\rho(\mu^k)^2, \qquad k \in \mathcal{K},
\end{aligned}$$

where the third step follows from (4.13) and Prop. 2.1 (1). By a symmetrical argument, we can show that the above bound also holds for $i \in N$. Plugging the above bound into left hand side of (4.12) and cancelling $(\mu^k)^2$ from both sides, we obtain

$$\frac{\sqrt{n}\left(\rho + c\epsilon^k\right)\rho}{\alpha} > \frac{c}{\sqrt{1+2\sqrt{2}n/\alpha}+1}, \quad \forall k \in \mathcal{K}$$

Letting $k \to \infty$ and using $\lim_{k\to\infty} \epsilon^k = 0$ yields

$$\frac{\sqrt{n}\rho^2}{\alpha} \ge \frac{c}{\sqrt{1+2\sqrt{2}n/\alpha}+1}$$

which contradicts (4.11). Hence, for sufficiently large c the safeguard in Algorithm 2 can be activated at most finite number of times.

Now suppose $\mathcal{K}$ is finite. Then, for sufficiently large k, the stepsize (3.2) will always be accepted and we have $(x^{k+1}, s^{k+1}) = (x^k, s^k) + \theta^k(d_x^k, d_s^k)$. By Theorem 4.2, the duality gap μ^k converges to zero quadratically. In light of Lemma 4.2, this further implies that $\|(x^{k+1}, s^{k+1}) - (x^k, s^k)\|$ converges to zero R-quadratically. Thus, the sequence $\{(x^k, s^k)\}$ is Cauchy and therefore (R-quadratically) convergent. It is obvious that the vector to which $\{(x^k, s^k)\}$ converges is an optimal primal-dual pair. **Q.E.D.**

Now we can use (4.11) to choose a constant $c > c^*$; then Theorems 4.1 and 4.2 imply that Algorithm 2 attains one-step quadratic convergence rate and has also $O(\sqrt{n}L)$ iteration complexity. Admittedly, the constant c^* is unknown and possibally large, thus making the result impractical. However, we can modify Algorithm 2 to dynamically estimate c^*. For example, we start with $c = c^0 := n$ and set $t = 1$. After $c\mu^k \leq .9$ and the predictor-corrector safeguard iteration is activated tn times, we update $c = c^t := 2c^{t-1}$. We repeat this updating process. Note that after $\log(c^*/n)$ updates, we should have

$$c^* \leq c.$$

For problems with a totally unimodular matrix (e.g., network flow problems), it is possible to obtain a polynomial bound on c. This is because in this case ρ can be bounded polynomially in n.

5. Concluding Remarks

In this paper we have proposed an one-step quadratically convergent $O(\sqrt{n}L)$ interior point algorithm for linear programming. At each iteration, the algorithm generates a new direction by solving a single linear system. This direction can be viewed as a convex combination of the predictor direction and the corrector direction. To achieve quadratic convergence, we let the weighting factor for the corrector direction go to zero in proportion to the duality gap μ. Therefore, the direction used to update iterates become more and more like the pure predictor step for which the quadratic reduction of duality gap is known to be possible. Our result is in agreement with the computational experiences reported in the literature [8, 7, 10] which suggested that large stepsizes can be used in the predictor steps while only few corrector steps are needed. Notice that the iterate sequence generated by the new algorithm is also convergent. This should come as no surprise in light of the general convergence results [5]. [Note that Algorithm 2 chooses the centering stepsize $\gamma^k = c\mu^k$, so it tends to zero at least linearly, thus satisfying the criterion of [5] for iterate convergence.]

References

1. C.C. Gonzaga and R.A. Tapia. On the convergence of the Mizuno-Todd-Ye algorithm to the analytic center of the solution set. Unpublished manuscript, 1993.
2. C.C. Gonzaga and R.A. Tapia. On the quadratic convergence of the simplified Mizuno-Todd-Ye algorithm for linear programming. Unpublished manuscript, 1993.

3. C.C. Gonzaga and M.J. Todd. An $O(\sqrt{n}L)$-iteration large-step primal-dual affine algorithm for linear programming. Technical Report 862, School of Operations Research and Industrial Engineering, Cornell University, 1989.
4. O. Güler and Y. Ye. Convergence behavior of some interior point algorithms. *Mathematical Programming* 60, pp. 215–228, 1993.
5. Z.-Q. Luo. Convergence analysis of primal-dual interior point algorithm for convex quadratic programs, Working paper, Communications Research Laboratory, McMaster University, Hamilton, Ontario, Canada L8S 4K1, 1993.
6. Z.-Q. Luo and S.-Q. Wu. A modified predictor-corrector method for linear programming, Working paper, Communications Research Laboratory, McMaster University, Hamilton, Ontario, Canada L8S 4K1; Accepted for publication in *Computational Optimization and Applications*, 1993.
7. I.J. Lustig, R.E. Marsten and D.F. Shanno. Computational experience with a primal-dual interior point method for linear programming. *Linear Algebra and Its Applications*, 152:191–222, 1991.
8. K.A. McShane, C.L. Monma and D.F. Shanno. An implementation of a primal-dual interior point method for linear programming. *ORSA J. Computing,* 1:70–83, 1989.
9. S. Mizuno, M.J. Todd and Y. Ye. On adaptive-step primal-dual interior-point algorithms for linear programming, Technical Report No. 944, School of Operations Research and Industrial Engineering, Cornell University, (Ithaca, New York), 1990, to appear in *Mathematics of Operations Research.*
10. M.J. Todd and J.-P. Vial. Todd's low-complexity algorithm is a predictor-corrector path following method. Technical Report No. 952, School of Operations Research and Industrial Engineering, Cornell University (Ithaca, NY.), 1990.
11. Y. Ye. On the Q-order of convergence of interior-point algorithms for linear programming. In Wu Fang, ed,. *Proceedings of the 1992 Symposium on Applied Mathematics* (Institute of Applied Mathematics, Chinese Academy of Sciences, 1992) pp. 534-543.
12. Y. Ye, O. Güler, R.A. Tapia and Y. Zhang. A quadratically convergent $O(\sqrt{n}L)$-iteration algorithm for linear programming. *Mathematical Programming* 59, pp. 151–162.
13. Y. Zhang and R.A. Tapia. A quadratically convergent polynomial primal-dual interior point algorithm for linear programming. *SIAM J. Optimization* 2, pp. 304–323.

A MODIFIED BARRIER FUNCTION METHOD FOR LINEAR PROGRAMMING

M.R. OSBORNE
Centre for Mathematics and its Applications
School of Mathematical Sciences
Australian National University
Canberra, A.C.T. 0200
Australia

Abstract. A barrier formulation of interior point methods for the linear programming problem is considered. It is shown that a dual feasible vector can be constructed provided the Newton iterate satisfies a slightly stronger condition than feasibility. The duality gap computed using this data is strictly smaller than that obtained by running the Newton iteration to convergence. Thus this condition provides a new stopping criterion for the barrier function based interior point methods for linear programming. An $O(nL)$ estimate for the number of steps to termination is derived. Numerical experience is reported for a method in which the barrier parameter is modified adaptively in an attempt to keep the number of Newton iterations per step fixed.

Key words: barrier function, interior point method, Newton's method, stopping criterion, reduced duality gap, polynomial complexity

1. Introduction

The purpose of this note is to consider a variant of the barrier function method when it is applied to the linear programming problem in active set form

$$\min_{\boldsymbol{x}\in X} \boldsymbol{c}^T\boldsymbol{x}; \quad X = \{\boldsymbol{x} : A\boldsymbol{x} \geq \boldsymbol{b}\} \tag{1}$$

where $A : R^p \to R^n$ is assumed to have rank $p < n$. Newton's method is used to minimize the successive barrier objectives, and it is well known that economising on the number of Newton iterations at each stage is essential if the method is to be competitive. A good general reference to this material is [3].

Recall that the dual linear program is

$$\min_{\boldsymbol{u}\in U} -\boldsymbol{b}^T\boldsymbol{u}; \quad U = \{\boldsymbol{u} : A^T\boldsymbol{u} = \boldsymbol{c};\ \boldsymbol{u} \geq 0\}. \tag{2}$$

If $\boldsymbol{x}$ is feasible for (1), and $\boldsymbol{u}$ is feasible for (2), then duality gives

$$\boldsymbol{c}^T\boldsymbol{x} - \boldsymbol{b}^T\boldsymbol{u} \geq 0. \tag{3}$$

If equality holds in (3) then $\boldsymbol{x}$ minimizes (1) and $\boldsymbol{u}$ minimizes (2).

The barrier function for (1) is

$$\mathcal{B}(\boldsymbol{x},\mu) = \boldsymbol{c}^T\boldsymbol{x} - \mu\sum_{i=1}^{n}\log r_i \tag{4}$$

Ding-Zhu Du and Jie Sun (eds.), Advances in Optimization and Approximation, 247–255.

where (provided $\boldsymbol{x} \in X$)

$$r_i = \boldsymbol{a}_i^T \boldsymbol{x} - b_i \geq 0, \ i = 1, 2, \cdots, n. \tag{5}$$

with $\boldsymbol{a}_i^T = \boldsymbol{e}_i^T A$, and $\boldsymbol{e}_i$ is the i'th coordinate vector. On occasion, the notation $r_i(\boldsymbol{x})$ will be used to make the dependence explicit. The necessary conditions for a minimum of $\mathcal{B}(\boldsymbol{x}, \mu)$ with respect to $\boldsymbol{x}$ for fixed $\mu > 0$ give

$$\nabla_x \mathcal{B} = \boldsymbol{c}^T - \sum_{i=1}^{n} \frac{\mu}{r_i} \boldsymbol{a}_i^T = 0. \tag{6}$$

An immediate consequence is that the vector $\boldsymbol{u}$ with components

$$u_i = \frac{\mu}{r_i} > 0, \ i = 1, 2, \cdots, n \tag{7}$$

is feasible for the dual problem (2). The main result is that

$$\boldsymbol{x} \rightarrow \boldsymbol{x}^*, \ \boldsymbol{u} \rightarrow \boldsymbol{u}^*, \ \text{as } \mu \downarrow 0, \tag{8}$$

where $\boldsymbol{x}^*$, $\boldsymbol{u}^*$ solve (1) and (2) respectively. Let

$$\mathcal{D}(\boldsymbol{u}, \mu) = -\boldsymbol{b}^T \boldsymbol{u} - \mu \sum_{i=1}^{n} \log u_i. \tag{9}$$

Then the appropriate barrier function formulation of (2) considers

$$\min_{\boldsymbol{u}, A^T \boldsymbol{u} = \boldsymbol{c}} \mathcal{D}(\boldsymbol{u}, \mu). \tag{10}$$

The necessary conditions for a minimum of (10) give

$$-\boldsymbol{b} = \mu \sum_{i=1}^{n} \frac{1}{u_i} \boldsymbol{e}_i - A\boldsymbol{\xi}, \tag{11}$$

where $\boldsymbol{\xi}$ is the vector of multipliers for the equality constraints. It follows by comparing (6), (11), (18), and (5) that the minimizers of (4) and (9) are related by

$$r_i = \frac{\mu}{u_i}, \ \boldsymbol{\xi} = \boldsymbol{x}. \tag{12}$$

The duality gap can be computed by multiplying (6) by $\boldsymbol{x}$ or (11) by $\boldsymbol{u}$. This gives

$$\boldsymbol{c}^T \boldsymbol{x} - \boldsymbol{b}^T \boldsymbol{u} = n\mu, \ \rightarrow 0, \mu \downarrow 0. \tag{13}$$

Newton's method is used to minimize (4). This computes a correction to the current $\boldsymbol{x}$ given by

$$\boldsymbol{h} = -\nabla_x^2 \mathcal{B}^{-1} \nabla_x \mathcal{B}^T \tag{14}$$

where the Hessian matrix is given by

$$\nabla_x^2 \mathcal{B} = \frac{1}{\mu} \sum_{i=1}^{n} \frac{\mu^2}{r_i^2} \boldsymbol{a}_i \boldsymbol{a}_i^T. \tag{15}$$

It follows from (15) that the condition number of the Hessian is bounded for the sequence of minima corresponding to the sequence of barrier parameters $\{\mu_k, \ k = 1, 2, \cdots\} \downarrow 0$ provided

1. $\{x_k\} \to x^*$ where x^* is a vertex of the feasible region, and
2. $\{u_k\} \to u^*$ where $u_i^* > 0$ for the constraints active at x^*.

These give conditions under which the Newton iteration can be expected to perform well. They require that the minimum is unique, but not that it is nondegenerate. In the case that the minimum is not unique then $\#\{u_i > 0\} < p$, the Hessian matrix (15) becomes singular as $\mu \downarrow 0$, and x^* is not a vertex of the feasible region in general. If the solution to the primal is not unique then the solution to the dual is degenerate, and it is preferable to solve the dual problem. The Newton correction k satisfies the system of equations

$$\begin{bmatrix} \nabla_x^2 \mathcal{D} & A \\ A^T & 0 \end{bmatrix} \begin{bmatrix} \frac{1}{\mu} k \\ \xi \end{bmatrix} = \begin{bmatrix} \sum_{i=1}^n \frac{\mu}{u_i} e_i + b \\ 0 \end{bmatrix}, \tag{16}$$

where

$$\nabla_x^2 \mathcal{D} = \sum_{i=1}^n \frac{\mu^2}{u_i^2} e_i e_i^T. \tag{17}$$

This system is well behaved if the Hessian, which is a diagonal matrix in this case, has $n - p$ diagonal elements bounded away from zero and if the complementary set of indices defines the rows of a nonsingular submatrix of A. It follows from the identification 12 that This condition can be expected to be satisfied when x^* is not a vertex of the feasible region. To simplify presentation uniqueness is assumed in this paper. However, a method for determining when to switch from primal to dual is suggested in section 4.

2. An Identity Associated with Newton's Method

The cost of interior point methods is determined by

1. the number of 'outer' iterations in which an approximate minimum of $\mathcal{B}(x, \mu_k)$ is sought for the sequence of values $\{\mu_k, k = 1, 2, \cdots\} \downarrow 0$, and
2. the number of Newton or 'inner' iterations performed for each value of μ.

These components are not independent as larger changes in μ_k at each stage lead to more inner iterations. Thus minimizing the cost requires that a balance be drawn. In this section a new method for terminating the inner iteration is developed.

Let h be the predicted correction at x when Newton's method is applied to find a zero of $f(x)$. Then Taylor's series gives

$$f(x + h) = f(x) + f^{(1)}(x : h) + \sum_{i=2}^{\infty} \frac{1}{i!} f^{(i)}(x : h, \cdots, h) = \sum_{i=2}^{\infty} \frac{1}{i!} f^{(i)}(x : h, \cdots, h).$$

In the case of (4)

$$f(x) = c - \sum_{i=1}^n \frac{\mu}{r_i(x)} a_i.$$

The derivatives are calculated readily and give

$$f(x + h) = c - \sum_{i=1}^n \frac{\mu}{r_i(x + h)} a_i,$$

$$= -\sum_{i=1}^{n}\sum_{j=2}^{\infty}(-1)^j \frac{\mu}{r_i}\left(\frac{\boldsymbol{a}_i^T\boldsymbol{h}}{r_i}\right)^j \boldsymbol{a}_i$$

$$= -\sum_{i=1}^{n}\frac{\mu}{r_i}\frac{\left(\frac{\boldsymbol{a}_i^T\boldsymbol{h}}{r_i}\right)^2}{1+\left(\frac{\boldsymbol{a}_i^T\boldsymbol{h}}{r_i}\right)}\boldsymbol{a}_i$$

provided $|\frac{\boldsymbol{a}_i^T\boldsymbol{h}}{r_i}| < 1$. This gives

$$\boldsymbol{c} = \sum_{i=1}^{n}\frac{\mu}{r_i(\boldsymbol{x}+\boldsymbol{h})}\{1-(\frac{\boldsymbol{a}_i^T\boldsymbol{h}}{r_i})^2\}\boldsymbol{a}_i. \tag{18}$$

Remark 1 The condition $|\frac{\boldsymbol{a}_i^T\boldsymbol{h}}{r_i}| < 1$ ensures that $\boldsymbol{x}+\boldsymbol{h}$ is feasible. This is a sensible condition for potential candidates for the optimal set. Typically, these are the constraints for which $\boldsymbol{a}_i^T\boldsymbol{h} < 0$. However, it is stronger than feasibility because it restricts the size of corrections to r_i as $\boldsymbol{x}$ moves away from the associated constraint. Typically, these are the constraints for which $\boldsymbol{a}_i^T\boldsymbol{h} > 0$.

Remark 2 The important feature of (18) is that it provides a dual feasible vector with components

$$\hat{u}_i = \frac{\mu}{r_i(\boldsymbol{x}+\boldsymbol{h})}\{1-(\frac{\boldsymbol{a}_i^T\boldsymbol{h}}{r_i})^2\},\ i = 1,2,\cdots,n.$$

Computing the duality gap gives:

$$\begin{aligned}\boldsymbol{c}^T(\boldsymbol{x}+\boldsymbol{h}) &= \sum_{i=1}^{n}\frac{\mu}{r_i(\boldsymbol{x}+\boldsymbol{h})}\{1-(\frac{\boldsymbol{a}_i^T\boldsymbol{h}}{r_i})^2\}\boldsymbol{a}_i^T(\boldsymbol{x}+\boldsymbol{h}),\\ &= \boldsymbol{b}^T\hat{\boldsymbol{u}} + \sum_{i=1}^{n}\mu\{1-(\frac{\boldsymbol{a}_i^T\boldsymbol{h}}{r_i})^2\}.\end{aligned} \tag{19}$$

Thus

$$\begin{aligned}\boldsymbol{c}^T(\boldsymbol{x}+\boldsymbol{h}) - \boldsymbol{b}^T\hat{\boldsymbol{u}} &= \sum_{i=1}^{n}\mu\{1-(\frac{\boldsymbol{a}_i^T\boldsymbol{h}}{r_i})^2\},\\ &< n\mu.\end{aligned} \tag{20}$$

This gives a more favourable duality gap than does iterating the Newton iteration to convergence. *It justifies stopping the Newton iteration as soon as a point satisfying* $|\frac{\boldsymbol{a}_i^T\boldsymbol{h}}{r_i}| < 1$ *is obtained.*

Remark 3 A similar argument applies to the minimization of (9). Here

$$-\boldsymbol{b} = \sum_{i=1}^{n}\frac{\mu}{u_i+k_i}\{1-(\frac{k_i}{u_i})^2\}e_i - A\boldsymbol{\xi}, \tag{21}$$

and the duality gap is given by

$$\boldsymbol{c}^T\boldsymbol{\xi} - \boldsymbol{b}^T(\boldsymbol{u}+\boldsymbol{k}) = \sum_{i=1}^{n}\mu\{1-(\frac{k_i}{u_i})^2\} < n\mu. \tag{22}$$

3. A Small Step Complexity Estimate

Here the rule

$$\mu_{k+1} = \zeta \mu_k \tag{23}$$

is considered for updating the barrier parameter μ_k. If the barrier reduction multiplier ζ is close enough to 1 then the stopping criterion

$$\max_{1 \le i \le n} |w_i| < 1; \; w_i = \frac{a_i^T h}{r_i}, \tag{24}$$

can be satisfied in the first Newton step.

First an estimate of ζ is derived so that (24) is satisfied in one Newton step starting from x a point on a central trajectory (that is x satisfies (6)) when $\mu \leftarrow \zeta\mu$. Here the step is defined by

$$\frac{\zeta}{\mu}\{\sum_{i=1}^{n}(\frac{\mu}{r_i})^2 a_i a_i^T\}h = -c + \sum_{i=1}^{n}\frac{\zeta\mu}{r_i}a_i. \tag{25}$$

This can be written

$$\frac{\zeta}{\mu}A^T U^2 A h = -(1-\zeta)A^T U e, \tag{26}$$

using (6), where the diagonal matrix U is defined by

$$U_i = \frac{\mu}{r_i}, \; i = 1, 2, \cdots, n. \tag{27}$$

It follows that

$$w_i = \frac{a_i^T h}{r_i} = -\frac{1-\zeta}{\zeta}\frac{\mu}{r_i}a_i^T (A^T U^2 A)^{-1} A^T U e.$$

Whence

$$\begin{aligned} w &= -\frac{1-\zeta}{\zeta} U A (A^T U^2 A)^{-1} A^T U e, \\ &= -\frac{1-\zeta}{\zeta} P e, \end{aligned} \tag{28}$$

where P is an orthogonal projection matrix. This gives the estimate

$$\max_i |w_i| \le \|w\|_2 \le \frac{1-\zeta}{\zeta}\sqrt{n}. \tag{29}$$

If $\zeta = 1 - \alpha/\sqrt{n}$ then

$$\max_i |w_i| \le \frac{\alpha}{1-\alpha/\sqrt{n}} < 1, \forall n \tag{30}$$

provided $\alpha < .5$. The optimality of an updating formula of this kind is discussed in [3].

Now assume that the current point is not on a central trajectory but has been reached after the first Newton step has been accepted. Then the first step predicted at the next inner iteration satisfies

$$\begin{aligned}\frac{\zeta}{\mu}A^T U^2 A\boldsymbol{h} &= -\boldsymbol{c}+\zeta\sum_{i=1}^{n}\frac{\mu}{r_i}\boldsymbol{a}_i,\\ &= -\sum_{i=1}^{n}(1-\zeta-w_i^2)\frac{\mu}{r_i}\boldsymbol{a}_i,\\ &= -A^T U\boldsymbol{z},\end{aligned} \tag{31}$$

where (18) has been used to substitute for $\boldsymbol{c}$, and $z_i = 1-\zeta-w_i^2$, $i = 1,2,\cdots,n$. It follows that

$$\boldsymbol{w} = -\frac{1}{\zeta}P\boldsymbol{z}, \tag{32}$$

so that

$$\max_i |w_i| \le \frac{1}{\zeta}\|\boldsymbol{z}\|_2. \tag{33}$$

The aim is to show that this step will be accepted also provided ζ is sufficiently close to 1. To this end consider

$$\max_{|w_i|\le 1}(1-\zeta-w_i^2)^2. \tag{34}$$

This has a minimum at $w_i^2 = 1-\zeta$ so that the maximum value is attained either at

$w_i^2 = 0$, giving the maximum value $max = (1-\zeta)^2$, or
$w_i^2 = 1$, giving the maximum value $max = \zeta^2$.

The case $w_i = 1$ leads to an infeasible estimate on the right hand side of (33), so attention is restricted to $0 < w_1^2 < 2(1-\zeta) < 1$. Substituting into (33) leads to the condition

$$\sqrt{2(1-\zeta)} \le \frac{1-\zeta}{\zeta}\sqrt{n}. \tag{35}$$

This can be satisfied in a consistent fashion by setting $\zeta = 1-\alpha/n$ where $\alpha = .45$ works for all n.

Remark 4 It is shown in [3] that a ζ satisfying a condition of this kind implies termination of the outer iteration in at most $O(nL)$ steps where L is the standard measure of problem complexity. This suffices to prove termination in polynomial time, but it is known that more favourable results are possible. More information can be found in the quoted reference.

4. A form of algorithm

Aspects of an algorithm implementing the the stopping criterion $|w_i| < 1$, $i = 1,2,\cdots,n$ are considered here. Termination at outer step $k-1$ of the algorithm at $\boldsymbol{x}^k$ implies

$$\boldsymbol{c} = \sum_{=1}^{n}\frac{\mu_{k-1}}{r_i(\boldsymbol{x}^k)}s_i^k\boldsymbol{a}_i \tag{36}$$

where

$$s_i^k = 1 - \left(\frac{a_i^T h^{k-1}}{r_i(x^k - h^{k-1})}\right)^2,$$

h^{k-1} is the last Newton step generated, $\mu_k = \zeta_k \mu_{k-1}$, and ζ_k defines the current updating of the barrier parameter. Let the current point be x_c. Then the Newton step h_c from the point x_c is found by solving the linear system

$$\begin{aligned} A^T U(x_c)^2 A h_c &= -\mu_k \{c - \sum_{i=1}^{n} \frac{\mu_k}{r_i(x_c)} a_i\}, \\ &= A^T U(x_c) z, \end{aligned}$$

where, substituting for c using (18),

$$z_i = \mu\{1 - \frac{s_i^k}{\zeta_k} \frac{r_i(x_c)}{r_i(x^k)}\}, \ i = 1, 2, \cdots, n.$$

It follows that h_c can be found at each step of the inner iteration by solving a least squares problem with matrix $U(x_c)A$. If the termination criterion (24) is not satisfied then a line search is carried out to determine the minimum with respect to λ of $\mathcal{B}(x_c + \lambda h_c, \mu_k)$. The line search described in [5] has proved satisfactory.

The number of steps in the inner iteration is controlled by ζ_k. This is chosen adaptively in an attempt to have the same number of Newton steps in each step of the outer iteration in the numerical experiments reported in the next section, The rule that has been used for adjusting ζ is:

- if $q < m$ then $\zeta \leftarrow \theta\zeta$, $\theta < 1$;
- if $q = m$ then ζ unchanged;
- if $q > m$ then $\zeta \leftarrow \zeta + \frac{1-\zeta}{\phi}$, $\phi > 1$;

where m is the target number of steps and q is the number taken in the just completed outer step.

The test that has been used to terminate the outer iteration is based on sorting the r_i in increasing order of magnitude and then testing the ratio $\frac{\sum_{i=1}^{p} r_{\nu(i)}}{p} / \frac{\sum_{i=p+1}^{n} r_{\nu(i)}}{n-p}$ against a prescribed tolerance. Here ν is an index set with elements pointing to the r_i in sorted order.This test is robust against degeneracy, but not against non–uniqueness. An alternative stopping procedure is provided by testing the duality gap against a prescribed tolerance. This test should be robust against non–uniqueness, so a comparison between the two procedures could provide a means for detecting non–uniqueness and triggering a switch to the dual barrier procedure.

Remark 5 The formula (18) can be used in other ways. For example, if the s_i^k are used as weights in the barrier function then a factor $(1 - \zeta)/\zeta$ can be taken out of the formula for the Newton correction in the first step. This means that ζ can be chosen close enough to 1 so that the first step is always accepted. Now the free parameter is $\max_i |w_i|$, and usually this determines how closely the likely active constraints are approached in the current step. But the numerical results have not proved too satisfactory in comparison with the procedure outlined above. First, it appears to be a characteristic of the method that the approach to the minimum is interrupted after a sequence of good steps ($\max_i w_i < 0$) by a sequence of steps in which $\max_i w_i > 0$ corresponding to a move away from the corresponding constraint. This suggests that the trajectory has approached too close to a non-optimal boundary point. This process is associated with a noticeable slowing down in the rate of convergence. Second, the process makes explicit use of (18), and this formula is susceptible to error build up through cancellation. This is very evident in this one step method with typically all figures being lost in the representation of c eventually. It has not proved as serious in the barrier method described above, and it's use to convert the calculation of the Newton step into a least squares step, the part of the calculation most directly affected, is an implementation ploy rather than an intrinsic feature of the method. A further alternative that has been considered has been to use (18) in conjunction with the accelerated barrier methods discussed in [5]. Formally, this approach suggests that second order convergence to zero of the barrier parameter should be possible. However, the practice appears different, and similar problems to those described above for the one step method appear.

5. Some numerical results

Results of some numerical experiments are reported in this sestion. The procedure used to set up the problems follows that described in [4] for generating random linear programming problems with known solutions. However, one variant is employed here. This notes that if $A_{ij} > 0$ and $-1 < x_i^* < 1$ then $\boldsymbol{x}_0 = \boldsymbol{e}$ provides a feasible initial starting point. Values chosen for the parameters include $m = 3$ for the target number of Newton iterations, $\theta = .5$, $\phi = 4$ for the parameters in the barrier adjustment procedure, and the stopping tolerance was set to 10^{-6}. The initial barrier parameter μ_0 is computed by

$$\mu_0 = arg \min_{\mu} \|\nabla_x \mathcal{B}(\boldsymbol{x}_0, \mu)\|_2 \tag{37}$$

following the suggestion of [1]. Initially $\zeta = .8$. A uniform random number generator is used to generate the problem data, and computations are carried out for each of five different seeds for the cases $n = 200$, $p = 40$, and $n = 1000$, $p = 100$. Results are presented in the following tables. These report the number of outer iterations, the cumulative total of Newton iterations, the final duality gap, the final values of μ and ζ, and the accuracy attained in $\boldsymbol{x}$ and $\boldsymbol{u}$. For comparison, numbers of steps for the projected gradient algorithm applied to the same problems is given also. The program used has been described in [2].

The results show a reasonable performance from the barrier function method. In particular, it is able to work with quite large steps in μ. The process for choosing ζ so that the number of Newton steps taken in each inner iteration is approximately constant worked remarkably well also. The test problems are not ones that would be expected to favour interior point methods because of the way the problems are generated. This is because all directions into the positive orthant are directions of recession so there is a tendency for vertices of the feasible region to cluster in the region where the generated solutions lie. This would be expected to make the termination test harder to satisfy. Also, A is a dense matrix so the full cost of each Newton iteration is incurred. Thus it could be encouraging that the C timing routine shows a factor of approximately 2 favouring the projected gradient method against the barrier method in both simulations.

200x40	seed1	seed2	seed3	seed4	seed5
outer iterations	13	12	12	13	12
Newton steps	34	31	32	34	32
duality gap	1.2(-5)	1.7(-5)	6.9(-6)	7.0(-6)	3.2(-6)
barrier parameter	6.0(-8)	8.5(-8)	3.5(-8)	3.5(-8)	1.7(-8)
final correction	8.1(-2)	8.1(-2)	8.1(-2)	4.1(-2)	4.1(-2)
x error	7.7(-5)	3.8(-5)	7.7(-5)	6.1(-6)	4.0(-6)
u error	1.2(-3)	3.4(-4)	4.6(-5)	4.5(-5)	3.7(-5)
PG iterations	67	65	60	58	62
1000x100	**seed1**	**seed2**	**seed3**	**seed4**	**seed5**
outer iterations	15	16	14	15	15
Newton steps	40	44	38	41	41
duality gap	4.7(-6)	8.8(-6)	2.8(-5)	1.2(-5)	4.9(-6)
barrier parameter	4.7(-8)	8.8(-9)	2.7(-8)	1.2(-8)	4.9(-9)
final correction	6.2(-2)	6.2(-2)	1.2(-1)	6.2(-2)	6.2(-2)
x error	2.5(-6)	2.7(-6)	7.4(-6)	6.2(-6)	2.3(-6)
u error	1.4(-3)	1.7(-4)	3.3(-3)	4.1(-4)	1.2(-4)
PG iterations	258	252	258	233	239

Table 1: numerical results from the two simulations.

References

1. A.V. Fiacco and G.P. McCormick, Nonlinear Programming, Wiley, 1968.
2. Karen George and M.R. Osborne, On degeneracy in linear programming and related problems, *Annals of Operations Research*, 1993 (to appear).
3. C.C. Gonzaga, Path-following methods in linear programming, *SIAM Review*, 34(2):167–224, 1992.
4. M.R. Osborne, Finite Algorithms in Optimization and Data Analysis, Wiley, 1985.
5. M.R. Osborne, Dual barrier functions with superfast rates of convergence for the linear programming problem, *J. Austral. Math. Soc. Ser. B*, 29(1):39–58, 1987.

A NEW FACET CLASS AND A POLYHEDRAL METHOD FOR THE THREE-INDEX ASSIGNMENT PROBLEM

LIQUN QI
School of Mathematics, The University of New South Wales, Kensington, N.S.W. 2033, Australia

EGON BALAS
Graduate School of Industrial Administration, Carnegie Mellon University, Pittsburgh, PA 15213, U.S.A.

and

GEENA GWAN
School of Mathematics, The University of New South Wales, Kensington, N.S.W. 2033, Australia

Abstract. Since the number of variables of the three-dimensional assignment problem of order n is n^3, an $O(n^3)$ separation algorithm for a class of facets of the three-dimensional assignment polytope is linear-time and its complexity is best possible. In [1] and [12], linear-time separation algorithms were given for four classes of facets. In this paper, we introduce a new class of facet-defining inequalities, with coefficients equal to 0 or 1 and right-hand side equal to a positive integer p. These inequalities are of Chvàtal rank 2. The special case when $p = 1$ has been identified by Balas and Saltzman in [2]. An $O(n^3)$ separation algorithm is given for facets with $p = 2$ in this class. Based upon the linear-time separation algorithms for the five classes of facets found in [1], [12] and this paper, we construct a polyhedral procedure for solving the three-index assignment problem. Computational results are given to show that this procedure is efficient.

1. Introduction

The three-index assignment problem of order n can be stated as a 0-1 programming problem as follows:

$$\begin{array}{ll} \min & \sum\{c_{ijk}x_{ijk} : i \in I, j \in J, k \in K\}, \\ \text{s.t.} & \sum\{x_{ijk} : j \in J, k \in K\} = 1, \ \ \forall i \in I, \\ & \sum\{x_{ijk} : i \in I, k \in K\} = 1, \ \ \forall j \in J, \\ & \sum\{x_{ijk} : i \in I, j \in J\} = 1, \ \ \forall k \in K, \\ & \quad x_{ijk} \in \{0,1\}, \quad \forall i,j,k, \end{array} \tag{1}$$

where I, J and K are disjoint sets with $|I| = |J| = |K| = n$. Let A be the coefficient matrix of the constraint set of (1). Then $R = I \cup J \cup K$ is the row index set of A. Let S be the column index set of A. Let G_A be the intersection graph of A, i.e., the graph that has a vertex for every column of A and an edge for every pair of non-orthogonal columns. Let

$$P = \{x \in \Re^{n^3} : Ax = e, x \geq 0\},$$

Ding-Zhu Du and Jie Sun (eds.), Advances in Optimization and Approximation, 256–274.

where $e = (1, ..., 1)^T \in \Re^{3n}$. Then

$$P_I = \text{conv}\{x \in \{0,1\}^{n^3} : x \in P\}$$

is the three-index assignment polytope of order n.

The three-dimensional assignment problem is well-known to be NP-hard [14]. For this type of problem, typically implicit enumeration methods are used. Among the algorithms in the literature are those of Vlach [20], Pierskalla [16] [17] and Leue [15]; a primal-dual algorithm described by Hansen and Kaufman [13]; a branch and bound algorithm using a Lagrangian dual and subgradient optimization implemented by Fröhlich [10] and discussed by Burkard and Fröhlich [4]. Also see Burkard and Rudolf [5]. More recently, Balas and Saltzman [3] developed a branch and bound algorithm that also uses facet-defining inequalities in a Lagrangian fashion with subgradient optimization.

Balas and Saltzman [2] started to study the facial structure of P_I. They gave an $O(n^4)$ procedure to detect whether there is a clique facet of P_I, violated by a given noninteger point x. In [1], Balas and Qi gave an $O(n^3)$ procedure to do this. Since the number of variables of (1) is n^3, an $O(n^3)$ separation algorithm for a facet class of P_I is linear-time and its complexity is best possible. Balas and Qi [1], Gwan and Qi [12] also gave linear-time separation algorithms for other two facet classes of P_I, identified in [2].

In Section 2, we identify a class of facets of P_I, whose left hand side coefficients are in $\{0,1\}$, whose right hand side coefficient is p, $1 \le p \le n-2$. For $p = 1$, this subclass of facets is a clique facet class discovered in [2]. Thus, we denote the subclass of facets for a fixed p, $1 \le p \le n-2$, by $\mathcal{P}(p)$ and the whole class by $\mathcal{P}$. We show that the facets in $\mathcal{P}$ are of Chvátal rank 2. We pay special attention to the subclass $\mathcal{P}(2)$. In Section 3, we discuss the properties of a facet in $\mathcal{P}(2)$ violated by a given noninteger $x \in P$, and in Section 4 we give an $O(n^3)$ procedure to detect whether there is such a facet in $\mathcal{P}(2)$, i.e., we give a linear-time separation algorithm for $\mathcal{P}(2)$.

Therefore, we now know five facet classes of P_I, for which we have linear-time separation algorithms. We call them linear-time separable facet classes. In Section 5, we discuss some features of these five facet classes. Based upon the linear-time separation algorithms for these five facet classes, in Section 6, we construct a polyhedral procedure for solving (1). Computational results are given in Section 7 to show that this procedure is efficient.

The results in Sections 2-4 were originally studied by Qi and Balas in an unpublished technical report [18]. Sections 6 and 7 were done by Gwan as a part of her Ph.D. thesis [11].

Other papers on the three-index assignment problem include [6] - [9] and [19].

2. Facet Class $\mathcal{P}$

We use the same symbols as [2]. We may specify $s = (i,j,k)$ for $i \in I$, $j \in J$, $k \in K$. For a set $Q \subseteq R$, we use Q_I, Q_J and Q_K to denote the parts of Q in I, J and K respectively. Furthermore, we also regard $s \in S$ as a three-element subset of R. So we may write $s \cap Q$ for $s \in S$ and $Q \subseteq R$. For $x \in \Re^{n^3}$ and $S' \subseteq S$, let

$$x(S') = \sum \{x_s : s \in S'\}.$$

In [2], Balas and Saltzman identified (among other facets) the following two classes of facet-defining inequalities for P_I:

1. Let $Q \subseteq R$, $|Q| = 2p+1$, $1 \le p \le n-1$, and $1 \le |Q_L| \le p$ for $L = I, J, K$. Let

$$S(Q) = \{s \in S : |s \cap Q| \ge 2\}.$$

Then

$$x(S(Q)) \le p, \tag{2}$$

defines a facet of P_I for $n \ge 3$. This facet is of Chvátal rank 1. The number of distinct inequalities (2) is $O(2^{3n})$. Denote the class of facet-defining inequalities (2) by $\mathcal{Q}(p)$ for $1 \le p \le n-1$, and let

$$\mathcal{Q} = \cup_{p=1}^{n-1} \mathcal{Q}(p).$$

2. Let $s, t \in S$ and assume that $s \cap t = \emptyset$. Let

$$C(s,t) = \{s\} \cup \{s' \in S : |s \cap s'| = 1, |s' \cap t| = 2\}.$$

Then $C(s,t)$ is a clique of G_A and

$$x(C(s,t)) \le 1, \tag{3}$$

defines a facet of P_I for $n \ge 4$. This facet is of Chvátal rank 2. The number of distinct inequalities (3) is $\frac{1}{4}n^3(n-1)^3$. Denote the class of facet-defining inequalities (3) by $\mathcal{P}(1)$.

For any $s, t \in S$, $s \cap t = \emptyset$, $C(s,t)$ is a clique of G_A. For $p = 1$, $S(Q)$ is also a clique of G_A. Balas and Saltzman [2] showed that $\mathcal{Q}(1)$ and $\mathcal{P}(1)$ include all clique facets of P_I. Balas and Qi [1] gave linear-time separation algorithms for $\mathcal{Q}(1)$, $\mathcal{P}(1)$ and in $\mathcal{Q}(2)$.

Consider $x \in P_I$ such that $x_{ijk} = x_{i'j'k'} = 1$, where (i,j,k) and $(i'j'k')$ are a pair of disjoint pair of triplets in S. Define x' by $x'_{ijk} = x'_{i'j'k'} = 0$, $x'_{i'jk} = x'_{ij'k'} = 1$ and $x'_t = x_t$ otherwise. Then $x' \in P_I$. As in [2], we call the construction of x' from x a first index interchange on the triplets (i,j,k) and (i',j',k') (second and third index interchanges are defined analogously). These three operations will be used several times in this section.

Theorem 2.1. *Suppose that $D, Q \subseteq R$, $D \cap Q = \emptyset$, $|D_L| + |Q_L| = p+1$ and $1 \le |Q_L| \le r$ for $L = I, J, K$, $|Q| = 2r+1$, $1 \le r \le p \le n-3$. Let*

$$C_1(D) = \{s \in S : s \subseteq D\},$$

$$C_2(D,Q) = \{s \in S : |s \cap D| = 1, |s \cap Q| = 2\},$$

$$C(D,Q) = C_1(D) \cup C_2(D,Q).$$

Then for $n \ge 4$,

$$x(C(D,Q)) \le p \tag{4}$$

defines a facet of P_I.

We first prove a lemma. Let $P_I^{C(D,Q)} = \{x \in P_I : x(C(D,Q)) = p\}$.

Lemma. Suppose that $s = (i_s, j_s, k_s) \in S$ and $t = (i_t, j_t, k_t) \in S$ are disjoint. Suppose that either (i) $s \subseteq D \cup Q$, $t \subseteq R \setminus (D \cup Q)$; or (ii) $|(s \cup t) \cap (D \cup Q)| \leq 2$, $|((s \cup t) \cap (D \cup Q))_L| \leq 1$ for $L = I, J, K$; or (iii) $s \subseteq D, t \subseteq Q$. Then there exists a 0-1 point $x \in P_I^{C(D,Q)}$ such that $x_s = x_t = 1$.

Proof: We prove this lemma by induction on $n - p, p - r$ and r. We first deal with the base case that $p = r = 1$ and $n = 4$. Then we show that every other case can be inductively reduced to this base case.

We now prove the base case: $p = r = 1$ and $n = 4$. Without loss of generality, assume that $D_L = \{1\}, Q_L = \{2\}$ for $L = I, J, K$. In the following, we specify $s' = (i_{s'}, j_{s'}, k_{s'})$ and $t' = (i_{t'}, j_{t'}, k_{t'})$.

In case (i), we have $i_s, j_s, k_s \in \{1,2\}$ and $i_t, j_t, k_t \in \{3,4\}$. Let $i_{s'} \in \{1,2\} \setminus \{i_s\}, j_{s'} \in \{1,2\} \setminus \{j_s\}, k_{s'} \in \{1,2\} \setminus \{k_s\}, i_{t'} \in \{3,4\} \setminus \{i_t\}, j_{t'} \in \{3,4\} \setminus \{j_t\}, k_{t'} \in \{3,4\} \setminus \{k_t\}$.

In case (iii), we have $s = (1,1,1)$ and $t = (2,2,2)$. Let $s' = (3,3,3)$ and $t' = (4,4,4)$.

We divide case (ii) into three subcases. (a). The subcase where $(s \cup t) \cap D = \emptyset$. Let $s' = (1,1,1)$ (b). The subcase where $|(s \cup t) \cap D| = 2$. Assume that $(s \cup t) \cap D_L = \emptyset$. If $L = I$, let $s' = (1,2,2)$; if $L = J$, let $s' = (2,1,2)$; if $L = K$, let $s' = (2,2,1)$. (c). The subcase where $|(s \cup t) \cap D| = 1$. Without loss of generality, assume that $|(s \cup t) \cap D_I| = 1$. We have $(s \cup t) \cap Q_L = \emptyset$ at least for $L = J$ or $L = K$. Without loss of generality, assume that $(s \cup t) \cap Q_J = \emptyset$. Let $s' = (2,2,1)$. In all three subcases (a), (b) and (c), let $i_{t'} \in \{1,2,3,4\} \setminus \{i_s, i_t, i_{s'}\}, j_{t'} \in \{1,2,3,4\} \setminus \{j_s, j_t, j_{s'}\}, k_{t'} \in \{1,2,3,4\} \setminus \{k_s, k_t, k_{s'}\}$.

In all three cases (i), (ii) and (iii), let $x \in \Re^S$ be defined by

$$x_s = x_t = x_{s'} = x_{t'} = 1$$

and $x_u = 0$ for $u \neq s, s', t, t'$. Then we have $x \in P_I^{C(D,Q)}$ and $x_s = x_t = 1$. This proves the base case.

We now show that every other case can be inductively reduced to this base case.

Suppose that $n - p \geq 4$. Pick $u = (i_u, j_u, k_u) \subseteq R \setminus (D \cup Q)$ such that u is disjoint from s and t. Fix $x_u = 1$ in the constraints which define P and P_I, i.e., remove the indices $i_u \in I, j_u \in J, k_u \in K$. Then we have a smaller problem with n reduced by 1 but D and Q remain the same, i.e., the value p remains the same. Hence by induction every other case can be reduced to the case that $n - p = 3$.

Suppose that $r \geq 2$. Without loss of generality, assume that $|Q_I| \geq |Q_K|$ and $|Q_J| \geq |Q_K|$. We have $|Q_I| \geq 2$, $|Q_J| \geq 2$ and $|Q_K| \leq (2r+1)/3$. Then $|D_K| \geq p + 1 - |Q_K| \geq r/3 + 2/3 \geq 4/3$. Since $|D_K|$ is integer, $|D_K| \geq 2$. Thus, in either the case (i), or the case (ii), or the case (iii), we have $|Q_I \setminus \{i_s, i_t\}| \geq 1, |Q_J \setminus \{j_s, j_t\}| \geq 1, |D_K \setminus \{k_s, k_t\}| \geq 1$. Pick $u = (i_u, j_u, k_u) \in S$ such that u is disjoint from s and t, $i_u \in Q_I, j_u \in Q_J$, and $k_u \in D_K$. Similarly, we may fix $x_u = 1$ and remove indices $i_u \in I, j_u \in J, k_u \in K$ from the problem. We have a smaller problem with r and p

reduced by 1. Hence by induction every other case can be reduced to the case that $r = 1$.

Suppose that $p - r \geq 1$. Then $|D_L| \geq 2$ for $L = I, J, K$. But in all the three cases of s and t, $|(s \cup t) \cap D_L| \leq 1$ for $L = I, J, K$. Thus, we may also pick $u = (i_u, j_u, k_u) \subseteq D$ such that u is disjoint from s and t. Remove indices i_u, j_u and k_u from I, J and K respectively. We have a smaller problem with $p - r$ reduced by 1. Therefore, every other case can be inductively reduced to the case that $p - r = 0$, i.e., $p = r$.

These show that every other case can be inductively reduced to the case that $p = r = 1$ and $n - p = 3$, i.e., the base case. This proves the lemma. ∎

Proof of Theorem 2.1: The inequality (4) can be obtained by adding the equations of $Ax = e$ indexed by $i \in D_I$, $j \in D_J$, $k \in D_K$, and twice the inequality

$$x(S(Q)) \leq r; \tag{5}$$

dividing the resulting inequality by 3 and rounding down all coefficients to the nearest integer. Inequality (5) has the form (2) which, by [2] defines a facet of P_I of rank 1. Thus, (4) is a valid inequality of Chvátal rank at most 2.

Without loss of generality, assume that

$$D_L \cup Q_L = \{1, 2, ..., p + 1\},$$

and the indices in Q_L are always higher than indices in D_L, for $L = I, J, K$. In particular, $(p+1, p+1, p+1) \notin C(D, Q)$. Let $x \in \Re^S$ be defined by

$$x_{1nn} = x_{n11} = x_{222} = x_{333} = \cdots = x_{n-1,n-1,n-1} = 1$$

and $x_{ijk} = 0$ for other (i, j, k). Then $x \in P_I$. Clearly, $(1, n, n), (n, 1, 1), (p+1, p+1, p+1), (p+2, p+2, p+2), ..., (n-1, n-1, n-1) \notin C(D, Q)$. Hence,

$$x(C(D, Q)) \leq x_{222} + x_{333} + ... + x_{ppp} \leq p - 1 < p.$$

Thus, the inequality (4) does not induce an improper face of P_I. This shows that $\dim P_I^{C(D,Q)} \leq \dim P_I - 1$.

To show that (4) defines a facet of P_I, i.e., that $\dim P_I^{C(D,Q)} = \dim P_I - 1$, we use the same approach as [2], i.e., given an equality $\alpha x = \alpha_0$ which is satisfied by all $x \in P_I^{C(D,Q)}$ we exhibit scalars $\lambda_i, i \in I, \mu_j, j \in J, \nu_k, k \in K$ and π such that

$$\alpha_{ijk} = \begin{cases} \lambda_i + \mu_j + \nu_k, & \text{if } (i, j, k) \in S \setminus C(D, Q), \\ \lambda_i + \mu_j + \nu_k + \pi, & \text{if } (i, j, k) \in C(D, Q), \end{cases} \tag{6}$$

and

$$\alpha_0 = \sum\{\lambda_i : i \in I\} + \sum\{\mu_j : j \in J\} + \sum\{\nu_k : k \in K\} + p\pi. \tag{7}$$

This demonstrates that any equation satisfied by all $x \in P_I^{C(D,Q)}$ is a linear combination of the equations in the system defining $P_I^{C(D,Q)}$; in other words, the addition of $x(C(D,Q)) = p$ to the constraints defining P increases the rank of the equality system of P by exactly one, i.e., dim $P_I^{C(D,Q)} =$ dim $P_I - 1$.

Define

$$\lambda_i = \alpha_{inn} - \alpha_{nnn}, \quad i \in I$$

$$\mu_j = \alpha_{njn} - \alpha_{nnn}, \quad j \in J$$

$$\nu_k = \alpha_{nnk}, \quad k \in K$$

Then we have to show that for $(i,j,k) \in S \setminus C(D,Q)$,

$$\alpha_{ijk} = \lambda_i + \mu_j + \nu_k = \alpha_{inn} + \alpha_{njn} + \alpha_{nnk} - 2\alpha_{nnn}. \tag{6a}$$

If at least two of the indices i, j, k are equal to n, (6a) obviously holds.

Now we show (6a) for $i = n$, $j \neq n$, $k \neq n$. Notice that if $x \in P_I^{C(D,Q)}$ with $x_{ijk} = x_{i'j'k'} = 1$ and $(i,j,k), (i',j',k') \in S \setminus C(D,Q)$, after an index interchange on the triplets (i,j,k) and (i',j',k'), we have $x' \in P_I^{C(D,Q)}$. Choose h and l such that $p+2 \leq h \leq n-1$, $l \neq k, l \neq n$ and $\max\{k,l\} \geq p+2$. Since $p+2 \leq h < n$, $j \neq n$ and $l \neq n$, by the lemma (case (ii)), there is an $x \in P_I^{C(D,Q)}$ such that $x_{nnn} = x_{hjl} = 1$. Performing a second index interchange on (n,n,n) and (h,j,l), we get x' which is still in $P_I^{C(D,Q)}$. By $\alpha x = \alpha x'$, we have

$$\alpha_{nnn} + \alpha_{hjl} = \alpha_{njn} + \alpha_{hnl}. \tag{8}$$

Similarly, since $p + 2 \leq h < n$, $j \neq n$, $l \neq k$ and $\max\{k,l\} \geq p+2$, by the lemma (case (ii)), there is an $x \in P_I^{C(D,Q)}$ such that $x_{njk} = x_{hnl} = 1$. Performing a second index interchange on (n,j,k) and (h,n,l), we get x' which is still in $P_I^{C(D,Q)}$. By $\alpha x = \alpha x'$, we have

$$\alpha_{njk} + \alpha_{hnl} = \alpha_{nnk} + \alpha_{hjl}. \tag{9}$$

Summing (8) and (9), we have

$$\alpha_{njk} = \alpha_{njn} + \alpha_{nnk} - \alpha_{nnn}. \tag{10}$$

This proves (6a) for $i = n$, $j \neq n$ and $k \neq n$. By symmetry, (6a) holds when one of i, j and k is n.

Suppose now that $i \neq n$, $j \neq n$, $k \neq n$ and $(i,j,k) \in S \setminus C(D,Q)$. By the lemma (case (i) if $(i,j,k) \in D \cup Q$, and case (ii) otherwise), we have an $x \in P_I^{C(D,Q)}$ such that $x_{nnn} = x_{ijk} = 1$. Performing a first index interchange on (n,n,n) and (i,j,k), we get $x' \in P_I^{C(D,Q)}$. Thus, $\alpha x = \alpha x'$. This yields

$$\alpha_{nnn} + \alpha_{ijk} = \alpha_{inn} + \alpha_{njk},$$

i.e.,

$$\alpha_{ijk} = \alpha_{inn} + \alpha_{njk} - \alpha_{nnn}. \tag{11}$$

Combining (11) and (10), we get (6a). This exhausts the cases $(i,j,k) \in S \backslash C(D,Q)$.

Next consider any $(i,j,k) \in C(D,Q)$. Define

$$\pi_{ijk} = \alpha_{ijk} - \lambda_i - \mu_j - \nu_k. \tag{12}$$

To prove the second case of (6), we have to show that all π_{ijk} are equal for $(i,j,k) \in C(D,Q)$.

Let $x \in P_I^{C(D,Q)}$ be such that $x_s = x_t = 1$, where $s = (i_s, j_s, k_s) \subseteq D, t = (i_t, j_t, k_t) \subseteq Q$. By the lemma (case (iii)), this can be realized. Define x' from x by a first index interchange on s and t. We have $x'_{s'} = x'_{t'} = 1, x'_s = x'_t = 0$, where $s' = (i_t, j_s, k_s), t' = (i_s, j_t, k_t)$. Thus, $s, t' \in C(D,Q)$, $s', t \notin C(D,Q)$. Hence, $x' \in P_I^{C(D,Q)}$ and $\alpha x = \alpha x'$. We obtain

$$\alpha_s + \alpha_t = \alpha_{s'} + \alpha_{t'}. \tag{13}$$

By (6),

$$\alpha_t = \lambda_{i_t} + \mu_{j_t} + \nu_{k_t}, \tag{14}$$

$$\alpha_{s'} = \lambda_{i_t} + \mu_{j_s} + \nu_{k_s}. \tag{15}$$

By (12),

$$\alpha_s = \pi_s + \lambda_{i_s} + \mu_{j_s} + \nu_{k_s}, \tag{16}$$

$$\alpha_{t'} = \pi_{t'} + \lambda_{i_s} + \mu_{j_t} + \nu_{k_t}. \tag{17}$$

Combining (13)- (17), we have $\pi_s = \pi_{t'}$, i.e.,

$$\pi_{i_s j_s k_s} = \pi_{i_s j_t k_t}.$$

By symmetry, we have

$$\pi_{i_s j_s k_s} = \pi_{i_s j_t k_t} = \pi_{i_t j_s k_t} = \pi_{i_t j_t k_s},$$

for any $i_s \in D_I, j_s \in D_J, k_s \in D_K, i_t \in Q_I, j_t \in Q_J, k_t \in Q_K$. Hence, all π_{ijk} are equal for $(i,j,k) \in C(D,Q)$. Let this number be π. Then the second case of (6) holds by (12).

Finally, let $x \in P_I^{C(D,Q)}$, $x \in \{0,1\}^{n^3}$. Sine $\alpha x = \alpha_0$, inequality (4) is satisfied as an equality and (6) holds, we obtain (7). This completes the proof of the theorem. ∎

We denote this class of facets of P_I by $\mathcal{P}$. Unlike facets in $\mathcal{Q}$, the facets in $\mathcal{P}$ are not of rank 1. Hence $\mathcal{P}$ and $\mathcal{Q}$ are disjoint.

Theorem 2.2. *Facets in $\mathcal{P}$ are of Chvátal rank 2.*

Proof: In the proof of Theorem 2.1, we have shown that (4) is of Chvátal rank at most 2. It suffices now to prove that it is not of rank 1. If (4) is of rank 1, then there exists ϵ, $0 < \epsilon < 1$, such that every point $x \in P$ satisfies

$$x(C(D,Q)) \leq p + 1 - \epsilon. \tag{4a}$$

We now show that there is $x \in P$ such that

$$x(C(D,Q)) = p+1, \tag{18}$$

which proves the theorem. We do this by induction on r. For $r=1$, without loss of generality, we may assume that $D_I = D_J = D_K = \{1,2,...,p\}$, $Q_I = Q_J = Q_K = \{p+1\}$. Let $x_{iii} = 1$ for $i = 1,2,...,p-1,p+2,...,n$. Let $x_{ppp} = x_{p,p+1,p+1} = x_{p+1,p,p+1} = x_{p+1,p+1,p} = \frac{1}{2}$, and $x_t = 0$ for all other t. Then $x \in P$ and (18) holds. Suppose that there is such an x for $r = 1,...,r_0$. For $r = r_0+1$, without loss of generality, assume that $|Q_K|$ is the smallest among $|Q_I|, |Q_J|$ and $|Q_K|$. Then $|Q_I|, |Q_J| \geq 2$. Choose $u = (i_u, j_u, k_u) \in S$ such that $i_u \in Q_I, j_u \in Q_J$, and $k_u \in D_K$. Delete i_u, j_u and k_u from I, J and K respectively. Then we have the case $r = r_0$. By our induction hypothesis, there is $\hat{x} \in P^{n-1}$, where P^{n-1} is $n-1$ order version of P, such that

$$\hat{x}(C(\hat{D},\hat{Q})) = p,$$

where $\hat{D}$ is obtained from D by deleting k_u from D_K, $\hat{Q}$ is obtained from Q by deleting i_u and j_u from Q_I and Q_J respectively. Let $x_u = 1$, $x_{ijk} = \hat{x}_{ijk}$ if $i \neq i_u, j \neq j_u$ and $k \neq k_u$, and $x_{ijk} = 0$ if either $i = i_u$ or $j = j_u$ or $k = k_u$ but $(i,j,k) \neq u$. Then $x \in P$ and (18) holds. This completes the proof. ∎

It turns out that different (D,Q) may give the same $C(D,Q)$:

Proposition 2.3. *Let $D, Q \subseteq R$ satisfy the hypothesis of Theorem 2.1. Moreover, let*

$$D_I = D_I^1 = Q_I^2 = Q_I^3, \; Q_I = Q_I^1 = D_I^2 = D_I^3,$$

$$D_J = Q_J^1 = D_J^2 = Q_J^3, \; Q_J = D_J^1 = Q_J^2 = D_J^3,$$

$$D_K = Q_K^1 = Q_K^2 = D_K^3, \; Q_K = D_K^1 = D_K^2 = Q_K^3.$$

Then $C(D,Q) = C(D^\nu, Q^\nu)$ for $\nu = 1,2,3$.

Proof: It suffices to show the case for $\nu = 1$. Then the other two cases hold by symmetry. First we have to show that D^1 and Q^1 satisfy the conditions for D and Q of Theorem 2.1, so that the set $C(D^1,Q^1)$ is well defined. Note

$$|Q^1| = |Q_I^1| + |Q_J^1| + |Q_K^1| = |Q_I| + |D_J| + |D_K| = |Q_I| - |Q_J| - |Q_K| + 2(p+1).$$

Thus $|Q^1|$ is odd. The other conditions on D and Q hold obviously. However, it is easy to see that $(i,j,k) \in C(D,Q)$ if and only if $(i,j,k) \in C(D^1,Q^1)$. This establishes the claim. ∎

Note that in the special case that $p = 1$, Theorem 2.1, Theorem 2.2 and Proposition 2.3 are in fact Theorem 3.7, Proposition 3.8 and Proposition 2.4 of [2] respectively.

Let the number of distinct inequalities (4) be $\kappa(\mathcal{P})$. Let $1 \leq p \leq n-3$, $M \subseteq R$, $|M_L| = p+1$ for $L = I,J,K$. Then, the number of distinct sets Q such that $Q \subseteq M$ and $|Q|$ is odd, is

$$\sigma_p \leq 2^{3(p+1)-1}.$$

Thus, by Proposition 2.3,

$$\kappa(\mathcal{P}) \le \frac{1}{4}\sum_{p=1}^{n-3}\binom{n}{p+1}^3 \sigma_p \le \sum_{p=1}^{n-3}\binom{n}{p+1}^3 2^{3p}.$$

3. Facets in $\mathcal{P}(2)$

Next we focus on the class $\mathcal{P}(2)$ and establish some properties that will enable us to give in Section 4 an efficient separation algorithm for this class.

From Theorem 2.1 and Proposition 2.3, a facet defining inequality in $\mathcal{P}(2)$ has the form

$$x(C(D,q)) \le 2, \tag{19}$$

where $D \subseteq R$, $q \in S$, $D \cap q = \emptyset$, $|D_I| = |D_J| = |D_K| = 2$,

$$C(D,q) = C_1(D) \cup C_2(D,q),$$

$$C_1(D) = \{t \in S : t \subseteq D\},$$

$$C_2(D,q) = \{t \in S : |t \cap D| = 1, |t \cap q| = 2\}.$$

Suppose that D and q are given as above and that $D_I = \{i_1, i_2\}$, $D_J = \{j_1, j_2\}$, $D_K = \{k_1, k_2\}$, $q = (i_q, j_q, k_q)$. Suppose that x is a given noninteger point in P and x violates (19), i.e.,

$$x(C(D,q)) > 2. \tag{20}$$

We also assume that x does not violate any clique facets of P_I, i.e., for any $s, t \in S$, $s \cap t = \emptyset$, the following inequalities hold:

$$x(S(s)) \le 1, \tag{21}$$

$$x(C(s,t)) \le 1, \tag{22}$$

For any index l in L, where L is one of I, J and K, use $S(l, L)$ to denote the collection of $t \in S$ whose element in L is l. Then

$$x(S(l,L)) = 1, \tag{23}$$

since $x \in P$.

Furthermore, for each l and L, define $S_x(l, L)$ to be the set that indexes the 18 largest components of x among those indexed by $S(l, L)$.

Let $C_x(D,q) = \{t \in C(D,q) : 1/10 < x_t < 1\}$.

The main results of this section is contained in the following theorem.

Theorem 3.1. *With the above definitions, there exists a triple $s, t, u \in S$ such that*

(a). $s \in C_1(D)$ and $1/8 < x_s < 1$;

(b). $t \in C(D,q)$, $t \cap s = \emptyset$ and

$$t \in \cup\{w : w \in C_x(D,q)\} \tag{24}$$

holds;

(c). $u \in C(D,q)$, $u \not\subseteq s \cup t$, *and* $u \in S_x(l,L)$ *for some pair* l, L *satisfying* $l \in (s \cup t) \cap L$.

Furthermore, if $|s \cup t \cup u| = 7$, *then*

(d). there exists $v \in C(D,q)$ *such that* $v \not\subseteq s \cup t \cup u$ *and* $v \in S_x(l,L)$ *for some pair* l, L *satisfying* $l \in (s \cup t) \cap L$.

To prove Theorem 3.1, we need several lemmas.

Lemma 3.2. *For any index* l *in* D_L, *where* $L = I, J$ *or* K, *there exists* $t \in C(D,q) \setminus S(l,L)$ *such that* $x_t > 1/9$.

Proof: By (20) and (23),

$$\sum \{x_t : t \in C(D,q) \setminus S(l,L)\} > 1.$$

But $|C(D,q) \setminus S(l,L)| = 9$. To see this, note that $|C_1(D)| = 8$, $|C_1(D) \cap S(l,L)| = 4$, $|C_2(D,q)| = 6$ and $|C_2(D,q) \cap S(l,L)| = 1$. The conclusion follows. ∎

Lemma 3.3. *For any* $s \in C(D,q)$, $x_s < 1$.

Proof: Since $x \in P$, $x_s \leq 1$. Assume that $x_s = 1$. Let

$$Z(s) = \{t \in S : t \cap s \neq \emptyset,\ t \neq s\}.$$

Then $x_t = 0$ for any $t \in Z(s)$.

(i). Assume that $s \in C_1(D)$. Without loss of generality, assume that $s = (i_1, j_1, k_1)$. Let $s' = (i_2, j_2, k_2)$. Then $s' = D \setminus s$. Hence

$$C(D,q) \subseteq \{s\} \cup Z(s) \cup C(s',q).$$

Thus,

$$x(C(D,q)) \leq x_s + x(C(s',q)) \leq 2.$$

where the second inequality is due to the assumption and (22). This contradicts (20).

(ii). Assume that $s \in C_2(D,q)$. Without loss of generality, assume that $s = (i_1, j_q, k_q)$. Then

$$C(D,q) \subseteq \{s\} \cup Z(s) \cup S(i_2, I).$$

Hence,

$$x(C(D,q)) \leq x_s + x(S(i_2, I)) \leq 2,$$

which also contradicts (20). This completes the proof. ∎

By Lemma 3.3, $C_x(D,q) = \{t \in C(D,q) : x_t > 1/10\}$.

Lemma 3.4. *(a). There exists* $s \in C_1(D)$ *such that* $x_s > 1/8$. *(b). There exists* $t \in C(D,q)$ *such that* $t \cap s = \emptyset$ *and (24) holds.*

Proof: By (21), $x(S(q)) \leq 1$. But, $C_1(D) = C(D,q) \setminus S(q)$. Thus, by (20), $x(C_1(D)) > 1$. Since $|C_1(D)| = 8$, there exists $s \in C_1(D)$ such that $x_s > 1/8$. This proves (a).

Without loss of generality, assume that $s = (i_1, j_1, k_1)$.

If there is $t \in C_x(D,q)$ such that $s \cap t = \emptyset$, then we have (b). Suppose that

$$s \cap t \neq \emptyset \qquad \text{for all } t \in C_x(D,q). \tag{25}$$

(i). Assume that $C_x(D,q) \cap C_2(D,q) \neq \emptyset$. Without loss of generality, assume that $s_1 = (i_1, j_q, k_q)$ and $s_1 \in C_x(D,q)$. By Lemma 3.2, there exists $s_2 \in C(D,q) \setminus S(i_1, I)$ such that $s_2 \in C_x(D,q)$. By (25),

$$s_2 \cap s \neq \emptyset, \quad s_2 \cap s_1 \neq \emptyset.$$

Without loss of generality, assume that $s_2 = (i_q, j_1, k_q)$. By (20) and (22),

$$x(C(D,q) \setminus C(s_1, q)) > 1.$$

Now $|C(D,q)| = 14$, $|C(s_1,q)| = 4$, $C(s_1,q) \subseteq C(D,q)$, hence $|C(D,q) \setminus C(s_1,q)| = 10$. Therefore, there exists $s_3 \in C_x(D,q) \setminus C(s_1,q)$. By (25),

$$s_3 \cap s \neq \emptyset, \quad s_3 \cap s_1 \neq \emptyset, \quad s_3 \cap s_2 \neq \emptyset.$$

Then $s_3 = (i_1, j_1, k_2)$. Since $s, s_1, s_2, s_3 \in C_x(D,q)$, (24) holds for $t = (i_q, j_q, k_2)$. We have (b).

(ii). Assume now that $C_x(D,q) \subseteq C_1(D)$. By Lemma 3.2, there exist $s_1 \in C_x(D,q) \setminus S(i_1, I)$, $s_2 \in C_x(D,q) \setminus S(j_1, J)$ and $s_3 \in C_x(D,q) \setminus S(k_1, K)$. Then $s_1, s_2, s_3 \in C_1(D)$, $i_2 \in s_1$, $j_2 \in s_2$, $k_2 \in s_3$. Hence, (24) holds for $t = (i_2, j_2, k_2)$. We also have (b).

This completes the proof of this lemma. ∎

Define $B(s,t) = \{w \in C(D,q) : w \subseteq (s \cup t)\}$.

Lemma 3.5. *Suppose that $s \in C_1(D)$, $t \in C(D,q)$ and $s \cap t = \emptyset$. If $t \in C_1(D)$, there exists $u \in C(D,q) \setminus B(s,t)$ such that*

$$x_u > (2 - x(B(s,t))/6; \tag{26}$$

if $t \in C_2(D,q)$, there exists $u \in C(D,q) \setminus B(s,t)$ such that

$$x_u > (2 - x(B(s,t))/10. \tag{27}$$

Furthermore, in both cases, we have

$$2 - x(B(s,t)) \geq 1 - x(B(s,t) \cap S(l, L)), \tag{28}$$

for some pair l, L satisfying $l \in (s \cup t) \cap u \cap L$.

Proof: If $t \in C_1(D)$, $|C(D,q) \setminus B(s,t)| = |C_2(D,q)| = 6$. If $t \in C_2(D,q)$, $|C(D,q) \setminus B(s,t)| = 10$. Since x violates (19), in the first case there exists $u \in$

$C(D,q)\setminus B(s,t)$ such that x_u satisfies (26), while in the second case there exists $u \in C(D,q)\setminus B(s,t)$ such that x_u satisfies (27). Notice that in both cases $u \cap (s \cup t) \neq \emptyset$.

We now discuss (28). Since $u \cap (s \cup t) \neq \emptyset$, it suffices to prove (28) for any index $l \in (s \cup t) \cap L$, $L = I, J, K$. Assume that $t \in C_1(D)$. Then $B(s,t) = C_1(D)$ and $C(D,q) \setminus B(s,t) = C_2(D,q)$. Thus,

$$x(C_1(D)) = x(C_1(D) \cap S(i_1, I)) + x(C_1(D) \cap S(i_2, I)) \leq 2,$$

where the last inequality follows from (23). Hence,

$$2 - x(C_1(D)) = (1 - x(C_1(D) \cap S(i_1, I))) + (1 - x(C_1(D) \cap S(i_2, I))).$$

By (23), each term in the brackets on the right-hand-side of the above equality is nonnegative. Thus, (28) holds for $l = i_1$ or i_2 in this case. By symmetry, (28) also holds for $l = j_1$ or j_2 or k_1 or k_2. The case that $t \in C_2(D,q)$ is similar. ∎

In Lemma 3.5, if $t \in C_1(D)$, then $|s \cup t \cup u| = 8$. There is only one index in (D,q) which is not in $s \cup t \cup u$. If $t \in C_2(D,q)$, then $|s \cup t \cup u| = 8$ or 7. We discuss the last case in the following lemma.

Define $B(s,t,u) = \{w \in C(D,q) : w \subseteq (s \cup t \cup u)\}$.

Lemma 3.6. *Let s, t and u be as in Lemma 3.5. Suppose that $t \in C_2(D,q)$ and $|s \cup t \cup u| = 7$. Then, there exists $v \in C(D,q) \setminus B(s,t,u)$ such that*

$$x_v > (2 - x(B(s,t,u))/8. \tag{29}$$

Furthermore,

$$2 - x(B(s,t,u)) \geq 1 - x(B(s,t,u) \cap S(l, L)), \tag{30}$$

for some pair l, L satisfying $l \in (s \cup t) \cap v \cap L$.

Proof: Without loss of generality, assume that $s = (i_1, j_1, k_1)$, $t = (i_2, j_q, k_q)$. Since $|C(D,q)\setminus B(s,t,u)| \leq 8$ and x violates (19), there exists $v \in C(D,q)\setminus B(s,t,u)$ such that (29) holds.

Let l_u be the unique index which is in u but not in $s \cup t$. There are two possibilities for l_u: (i). it is in D; (ii). it is in q. We now discuss (30) in cases (i) and (ii) separately.

(i). In this case $l_u = j_2$ or k_2. Assume l_u is j_2. The case when l_u is k_2 follows by symmetry. Then

$$x(B(s,t,u)) = x(B(s,t,u) \cap S(i_1, I)) + x(B(s,t,u) \cap S(i_2, I)) \leq 2.$$

$$x(B(s,t,u)) = x(B(s,t,u) \cap S(k_1, K)) + x(B(s,t,u) \cap S(j_q, J)) \leq 2.$$

$$x(B(s,t,u)) = x(B(s,t,u) \cap S(k_1, K)) + x(B(s,t,u) \cap S(k_q, K)) \leq 2.$$

The last inequality in each of the above three formulas follows from (23). By an argument similar to that in the proof of Lemma 3.5, we see that (30) holds for $l = i_1, i_2, k_1, j_q, k_q$. These five indices are in $s \cup t$. Since $v \in C(D,q)$, at least one of these five indices is in v. Thus, (30) holds for some pair l, L satisfying $l \in (s \cup t) \cap v \cap L$.

(ii). In this case $l_u = i_q$ and we have

$$x(B(s,t,u)) = x(B(s,t,u) \cap S(j_1, K)) + x(B(s,t,u) \cap S(j_q, J)) \leq 2.$$

$$x(B(s,t,u)) = x(B(s,t,u) \cap S(k_1, K)) + x(B(s,t,u) \cap S(k_q, K)) \leq 2.$$

$$x(B(s,t,u)) = x(B(s,t,u) \cap S(i_2, I)) + x(B(s,t,u) \cap C(s,q)) \leq 2.$$

$$x(B(s,t,u)) = x(B(s,t,u) \cap S(i_1, I)) + x(B(s,t,u) \cap C(s',q)) \leq 2,$$

where $s' = (i_2, j_1, k_1)$. The last inequality in each of the first two formulas follows from (23). The last inequality in each of the next two formulas follows from (22) and (23). Then (30) holds for all indices in $s \cup t$. Since $v \in C(D,q)$, $v \cap (s \cup t) \neq \emptyset$. Thus, (30) holds for some pair l, L satisfying $l \in (s \cup t) \cap v \cap L$. ∎

Proof of Theorem 3.1: Combining Lemmas 3.3 and 3.4, we have conclusions (a) and (b). In Lemma 3.5, notice that in both cases we have $u \not\subseteq s \cup t$ and $u \in C(D,q)$. By (26), (27) and (28),

$$x_u > (1 - x(B(s,t) \cap S(l,L)))/10,$$

for some pair l, L satisfying $l \in (s \cup t) \cap u \cap L$. However, $u \in S(l,L) \setminus B(s,t)$. By (23), the number of elements in $S(l,L) \setminus B(s,t)$, which satisfy the above inequality, is no more than 10 for each l. Notice $|B(s,t)| \leq 8$. Thus, x_u must be among the 18 largest members of $\{x_w : w \in S(l,L)\}$, i.e., $u \in S_x(l,L)$. This proves conclusion (c). Similarly, in Lemma 3.6, in both cases we have $v \not\subseteq s \cup t \cup u$ and $v \in C(D,q)$. Besides, there always exists an $l \in v \cap (s \cup t)$ such that (30) holds. By (29) and (30),

$$x_v > (1 - x(B(s,t,u) \cap S(l,L)))/8.$$

Similarly, we have $v \in S(l,L) \setminus B(s,t)$. By (23), the number of elements in $S(l,L) \setminus B(s,t,u)$, which satisfy the above inequality, is no more than 8 for each l. Notice $|B(s,t,u)| \leq 6$. Thus, x_v must be among the 14 largest members of $\{x_w : w \in S(l,L)\}$, i.e., $v \in S_x(l,L)$. This proves conclusion (d). ∎

4. A Linear-Time Separation Algorithm for $\mathcal{P}(2)$

From Theorem 3.1, a linear-time separation algorithm for $\mathcal{P}(2)$ follows in a straightforward manner.

Suppose that x is a noninteger point in P. We may use the $O(n^3)$ algorithm in [1] to detect whether there is any clique facet violated by x. Hence, we may assume that x does not violate any clique facets of P_I, i.e., (21) and (22) hold.

We now describe the algorithm.

Algorithm 4.1. Suppose that x is a noninteger point in P. Suppose that (21) and (22) hold for all $s, t \in S$ such that $s \cap t = \emptyset$.

Step 1. Form $S_x = \{s \in S : 1/10 < x_s < 1\}$. For each index $l \in L$, where $L = I, J$ or K, form $S_x(l,L)$ as the set that indexes the 18 largest components of x among those indexed by $S(l,L)$.

Step 2. If there exists a pair $s, t \in S_x$ satisfying $s \cap t = \emptyset$ and $x_s > 1/8$, go to Step 4.

Step 3. For each $s \in S_x$, let

$$L_s = (\cup\{u \in S_x : u \cap s \neq \emptyset\}) \setminus s.$$

If there exists a pair (s,t) such that $s \in S_x$, $t \in S$ and $t \subseteq L_s$, go to Step 4.

Step 4. For each pair s,t in Steps 2 and 3, let $G_1 = s \cup t$. If there exists a $u \in S_x(l, L)$ such that $l \in G_1$ and $u \not\subseteq G_1$, let $G_2 = G_1 \cup u$.

Step 5. For each possible triple s, t, u in Step 3, if $|G_2| = 8$, let $G_3 = G_2$ and go to Step 6. Otherwise, for each $l \in G_1$, if there exists a $v \in S_x(l, L)$ such that $v \not\subseteq G_2$ and $|(G_2 \cup v)_{L'}| \leq 3$ for $L' = I, J, K$, let $G_3 = G_2 \cup v$. If $|G_3| = 9$, let $G = G_3$ and go to Step 7. Otherwise, go to Step 6.

Step 6. For each G_3 and each index $l \in L' \setminus (G_3)_{L'}$ satisfying $|(G_3)_{L'}| = 2$, add l to G_3 and call it G now.

Step 7. For each G, let $D = s \cup t$ and $q = G \setminus D$. Check whether (19) is violated or not. If (19) is violated, stop.

Step 8. For each G and each $q \subset G \setminus s$ satisfying $q \in S$ and $|q \cup t| = 1$, let $D = G \setminus q$ and check whether (19) is violated or not. If (19) is violated, stop.

If all pairs (s,t) in Step 2 have been considered, go to Step 3. After all pairs (s,t) in Step 3 have been considered, stop. Similarly, in Steps 4 and 5, each possible u and v should be considered to form G such that (19) is checked in Steps 7 and 8. ∎

Theorem 4.2. *Algorithm 4.1 determines in $O(n^3)$ steps whether a given $x \in P$ violates a facet defining inequality (19).*

Proof: By Theorem 3.1, Algorithm 4.1 checks all possible situations where (19) may be violated. We now discuss the complexity of the procedure. Clearly, the complexity of Step 1 is $O(n^3)$. Since

$$x(S) = \sum\{x(S(i, I) : i \in I\} = n,$$

the number of $s \in S$ such that $x_s > 1/\lambda$ is no more than λn for any $\lambda > 0$. Thus, $|S_x| \leq 10n$ and the number of candidate pairs of s and t in Step 2 is no more than $80n^2$. By (23), for each s in Step 3, $|S_s| \leq 30$. Hence, the number of candidate pairs of s and t in Step 3 is no more than $240n$. Since $|S_x(l, L)| = 18$ for each l, the number of sets G_2 and G_3 in Steps 4 and 5 is $O(n^2)$. In Step 6, for each G_3, there are $n - 2$ indices which can be chosen to be added to G_3. So the number of sets G in Step 6 is $O(n^3)$. For each G, there are one way to form (D, q) in Step 7 and three ways to form (D, q) in Step 8. The testing time for (19) is $O(1)$. Therefore, the overall complexity of the algorithm is $O(n^3)$. This completes the proof. ∎

Remark 4.3. There are some ways to reduce the coefficient function of the complexity of this algorithm, but the order of complexity is already best possible. ∎

5. Facet Classes with Linear-Time Separation Algorithms

By [1] and this paper, four facet classes of the three index-assignment polytope, namely, $\mathcal{P}(1)$, $\mathcal{Q}(1)$, $\mathcal{P}(2)$ and $\mathcal{Q}(2)$, have linear-time separation algorithms. It seems that the complexity of separation algorithms for $\mathcal{P}(p)$ and $\mathcal{Q}(p)$ where $p \geq 3$ will be higher than $O(n^{p+1})$.

In [12], Gwan and Qi gave a linear-time separation algorithm for the facet class defined by

$$2x_s + x(F(Q,s)) \leq 2, \tag{31}$$

where $s \in S, s \subseteq Q, |Q| = 5, |Q_I| \leq 2$ for $L = i, J, K$, and

$$F(Q,s) = \{t \in S : |t \cap Q| \geq 2, 2 \geq |t \cap s| \geq 1\}.$$

This facet class was first identified in [2]. But it was in [12] that it was recognized that its Chvátal rank is 1. In fact, the inequality (31) can be obtained by multiplying the three equations of $Ax = e$ indexed by the indices of s by $\frac{2}{3}$, and multiplying the two equations of $Ax = e$ indexed by indices in $Q \setminus s$ by $\frac{1}{3}$, and adding them together and rounding down all coefficients to their nearest integers. Hence the Chvátal rank of (31) is at most 1. See [12] for details.

Therefore, we now know five linear-time separable facet classes of P_I. Two of them are clique facets for which the right-hand sides of the facet-defining inequalities are 1. The right-hand sides of the facet-defining inequalities for the other three linear-time separable facet classes are 2. Are the right-hand sides of the facet-defining inequalities for all the linear-time separable facet classes 1 or 2? Do all the facet classes defined by inequalities with right-hand sides 2 have linear-time separation algorithms? In [12], another facet class, called two-tooth comb facet class, has been identified. The right-hand sides of the corresponding facet-defining inequalities are also 2. Their Chvátal rank is 2. Are there other facet classes such that the right-hand sides of their defining inequalities are 2 and their Chvátal rank is 1 or 2? Are there other facet classes such that the right-hand sides of their defining inequalities are 2 and their Chvátal rank is more than 2? These are some of the questions left open at this point.

6. An Algorithm Using a Polyhedral Tightening Procedure.

In this section we describe an algorithm for the three-index assignment problem, based on a procedure for tightening the linear programming relaxation of the problem. The procedure uses the five known linear-time separation algorithms to generate cutting planes, namely facet defining inequalities in the above five classes, that are violated by the current optimal solution. If at some point the solution becomes integer, it is optimal. Otherwise, when no more violated inequalities can be found in either of the five classes, we solve by branch and bound the resulting problem, whose linear constraint set has been tightened by the cut generating procedure.

The resulting algorithm for the three-index assignment problem can be stated as follows.

Let (LP) be the linear programming relaxation of the three-index assignment problem, i.e. the problem obtained from (1) by replacing the condition $x_{ij} \in \{0,1\}$ with $x_{ij} \geq 0$.

Step 1. Solve (LP). Let $\bar{x}$ be the optimal solution found.

If $\bar{x} \in \{0,1\}^{n^3}$, stop: $\bar{x}$ is optimal.

Step 2. Apply the separation algorithm for clique facets of the class $\mathcal{P}(1)$.

If a violated inequality is found, add it to the constraint set of (LP) and go to Step 1.

Steps 3, 4, 5, 6 are analogous to Step 2, with facet class $\mathcal{P}(1)$ replaced by the classes $\mathcal{Q}(1)$, $\mathcal{Q}(2)$, $\mathcal{P}(2)$, and (31).

Step 7. Use a branch and bound algorithm to solve the current (LP).

The practical usefulness of an algorithm of this type hinges on the question whether the computational effort spent on tightening the linear constraint set is less than the effort that it saves by making the branch and bound procedure more efficient on the tightened problem. In the next section we bring ample computational evidence to the effect that using the tightening procedure before applying branch and bound does indeed pay off. We also show that the new procedure compares favorably with earlier procedures and thus advances the state of the art.

The more ambitious task of developing a branch and cut algorithm, in which cutting planes are generated at different nodes of the search tree rather than just at the root node, is left for a future project.

7. Computational Results

To implement our algorithm, we used two routines of the FORTRAN Library developed by NAG (Numerical Algorithm Group, Ltd.) available at the University of New South Wales. Namely, the simplex routine E04MBF was used to solve the linear programs of Step 1, and the branch and bound code H02BBF was used in Step 7. The overall program was written in FORTRAN and the resulting code was run first on the VAX 6000-30 multiprocessor, then on the Sequent Hydra multiprocessor (with DYNIX). The programming attempted to make use of the parallelism made possible by the multiprocessors. Since memory allocation turned out to be a problem, a memory scheduling software package of the information service department of Pioneer International Ltd., Australia, was used for the runs on the Hydra.

A first set of three-index assignment problems, with n set equal to every even number between 4 and 26, was generated in the way described in [3]. In other words, the cost coefficients c_{ij} were randomly drawn from a uniform distribution of the integers between 0 and 100. The resulting 12 problems were run on the VAX 6000-30 with the outcome shown in Table 1.

Here the column giving the number of iterations also gives the number of cuts generated, since every iteration of Steps 1-6 ends either by finding a facet defining inequality violated by the current solution, or by terminating the cut generation and going to the final branch and bound step. The small number of search tree nodes needed by the branch and bound procedure to solve the problem after the addition of the cuts, as reflected in the entries of the last column, attests to the efficiency of the tightening procedure.

TABLE I

Problem	n	CPU times (seconds)	Iterations of Steps 1-6	Nodes of the branch and bound tree
1	4	0.05	3	0*
2	6	0.25	5	0*
3	8	0.68	10	9
4	10	1.33	7	11
5	12	2.23	15	5
6	14	7.19	22	3
7	16	27.68	42	3
8	18	53.67	52	13
9	20	87.22	57	0*
10	22	132.76	74	7
11	24	286.46	179	11
12	26	449.32	357	5

*The branch and bound code was not used.

Subsequently, 9 additional problems were generated for each class, plus for the class $n = 28$. Because of the above mentioned difficulties with memory allocation, these subsequent tests were run on a different computer, namely a Sequent Hydra (with DYNIX). Also, in the new computational environment, a new random number generator was used. However, the cost coefficients were still drawn from the same interval i.e. from the integers between 0 and 100 Table 2 compares the computing times needed by our algorithm with those needed by the branch and bound routine when run without the tightening procedure. Each entry of the table represents the average time for 9 problems.

The discrepancy between the computing times reported in Table 1 and Table 2 is due to differences in the computing environment. The effect of the tightening procedure on the overall computational effort is quite dramatic.

Finally, Table 3 provides a comparison of our computing effort to that involved in two earlier procedures, those of Balas and Saltzman [3] and Hansen and Kaufman [13], respectively. Since our results were obtained on a Sequent Hydra, whereas the other two research groups used a VAX 8650 and a CDC 6400, respectively, the computing times are hard to compare. However, the increase in computing time from one problem class to the next, given in the form of the ratios shown in columns 4, 6 and 8, gives a good idea of the efficiency of the procedures. Under this criterion, the performance of our procedure is comparable to that of Balas and Saltzman [3]. This suggests that a branch and cut approach, that would generate cuts not only at the root node, might be superior both to our current approach and to earlier algorithms.

TABLE II

Problem class	n	CPU time (seconds) Algorithm of Section 6	Branch and Bound Without Tightening
1	4	0.05	0.3
2	6	0.28	0.77
3	8	0.88	10.69
4	10	1.34	70.01
5	12	2.30	400.05
6	14	10.41	612.02
7	16	37.54	765.13
8	18	61.70	988.83
9	20	164.01	1262.07
10	22	302.03	4324.27
11	24	469.69	8534.17
12	26	591.37	37335.38
13	28	1966.86	unfinished

TABLE III

Problem	n	Algorithm of Section 6		Balas-Saltzman		Hansen-Kaufman	
		CPU seconds (Sequent Hydra)	CPU ratio* n versus $n-2$	CPU seconds (VAX 8650)	CPU ratio* n versus $n-2$	CPU seconds (CDC 6400)	CPU ratio* n versus $n-2$
1	4	0.05	–	0.03	–	0.59	–
2	6	0.28	5.6	0.16	5.3	2.75	4.7
3	8	0.88	3.1	0.84	5.3	10.59	3.9
4	10	1.34	1.5	1.36	1.6	60.27	5.7
5	12	2.30	1.7	2.19	1.6	359.77	5.9
6	14	10.41	4.5	11.95	5.5	–	–
7	16	37.54	3.6	39.80	3.3	674.69	–
8	18	61.70	1.6	55.30	1.4	–	–
9	20	164.01	2.6	169.29	3.1	–	–
10	22	302.03	1.8	371.52	2.2	–	–
11	24	469.69	1.5	514.52	1.3	–	–
12	26	591.37	1.2	624.00	1.2	–	–
13	28	1966.86	3.3	–	–	–	–

*Computing time for n divided by computing time for $n-2$.

Acknowledgements

The first author's work is supported by the Australian Research Council. The second author's work is supported by the National Science Foundation through Grant No.

DDM-8901495 and the Office of Naval Research through Contract N00014-85-K-0198. The third author's work is a part of her Ph.D. thesis [11].

References

[1] E. Balas and L. Qi, "Linear-time separation algorithms for the three-index assignment polytope", *Discrete Applied Math.* 43 (1993) 1-12.

[2] E. Balas and M. J. Saltzman, "Facets of the three-index assignment polytope", *Discrete Applied Math.* 23 (1989) 201-229.

[3] E. Balas and M. J. Saltzman, "An algorithm for the three-index assignment problem", *Oper. Res.* 39 (1991) 150-161.

[4] R. Burkard and K. Fröhlich, "Some remarks on three-dimensional assignment problem", *Method of Oper. Res.* 36 (1980) 31-36.

[5] R. Burkard and R. Rudolf, "Computational investigations on 3-dimensional axial assignment problems", to appear in: *Belgian Journal of Operations Research, Statistics and Computer Science.*

[6] R. Burkard, R. Rudolf and G. Woeginer, "Three-dimensional axial assignment problems with decomposable cost-coefficients", Report 238, Technische Universität Graz, Austria, 1992.

[7] R. Euler, "Odd cycles and a class of facets of the axial 3-index assignment polytope", *Applicationes Mathematicae (Zastosowania Matematyki)* XIX (1987) 375-386.

[8] A. M. Frieze, "Complexity of a 3-dimensional assignment problem", *European J. Oper. Res.* 13 (1983) 161-164.

[9] A. M. Frieze and J. Yadegar, "An algorithm for solving 3-dimensional assignment problem with application to scheduling a teaching practice", *J. Oper. Res. Soc.* 32 (1981) 989-995.

[10] K. Fröhlich, "Dreidimensionale Zuordnungsprobleme", Masters Thesis, Math. Institut, Universität Köln, 1979.

[11] G. Gwan, "A polyhedral method for the three index assignment problem", Ph.D. Thesis, School of Mathematics., University of New South Wales, Australia, 1993.

[12] G. Gwan and L. Qi, "On facet of the three index assignment polytope", *Australasian Journal of Combinatorics* 6 (1992) 67-87.

[13] P. Hansen and L. Kaufman, "A primal-dual algorithm for the three-dimensional assignment problem", *Cahiers du CERO* 15 (1973) 327-336.

[14] R. Karp, "Reducibility among combinatorial problems", *Proc. Complexity of Computer Computations*, eds., by R.E. Miller and J.W. Thatcher, (Plenum Press, 1972) pp.85-104.

[15] O. Leue, "Methoden zur Lösung dreidimensionaler Zuordnungsprobleme", *Angew. Inform.* (1972) 154-162.

[16] W. P. Pierskalla, "The tri-substitution method for the three-dimensional assignment problem", *CORS J.* 5 (1967) 71-81.

[17] W. P. Pierskalla, "The multidimensional assignment problem", *Oper. Res.* 16 (1968) 422-431.

[18] L. Qi and E. Balas, "A new class of facet-defining inequalities for the three-index assignment polytope", Management Science Research Report 563, Graduate School of Industrial Administration, Carnegie Mellon University, Pittsburgh, PA, 1990.

[19] R. Rudolf, "Dreidimensionale axiale Zuordnungsprobleme", Masters Thesis, Technische Universität Graz, Austria, 1991.

[20] M. Vlach, "Branch and bound method for the three index assignment problem", *Ekonomicko-Matematicky Obzor* 2 (1967) 181-191.

A FINITE SIMPLEX-ACTIVE-SET METHOD FOR MONOTROPIC PIECEWISE QUADRATIC PROGRAMMING *

R. T. ROCKAFELLAR
Department of Applied Mathematics, University of Washington, Seattle, WA 98195, USA

and

J. SUN
Department of Decision Sciences, National University of Singapore, Singapore 0511, Republic of Singapore

Abstract. An algorithm for linearly constrained convex programs with separable piecewise quadratic costs is developed. It can be used to solve extended linear-quadratic programs of box-diagonal form. The algorithm converges to an exact solution in finitely many iterations and can be thought of as an extension of the simplex method for convex programming and the active set method for quadratic programming. Pivoting operations and line searches are used to reduce the size of quadratic subproblems and to take advantage of the case where the cost function has some linear pieces. Computational results are reported which show that solving such a piecewise quadratic program is not much harder than solving a quadratic program of the same number of variables.

Key words: Active Set Methods, Monotropic Programming, Nonsmooth Optimization, Piecewise Quadratic Programming, Simplex Methods

1. Introduction

This paper presents an algorithm for monotropic piecewise quadratic programs (MPQP for short) which often arise from certain nonsmooth optimization problems in control and decision sciences.

By MPQP we mean the following mathematical problem

$$(\text{MPQP}) \begin{cases} \text{minimize} & F(x) = \sum_{j=1}^{n} f_j(x_j) \\ \text{subject to} & Ax = b \\ & c^- \le x \le c^+ \end{cases}$$

where the vector inequality is understood coordinatewise. Note that we have $A \in R^{m \times n}, b \in R^m, c^-, c^+$ *and* $x \in R^n$ with the exceptions that some components of c^- could be $-\infty$ and that some components of c^+ could be $+\infty$, meaning that there are actually no lower and upper bounds for the corresponding components of x. Each f_j is a convex piecewise quadratic function of a single variable x_j, i.e.

* This research is supported in part by the Air Force Office of Scientific Research and the National Science Foundation.

Ding-Zhu Du and Jie Sun (eds.), Advances in Optimization and Approximation, 275–292.

$$f_j(x_j) = \begin{cases} +\infty, & \text{if } x_j < c_j^- := c_{j0} \\ \frac{1}{2}p_{j1}x_j^2 + q_{j1}x_j + r_{j1}, & \text{if } c_{j0} \leq x_j \leq c_{j1} \\ \cdots & \\ \frac{1}{2}p_{jk_j}x_j^2 + q_{jk_j}x_j + r_{jk_j}, & \text{if } c_{jk_j-1} \leq x_j \leq c_{jk_j} \\ +\infty, & \text{if } x_j > c_j^+ := c_{jk_j} \end{cases}$$

where for each j, we call the numbers $c_{j0}, c_{j1}, \ldots, c_{jk_j}$ the *breakpoints* of f_j; they satisfy

$$-\infty \leq c_{j0} < c_{j1} < \ldots < c_{jk_j} \leq +\infty.$$

The interest of studying problem (MPQP) comes primarily from a new area of optimization called extended linear-quadratic programming (ELQP for short). Recently it has been found [5][6] that the following ELQP model is especially useful in solving certain large-scale decision-making problems of deterministic or stochastic type

$$\text{(ELQP)} \begin{cases} \text{minimize} & c^T x + \frac{1}{2}x^T P x + \sup_{y \in Y}\{(b - Ax)^T y - \frac{1}{2}y^T Q y\} \\ \text{subject to} & x \in X, \end{cases}$$

where $X \subset R^n, Y \subset R^m$ are convex polyhedra, $x, c \in R^n$, $y, b \in R^m$, $A \in R^{m \times n}$, $P \in R^{n \times n}, Q \in R^{m \times m}$, and the matrices P and Q are positive semidefinite.

It has been shown in [5] that, if P and Q are diagonal and if X and Y are boxes (hyperrectangles), then problem (ELQP) and its dual are MPQP. In addition, the expressions of $f_j(x_j)$ may be linear instead of quadratic between some of its breakpoints.

As another example of large-scale MPQP, let us consider a stochastic transportation problem in which a shipment vector x and a demand vector z must be chosen optimally with respect to present costs and constraints as well as certain expected penalties due to the discrepancies between z and a random demand w in the future. The cost function to be minimized subject to flow constraints is

$$\begin{aligned} \phi(x, z) &= \sum_{j=1}^{n} t_j(x_j) + \Big\{\sum_{i=1}^{m} \mathcal{E}_{w_i} F_i(z_i, w_i)\Big\} \\ &= \sum_{j=1}^{n} t_j(x_j) + \sum_{i=1}^{m} u_i(z_i), \end{aligned}$$

where $\mathcal{E}_{w_i} F_i(z_i, w_i)$ stands for the expected penalty with respect to the random demand w_i at the demand node i. It can be shown that, if $F_i(z_i, w_i)$ is piecewise linear in $z_i - w_i$ and if the distribution of w_i is of "histogram type" (both assumptions are fairly natural), then the function $u_i(z_i)$ is convex piecewise quadratic. Moreover, the number of breakpoints of $u_i(z_i)$ could be very large. For a relatively detailed formulation of this problem, see [10].

In addition to those direct applications, MPQP may appear as subproblems in aggregation algorithms for ordinary linear-quadratic programming [7][9] and in approximate methods for monotropic programming (i.e. linearly constrained convex separable programming), etc. It is worth mentioning that in general the dual of convex quadratic programming can be reduced to this form [4].

Notice that the (effective) domain of the objective function in problem (MPQP) is a box in R^n and is a union of finitely many smaller boxes on each of which the objective function is given by a quadratic (or linear) formula. This fact makes function $F(x)$ a special case of the so-called convex piecewise quadratic function [8]. Since $f_j(x_j)$ may not be differentiable at its breakpoints, MPQP can be generally categorized as a nonsmooth optimization problem. An intuitive idea would be to find out suboptima in all such small boxes, which corresponds to solving a series of quadratic programs, and then to choose the optimal solution among them. This is unrealistic because of the possible huge number of these small boxes in many practical situations.

The algorithm to be introduced below is essentially a simplex method. Therefore, many extensively studied techniques in numerical linear algebra, such as those used in MINOS [3], could be adopted to improve its efficiency and robustness. However, for clear statements of the basic ideas, we will use simplex tableaux for elaboration. Since simplex methods generally do not ensure finite termination, also for the sake of accelerating local convergence, we combine a Newton step restricted on certain subspaces with the simplex method. Various techniques such as pivoting and line searches are used to reduce the size of quadratic subproblems and to take advantage if some f_j's in the objective function are piecewise linear. As a result, the algorithm effectively converges to a solution in finitely many iterations. In the following, the framework of the algorithms is presented in Section 2; its implementation is discussed in Section 3; the degeneracy processing is the topic of Section 4 and some computational results are reported in Section 5.

2. Framework of a Finitely Convergent Algorithm

In this section we introduce a model algorithm and clarify its convergence.

Definition (2.1). Let $\bar{x}$ be a feasible solution of problem (MPQP). Denote by $I(\bar{x})$ the index set

$$\{i \in \{1, 2, \cdots, n\} \mid \bar{x}_i \text{ is a breakpoint of } f_i\}.$$

Then x is called a *quasi-optimal solution* to (MPQP) if it is the optimal solution to the following subproblem:

$$(\mathrm{SP1}) \begin{cases} \text{minimize} & F(x) \\ \text{subject to} & Ax = b \\ & c^- \leq x \leq c^+ \\ & x_j = \bar{x}_j \quad \forall\, j \in I(\bar{x}) \text{ for some } \bar{x}. \end{cases}$$

If x is a quasi-optimal solution, then $F(x)$ is called a *quasi-optimal value*.

According to this definition, the optimal solutions to MPQP must be quasi-optimal. In addition there is only a finite number of quasi-optimal values. To see this point, notice that the total number of different sets $I(\bar{x})$ is finite as $\bar{x}$ ranges over all feasible solutions to the given problems. Hence there is only a finite number of different subproblems (SP1). In addition, if we can strictly descend from one quasi-optimal solution to another, then after finitely many steps we will end up with an optimal solution.

The framework of this type of algorithms consists of three steps.

Algorithm (2.2).

Step 0. Find a feasible solution or determine the infeasibility of problem (MPQP).

Step 1. Move from a feasible solution to a quasi-optimal solution without increasing the objective and go to Step 2.

Step 2. Verify whether this quasi-optimal solution is optimal. If not, find a descent direction , decrease the objective function and go to Step 1.

If each step takes finitely many operations, then any algorithm which can be embedded in this framework will have to converge in finitely many iterations.

3. Implementation

In this section we consider the question of how to realize the model algorithm.

3.1. Initialization

Step 0 can be accomplished by any linear programming algorithm if we replace the objective function by zero. Since some algorithms for piecewise linear programming have been implemented (e.g. [1][10]), a better idea is to use a piecewise linearized approximation of the objective function in order to get a starting point closer to the optimum of (MPQP). Actually in certain case a quantitative estimate can be made about the distance between the optimal solutions of the piecewise linear approximation and the optimal solutions of problem (MPQP), as indicated by the following proposition.

Proposition (3.1.1). Suppose that $c^- > -\infty$ and $c^+ < \infty$ for all j. Let $\bar{f}_j(x_j)(j = 1, ..., n)$ be defined as

$$\bar{f}_j(x_j) = \begin{cases} +\infty, & \text{if } x_j < c_j^- = g_{j0}; \\ f_j(g_{j0}) + f_j'(\frac{g_{j0}+g_{j1}}{2})(x_j - g_{j0}), & \text{if } g_{j0} \le x_j \le g_{j1}; \\ \cdots & \\ f_j(g_{j,m_j-1}) + f_j'(\frac{g_{j,m_j-1}+g_{jm_j}}{2})(x_j - g_{j,m_j-1}), & \text{if } g_{j,m_j-1} \le x_j \le g_{jm_j} \\ +\infty, & \text{if } x_j > c_j^+ = g_{jm_j}. \end{cases}$$

This is a piecewise linear approximation of $f_j(x_j)$. Suppose that the function $f_j(x_j)$ can be expressed as $\frac{1}{2}p_{ji}x_j^2 + q_{ji}x_j + r_{ji}(p_{ji} > 0)$ between the grid points $g_{j,i-1}$ and g_{ji} for $i = 1, ..., m_j$ and $j = 1, ..., n$. Then the distance between the optimal solution x^* of $\bar{F}(x) = \sum_{j=1}^n \bar{f}_j(x_j)$ and the optimal solution x^{**} of $F(x) = \sum_{j=1}^n f_j(x_j)$ on the set $S = \{x \in R^n | Ax = b, c^- \le x \le c^+\}$ satisfies $||x^* - x^{**}|| \le Md$, where $|| \cdot ||$ is the Euclidean norm of $R^n, d = \max\{g_{ji} - g_{j,i-1} | i = 1, ..., m_j, j = 1, ..., n\}$, and M is a constant independent from d.

Proof. Since S is a compact set, both x^* and x^{**} exist. Let

$$\lambda = \min\{p_{ji} | i = 1, ..., m_j, j = 1, ..., n\}.$$

By the assumption that f_j is strictly quadratic we have $\lambda > 0$. It is not hard to see that geometrically the graph of $\bar{f}_j$ is the brokenline connecting $(g_{j0}, f_j(g_{j0})), (g_{j1}, f_j(g_{j1}))$, $\cdots$, and $(g_{jm}, f_j(g_{jm}))$. Thus we have

$$\begin{aligned} &\bar{F}(x^*) - F(x^{**}) \geq F(x^*) - F(x^{**}) \\ = &\sum_{j=1}^{n}(f_j(x_j^*) - f_j(x_j^{**})) \geq \sum_{j=1}^{n} f_j'(x_j^{**}, x_j^* - x_j^{**}) + \frac{1}{2}\lambda \sum_{j=1}^{n}(x_j^* - x_j^{**})^2 \\ = &F'(x^{**}, x^* - x^{**}) + \frac{1}{2}\lambda ||x^* - x^{**}||^2 \end{aligned}$$

where $f'(\cdot,\cdot)$ and $F'(\cdot,\cdot)$ are the directional derivatives. Since $x^* - x^{**}$ is an ascent direction of F at x^{**},

$$F'(x^{**}, x^* - x^{**}) \geq 0.$$

Therefore

$$\bar{F}(x^*) - F(x^{**}) \geq \frac{\lambda}{2}||x^* - x^{**}||^2.$$

On the other hand, we have

$$\bar{F}(x^*) - F(x^{**}) \leq \bar{F}(x^{**}) - F(x^{**}) = \sum_{j=1}^{n}[\bar{f}_j(x_j^{**}) - f_j(x_j^{**})].$$

Assume that $g_{j,i_j-1} \leq x_j^{**} \leq g_{ji_j}$, for $j = 1, ..., n$. Then

$$\begin{aligned} &\sum_{j=1}^{n}[\bar{f}_j(x_j^{**}) - f_j(x_j^{**})] \\ = &\sum_{j=1}^{n}\{f_j(g_{ji_j-1}) + [p_{ji_j}(\frac{g_{ji_j-1} + g_{ji_j}}{2}) + q_{ji_j}](x_j^{**} - g_{ji_j-1}) \\ &-[f_j(g_{ji_j-1}) + (p_{ji_j}g_{ji_j-1} + q_{ji_j})(x_j^{**} - g_{ji_j-1}) + (\frac{1}{2}p_{ji_j}(x_j^{**} - g_{ji_j-1})^2]\} \\ = &\sum_{j=1}^{n} p_{ji_j}(x_j^{**} - g_{ji_j-1})(\frac{g_{ji_j-1} + g_{ji_j}}{2} - x_j^{**}) \\ \leq &\sum_{j=1}^{n} p_{ji_j}(q_{ji_j-1} - g_{ji})^2/16 \\ \leq &\frac{\mu}{16}d^2 (\mu > 0) \end{aligned}$$

where $\mu = \max\{\sum_{j=1}^{n} p_{ji} | i_j = 1, ..., m_j, j = 1, ..., n\}$.

Thus

$$\frac{\lambda}{2}||x^* - x^{**}||^2 \le \frac{\mu}{16}d^2$$

i.e.

$$||x^* - x^{**}|| \le (\frac{\mu}{8\lambda})^{1/2}d.$$

$(\mathcal{Q}.\mathcal{E}.\mathcal{D}.)$

Proposition (3.1.1) says by choosing relatively fine grid points we can solve (MPQP) approximately. The trade-off is that we will have to solve a piecewise linear program with many pieces. However, if we just want a good initial point for (MPQP), then using the original breakpoints of f_j as the breakpoints of $\bar{f}_j$ is often a good compromise.

3.2. Algorithm for Finding a quasi-optimal Solution

Now we discuss Step 1. Suppose that $\bar{x}$ is a feasible solution. In order to get a quasi-optimal solution, we need to solve (SP1). However, (SP1) is again a monotropic piecewise quadratic programming problem. Let $\bar{c}^-(\bar{x})$ and $\bar{c}^+(\bar{x})$ be the closest lower and upper breakpoint vectors for $\bar{x}$, that is

$$\bar{c}^-(\bar{x}) = [\bar{c}_1^-(\bar{x}), \cdots, \bar{c}_n^-(\bar{x})]^T, \bar{c}^+(\bar{x}) = [\bar{c}_1^+(\bar{x}), \cdots, \bar{c}_n^+(\bar{x})]^T,$$

where for $j = 1, \cdots, n$,

$$\begin{aligned} \bar{c}_j^-(\bar{x}) &= \max\{c_{jl} | c_{jl} \le \bar{x}_j \quad l = 0, 1, \cdots, k_j\} \\ \bar{c}_j^+(\bar{x}) &= \min\{c_{jl} | c_{jl} \ge \bar{x}_j \quad l = 0, 1, \cdots, k_j\} \end{aligned}$$

(Recall that the c_{jl}'s are the breakpoints of f_j; c_{j0} is c_j^- and c_{jk_j} is c_j^+.) Both $\bar{c}^-(\bar{x})$ and $\bar{c}^+(\bar{x})$ depend on $\bar{x}$. Then we have the following.

Proposition (3.2.1). Any optimal solution x^* to

$$\text{(SP2)} \begin{cases} \text{minimize} & F(x) \\ \text{subject to} & Ax = b \\ & \bar{c}^-(\bar{x}) \le x \le \bar{c}^+(\bar{x}) \end{cases}$$

is a quasi-optimal solution to (MPQP) and satisfies $F(x^*) \le F(\bar{x})$, where $\bar{x}$ is any fixed feasible solution.

Proof. Since x^* is optimal to (SP2) but $\bar{x}$ is merely feasible, we have $F(x^*) \le F(\bar{x})$. Observe that x^* is a local minimizer of the problem

$$\text{(SP3)} \begin{cases} \text{minimize} & F(x) \\ \text{subject to} & Ax = b \\ & c^- \le x \le c^+ \\ & x_j = x_j^*, \quad \forall\, j \in I(x^*), \end{cases}$$

so by convexity it is also a global minimizer of (SP3), and hence it is quasi-optimal by definition.

(Q.E.D.)

Proposition (3.2.1) says we can get a quasi-optimal solution by solving a quadratic program (SP2). Yet we can make the procedure even simpler if we have a simplex tableau of $Ax = b$ on hand. To simplify the statement of algorithms, let us first introduce some terminology.

Suppose that $\tilde{c}^- = [\tilde{c}_1^-, \cdots, \tilde{c}_n^-]^T$ and $\tilde{c}^+ = [\tilde{c}_1^+, \cdots, \tilde{c}_n^+]^T$ are two vectors whose jth components are certain breakpoints of f_j. They satisfy $\tilde{c}^- \leq \tilde{c}^+$. Call the set

$$[\tilde{c}^-, \tilde{c}^+] = \{x | \tilde{c}^- \leq x \leq \tilde{c}^+\}$$

a *piece* if for $j = 1, \cdots, n$ there is no other breakpoint of f_j between $\tilde{c}_j^-$ and $\tilde{c}_j^+$. Given a feasible solution $\bar{x}$ and a feasible direction $\bar{y}$ at $\bar{x}$(i.e. $\bar{x} + \epsilon\bar{y}$ is feasible for some $\epsilon > 0$); we say that the piece $[\tilde{c}^-, \tilde{c}^+]$ is associated with $(\bar{x}, \bar{y})$ if there exists a positive number $\bar{\epsilon}$ such that for all ϵ in $[0, \bar{\epsilon}]$, $\bar{x} + \epsilon\bar{y}$ is in $[\tilde{c}^-, \tilde{c}^+]$.

Notice that $\tilde{c}^-$ and $\tilde{c}^+$ depend on $\bar{x}$ and $\bar{y}$, and once $\bar{x}$ and $\bar{y}$ are given, it is not hard to determine the unique piece that is associated with them, if we adopt the regulation that $\bar{y}_j = 0$ and $\bar{x}_j = c_{jk} \Rightarrow \tilde{c}_j^- = \tilde{c}_j^+ = c_{jk}$.

Definition (3.2.2). Given $\bar{x}$ and $\bar{y}$, the procedure for finding a minimizer of $F(\bar{x}+\epsilon\bar{y})$ in the piece associated with $(\bar{x}, \bar{y})$ is called *one-piece line search from $\bar{x}$ in the direction of $\bar{y}$*, or is simply called "the one-piece line search with $(\bar{x}, \bar{y})$". The procedure of finding the global minimizer of $F(\bar{x} + \epsilon\bar{y})$ is called *multi-piece line search from $\bar{x}$ in the direction of $\bar{y}$*, or "the multi-piece line search with $(\bar{x}, \bar{y})$".

For any $\bar{x}$ and $\bar{y}$, $F(\bar{x} + \epsilon\bar{y})$ is a piecewise quadratic function of ϵ. Suppose $F(x)$ is $\sum_{j=1}^n \frac{1}{2}(p_j x_j^2 + q_j x_j + r_j)$ on $[\tilde{c}^-, \tilde{c}^+]$ (if $\tilde{c}_j^- = \tilde{c}_j^+$, simply take $p_j = q_j = 0, r_j = f_j(\tilde{c}_j^-)$). We make the following convention: our one-piece line search always ends up at a relative boundary point of $[\tilde{c}^-, \tilde{c}^+]$ if this point is one of the minimizers. Namely, if $\bar{y}$ is an ascent vector (i.e. $F'(\bar{x}, \bar{y}) > 0$), then the one-piece line search stops at $\bar{x}$; otherwise the one-piece line search will produce $\bar{x} + \epsilon_0\bar{y}$, where

$$\epsilon_0 = \min \begin{cases} \frac{\tilde{c}_j^+ - \bar{x}_j}{\bar{y}_j} & \text{for } j \text{ with } \bar{y}_j > 0; \\ \frac{\tilde{c}_j^- - \bar{x}_j}{\bar{y}_j} & \text{for } j \text{ with } \bar{y}_j < 0; \\ \frac{-\sum_{j=1}^n \bar{y}_j(p_j\bar{x}_j + q_j)}{\sum_{j=1}^n \bar{y}_j p_j^2} & (\text{or } +\infty, \text{ if all } \bar{y}_j p_j^2 = 0). \end{cases}$$

If $\epsilon_0 = +\infty$ and $F'(\bar{x}, \bar{y}) < 0$, this means that piece $[\tilde{c}^-, \tilde{c}^+]$ contains the half line $\bar{x} + \epsilon\bar{y}$ $(\epsilon \geq 0)$ and along this half line, $F(x)$ is linear and decreasing. In this case, the one-piece line search halts.

The algorithm for multi-piece line search consists of repeated use of the one-piece line search. Specifically we can express this as follows, where inf(MPQP) stands for the infimum of problem (MPQP).

Algorithm (3.2.3) (The Multi-Piece Line Search).

Step 1. Do one-piece line search starting with a feasible $\bar{x}$ and a vector $\bar{y}$. If ϵ_0 is $+\infty$, then inf(MPQP) $= -\infty$; stop. Otherwise go to Step 2.

Step 2. If $\epsilon_0 > 0$ but $\bar{x} + \epsilon_0 \bar{y}$ is not on the relative boundary of $[\tilde{c}^-, \tilde{c}^+]$, replace $\bar{x}$ by $\bar{x} + \epsilon_0 \bar{y}$; stop. Otherwise replace $\bar{x}$ by $\bar{x} + \epsilon_0 \bar{y}$; go to Step 3.

Step 3. Check if $\bar{y}$ is a descent vector at $\bar{x}$. If yes, go to Step 1; if no, stop.

Now we are ready to state our algorithm for finding quasi-optimal solutions. The basic idea is, instead of solving the quadratic program (SP2) that corresponds to a given $\bar{x}$, we first make a series of pivoting steps and one-piece line searches until (SP2) becomes a strictly convex quadratic program with minimal dimension. Then we treat (SP2) as if it is an unconstrained quadratic program with respect to the variables x_j, $j \notin I(\bar{x})$. We take the Newton direction vector as $\bar{y}$ and do one-piece line search with $(\bar{x}, \bar{y})$. The whole procedure (reduction of dimension and one-piece line search) will be repeated until the one-piece line search no longer ends up with a point on the relative boundary of $[\tilde{c}^-, \tilde{c}^+]$. This is much like the first step of the classical "active set method" of quadratic programming, with the concept of "active constraints" being replaced by that of "breakpoints". To be sure that the algorithm can choose one quasi-optimal solution when there are infinitely many of them, we need the following convention.

Breakpoint Convention: If $f_j(x_j)$ is linear on $(-\infty, +\infty)$, then zero is regarded as a breakpoint of f_j .

Algorithm (3.2.4). (Finding a Quasi-Optimal Solution)

Step 0. Suppose that $\bar{x}$ is a feasible solution and that $x_B = Tx_N + d$ is a simplex tableau of the system $Ax = b$. Partition the set B into $FB \cup IB$, where FB(the set of *free* basic indices)$= B \backslash I(\bar{x})$ and IB(the set of *imprisoned* basic indices) $= B \cap I(\bar{x})$. Likewise partition $N = FN \cup IN$(the sets of free and imprisoned nonbasic indices). Let T_{FI} be the submatrix of T consisting of t_{ij} with $i \in FB$ and $j \in IN$. Similarly define T_{IF}, T_{II} and T_{FF}. (See Figure 1.)

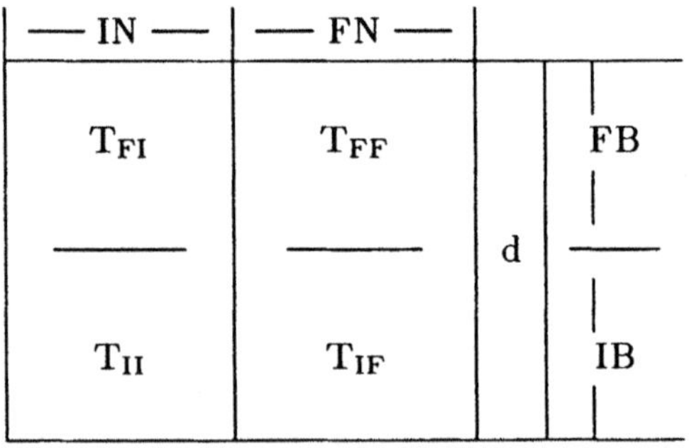

Figure 1. Partition of the Simplex Tableau Relative to Free and Imprisoned Variables

Step 1. If there is a nonzero $t_{ij} \in T_{IF}$, pivot on t_{ij} and repeat this until either $T_{IF} = \emptyset$ or $T_{IF} = 0$ (see Figure 2). Go to Step 2.

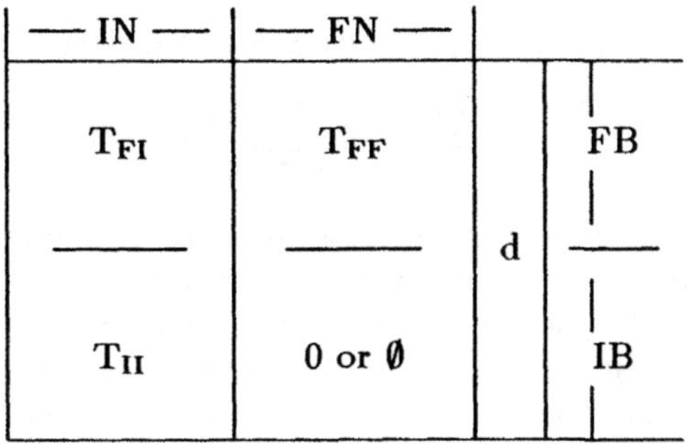

Figure 2. After Pivoting, $T_{IF} = 0$ or $\emptyset$

Step 2. If $FN = \emptyset$, stop; $\bar{x}$ is a quasi-optimal solution. Otherwise partition FN into $LFN \cup QFN$ (the sets of linear free nonbasic indices and quadratic free nonbasic indices), where

$$\begin{aligned} LFN &= \{j \in FN \mid f_j \text{ is linear on the interval which includes } \bar{x}_j\}, \\ QFN &= FN \backslash LFN. \end{aligned}$$

Similarly partition FB into $LFB \cup QFB$. Let T_{LL} be the submatrix of T_{FF} consisting of t_{ij} with $i \in LFB$ and $j \in LFN$. Similarly define T_{LQ}, T_{QL} and T_{QQ}. Pivot on nonzero entries $t_{ij} \in T_{QL}$ until either $T_{QL} = \emptyset$ or $T_{QL} = 0$ (See Figure 3). Go to Step 3.

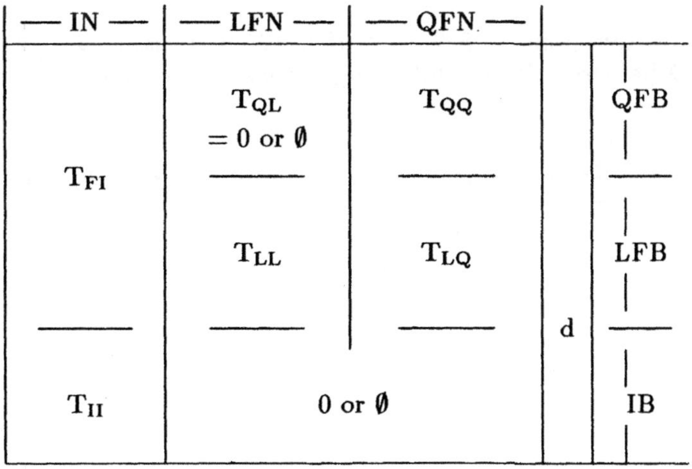

Figure 3. Further Partition of T_{FF} relative to Linear and Quadratic Variables

Step 3. If $LFN = \emptyset$, go to Step 4. Otherwise for each $j \in LFN$ calculate

$$\delta_j = f_j'(\bar{x}_j) + \sum_{i \in LFB} t_{ij} f_i'(\bar{x}_j).$$

Let $\bar{y}$ be the elementary vector corresponding to j. That is, its jth element is 1 and its basic variables are given by the jth column of T; all other components are zero. If $\delta_j < 0$, do one-piece line search with $(\bar{x}, \bar{y})$; if $\delta_j > 0$, do one-piece line search with $(\bar{x}, -\bar{y})$; if $\delta_j = 0$ and $\epsilon_0 \neq +\infty$, do one-piece line search with $(\bar{x}, \bar{y})$; if $\delta_j = 0$ and $\epsilon_0 = +\infty$, do one piece line search with $(\bar{x}, -\bar{y})$. Repeat Step 3 until either the one-piece line search indicates that inf(MPQP)$=-\infty$ or until $LFN = \emptyset$. In the former case, stop; in the latter case (see Figure 4), go to Step 4.

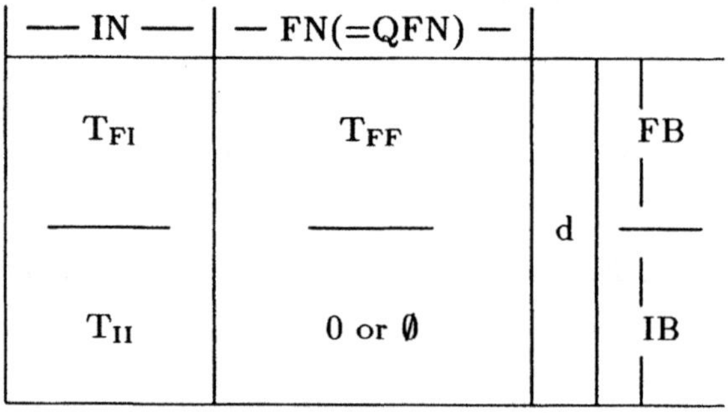

Figure 4. After Step 3, FN=QFN (LFN=$\emptyset$)

Step 4. At this point, $LFN = \emptyset$. If $QFN = \emptyset$, stop; $\bar{x}$ is a quasi-optimal solution. Otherwise solve the following unconstrained quadratic program:

$$\begin{aligned}
\text{minimize} \quad & \frac{1}{2} y_{FN} \cdot P_{FN}\, y_{FN} + q_{FN} \cdot y_{FN} + \frac{1}{2} y_{FB} \cdot P_{FB}\, y_{FB} + q_{FB} \cdot y_{FB} \\
\text{subject to} \quad & T_{FF}\, y_{FN} = y_{FB} \\
\text{where} \quad & y_{FN} = \{y_j \,|\, j \in FN\} \qquad y_{FB} = \{y_i \,|\, i \in FB\} \\
& P_{FN} = \operatorname{diag}[f_j''(x_j)]_{j \in FN} \qquad P_{FB} = \operatorname{diag}[f_i''(x_i)]_{i \in FB} \\
& q_{FN} = [f_j'(x_j)]_{j \in FN}^T \qquad q_{FB} = [f_i'(x_i)]_{i \in FB}^T
\end{aligned}$$

Calculate $y_{FB} = T_{FF}\, y_{FN}$, then set

$$\bar{y} = [\bar{y}_1, \cdots, \bar{y}_k, \cdots, \bar{y}_n]^T, \quad \text{where} \quad \bar{y}_k = \begin{cases} y_k & \text{if } k \in FB \cup FN, \\ 0 & \text{otherwise.} \end{cases}$$

Go to Step 5.

Step 5. Do a one-piece line search with (x^k, y^k). If the line search ends up with a point that is not on the relative boundary of the piece, stop; the point reached by the line search is a quasi-optimal solution. If the line search shows the infimum of problem (MPQP) is $-\infty$, stop; otherwise go to Step 1.

Justification of the algorithm

The algorithm starts with a feasible $\bar{x}$. After Step 1, we get the following simplex tableau:

$$\begin{pmatrix} x_{FB} \\ x_{IB} \end{pmatrix} = \begin{pmatrix} T_{FI} & T_{FF} \\ T_{II} & 0 \text{ or } \emptyset \end{pmatrix} \begin{pmatrix} x_{IN} \\ x_{FN} \end{pmatrix} + d.$$

Thus if $FN = \emptyset$, the values of all basic variables are uniquely determined by those of the imprisoned nonbasic variables. Then x is the unique feasible solution of (SP1), so it is quasi-optimal.

After Step 2 we get the simplex tableau

$$\begin{pmatrix} x_{LFB} \\ x_{QFB} \\ x_{IB} \end{pmatrix} = \begin{pmatrix} (T_{FI}) & \begin{pmatrix} T_{LQ} & T_{LL} \\ T_{QQ} & 0 \text{ or } \emptyset \end{pmatrix} \\ (T_{II}) & (0 \text{ or } \emptyset) \end{pmatrix} \begin{pmatrix} x_{IN} \\ x_{QFN} \\ x_{LFN} \end{pmatrix} + d$$

Therefore in Step 3 we have $F'(\bar{x}, \pm\bar{y}) = \pm\delta_j$ (cf. Proposition(3.3.2)). Since only linear functions are involved in the one-piece line search procedures in Step 3, if $\epsilon_0 \neq +\infty$ and $\delta_j \neq 0$, the line search must finish at some relative boundary point of $[\tilde{c}^-, \tilde{c}^+]$ and thus decrease the size of $LFB \cup LFN$ by at least 1. In this case we go back to Step 3 to repeat the procedure. The iteration can happen only finitely many times before we go to Step 4. If $\epsilon_0 = +\infty$ and $\delta_j \neq 0$, we have $\inf(\text{MPQP}) = \inf(\text{MPQP}) = -\infty$ (cf. the statement of one-piece line search); if $\delta_j = 0$, then by the breakpoint convention, either along $\bar{y}$ or along $-\bar{y}$ we will find a relative boundary point. Therefore Step 3 will produce a feasible solution whose simplex tableau has $LFN = \emptyset$ and whose objective value is not greater than $F(\bar{x})$.

If $QFN = \emptyset$ in Step 4, then $FN = \emptyset$ and $\bar{x}$ is quasi-optimal. Otherwise the simplex tableau will look like

$$\begin{pmatrix} x_{FB} \\ x_{IB} \end{pmatrix} = \begin{pmatrix} T_{FI} & T_{FF} \\ T_{II} & 0 \text{ or } \emptyset \end{pmatrix} \begin{pmatrix} x_{IN} \\ x_{FN} \end{pmatrix} + d,$$

where for each $j \in FN, f_j$ is quadratic around $\bar{x}_j$. The one-piece line search along $\bar{y}$ in Step 5 will give the minimizer of problem (SP1) if it does not stop at a relative boundary point of $[\tilde{c}^-, \tilde{c}^+]$, because $\bar{y}$ is the Newton direction for problem (SP1). If the line search ends at a relative boundary point of $[\tilde{c}^-, \tilde{c}^+]$, then by going back to Step 1, we will move at least one index from $FB \cup FN$ to $IB \cup IN$. Thus after finitely many iterations, we will find that either the set FN is empty or line search along $\bar{y}$ does not stop at any relative boundary points of $[\tilde{c}^-, \tilde{c}^+]$. Both cases yield a quasi-optimal solution where corresponding objective value is not greater than the value at the starting point.

$(\mathcal{Q.E.D.})$

3.3. Algorithm for Finding a Descent Vector

Now we turn to Step 3 of the model algorithm. The question is how to descend from a given feasible solution x. Obviously, either a one-piece or multi-piece line search in a descent direction will reach another feasible solution with lower objective value. Hence the problem reduces to getting a descent vector. The accurate definition of a descent vector is:

Definition (3.3.1). Suppose x is a feasible solution to (MPQP), y satisfies $Ay = 0$ and $F'(x, y) < 0$. Then y is a *descent vector* at x.

Now suppose we have a simplex tableau $x_B = Tx_N + d$ on hand. For simplicity of notations, suppose $B = \{1, ..., m\}$ and $N = \{m + 1, ..., n\}$. The j-th column of T, denoted by $[t_{1j}, ..., t_{mj}]^T$, corresponds to the nonbasic variable x_{m+j}, while the i-th row of T, denoted by $[t_{i1}, ..., t_{i,n-m}]$, corresponds to basic variable x_i.

Proposition (3.3.2). A necessary and sufficient condition for the elementary vector

$$[t_{1j}, ..., t_{mj}, 0, ..., 0, \underset{\substack{\uparrow \\ (m+j)-th\ component}}{1}, 0, ..., 0]^T$$

to be a descent vector is

$$\delta_j^+ = f_{m+j}^+(x_{m+j}) + \sum_{i=1}^{m} \max\{t_{ij} f_i^-(x_i), t_{ij} f_i^+(x_i)\} < 0.$$

Similarly, the elementary vector $[-t_{ij}, ..., -t_{mj}, 0, ..., 0, -1, 0, ..., 0]^T$ is a descent vector if and only if

$$\delta_j^- = f_{m+j}^-(x_{m+j}) + \sum_{i=1}^{m} \min\{t_{ij} f_i^-(x_i), t_{ij} f_i^+(x_i)\} > 0.$$

Proof. All the elementary vectors satisfy the system $Ay = 0$. Let the first elementary vector be z. Then by the definition of directional derivatives we have

$$\begin{aligned} F'(x, z) &= \lim_{t \downarrow 0} \frac{\sum_{j=1}^{n} f_j(x_j + tz_j) - \sum_{j=1}^{n} f_j(x_j)}{t} \\ &= \sum_{j=1}^{n} \lim_{t \downarrow 0} \frac{f_j(x_j + tz_j) - f_j(x_j)}{t} (\text{Note : No limit in this sum can be } -\infty.) \\ &= \sum_{j=1}^{n} f_j'(x_j, z_j) = \sum_{z_j > 0} z_j f_j^+(x_j) - \sum_{z_j < 0} z_j f_j^-(x_j) \\ &= \sum_{j=1}^{n} \max\{z_j f_j^-(x_j), z_j f_j^+(x_j)\} = \delta_j^+. \end{aligned}$$

The similar calculation gives

$$F'(x, -z) = f_{m+j}^-(x_{m+j}) + \sum_{i=1}^{m} \min\{t_{ij} f_i^+(x_i), t_{ij} f_i^-(x_i)\} = \delta_j^-.$$

The conclusion of the proposition follows.

$(\mathcal{Q.E.D.})$

According to Proposition (3.3.2), the determination of an elementary descent vector seems to be a quite simple issue. The complication comes from the so-called degeneracy. This will be discussed in the next section.

4. Degeneracy Processing

There is a problem left in Section 3. What if we have $\delta_j^- \le 0 \le \delta_j^+$ for all $j \in N$ in a given simplex tableau at a given point x? Can we find a descent direction by pivoting to another simplex tableau or can we declare the present x is an optimal solution?

To answer these questions, let us define the degeneracy of a simplex tableau at a given point x and look at the optimality condition of MPQP, as stated in the following.

Definition (4.1). A simplex tableau is degenerate at x if there exists an index $i_0 \in B$ such that the subdifferential $\partial f_{i_0}(x_{i_0})$ is not a singleton.

In other words, if one of the basic variables x_{i_0} happens to be at a breakpoint of f_{i_0} and f_{i_0} is first order discontinuous at this point, then the corresponding simplex tableau is degenerate at x.

Proposition (4.2). Suppose that $x_B = TX_N + d$ is a simplex tableau of $Ax = b$. Then x is optimal to (MPQP) if and only if there exist $v_i \in \partial f_i(x_i)$ for all $i \in B$ such that

$$-\sum_{i \in B} t_{ij} v_i \in \partial f_j(x_j) \ \textit{for all}\ j \in N.$$

Proof. Since any vector (v_B, v_N) in the range space of A^T satisfies $v_N = -Tv_B$ and vice versa, the condition of the proposition is equivalent to that of the vector $v = (v_i, i \in B, -\sum_{i \in B} t_{ij} v_i, j \in N)$ which is in the range space of A^T and satisfies $v_k \in \partial f_k(x_k)$ for all k, or $v \in \partial F$. This means there exists a vector u such that $-A^T u \in \partial F(x)$, or $0 \in \partial F(x) + A^T u$, so the optimality condition is satisfied and x is an optimal solution to problem (MPQP) (see the equilibrium theorem in [4]).

$(\mathcal{Q.E.D.})$

Corollary (4.3). If the simplex tableau is nondegenerate at x and $\delta_j^- \le 0 \le \delta_j^+$ for all $j \in N$, then x is optimal.

Proof. The nondegeneracy indicates that there exist $\{v_i\} = \partial f_i(x_i)$ for all $i \in B$. Now the condition $\delta_j^- \le 0 \le \delta_j^+$, for all $j \in N$ can be written as

$$f_j^-(x_j) + \sum_{i \in B} t_{ij} v_i \le 0 \le f_j^+(x_j) + \sum_{i \in B} t_{ij} v_i$$

for all $j \in N$. Hence the condition of Proposition (4.2) is satisfied.

$(\mathcal{Q.E.D.})$

When the simplex tableau $x_B = Tx_N + d$ is degenerate at x, the following algorithm will determine if x is optimal. If x is not optimal, it then finds an elementary descent direction. Thus our model algorithm can go on even if degeneracy is present.

Algorithm (4.4).

Start with a feasible solution x and a simplex tableau which is degenerate at x.

Step 0. Arbitrarily choose $v_i = f_i^-(x_i) > -\infty$ or $v_i = f_i^+(x_i) < +\infty$ for all $i \in B$. For the exceptional case that $f_i^-(x_i) = -\infty$ and $f_i^+(x_i) = +\infty$, variable x_i can be actually disregarded from the beginning of our consideration, so we assume this case does not arise.

Step 1. Check for all $j \in N$ if $\delta_j^- \leq 0 \leq \delta_j^+$. If not, then a descent direction is found; return to main algorithm. Otherwise go to Step 2.

Step 2. Check for all $j \in N$ if $-\sum_{i\in B} t_{ij} v_i \in \partial f_j(x_j)$. If yes, then x is optimal; stop. Otherwise choose the smallest index j_0 such that the above inclusion is not valid. This x_{j0} will be the entering variable. Go to Step 3.

Step 3. If

$$f_{j0}^-(x_{j0}) + \sum_{i\in B} t_{ij} v_i > 0 \tag{$*$}$$

in Step 2, then find the smallest index i_0 such that

$$f_{j_0}^+(x_{j_0}) + \sum_{i\in B, i<i_0} \max\{t_{ij} f_i^-(x_i), t_{ij} f_i^+(x_i)\} + \sum_{i\in B, i\geq i_0} v_i t_{ij_0} \leq 0;$$

and if

$$f_{j0}^+(x_{j0}) + \sum_{i\in B} t_{ij} v_i < 0 \tag{$**$}$$

in Step 2, then find the smallest index i_0 such that

$$f_{j_0}^- + \sum_{i\in B, i<i_0} \min\{t_{ij} f_i^-(x_i), t_{ij} f_i^+(x_i)\} + \sum_{i\in B, i\geq i_0} v_i t_{ij_0} \geq 0.$$

The variable x_{i_0} will be the leaving variable. Pivot on (i_0, j_0), set $v_{j_0} = f_{j_0}^-(x_{j_0})$ or $f_{j_0}^+(x_{j_0})$ depending on the former or latter case and go to Step 1.

Justification of the Algorithm

The descent direction in Step 1 and the optimality in Step 2 can be justified by Propositions (3.3.2) and (4.2), respectively. Hence what we need to show is the non-cycling property of the pivoting operation in Step 3.

Imagine our objective function f_j is replaced by a piecewise linear function:

$$\bar{f}_j(y_j) = \begin{cases} f_j(x_j) + f_j^+(x_j)(y_j - x_j) & \text{if } y_j \geq x_j; \\ f_j(x_j) + f_j^-(x_j)(y_j - x_j) & \text{if } y_j \leq x_j. \end{cases}$$

then the algorithm (4.4) will identify to a non-cyclic algorithm for the piecewise linear programming problem with the objective function $\sum \bar{f}_j$.(See [2] for detail.)

Hence cycling is impossible. Another thing that needs to be mentioned is the choice of v_{j_0} is always possible in Step 3 because the conditions (*) or (**) ensures the value $f_{j_0}^+(x_{j_0})$ or $f_{j_0}^-(x_{j_0})$ is finite when it is chosen.

$(\mathcal{Q}.\mathcal{E}.\mathcal{D}.)$

5. Computational Experiment

In this section we describe the computational program and the results of our numerical experiments.

5.1. Experiment Design and Program

Since there seems to be no algorithm previously reported in the literature for MPQP, the primary goal of the experimental work has not been to compare the effectiveness of our algorithm with others. Instead, we have mostly been interested in the following questions: Will the algorithm be able to solve small dense problems? If the numbers of variables and constraints are fixed, will the increase in the number of pieces of f_j substantially increase the computing time?

Let n, m and k_j be the number of variables, the number of constraints and the number of pieces of the j-th variable, respectively. We refer to n as the *dimension* of the problem. Since the algorithm directly compute the inverse of the Newton equation in finding a quasi-optimal solution, we restrict ourselves to test the problems of small dimension. On the other hand, we need to drive the numbers k_j high enough to get meaningful information. For simplicity, in our test all the k_j's are equal, and $m = \lfloor \frac{n}{2} \rfloor$. Let k be the common number of pieces. Then the pair (n, k) completely specifies the size of a tested problem. Let $T(n, k)$ denote the CPU time.

Roughly speaking, our program consists of two parts: the calculation of quasi-optimal solutions and multi-piece line search (including the determination of optimality and otherwise a descent direction) as we described in Section 3. Within this framework we find that if a quasi-optimal solution is calculated after several multi-piece line searches instead of after every such search, then considerable computation time can be saved. This is likely because the procedure of computing a quasi-optimal solution, which includes finding several Newton directions and making several pivotings, needs much more time than the procedure of multi-piece line searches. In our tested program code, the number of multi-piece line searches between two consecutive quasi-optimal solutions is taken to be 5. In computing the quasi-optimal solutions we strictly follow the steps described on Algorithm(3.2.4), i.e.

$$[\text{making } T_{IF} = 0 \text{ or } \emptyset] \rightarrow [\text{making } T_{QL} = 0 \text{ or } \emptyset] \rightarrow [\text{making } LFN = \emptyset] \rightarrow$$
$$[\text{calculating } M = P_{FN} + T_{FF}^T P_{FB} T_{FF}] \rightarrow [\text{calculating the Newton direction } y].$$

5.2. Tested Problems

All tested problems were generated randomly by a program called PQPGEN. We input the values of m, n and k. PQPGEN then generates k breakpoints and a characteristic curve for each x_j such that $0 \in \text{dom } F$. (Recall that F denotes the objective function of MPQP.) After this, PQPGEN produces an initial simplex tableau $x_B = Tx_N$, where $B = \{1, \cdots, m\}$ and $N = \{m+1, \cdots, n\}$. Since all entries in T come from a pseudo-random number subroutine, T is usually dense. Because all breakpoints are finite numbers, the generated problems always have optimal solutions.

The feasibility of the zero vector and the homogeneity of the Tucker representation are not restrictions, since a simple transformation will make any MPQP (if feasible) satisfy these conditions. In contrast, the suppression of the initial step (finding a feasible solution) makes the experimental results more accurately reflect the computation time that we wish to know.

5.3. Computational Results

The computational experiments ware conducted on VAX750/ UNIX System. The numerical results are listed in Table 1.

Although the code is written in a nonprofessional way, all the tested problems were still solved in reasonable time. The iterative sequence approaches an optimal solution in the following pattern: The simplex descent step first moves the initial solution to a point closer to the minimum. The first multi-piece line search usually crosses fairly many pieces. Then the algorithm spends lots of time locating the first quasi-optimal solution. For example, if the number of variables is 100 and the number of constraints is 50, then Algorithm (3.2.4) (the subprogram of finding quasi-optimal solutions) starts usually with a feasible solution with 50 free nonbasic variables. This leads to solving a linear equation system with 50 unknowns. After this a one-piece line search, probably including a pivoting operation in a 50×100 matrix, is performed and typically ends up with a new feasible solution having 49 free nonbasic variables. This procedure is subsequently repeated up to 49 times, each time with a problem of dimension one less, until a quasi-optimal solution is obtained. The algorithm then needs much less time to make one iteration because most of the nonbasic variables stay imprisoned and consequently the linear equation systems encountered in the calculation of the Newton directions have very small dimension (1 or 2, likely). Finally, when the solution sequence comes to the pieces where the optimal solution resides, a reverse mode of operation is experienced: The line search along the Newton direction almost always ends up with a relative interior point of a certain piece. Then the simplex descent step moves more and more imprisoned nonbasic variables out of breakpoints, and therefore the linear equation systems solved for finding Newton directions get larger and larger until an optimal solution is reached. Another phenomenon observed is that the amount of pivoting is low, although pivoting theoretically can happen after each one-piece line search.

n	m	k	T(n,k)(sec.)	n	m	k	T(n,k)(sec.)
5	2	1	0.3	15	7	1	2.3
5	2	10	0.7	15	7	10	4.1
5	2	20	1.3	15	7	20	5.9
5	2	30	1.4	15	7	30	6.4
5	2	40	2.2	15	7	40	7.8
5	2	50	2.2	15	7	50	8.6
5	2	60	2.8	15	7	60	10.9
5	2	70	3.0	15	7	70	12.1
5	2	80	3.8	15	7	80	16.1
5	2	90	3.7	15	7	90	13.7
5	2	100	4.1	15	7	100	14.2
5	2	200	9.1	15	7	200	26.0
5	2	300	12.8	15	7	300	37.3
5	2	400	15.5	30	15	1	16.0
5	2	500	19.7	30	15	5	19.6
5	2	600	24.2	30	15	10	23.2
5	2	700	28.1	30	15	15	18.7
5	2	800	32.0	30	15	30	21.1
5	2	900	36.5	30	15	50	28.9
5	2	1000	39.0	30	15	70	32.3
10	5	1	0.9	30	15	100	49.3
10	5	10	1.9	30	15	150	59.6
10	5	20	2.6	50	25	1	81.1
10	5	30	4.9	50	25	10	79.9
10	5	40	5.5	50	25	20	68.5
10	5	50	5.6	50	25	30	71.3
10	5	60	6.7	50	25	40	83.7
10	5	70	7.6	50	25	50	131.4
10	5	80	7.4	50	25	60	100.2
10	5	90	9.0	50	25	70	99.5
10	5	100	9.2	50	25	80	80.9
10	5	200	17.4	50	25	90	94.1
10	5	300	24.6	50	25	100	99.9
10	5	400	33.7	100	50	1	734.2
10	5	500	43.1	100	50	10	1420.3

Table 1. The Computational Time $T(n,k)$

The computational results show no sharp increase in $T(n,k)$ when k increases drastically. In fact, the quantity $T(n,k)$ increases roughly linearly with k. One can see that

$$2T(n,k) \le T(n,10k) \le 10T(n,k) \quad \text{for all tested } n,k.$$

On the other hand, this result apparently indicates that solving a piecewise quadratic problem is somewhat like solving a quadratic program of the same dimension. This point is supported by comparing all the tested pairs $T(n,k)$ with $T(n,1)$. The computational efficiency so demonstrated should at least encourage more research in dual methods of quadratic programming because, as we mentioned before, the dual of a quadratic program is in general piecewise quadratic. Besides, the insensitivity of computational time versus the number of pieces is good news, since some practical models, like the stochastic transportation model introduced in Section 1, generate MPQP with a large number of pieces.

6. Conclusions

We offer a framework of a class of finitely convergent algorithms for MPQP. One of them is described in detail. It is an extension of simplex methods of convex programming and active set methods of quadratic programming. A computational experiment is conducted based on this algorithm and randomly generated dense problems. The results of the experiment show that solving such a piecewise problem is not much harder than solving a non-piecewise problem of the same number of variables. This, we wish, will stimulate more research in the theory, application and software of MPQP.

References

1. R. Fourer, "A simplex algorithm for piecewise linear programming I: derivation and proof," *Mathematical Programming* 33 (1985), 204-233.
2. R. Fourer, "A simplex algorithm for piecewise linear programming II: finiteness, feasibility and degeneracy," *Mathematical Programming* 41 (1988) 281-316.
3. B.A. Murtagh, *Advanced Linear Programming,* McGraw- Hill Inc. New york (1981).
4. R. T. Rockafellar, *Network Flow and Monotropic Optimization,* John Wiley and Sons, New York (1984).
5. R. T. Rockafellar, "Linear-quadratic programming and optimal control," *SIAM Journal of Control and Optimization* 25 (1987) 781-814.
6. R. T. Rockafellar and R. J.-B. Wets, "Generalized Linear-quadratic problems of deterministic and stochastic optimal control in discrete time," *SIAM J. Control and Optimization* 28 (1990) 810-822.
7. J. Sun, "Tracing the characteristic curve of a quadratic black box," *Networks* 19 (1989) 637-650.
8. J. Sun, "On the structure of convex piecewise quadratic functions," *Journal of Optimization Theory and Applications* 72 (1992) 499-510.
9. J. Sun, "An aggregation scheme for parallel solution of network linear programs," submitted to *Journal of Chinese Institute of Industrial Engineers,* (1993).
10. J. Sun, K.-H. Tsai and L. Qi, "A simplex method for network programs with convex separable piecewise linear costs and its application to stochastic transshipment problems," in: D.-Z. Du and P.M. Pardalos eds. *Network Optimization Problems: Theory, Algorithms and Complexity,* World Scientific Publishing Co. London (1993), pp. 281-300.

A NEW APPROACH IN THE OPTIMIZATION OF EXPONENTIAL QUEUES

SUSAN H. XU
Department of Management Science and Information Systems
Smeal College of Business Administration
Pennsylvania State University
University Park, PA 16802

Abstract. Queueing theory is often a vital device in the analysis and optimization of diverse systems such as communication networks, job shops, flexible manufacturing systems, and transportation systems. Most queueing control results are derived via the theory of dynamic programming and induction techniques. In this paper we apply a new approach to derive the optimal scheduling and/or flow control policies for several exponential queueing systems. The underlying idea of this approach is to convert a system into its dual: A system that uses a different service discipline yet is stochastically identical to the original system. We show that the individually optimal policy in the dual system is the socially optimal policy in the original system. In contract with the dynamic programming technique that considers the system as a whole, we adopt the viewpoint of individual jobs and analyze the impact of their behaviors on the social outcome. This approach substantially simplifies analyses and reveals a better insight into the structural properties of the optimal control policies. Besides being of analytical interest, our duality approach also provides computational advantages.

Key words: Ordered-entry queues, scheduling, admission, optimal control, FCFS, LCFS, individual optimum, social optimum.

1. Introduction

In this paper we present some recent results in the optimal admission and scheduling control of some exponential queueing systems using an approach that is significantly different from the conventional dynamic programming technique.

We consider an $M/M/s$ system with heterogeneous servers that services a common buffer or a queue. Processor i has an exponential processing time distribution with rate μ_i, where $\mu_i \neq \mu_j$, in general. The queue is fed by jobs that arrive to the system in a Poisson stream with rate λ. The service discipline is *nonpreemptive first-come first-served* (FCFS).

We wish to tackle various optimization problems:

Model I. *Admission control of an ordered-entry queue.*

Admission control decides on whether an incoming job should be permitted or prohibited to enter the system. Suppose that the system receives a constant reward r from each job upon its service completion and pays a cost c per job per unit time. Let the service discipline be the *ordered-entry* that places an admitted job to the end of the queue if all processors are busy, and assigns the job to the lowest indexed available processor if at least one processor is available. The objective is to maximize the expected discounted or long-run average profit

Ding-Zhu Du and Jie Sun (eds.), Advances in Optimization and Approximation, 293–312.

(reward minus cost) of the system.

Model II. *Scheduling control of an $M/M/2$ queue.*

The objective of the scheduling control is to decide how to utilize available processors so as to minimize the expected sojourn time of a job. We restrict our attention to the two processor system ($s = 2$). This problem is studied by Lin and Kumar (1984) and Walrand (1985), but we shall use a different approach.

Model III. *Admission and scheduling control of an $M/M/2$ queue.*

As in Model II, we assume $s = 2$. Let the cost and reward of the system be the same as in Model I. Suppose the system is subject to both *admission and scheduling control.* The objective is to maximize the expected discounted or long-run average profit.

We refer the objective in each of the above models to as the *social optimum*, as we aim at finding the policy that optimizes the system's performance as a whole. In contrast with the social optimum is the *individual optimum* under which each job optimizes its own objective function. For instance, in model II the objective of an individual job is to minimize its expected discounted (undiscounted) sojourn time. In Model I or Model 2 the objective of an individual job is to maximize its expected discounted (undiscounted) profit $rE[e^{-\beta D}] - cE[\int_{t=0}^{D} e^{-\beta t} dt]$ $(r - cE[D])$, where D is the sojourn time of a job and $\beta, \beta \geq 0$, is the discounting rate (D of course depends upon the state of the system at the job's arrival epoch). We assume that a job is risk neutral in the sense that it will remain in the system if and only if its expected profit is strictly positive. To avoid the trivial situation where an arriving job may not join the system even when there are processors idling, in Models I and III we will assume $c > 0$ and

$$r - c/\mu_i > 0, \quad i = 1, \ldots, s. \tag{1}$$

An oft-adopted method in studying queueing control problems such as the aforementioned models is dynamic programming (DP) and inductive techniques (Stidham 1988). While DP plays a fundamental role in solving optimization problems, one sometimes finds that it is analytically cumbersome or even intractable. The approach we adopt to derive the solutions for Models I-III is significantly different from DP. Briefly, our solution procedure is as follows: We convert each system into a *dual system* that is stochastically isomorphic to the original system. We show that the individually and socially optimal policies in the dual system are identical and are of a threshold type. Finally, we prove that the optimal control in the dual system also optimizes the original system. This approach has the merits of providing a deeper insight into the structural properties of the optimal policy and of reducing the algebraic manipulation to a minimum. This approach also enables us to develop an accurate and robust approximation method for each model.

The optimal control of admission, servicing, routing, and scheduling in a queueing system has been a heated subject of research starting from the pioneering work of Naor (1969). Naor considered a simple $M/M/1$ queue with control of admission of customers, a special case of Model I with $m = 1$. Naor shows quantitatively that the optimal admission control policy that maximizes the expected profit of the system is of a threshold. He observes that the individually optimal policy also has a threshold structure, though the critical numbers which characterize the two policies are not necessarily equal. Typically, the socially optimal critical number is strictly less than its individual counterpart. Naor's work attracted considerable interest among economists and operations researchers, and research along this line is continuing to the present. For detailed survey on flow, scheduling, and routing control, see Stidham (1984, 1988) and references therein.

The discrepancy between the socially and individually optimal admission policies has been widely observed in many admission control systems (e.g., in $GI/M/1$, $M/M/s$, $GI/M/s$, and $M/D/1$ queueing systems). Economists refer this phenomenon to as the *external effect*: The decrease in utility imposed on future customers by an arriving customer's decision to enter the system. It is understood that, due to the presence of the externality, the policy implemented by self-interested individuals does not in general lead to the best social outcome. Bell and Stidham (1983) and Hassin (1986) proposed an interesting way to get around the externality. In particular, Hassin uses the *preemptive last-come-first-served* (LCFS-P) service discipline in an $M/M/1$ queue. He argues verbally that LCFS-P makes the externality vanish and thus brings the social and individual optimality together.

Our approach is motivated by Hassin's observation. We shall demonstrate that, not only does the LCFS-P service discipline bridge the gap between social and individual optimum for the $M/M/1$ queueing system, but it also opens the possibility for an innovative approach to the study of more general queue control problems.

The rest of the paper is organized as follows. Sections 2-4 study Models I-III, respectively. Section 5 considers the computational aspect of the models.

2. Model I: Admission Control of an Ordered-Entry Queue

2.1. The Dual System of Model I

We first define the LCFS-P ordered-entry service discipline.

Definition 2.1. A service discipline is said to be LCFS-P ordered-entry if:

a. The system is non-idling: no processor is allowed to be idle if the queue is nonempty.

b. The system always activates the lowest indexed available processor at an arrival epoch of a new job.

c. The jobs under service must have arrived later than the jobs in the queue. In addition, the job arrived later is placed ahead of the job arrived earlier in the queue.

d. The jobs who arrived earlier will have no effect on the dynamics of the jobs who arrived later.

A system is said to be subject to *expulsion control* if the central controller, who cannot deny entry to an incoming job, may expel an entered job from the system, with the restriction that he/she can only expel jobs from the end of the queue. We call the system subject to expulsion control under LCFS-P service discipline the dual system of Model I.

To be specific, we implement the LCFS-P ordered-entry service discipline as follows. Let the central controller, in addition to controlling the congestion level of the system by expelling jobs, also intervene in the system at arrival and service completion epochs. At an *arrival epoch* of a new job, the central controller will immediately assign the job to processor 1, the lowest indexed processor. This may preempt the job, if any, who was originally being served by processor 1. If the preemption indeed occurs, the central controller will move the preempted job (from processor 1) to the lowest indexed processor, say i, that is idle or has a lower priority job (i.e., a job who arrived earlier than the preempted job); if processor i was idle, then the preemption/reassignment will stop there; otherwise, the job who is preempted from processor i will be assigned to the next lowest indexed processor that is either idle or has a lower priority job. Such a preemption/reassignment procedure terminates after either the job preempted from processor i is reassigned to an idle processor, or the job preempted from processor i is placed at the head of the queue. Now consider a *service completion epoch*. If the queue is empty at a service completion epoch, say service completed by processor i, the central controller will take no action. If the queue is nonempty, then processor i must be the only idle processor in the system (since no idling is allowed when there are jobs in the queue). In such a case the central controller will assign the job at the head of the queue to processor i.

Clearly, this procedure satisfies the requirement for the ordered-entry system since the lowest indexed available processor, if any, will be activated at the arrival epoch of an arriving job, and no idling will occur when the queue is nonempty. Furthermore the preemption/reassignment policy and the job priorities are such that whether we remove a lower priority job from the system at any time or not, the dynamics of the higher priority jobs are unaffected. Note that in the original system jobs are given priorities in ascending order of their arrival times, and in the dual system it is vice versa.

We start with individual optimality in the dual system. Suppose that an individual job's decisions include: (i) whether or not to *enter* the system at its arrival epoch; and (ii) when to *renege* after it enters the system. While a job remains in the system, the LCFS-P ordered-entry service discipline (implemented by the central controller), will be imposed on it. Let us, at any given time, label the jobs in the system in descending order of their arrival times, i.e., the most recently admitted job will be called job 1 at that time, etc. The following theorem gives the structure of individual optimum. We sketch the proof; details can be found in Xu and Shanthikumar (1993).

Theorem 2.1. The individually optimal policy, for either the expected discounted ($\beta > 0$) or undiscounted ($\beta = 0$) profit criterion, has the following structure:

1. An arriving job will always join the system.
2. There exists a threshold n^*_β so that a job will renege if and only if its index is greater than n^*_β.

Proof (Sketch). Since a future job has a higher priority to receive service than a present job, each arriving job views the system as if it were empty, and, consequently, always joins the system. On the other hand, since all the jobs are 'identical' in terms their processing times and cost/reward structure, they will follow the same optimal reneging policy. It is easy to see that this policy can be nothing else but of a threshold type, with some finite threshold n^*_β.

Using stochastic coupling argument (Ross 1983), it can be shown that n^*_β is nonincreasing in β. Let $n^* = \lim_{\beta\to 0} n^*_\beta$. Then n^* will be the individually optimal threshold corresponding to the undiscounted profit criterion. ||

Let π_n be the threshold policy under which a job will renege as soon as it becomes the $n+1$ ^st^ job. We define

$f_j(n)$: the ***first passage probability*** that, under policy π_n, the jth job will be served;

$\mathcal{W}_j(n)$: the sojourn time of job j under π_n;

$W_j(n) := E[\mathcal{W}_j(n)]$.

The optimal threshold for each individual job, n^*_β, is the largest integer n satisfying

$$rE\big[e^{-\beta\mathcal{W}_n(n)}\big]f_n(n) > cE\Big[\int_0^{\mathcal{W}_n(n)} e^{-\beta t}dt\Big]. \tag{2}$$

When $\beta = 0$, n^* is the largest integer n so that

$$rf_n(n) > cW_n(n). \tag{3}$$

Since the expected profit of a job ahead of job n^*_β in the queue cannot be worse than that of job n^*_β, one also has

$$rE\big[e^{-\beta\mathcal{W}_j(n^*_\beta)}\big]f_j(n^*_\beta) > cE\Big[\int_0^{\mathcal{W}_j(n^*_\beta)} e^{-\beta t}dt\Big], \quad \text{for } s \le j \le n^*_\beta. \tag{4}$$

We now turn our attention to the socially optimal policy in the dual system. First note that policy $\pi_{n^*_\beta}$ can be implemented as a social policy in the dual system: The central controller will expel the lowest priority job if and only if the number of jobs present is greater than n^*_β. Observe that if the initial number of jobs in the system is greater than n^*_β, multiple expulsions may occur (*batch disposal*) under π_n. However, it would be of particular interest (when we address the admission control policy in Model I) to implement $\pi_{n^*_\beta}$ under the constraint that the central controller can only expel jobs at an arrival epoch and can expel *at most one* job at such an instant (*single disposal*).

Theorem 2.2. Policy $\pi_{n^*_\beta}$ (π^*) maximizes the expected discounted (long-run average) profit of the dual system subject to the single or batch disposal constraint.

We only prove the discounted case under the single disposal constraint. The argument for the batch disposal is similar. Under certain conditions the long-run average optimal policy can be derived from the discounted optimal policy by letting β approach zero. The conditions required for the convergence turn out to be satisfied in Model I (Xu and Shanthikumar).

Proof (Sketch). Since the system possesses Markovian property, by standard uniformization technique the problem is equivalent to a discrete-time, discounted dynamic programming problem (see, e.g., Lippman 1975, Walrand 1989, Bertsekas 1987).

We establish the result via policy iteration (Ross 1983, Bertsekas 1987). The theory of policy iteration for the discounted dynamic programming states the following: Let f be any stationary policy. Let π be the policy that improves f by taking the best action at current time and following f from next time. If π is not strictly better than f for at least one initial state, then both π and f are optimal policies.

More specifically, we prove that if $\pi_{n_\beta^*}$ is our initial policy which will be adopted from the next step onwards, then an improved policy, π, constructed by taking the best action at this step and then switching to $\pi_{n_\beta^*}$ afterwards, is identical to $\pi_{n_\beta^*}$.

Let Π^* correspond to the system that uses policy $\pi_{n_\beta^*}$ throughout. Let Π correspond to the system that uses policy π, where π agrees with $\pi_{n_\beta^*}$ in the future, but disagrees with $\pi_{n_\beta^*}$ at the present decision epoch, time 0. We prove, if π deviates from $\pi_{n_\beta^*}$ at time 0 for at least one state, then the expected profit of system Π^* is no less than that of system Π. This signifies that $\pi_{n_\beta^*}$ must be the optimal policy. Let j be the number of jobs in the system at time 0, including the job arrived at time 0. We consider $j \le s$, $s < j \le n_\beta^*$, and $j > n_\beta^*$.

For $j \le s$, the reader should be easily convinced, from Eq. (1), that Π^* is superior to Π, because no job should be expelled when the system has idle processors.

Let $s < j \le n_\beta^*$. That $\pi \ne \pi_{n_\beta^*}$ indicates that Π will expel the jth job, the *tagged job*. Let σ be the time at which the tagged job in system Π^* becomes costless (i.e., either completes service or becomes the $n_\beta^* + 1^{st}$ job and hence leaves Π^*). Since during $(0, \sigma)$ systems Π^* and Π will follow $\pi_{n_\beta^*}$, under which no jobs are expelled, both systems incur the same cost and reward, except for the tagged job. The expected discounted profit system Π earns from the tagged job is zero (who is expelled at time 0). The expected discounted profit Π^* earns from the tagged job is

$$rE\left[e^{-\beta W_j(n_\beta^*)}\right] f_j(n_\beta^*) - cE\left[\int_0^{W_j(n_\beta^*)} e^{-\beta t} dt\right] > 0,$$

where the last inequality follows Eq. (4). Thus Π^* outperforms Π until time σ. After time σ the two systems are coupled and henceforth generate the same expected discounted profit. This is because, if at σ the tagged job completes its service, then both systems are empty at that time. If at σ the tagged job in Π^* becomes the $n_\beta^* + 1^{st}$ job, then the tagged job will leave Π^*. However, no job will leave Π, as at that time Π has exactly n_β^* jobs.

Finally, let $j > n_\beta^*$. In this case Π^* will expel job j, the tagged job, at time 0 and earns zero profit from it. Since the tagged job in system Π will remain in the system, which is not individually optimal, it will generate negative profit for Π. Next we prove that the expected discounted profit of *any* other job in Π can be no greater than its counterpart in Π^*. The expected profit of a job, say n, in the two systems may differ if different expulsion decisions are made on job n, which can occur only if, at the arrival instant of a new job, job n is the lowest priority job. Let us choose an arbitrary future arrival time t', and consider the lowest priority job n (it is possible that job n is our tagged job), in system Π^* at that time. Since π can only differ from $\pi_{n_\beta^*}$ for one step, and since only one expulsion can occur at a time, a little reflection tells us that the number of jobs in system Π is either n or $n+1$. Additionally, job n in both systems is the *same* job. The expected profit of job n will not alter as long as the numbers of jobs in both systems are the same at time t'. If at time t' job n is the second-to-the-last job in Π and if $n > n_\beta^*$, then job n in Π^* will be expelled and thus become costless from t'. However, in system Π, after the new job enters and job $n+1$ leaves, job n will be renumbered as job $n+1$. Since $n+1 > n_\beta^* + 1$, by Eq. (2), its expected discounted profit must be negative from time t' onwards.

We have thus verified that system Π^* gets a better total profit than system Π does. ||

2.2. Admission Control of Model I

Let us now consider the socially optimal policy of Model I.

Theorem 2.3. Policy $\pi_{n_\beta^*}$ (π_{n^*}), which admits an arriving job if and only if the number of jobs in the system is less than n_β^* (n^*), where n_β^* (n^*) is the optimal discounted (long-run average) threshold in the dual system, maximizes the discounted (long-run average) profit of Model I.

Proof. Let π be any admission control policy used by Model I. Let us implement π in the dual system subject to the single disposal constraint as the following: For any given state, policy π in the dual system will expel the last priority job if and only if, at that state, policy π in Model I will reject an arriving job. Because the newly rejected job in Model I has the same service time distribution as the newly expelled job in the dual system, the number of jobs in the system at time t and the number of jobs serviced by time t in Model I, $(N_1(\pi,t), S_1(\pi,t))$, and its counterpart in the dual system, $(N_2(\pi,t), S_2(\pi,t))$, are *stochastically identical*:

$$\{(N_1(\pi,t), S_1(\pi,t)), t \geq 0\} =_{st} \{(N_2(\pi,t), S_2(\pi,t)), t \geq 0\}.$$

where "$=_{st}$" means equal in distribution. Therefore, the socially optimal policies in Model I and its dual system must be identical. ||

3. Model II: Scheduling Control of an $M/M/2$ Queue

In Model II all arriving jobs must be accepted, regardless of the system's congestion level. Our objective is to find the optimal policy that minimizes the discounted (or long-run average) sojourn time of a job. Let $\mu_1 + \mu_2 > \lambda$ and $\mu_1 \geq \mu_2$.

3.1. Individual Optimum under LCFS-P

Let us define a service discipline for each individual job.

Definition 3.1. A service discipline is called LCFS-P in individual context if it grants priorities to jobs as follows:

a. At any time jobs in the system are numbered in descending order of their arrival times. A job with a lower index possesses a higher priority to access processors.

b. A lower indexed (higher priority) job, while waiting in the queue, can preempt the service of a higher indexed (lower priority) job. A job cannot abandon its processor unless it is preempted by a lower indexed job.

We let each job under this priority scheme decide whether to utilize a processor that is either idle or processing a higher indexed job. Since a lower indexed job can preempt the service of a higher indexed job, the sojourn time of a lower indexed job is not affected by the decisions of higher indexed jobs. Consequently, when an individual job makes its decision, it only takes into account of the jobs who arrive after itself.

Since in order for a job to make an individual decision, it needs to know the indices of the jobs under service and its own index, we let $\mathbf{k} = (k_1, k_2)$ be the state of the processors, where k_i is the index of the job receiving service from processor i, and $k_i = 0$ if processor i is empty. Let π_m be the threshold scheduling policy under which a job will accept processor 1 whenever possible, will use processor 2 if its index is $m + 1$ and processor 2 is either idle or with a higher indexed job. Let $\mathcal{W}_j(\mathbf{k}, m)$ be the sojourn time job j in the system under policy π_m, with processor state $\mathbf{k}$.

Theorem 3.1. The individually optimal policy minimizing the discounted (undiscounted) sojourn time of an individual job has the following structure:

1. A newly arrived job will always utilize processor 1 immediately.
2. There exists a *finite* critical number m^*_β (m^*) so that job $m^*_\beta + 1$ $(m^* + 1)$ will utilize processor 2 when it is idle or serving a higher indexed job.

Proof (Sketch). We only consider the discounted profit, the proof for the undiscounted profit is similar, with discount factor β being 0.

(1). First note that, as all the jobs are subject to the same cost and reward, the individually optimal policy of any job must be the same and depends only on the index of the job. Suppose that under the individually optimal policy a new jobs will not use processor 1 immediately. Then it will either use processor 2 immediately or wait in the queue. Consider the formal case first. Let m be the index so that the job will use processor 1 when it becomes job $m+1$, where $m = \infty$ implies that processor 1 will never be used under the individually optimal policy. Denote this policy by $\pi_2(m)$. Let $\pi_1(m)$ be the individual policy under which each job uses processor 1 upon its arrival, and uses processor 2 when it becomes job $m + 1$. It can be shown that the expected sojourn time of a job under $\pi_1(m)$ is no greater than that under $\pi_2(m)$. Thus $\pi_2(m)$ cannot be individually optimal. It can also be shown that the individual policy that starts a job's service immediately at processor 1 is always better than delaying the starting time of its service.

(2). We derive the result via contradiction. If no such a critical number, m^*_β, exists, then under the individually optimal policy all the jobs in the queue will wait to be processed by processor 1. Denote this policy by π_1. Then under LCFP-P the sojourn time of job j is $\sum_{i=1}^{j} B_i$, where B_i is the *busy period* of an $M/M/1$ system with arrival rate λ and service rate μ_1, and B_i, $i = 1, 2, \ldots, j$, are iid random variables. Thus the expected discounted sojourn time of job j is

$$E[\int_0^{\sum_{i=1}^{j} B_i} e^{-\beta t} dt] = \frac{1 - E(e^{-\beta \sum_{i=1}^{j} B_i})}{\beta}$$

Let m be the smallest integer j such that

$$\frac{1 - E(e^{-\beta \sum_{i=1}^{i} B_i})}{\beta} > \frac{1}{\beta + \mu_2} \tag{5}$$

The right hand side of the above expression is understood as the expected discounted processing time of processor 2. Since $\lim_{j \to \infty} E[e^{-\beta \sum_{i=1}^{j} B_i}] = 0$, m must be finite. Let us define $\pi_1(m)$ as the individual policy under which each job uses processor 1 upon its arrival, and uses processor 2 when its index is greater than m and processor 2 is either available or processing a lower priority job. We argue that each job will be better off under $\pi_1(m)$ than under π_1: Under $\pi_1(m)$, if the job is serviced by processor 1, then because some of higher indexed jobs (if any) may be routed to processor 2, its sojourn time cannot be longer than its counterpart under π_1; if the job is served by processor 2, then Eq. (5) ensures that the sojourn time under $\pi_1(m)$ is smaller than that under π_1. Thus π_1 cannot be individually optimal. This contradiction implies that the individually optimal scheduling policy is characterized by a finite threshold, say m^*_β. ||

3.2. Social Optimum of Model II and Its Dual System

In this subsection we define a *dual system* associated with Model II and derive the socially optimal policy of Model II via that of the dual system.

Definition 3.2. A system is said to be the dual of Model II if it implements a service discipline that satisfies the following (but otherwise *arbitrary*): At a scheduling instant, any job in the queue may be assigned to an available processor. A job under service may be preempted, as long as it is replaced immediately by another job in the queue.

We intend to make the service discipline in the dual system as flexible as possible. The only restriction is that, at any time, a job under service can leave its processor only when it is preempted by another job. Preemptions of this kind, however, do not matter from *the system point of view.* By memoryless property of exponential distribution, any two such service disciplines are *equivalent* since they induce stochastically identical departure processes and numbers of jobs in the systems, under any stationary policy (Wolff 1989). The proof of the following theorem resembles that of Theorem 2.3.

Theorem 3.2. The optimal expected discounted (long-run average) policies in Model II and its dual systems are identical.

Let π_m be the policy for the dual system under which the system controller will keep processor 1 busy whenever possible, and utilize processor 2 when the system contains more than m jobs. We shall prove that policy $\pi_{m^*_\beta}$ (π_{m^*}) is the optimal discounted (long-run average) scheduling policy for the dual system and, consequently, Model II.

Theorem 3.3. Policy $\pi_{m^*_\beta}$ (π_{m^*}) minimizes the expected discounted (long-run average) sojourn time of the dual system. More specifically:

(1). Processor 1 is used whenever possible.

(2). Processor 2 is assigned a new job if and only if there are more than m^*_β (m^*) jobs in the system.

Proof (Sketch). In this proof we only consider the discounted criterion. The proof for the long-run average criterion may be found in Xu.

We scale parameters so that $\lambda+\mu_1+\mu_2 = 1$. The proof is done by policy iteration. We assume that $\pi_{m^*_\beta}$ will be adopted from next time. At the current time, time 0, policy π will improve $\pi_{m^*_\beta}$ by taking the best action that minimizes the expected total discounted sojourn time. We show, if $\pi \neq \pi_{m^*_\beta}$ for at least one initial state, then a better decision may be made at time 0 under which the expected sojourn time of *every job* can be no less than its counterpart (the job with the same index) under π. As the total sojourn time of the *system* is the sum of the sojourn times of all *individual jobs*, it indicates that $\pi_{n^*_\beta}$ cannot be improved upon.

To facilitate our proof, we make the following conventions.

(i). We index jobs at time 0 as follows. The job with processor 1, if any, is numbered as job 1. The job with processor 2, if any, is numbered as job 1 (2) if processor 1 is idle (busy). Jobs in the queue are numbered in ascending order of their positions in the queue, excluding the numbers used for jobs under service. This index rule ensures that, at time 0, the highest indexed job will be at the back of the queue, as long as the queue is nonempty.

(ii) From time 1 onwards, the jobs in the system at any time is numbered in descending order of their arrival times.

(iii). From time 1 onwards, policy $\pi_{m^*_\beta}$ will be implemented by each individual jobs.

Let the number of jobs in the system at time 0 be j. We call job j, the highest indexed job, the *tagged job*.

(1). Let processor 1 be idle and the queue size be greater than 1 (the proof can be modified accordingly to show (1) holds when the queue size equals 1). Let π reject processor 1 at time 0. In the case that π calls for the use of idle processor 2 at time

0, let π assign the lowest indexed job, the job at the head of the queue, to processor 2 (this is a legitimate assignment since any job can be assigned to processor 2 in the dual system). By induction hypothesis, π will follow $\pi_{m^*_\beta}$ from next time. Let $\bar{\pi}$ be the policy (call the system under this policy $\bar{\Pi}$) which at time 0 follows π but assigns the tagged job (job j) to processor 1, and follows $\pi_{m^*_\beta}$ afterwards.

Now we compare the expected sojourn time of each job in systems Π and $\bar{\Pi}$. If at time 1 either a new job arrives or processor 2 completes its service, then the tagged job in system $\bar{\Pi}$ will be preempted from processor 1 by the new arrival or the job at the head of the queue and, subsequently, return to the back of the queue. Thus systems Π and $\bar{\Pi}$ from time 1 onwards are coupled, with the sojourn time of every job identical. If at time 1 processor 1 of system $\bar{\Pi}$ finishes the tagged job, then clearly the tagged job is better off in $\bar{\Pi}$ than in Π. The sojourn times of all other jobs are coupled.

(2). Suppose that processor 1 is busy and processor 2 is idle. First consider $j \leq m^*_\beta$. Then $\pi_{m^*_\beta}$ will reject processor 2 and π ($\neq \pi_{m^*_\beta}$) will use processor 2. As any job can be assigned to processor 2, let π assign the tagged job to processor 2 at time 0. By the induction hypothesis, both systems will use $\pi_{m^*_\beta}$ afterwards. Hence the jobs with the same index in the two systems, except the tagged job, will have the same sojourn time, as the decision made for the tagged job (who has the lowest priority) has no effect on the dynamics of lower indexed jobs. As for the tagged job, it is better off by stalling in the queue rather than using processor 2, since $\pi_{m^*_\beta}$ is the individually optimal policy.

Now let $j = m^*_\beta + 1$. Then π will reject processor 2 at time 0 and follow $\pi_{m^*_\beta}$ afterwards, whereas $\pi_{m^*_\beta}$ will assign job $m^*_\beta + 1$ to processor 2. Applying the similar argument as in the previous paragraph, we can show that system Π^* outperforms system Π.

Finally, let $j > m^*_\beta + 1$. Suppose that π will reject processor 2 at time 0. Let $\bar{\pi}$ be the policy which assigns the tagged job to processor 2 at time 0 and follows $\pi_{m^*_\beta}$ afterwards. Let us consider the expected sojourn time of a job with the same index in Π and $\bar{\Pi}$. If at time 1 a new job arrives or the job with processor 1 finishes, then, by the induction hypothesis, job $m^*_\beta + 1$ at time 1 will use processor 2 in both systems. Thus the two systems are coupled from time 1 onward. If at time 1 processor 2 in $\bar{\Pi}$ completes the tagged job, then following the same argument as in the proof of Theorem 3.3 (1), we see that every job, including the tagged job, is better off in $\bar{\Pi}$ than its counterpart in Π. ||

The last term in Eq. (3)

4. Model III: Admission and Scheduling Control of an $M/M/2$ Queue

In this section we study an $M/M/2$ queue which is subject to both admission and scheduling controls. Applying both controls simultaneously will further improve the system performance, as disregarding one poses only a suboptimal solution. Our objective is to maximize the expected discounted or long-run average profit of Model III.

4.1. Individual Optimal Under LCFS-P

Suppose that the service discipline for individual jobs, as defined in Definition 3.1, applies here. We let each job under this priority scheme decide: (i) Whether to join the system and, once joined, when to leave the system, and (ii) whether to utilize a processor that is either idle or processing a higher indexed job. Observe that the profit of a lower index job is not affected by the decisions of higher indexed jobs. Consequently, when an individual job makes its decision, it only takes into account of the jobs who arrive after itself.

Let $\mathbf{k}$, the state of the processors, be defined as in Section 3. Let $\pi_{m,n}$ be the threshold policy under which a job will accept processor 1 whenever possible, will use processor 2 if its index is $m+1$ and processor 2 is either idle or with a higher indexed job, and will renege if its index exceeds n. Define

$f_j(\mathbf{k}, m, n)$: *first passage probability* that job j will be served under $\pi_{m,n}$ and processor state $\mathbf{k}$;

$\mathcal{W}_j(\mathbf{k}, m, n)$: the sojourn time of job j under $\pi_{m,n}$.

Theorem 4.1. The individually optimal policy maximizing the discounted (undiscounted) profit has the following structure:

(1). A newly arrived job will always join the system and utilize processor 1 immediately.

(2). There exists a critical number, m^*_β (m^*), so that job $m^*_\beta + 1$ ($m^* + 1$) will utilize processor 2 when it is idle or serving a higher indexed job.

(3). There exists a critical number, n^*_β (n^*), so that a job will renege if and only if its index exceeds n^*_β (n^*), where $m^*_\beta < n^*_\beta$ ($m^* < n^*$).

Proof (Sketch). We only consider the discounted profit.

(1). The reader can be easily convinced that if a job decides to enter the system, it will choose processor 1 rather than processor 2 or wait in the queue. To see that using processor 1 is a better decision than reneging immediately, let the job utilize processor 1 but leave the system when either the next job arrives or it is served by processor 1, whichever occurs first. Let Y and X_i be the time to the next arrival and the service time of processor i, $i = 1, 2$, respectively. The discounted profit of the new job is

$$rP(X_1 < Y)E[e^{-\beta \min\{X_1, Y\}}] - cE[\int_0^{\min\{X_1,Y\}} e^{-\beta t} dt] = \frac{\mu_1}{\beta + \lambda + \mu_1}(r - \frac{c}{\mu_1}) > 0, \tag{6}$$

where the last inequality is due to Eq. (1).

(2). The result may be derived via contradiction. Because Eq. (6) is still valid if X_1 and μ_1 are replaced by X_2 and μ_2, respectively, a job will not renege when it can use processor 2. If no such a critical number, m^*_β, exists, then under the individually optimal policy all the jobs in the queue will wait to be processed by processor 1. Denote this policy by π_1. Thus the expected profit of job j under π_1 is:

$$rE(e^{-\beta \sum_{i=1}^{j} B_i}) - cE[\int_0^{\sum_{i=1}^{j} B_i} e^{-\beta t} dt] = E(e^{-\beta \sum_{i=1}^{j} B_i})[r + \frac{c}{\beta}] - \frac{c}{\beta} \tag{7}$$

where B_i is the busy period of an $M/M/1$ queue with arrival rate λ and service rate μ_1. The above expression will be negative when j is sufficiently large. Thus π_1 cannot be individually optimal. This contradiction implies that there must exist a finite index, say m^*_β, so that job m^*_β will use processor 2 when it is idle or processing a higher indexed job.

(3). Since a job will not renege as long as it is being served, let both processors be busy. It can be shown (see Xu for details) that there exists a *unique number* n^*_β, $n^*_\beta > m^*_\beta$, independent of $\mathbf{k}$ (as long as $k_1 > 0, k_2 > 0$), so that

$$V_j(\mathbf{k}, m^*_\beta, n^*_\beta) > 0 \Longleftrightarrow 1 \le j \le n^*_\beta. \tag{8}$$

This proves (3). ||

In fact, π_{m^*,n^*} is the limiting policy of $\pi_{m^*_\beta,n^*_\beta}$ as β approaches 0. That is,

$$\lim_{\beta \to 0}(m^*_\beta, n^*_\beta) = (m^*, n^*) \tag{9}$$

4.2. Social Optimum of Model III and Its Dual System

Definition 4.1. A system is said to be the dual of Model III if it is subject to *expulsion/scheduling* control: While an incoming job must be accepted, one job in the system may be expelled at the arrival epoch of a new job. Scheduling control is the same as in Model III. The service discipline satisfies the following but otherwise *arbitrary*:

a. At a *scheduling instant*, any job in the queue may be assigned to an available processor. A job under service may be preempted, as long as it is replaced immediately by another job in the queue.

b. At an *expulsion instant*, any job in the system may be expelled, with the constraint that if a job with a processor is expelled, then another job in the queue must be assigned to that processor immediately.

Let the system controller implement policy $\pi_{m^*_\beta,n^*_\beta}$ (π_{m^*,n^*}) in the dual system as follows: He/she will keep processor 1 busy whenever possible, utilize processor 2 when the system contains more than m^*_β (m^*) jobs and, upon the arrival of a new job, expel one job if the system has more than n^*_β (n^*) jobs, including the new arrival. Observe that policy $\pi_{m^*_\beta,n^*_\beta}$ (π_{m^*,n^*}) will be implemented if each individual job implements $\pi_{m^*_\beta, n^*_\beta}$ (π_{m^*, n^*}), under the constraint that reneging can only occur at the arrival epoch and only one job, the highest indexed job in the queue, can renege at such an epoch (call it the *single disposal constraint*). It is clear that individual implementation of $\pi_{m^*_\beta, n^*_\beta}$ (π_{m^*, n^*}) will induce a service discipline satisfying (a)-(b) of Definition 4.1.

Theorem 4.2. Policy $\pi_{m^*_\beta, n^*_\beta}$ (π_{m^*, n^*}) maximizes the expected discounted (long-run average) profit of the dual system. More specifically:

(1). Processor 1 is used whenever possible.

(2). Processor 2 is assigned a new job if and only if there are more than m^*_β (m^*) jobs in the system.

(3). A job in the queue is expelled immediately after a new arrival if and only if there are more than n^*_β (n^*) jobs in the system.

Proof (Sketch). The proofs of (1) and (2) are similar to that of (1) and (2) of Theorem 3.3, with some appropriate modifications. The proof of (3) is similar to that of Theorem 2.3. Details can be found in Xu. ||

Finally, the following theorem connects the dual system with Model III.

Theorem 4.3. The optimal expected discounted (long-run average) policies of Model III and its dual system are identical.

Remark. The two controls in model III are interrelated in the sense that a fixed scheduling threshold uniquely determines the optimal admission threshold, and vise versa. In addition, each optimal threshold in Model III is generally different from its counterpart in Model I or Model II, though our computational experience shows that the difference is often surprisingly small, if not equal to zero.

5. Approximate Solution

The consistency between the individual and social optimality under LCFP-P in Models I-III also offers computational advantages. Although the optimal policies of Models I–III exhibit simple forms, they induce rather complex stochastic processes so that the task of searching for the actual optimal thresholds is a difficult undertake (Xu and Shanthikumar, Xu). One way to circumvent computational obstacles is to use the individually optimal threshold (under FCFS) as an upper bound for its social counterpart (see, e.g., Bartroli and Stidham 1990). Unfortunately, since such a bound is not a function of arrival rate (a selfish job does not care about future arrivals), the gap between the individually and socially optimal thresholds can be rather large, even with a moderate arrival rate. It turns out that LCFP-P is a powerful tool in correcting this discrepancy. In this section we approximate individually optimal thresholds under LCFP-P for Models I-III. Computational efforts are near trivial, yet our test results demonstrate that each approximation is consistently very close to the true solution, over all parameter configurations selected. We only consider the long-run average criterion for each model. To avoid confusion, in this section we use n_a, m_s and (m^*, n^*) to represent the optimal thresholds associated with Models I, II, and III, respectively.

5.1. Approximate Solution of Model I

The approach we propose to derive an approximation of n_a is based on Eq. (3), which characterizes the condition that n_a must satisfy in the dual system. Recognizing that the major complication in computing $W_j(n)$ is caused by preemptions, we consider, instead, the optimal threshold under the *non-preemptive* LCFS service discipline.

The following result will be useful to derive our approximation method. Consider a gambler's ruin model in which a gambler, with j units at hand, has probability λ of winning one unit and probability $\mu = 1 - \lambda$ of losing one unit for each bet. The gambler will quit playing as soon as s(he) is bankrupted or her/his fortune attains $n+1$. Let $P_j(n)$ and $T_j^q(n)$ be the first passage probability that the gambler will go bankruptcy and the expected duration of the game, respectively. Let $\rho = \lambda/\mu$. It is derived that (see, e.g., Parzen 1962, Ross 1983)

$$P_j(n) = \begin{cases} \frac{1-\rho^{n+1-j}}{1-\rho^{n+1}} & \text{if } \rho \neq 1 \\ \frac{n+1-j}{n+1}, & \text{if } \rho = 1 \end{cases} \tag{10}$$

and

$$T_j^q(n) = \begin{cases} \frac{j}{\mu(1-\rho)} - \frac{(n+1)\rho^{n+1-j}(1-\rho^j)}{\mu(1-\rho)(1-\rho^{n+1})} & \text{if } \rho \neq 1 \\ j(n+1-j), & \text{if } \rho = 1. \end{cases} \tag{11}$$

Let $\mu = \sum \mu_i$ and $\lambda + \mu = 1$. Consider the nth job in the *non-preemptive* LCFS dual system who uses threshold policy π_n. That is, the job will renege when it becomes the $(n+1)^{st}$ job in the system. If, as the job is waiting at the head of the queue, a processor becomes free, the job will be assigned to that processor and will not be preempted by new arrivals.

Let $T_n(n)$ be the expected sojourn time of job n, $s < n$, under π_n. Then $T_n(n)$ comprises the job's expected waiting time in the queue and the job's service time, if the job will ever be assigned to a processor. The expected time job n spent in the queue is $T_{n-s}^q(n-s)$, with $\rho = \lambda/\mu$, because one may view the expected waiting time of job n in the queue as the mean duration of the gambler's game, given that the gambler starts with $n-s$ units (since job n is the $(n-s)^{th}$ job in the queue) and will quit playing when the gambler's fortune reaches either $n-s+1$ (if it becomes the $(n+1)^{st}$ job and reneges) or 0 (if it is served by a processor). The probability that the job is served is $P_{n-s}(n-s)$. The expected profit of job n is

$$\begin{aligned} &rP_{n-s}(n-s) - cL_n(n) \\ &= rP_{n-s}(n-s) - c[T_{n-s}^q(n-s) + \frac{sP_{n-s}(n-s)}{\mu}] \\ &= (r - \frac{cs}{\mu})P_{n-s}(n-s) - cT_{n-s}^q(n-s) \end{aligned}$$

Substituting (10) and (11) into the previous expression, one sees that job n will remain in the system if and only if

$$\begin{cases} \frac{r\mu}{c} - \frac{(n-s)(1-\rho)-\rho(1-\rho^{n-s})}{(1-\rho)^2} - s > 0 & \text{if } \rho \neq 1 \\ \frac{r\mu}{c} - \frac{(n-s)(n-s+1)}{2} > 0 & \text{if } \rho = 1. \end{cases}$$

The largest integer $\tilde{n}_a$ that satisfies the above expression, may be used as an approximation for n_a.

5.2. Approximate Solution of Model II

Consider an individual job under LCFP-P who wishes to minimize its expected sojourn time. As the major complication in obtaining the exact value of m_s is caused by possible preemptions from processor 2, we assume that preemptions are forbidden from processor 2. In other words, the service of a job may be interrupted (by a lower indexed job) only when it is with processor 1, but not with processor 2.

Let $\rho_1 = \lambda/\mu_1$. Suppose that processor 1 is busy and processor 2 is offered to job m, the last job in the buffer. If job m accepts processor 2, its expected completion time is $\frac{1}{\mu_2}$; if job m rejects processor 2 now but will accept processor 2 as soon as it becomes job $m+1$, its expected sojourn time is

$$T_m^q(m) + (1 - P_m(m))\frac{1}{\mu_2}$$

where $T_m^q(m)$ and $P_m(m)$ are defined as in (10) and (11), with ρ replaced by ρ_1 and μ by μ_1. Thus job m will reject processor 2 if and only if

$$\frac{1}{\mu_2} - \frac{T_m^q(m)}{P_m(m)} > 0.$$

Substituting (10)-(11) into the above expression yields

$$\begin{cases} \frac{\mu_1}{\mu_2} - \frac{m(1-\rho_1)-\rho_1(1-\rho_1^m)}{(1-\rho_1)^2} > 0 & \text{if } \rho_1 \neq 1 \\ \frac{1}{\mu_2} - m(m+1) > 0 & \text{if } \rho_1 = 1. \end{cases} \tag{12}$$

The largest integer m satisfying (12), $\tilde{m}_s$, can be used as an approximation of m_s.

5.3. Approximate Solution of Model III

When the system is subject to admission/scheduling control, the optimal thresholds m^* and n^* are interrelated and both are the functions of reward/cost ratio r/c. In general, $m^* \neq m_s$, as m_s is not a function of r/c. We shall decompose the problem into two one-dimensional problems and find the approximate solutions of m^* and n^* in a sequel.

We first consider the approximation for m^*, say $\tilde{m}$. Observe that while r/c is one of the major factors in determining n^*, the dependency of m^* on r/c is rather weak (mainly through the interrelation between m^* and n^*), because the primary goal of scheduling control is to minimize the sojourn times of admitted jobs, which, in turn, will maximize their expected profits. This suggests that $\tilde{m}_s$, defined as in (12), is a plausible approximation for m^*. Let $\tilde{m} = \tilde{m}_s$.

Now consider the reneging decision of job n, when scheduling threshold $\tilde{m}$ is used, $n > \tilde{m}$. Note that in order for job n to be served, it must become job $\tilde{m}+1$ first. The expected waiting time of job n until it either becomes job $n+1$ (and thus departing the system without completing service) or job $\tilde{m}+1$ is $T^q_{n-\tilde{m}-1}(n-\tilde{m}-1)$, which is defined by (11), with $\rho = \lambda/(\mu_1+\mu_2)$ and $\mu = \mu_1+\mu_2$. Similarly, the probability that job n becomes job $\tilde{m}+1$ before it becomes job $n+1$, $P_{n-\tilde{m}-1}(n-\tilde{m}-1)$, can be obtained from (10). We shall approximate the profit of job $\tilde{m}+1$ by $r - c/\mu_2$. So job n will not leave the system as long as

$$P_{n-\tilde{m}-1}(n-\tilde{m}-1)(r-c/\mu_2) - cT^q_{n-\tilde{m}-1}(n-\tilde{m}-1) > 0.$$

Using (10) and (11), this is equivalent to

$$\begin{cases} \frac{r\mu}{c} - \frac{\mu}{\mu_2} - \frac{(n-\tilde{m})(1-\rho)-\rho(1-\rho^{n-\tilde{m}})}{(1-\rho)^2} > 0 & \text{if } \rho \neq 1 \\ \frac{r\mu}{c} - \frac{\mu}{\mu_2} - (n-\tilde{m})(n-\tilde{m}+1) > 0 & \text{if } \rho = 1. \end{cases}$$

The largest integer satisfying the previous expression, $\tilde{n}$, can be used as an approximation for n^*.

5.4. Computational Results

To attain level of accuracy for those remarkably simple approximations, we compute the exact and approximate solutions of Models I–III for the long-run average profits. In all three models we assume that $s = 2$ and $\lambda+\mu_1+\mu_2 = 1$. For all three models and over extensive ranges of parameter combinations, our approximations yield solutions which is surprisingly robust and close to the true optimal solutions.

The numerical results of Model I are presented in Table 1. In this model the system configuration is determined by parameters ρ, $\gamma := \mu_2/\mu_1$, and r/c. We let ρ range from 0.1 to 0.9, γ from 0.1 to 0.4 (for $\gamma > 0.4$, the exact and approximate solutions are the same), and r/c equal 15 and 45. We report n_a and $\tilde{n}_a$, the exact and approximate optimal thresholds, $L_a(n_a)$, the expected number of jobs in the system under admission threshold n_a, $V_a(n_a)$, the expected profit per unit time under threshold n_a, $V_a\% = 100(1-V_a(\tilde{n}_a)/V_a(n_a))$, the relative error of the expected profit per unit time, and the arerage relative error. Table 1 shows that $\tilde{n}_a$ differs from n_a for at most 1, the relative error of the expected profit per unit time is below 3.3% for all sets of parameters, and the average relative error is 0.546%.

Table 2 reports numerical results for Model II. Here the system configuration is determined by $\rho = \lambda/(\mu_1+\mu_2)$ and γ. We let ρ vary from 0.1 to 0.9 and γ from 0.1 to 0.4 (again for $\gamma > 0.4$, the exact and approximate solutions are the same). We report m_s and $\tilde{m}_s$, the exact and approximate optimal scheduling thresholds, $L_s(m_s)$, the expected number of jobs in the system using threshold m_s, $L_s\% = 100(1 - L_s(\tilde{m}_s)/L_s(m_s))$, the relative error of the expected number of jobs in the system, and the average relative error. Table 2 shows that the difference between m_s and $\tilde{m}_s$ is no more than 1. The relative error of the number of jobs in the system under approximation is no more than 0.3%, and the average relative error is 0.034%.

Finally, Table 3 reports the computational results for Model III. Parameter ranges are the same as for model I. The entry with '*' indicates that Eq. (1) is violated. We find that $|\tilde{m} - m^*| \leq 1$ and $|\tilde{n} - n^*| \leq 2$, with the relative error of the expected profit, $V\% = 100(1 - V(\tilde{m}, \tilde{m})/V(m^*, n^*))$, not exceeding 3.2%, where $V(m, n)$ is the expected profit under policy $\pi_{m,n}$, and $L(m^*, n^*)$ is the expected number of jobs in the system under policy π_{m^*,n^*}. The average relative error is 0.22%.

r/c	γ	ρ	n_a	$L_a(n_a)$	$V_a(n_a)$	$\tilde{n}_a$	$V_a\%$
15	0.1	0.1	12	0.20	1.17	12	0.00
		0.5	6	1.48	3.45	6	0.00
		0.9	4	2.20	3.50	5	2.61
	0.2	0.1	12	0.17	1.20	12	0.00
		0.5	6	1.34	3.61	6	0.00
		0.9	4	2.12	3.64	5	2.98
	0.3	0.1	12	0.16	1.20	12	0.00
		0.5	6	1.28	3.67	6	0.00
		0.9	4	2.07	3.71	5	3.15
	0.4	0.1	12	0.17	1.20	12	0.00
		0.5	6	1.25	3.70	6	0.00
		0.9	4	2.05	3.75	5	3.24
45	0.1	0.1	37	0.20	3.90	37	0.00
		0.5	16	1.56	13.44	16	0.00
		0.9	7	3.41	15.83	8	0.08
	0.2	0.1	37	0.17	3.92	37	0.00
		0.5	16	1.41	13.59	16	0.00
		0.9	7	3.32	15.98	8	0.14
	0.3	0.1	37	0.16	3.93	37	0.00
		0.5	16	1.35	13.65	16	0.00
		0.9	7	3.26	16.05	8	0.18
	0.4	0.1	37	0.17	3.93	37	0.00
		0.5	16	1.32	13.68	16	0.00
		0.9	7	3.25	16.10	8	0.18
Average relative error = 0.547							

Table 1. Exact and Approximate Solutions for Model I

	$\gamma = 0.1$				$\gamma = 0.2$			
ρ	m_s	$L_s(m_s)$	$\tilde{m}_s$	$L_s\%$	m_s	$L_s(m_s)$	$\tilde{m}_s$	$L_s\%$
0.1	9	0.12	9	0.00	4	0.14	4	0.00
0.5	5	1.19	5	0.00	3	1.28	3	0.00
0.9	3	9.56	4	0.28	2	9.54	2	0.00
	$\gamma = 0.3$				$\gamma = 0.4$			
ρ	m_s	$L_s(m_s)$	$\tilde{m}_s$	$L_s\%$	m_s	$L_s(m_s)$	$\tilde{m}_s$	$L_s\%$
0.1	3	0.15	3	0.00	2	0.16	2	0.00
0.5	2	1.30	2	0.00	1	1.32	1	0.00
0.9	1	9.53	2	0.17	1	9.49	1	0.00
Average relative error = 0.034								

Table 2. Exact and Approximate Solutions for Model II

r/c	γ	ρ	(m^*, n^*)	$L(m^*, n^*)$	$V(m^*, n^*)$	$(\tilde{m}, \tilde{n})$	$V\%$
15	0.1	0.1	(9,12)	0.12	1.24	(9,12)	0.00
		0.3	(7,8)	0.49	2.97	(7,8)	0.00
		≥ 0.5	*	*	*	*	*
	0.2	0.1	(4,12)	0.14	1.23	(4,11)	0.00
		0.5	(3,6)	1.21	3.74	(3,6)	0.00
		0.9	(2,4)	2.06	3.72	(2,4)	0.00
	0.3	0.1	(3,12)	0.15	1.21	(3,12)	0.00
		0.5	(2,6)	1.23	3.71	(2,6)	0.00
		0.9	(2,4)	2.06	3.71	(2,5)	3.18
	0.4	0.1	(2,12)	0.16	1.20	(2,12)	0.00
		0.5	(1,6)	1.25	3.70	(1,6)	0.00
		0.9	(1,4)	2.05	3.75	(1,4)	0.00
45	0.1	0.1	(9, 37)	0.12	3.97	(9,37)	0.00
		0.5	(5,16)	1.19	13.80	(5,16)	0.00
		0.9	(3,7)	3.28	16.04	(4,9)	0.78
	0.2	0.1	(4,36)	0.14	3.95	(4,36)	0.00
		0.5	(3,16)	1.28	13.72	(3,16)	0.00
		0.9	(2,7)	3.27	16.06	(2,8)	0.18
	0.3	0.1	(3,37)	0.15	3.94	(3,37)	0.00
		0.5	(2,16)	1.30	13.70	(2,16)	0.00
		0.9	(1,7)	3.28	16.05	(2,9)	0.77
	0.4	0.1	(2,36)	0.16	3.93	(2,36)	0.00
		0.5	(1,16)	1.32	13.68	(1,16)	0.00
		0.9	(1,7)	3.25	16.10	(1,8)	0.20
Average relative error = 0.22							

Table 3. Exact and Approximate Solutions for Model III

References

1. Bartroli, M. and S. Stidham. (1990). Admission to a General Stochastic Congestion System: Comparison of Individually and Socially Optimal Policies. Technical Report, Department of

Operations Research, University of North Carolina, Chapel Hill, NC, 27599.
2. Bell, C.E. and S. Stidham. (1983). Individual versus Social Optimization in the Allocation of Jobs to Alternative Processors. *Mgt. Sci.* , **29**, 831-839.
3. Bertsekas, D. (1987). *Dynamic Programming*, Prentice Hall, Englewood Cliffs, N.J.
4. Hassin, R. (1985). On the Optimality of First Come Last Served Queues. *Econometrica*, **53**, 201-202.
5. Lin, W. and P.R. Kumar (1984), Optimal Control of a Queueing System with Two Heterogeneous Servers. *IEEE Trans Autom. Control*, **29**, 211-216
6. Lippman, S.A. (1975). Applying a New Device in the Optimization of Exponential Queueing Systems. *Oper. Res.* **23**, 687-710.
7. Naor, P. (1969). The Regulation of Queue Size by Levying Tolls. *Econometrica, 37*, 15-24.
8. Parzen, E. (1962). *Stochastic Processes*, Holden-Day.
9. Ross, S. (1983). *Introduction to Stochastic Dynamic Programming*, Academic Press, New York, NY.
10. Stidham, S., Jr. (1984). Optimal Control of Admission, Routing, and Service in Queues and Networks of Queues: A Tutorial Review. *Proc. ARO workshop: Analytical and Computational Issues in Logistics R&D.* George Washington University, 330-377.
11. Stidham, S., Jr. (1988). Scheduling, Routing, and Flow Control of Stochastic Network: *Stochastic Differential Systems, Stochastic Control Theory and Applications.* IMA Volume 10, 529-554, W. Fleming and P.L. Lions (ed.), Springer-Verlag, New York.
12. Walrand, J. (1984). A Note on 'Optimal Control of a Queueing System with Two Heterogeneous Processors', *Syst. & Cont. Lett.* **4**, 131-134.
13. Walrand, J. (1989). *Introduction to Queueing Networks*, Prentice Hall, Englewood Cliffs, NJ.
14. Wolff, R.W. (1989). *Stochastic Modeling and the Theory of Queues*, Prentice Hall, Englewood Cliffs, NJ.
15. Xu, S.H. and J.G. Shanthikumar. (1993). Optimal Expulsion Control – A Dual Approach to Admission Control of an Ordered-Entry System. *Oper. Res.*, forthcoming.
16. Xu, S.H. (1993). A Duality Approach to Admission and Scheduling Controls of Queues. Preprint.

THE EUCLIDEAN FACILITIES LOCATION PROBLEM

GUOLIANG XUE
Department of Computer Science and Electrical Engineering
The University of Vermont Burlington, VT 05405, USA

and

CHANGYU WANG
Institute of Operations Research, Qufu Normal University
Qufu, Shandong, PRC

Abstract. The facilities location problem is one of locating new facilities with respect to existing facilities, the locations of which are known. The problem consists of finding locations of new facilities which will minimize a total cost function which consists of a sum of costs directly proportional to the distances between the new facilities, and costs directly proportional to the distances between new and existing facilities. This is an old and yet thriving problem which has important applications in transportation and logistics.

The purpose of this paper is to provide a starting point for those who want to do research in the field of facilities location. We will introduce the problems, address recent advances, and highlight unsolved problems and research directions in this field. Several new results on the Euclidean facilities location-allocation are also presented.

Key words: Euclidean facilities location and allocation, spherical facility location, transportation, algorithms, convergence, open problems.

1. Introduction

The facilities location problem is one of locating new facilities with respect to existing facilities, the locations of which are known. The problem consists of finding locations of new facilities which will minimize a total cost function which consists of a sum of costs directly proportional to the distances between the new facilities, and costs directly proportional to the distances between new and existing facilities. This problem has important applications in transportation and logistics.

The facilities location problem trace back to a problem posed by Fermat early in the 17th century. At the end of his celebrated essay on maxima and minima, in which he presented pre-calculus rules for finding tangents to a variety of curves, he threw out the challenge: "Let he who does not approve of my method attempt the solution of the following problem: Given three points in the plane, find a fourth point such that the sum of its distances to the three given points is a minimum!" (This story is quoted from [19]). The solution to the original Fermat problem is either the Torricelli point (an interior point which opens an angle of 120° to each of the three sides of the triangle) or one of the given points whose angle is no less than 120°. For this reason, facilities location problems are sometimes called the generalized Fermat problem. If there are m existing facilities and only one new facility, the problem is called an Euclidean Single Facility Location (ESFL) problem. If there are more

Ding-Zhu Du and Jie Sun (eds.), Advances in Optimization and Approximation, 313–331.

than one new facilities, the problem is called an Euclidean Multifacility Location (EMFL) problem.

For the general ESFL problem, Weiszfeld [36] gave a simple closed form iterative algorithm in 1937. Later, it was proved by numerous authors [20, 28, 34] that the algorithm converges globally and, under certain conditions, linearly. [5, 6] exhibited a solution to the strong separation problem associated with the ESFL problem which shows that an ϵ-approximation solution to the ESFL problem can be constructed in polynomial time using the Ellipsoid method. [35, 37] studied the ESFL problem subject to convex constraints. [11] studied the problem in general spaces. Both sequential and parallel computational studies were reported in [30]. There has been a large literature on the Euclidean facility location problems [14]. For more details, see the books by Francis, McGinnis and White [15] and by Love, Morris and Wesolowsky [22].

For the ESFL problem, the objective function is nondifferentiable only at a finite number of points (where the new facility coincides with one of the existing facilities). Therefore, the ESFL problem is smooth in nature. For the EMFL problem, the objective function is nondifferentiable on an exponential number of different linear manifolds (where some new facilities coincide). This makes the EMFL problem more difficult because we cannot avoid nondifferentiability easily.

Miehle [24] was the first to propose an extension of the Weiszfeld's algorithm for ESFL to solve EMFL problems. Ostresh [27] proved that Miehle's algorithm is a descent one. However, Miehle's algorithm may converge to a nonoptimal point [32, 39]. Eyster, White and Wierwille [12] proposed a hyperboloid approximation procedure (HAP) for solving the perturbed EMFL problem. [27] proved that the HAP is a descent algorithm. [39, 31] proved that the HAP always converges from any initial point. In 1980, Calamai and Conn proposed a projected gradient algorithm [2]. In 1982, they proposed a projected Newton algorithm [3]. In 1983, Overton [29] proposed a projected Newton algorithm and proved that the algorithm has a quadratic rate of convergence provided the sequence of points generated by the algorithm converges to a strong minimizer plus some other conditions. [9] studied the optimality conditions for the EMFL problem. In 1987, Calamai and Conn [4] proved that their projected Newton algorithm converges globally and quadratically under a nondegeneracy assumption and a strong minimizer assumption. In [33], Rosen and Xue proposed an algorithm for EMFL which, from any initial point, generates a sequence of points which converges to the closed convex set of optimal solutions of EMFL.

As we have seen, there have been many research going on this old problem. The purpose of this paper is to provide a starting point for those who want to do research in this field. In the next four sections, we will introduce the Euclidean single facility location problem, the Euclidean multifacility location problem, the Euclidean location-allocation problem and the spherical facility location problem. For each topic, we will introduce the problem, address recent advances, and highlight research directions and open problems. In the jargon of facility locations, existing facilities are called **peak** points and all other points are called **smooth** points. We will use these terms throughout the paper.

2. The Euclidean Single Facility Location Problem

2.1. The Problem

Let $a_1, a_2, \cdots, a_m$ be m points in R^d, the d-dimensional Euclidean space, which represent the respective locations of m clients in a given region. Let $c_1, c_2, \cdots, c_m$ be m positive weights, which represent the respective amount of service requests of the clients. A service center to provide service to these m clients is going to be set up. If the center is located at $x \in R^d$, then the operation cost is defined as

$$f(x) = \sum_{j=1}^{m} c_j ||x - a_j||, \tag{1}$$

which is the sum of weighted Euclidean distances from the center to each of the clients. The so-called *Euclidean single facility location* (ESFL) problem is to find a point $\overline{x}$ such that the function $f(x)$ is globally minimized.

$$\min\{f(x)|x \in R^d\}. \tag{2}$$

2.2. Optimality Conditions

The objective function $f(x)$ is continuous in R^d. $f(x)$ is differentiable at a point $x \in R^d$ if and only if x is different from the locations of the clients. For this reason, the points $a_1, a_2, \cdots, a_m$ are called *peak points*, every other point is called a *smooth point*.

At a smooth point x, the gradient of $f(x)$ is given by

$$\nabla f(x) = \sum_{j=1}^{m} c_j \frac{x - a_j}{||x - a_j||}, \tag{3}$$

and the Hessian is given by

$$\nabla^2 f(x) = \sum_{j=1}^{m} c_j \frac{1}{||x - a_j||^3}(||x - a_j||^2 I - (x - a_j)(x - a_j)^T). \tag{4}$$

For $j = 1, 2, \cdots, m$ and $x \neq a_j$, define

$$g_j(x) = c_j \frac{x - a_j}{||x - a_j||}, H_j(x) = c_j \frac{||x - a_j||^2 I - (x - a_j)(x - a_j)^T}{||x - a_j||^3}, \lambda_j(x) = \frac{c_j}{||x - a_j||}. \tag{5}$$

For any $x \in R^d$ (peak point or smooth point), define

$$g(x) = \sum_{x \neq a_j} g_j(x), H(x) = \sum_{x \neq a_j} H_j(x), \lambda(x) = \sum_{x \neq a_j} \lambda_j(x). \tag{6}$$

It is clear from the definitions that if x is a smooth point, then $g(x)$ and $H(x)$ are the gradient and Hessian matrix of $f(x)$ respectively. When x coincide with one of the clients, say $x = a_i$, then $g(x)$ and $H(x)$ are the gradient and Hessian matrix of $(f(x) - c_i||x - a_i||)$, respectively.

Theorem 2.1 (Overton [29], Xue [38]) Suppose that the existing facilities are non-collinear. Then function $f(x)$ is continuous and strictly convex and is differentiable everywhere except at those peak points. $g(x)$ is continuous at smooth points, discontinuous at but bounded around peak points. $H(x)$ is positive definitive everywhere, continuous at smooth points, discontinuous at and unbounded around peak points. definite at each smooth point. □

Theorem 2.2 (Kuhn [20], Wang [34]) Suppose that the existing facilities are non-collinear. Then problem (2) has a unique optimal solution. For a smooth point x to be optimal, it is necessary and sufficient that $g(x) = 0$. For a peak point a_i to be optimal, it is necessary and sufficient that $||g(a_i)|| \leq c_i$. □

What happens when the existing facilities are collinear? In this case, the objective function is not strictly convex. However, this does not completes the solution of the problem. The following theorem shows that the set of minimizers can be found in finite time if the exsiting facilities are collinear.

Theorem 2.3 (Kuhn [20], Wang [34]) If all of the existing facilities lie on a straight line, then one of the existing facilities must be an optimal solution of the ESFL problem. Suppose that the existing facilities line on the real line from left to right in the order $a_1, a_2, \cdots, a_m$, then a_t is an optimal solution if

$$\sum_{i=1}^{t-1} c_i < \frac{1}{2}\sum_{i=1}^{m} c_i \leq \sum_{i=1}^{t} c_i. \tag{7}$$

In addition, if the second inequality in 7 is strict, then a_t is the only optimal solution of the ESFL problem; otherwise the closed interval $[a_t, a_{t+1}]$ is the set of all the optimal solutions of the ESFL problem. □

2.3. The Weiszfeld Algorithm

In order to find a point $x \in R^d$ where the gradient of the objective function is zero, Weiszfeld [36] tried to solve the following system of nonlinear equations.

$$\sum_{j=1}^{m} c_j \frac{x - a_j}{||x - a_j||}. \tag{8}$$

By changing the above equation to

$$\sum_{j=1}^{m} c_j \frac{x^{k+1} - a_j}{||x^k - a_j||}, \tag{9}$$

Weiszfeld proposed the following closed form iteration formula.

$$x^{k+1} = \frac{\sum_{j=1}^{m} \frac{c_j a_j}{||x^k - a_j||}}{\sum_{j=1}^{m} \frac{c_j}{||x^k - a_j||}}. \tag{10}$$

Note that (10) is well defined and is equivalent to

$$x^{k+1} = x^k - \frac{g(x^k)}{\lambda(x^k)} \tag{11}$$

if x^k is a smooth point, and is undefined when x^k is a peak point. Wang [34] augmented (10) by letting

$$x^{k+1} = x^k - \frac{g(x^k)}{\lambda(x^k)} \frac{||g(x^k)|| - c_{i_k}}{||g(x^k)||} \tag{12}$$

when x^k is a nonoptimal peak point a_{i_k}. This modified Weiszfeld algorithm can be summarized as follows.

Algorithm 2.1{Modified Weiszfeld Algorithm}

Step_1 Randomly choose a initial point $x^1 \in R^d$, let $k = 1$.

Step_2 Compute $g(x^k)$ and $\epsilon_k = ||x^k - a_{j_k}|| = \min\{||x^k - a_j|| | j = 1, 2, \cdots, m\}$ If $\epsilon_k > 0$ then goto Step_3; otherwise goto Step_4;

Step_3 If $g(x^k) = 0$, then stop, x^k is a smooth optimal solution; otherwise compute $\lambda_k(x^k)$, set $x^{k+1} = x^k - g(x^k)/\lambda_k(x^k)$. Replace k by $k + 1$ and go to Step_2.;

Step_4 If $||g(x^k)|| \leq c_{j_k}$, stop, x^k is a peak optimal solution; otherwise compute $\lambda_k(x^k)$, set $x^{k+1} = x^k - [g(x^k)/\lambda_k(x^k)][(||g(x^k)|| - c_{i_k})/||g(x^k)||]$. Replace k by $k + 1$ and go to Step_2.;

Theorem 2.4 (Kuhn [20], Wang [34]) For any initial point x^1, Algorithm 2.1 either stops at the optimal solution $\overline{x}$ or generates a sequence $\{x^k\}$ which converges to $\overline{x}$. If $\overline{x}$ is a smooth point, then the rate of convergence is linear. If $\overline{x}$ is peak point a_j and $||g(\overline{x})|| < c_j$, then the rate of convergence is superlinear. If $\overline{x}$ is peak point a_j and $||g(\overline{x})|| = c_j$, then the rate of convergence is sublinear. □

Weiszfeld's algorithm is very simple. However, its rate of convergence is not very attractive. In [3], Calamai and Conn first proposed a second order method to solve the Euclidean multifacility location problem. In [29], Overton proposed a second order method to solve the Euclidean multifacility location problem and proved quadratic local convergence under certain conditions. In [4], Calamai and Conn proposed a similar second order method and proved both global convergence and local quadratic convergence under certain conditions. In [37, 38], Xue proposed a second order method to solve the constrained Euclidean single facility location problem and proved that for any initial point, the algorithm either stops at the optimal solution or generates a sequence which converges to the optimal solution at a quadratic rate. Because of space limitations, will choose not to discuss these algorithms here. Interested readers are referred to the specific references.

3. The Euclidean Multifacility Location Problem

3.1. The Problem

Let $a_1, a_2, \ldots, a_m$ be m points in R^d, the d-dimensional Euclidean space. Let w_{ji}, $j = 1, 2, \ldots, n$, $i = 1, 2, \ldots, m$, and v_{jk}, $1 \leq j < k \leq n$ be given nonnegative numbers.

Find a point $x = (x_1^T, \ldots, x_n^T)^T \in R^{n\times d}$ that will minimize

$$f(x) = \sum_{j=1}^{n}\sum_{i=1}^{m} w_{ji}||x_j - a_i|| + \sum_{1\le j<k\le n} v_{jk}||x_j - x_k||, \tag{13}$$

where $|| \bullet ||$ is the Euclidean norm. This is the so-called Euclidean Multifacility Location problem (EMFL).

In the EMFL problem, $a_1, a_2, \ldots, a_m$ represent the locations of m existing facilities; $x_1, x_2, \ldots, x_n$ represent the locations of n new facilities; the objective function $f(x)$ is the sum of weighted Euclidean distances from each new facility to each existing facility and those between each pair of new facilities; our goal is to find optimal locations for the new facilities, i.e., to minimize $f(x)$.

A new facility x_j and an existing facility a_i are said to have an *exchange* whenever $w_{ji} > 0$. Two new facilities x_j and x_k are said to have an *exchange* whenever $v_{jk} > 0$. A new facility x_{j_0} is said to be *chained* if there exist $j_1, j_2, \cdots, j_l \in \{1, 2, \cdots, n\}$ and $i_0 \in \{1, 2, \cdots, m\}$ such that $v_{j_0j_1} \cdots v_{j_{l-1}j_l} w_{j_l i_0} \neq 0$. A variable x_j which is not chained is said to be a *free* variable. Let f and C be the index sets for free variables and chained variables, respectively. We can rewrite $f(x)$ as

$$f(x) = \sum_{j\in C}\sum_{i=1}^{m} w_{ji}||x_j - a_i|| + \sum_{j,k\in C,\ 1\le j<k\le n} v_{jk}||x_j - x_k|| + \sum_{j,k\in \mathcal{F},\ 1\le j<k\le n} v_{jk}||x_j - x_k||. \tag{14}$$

Due to the nonnegativity of the v_{jk}'s, for any $x \in R^{n\times d}$ and $\overline{x}_{free} \in R^d$, if we change all the free variables in x to have the value $\overline{x}_{free}$, the function value (2.1) will not increase. Furthermore, the first two terms in (2.1) make up another EMFL problem with no free variable. For this reason, with no loss of generality, we will assume in the rest of this paper that there is no free variable in EMFL problem. For convenience, we will also assume that $v_{jk} = v_{kj}$ for $n \ge j > k \ge 1$.

3.2. Optimality Conditions

It is clear that the objective function 13 is continuous and is convex. Francis and Cabot [13] obtained the following necessary and sufficient condition for the objective function to be strictly convex.

Theorem 3.1 (Francis and Cabot [13]) A necessary and sufficient condition for the objective function 13 to be strictly convex is that the existing facilities are noncollinear. □

We challenge the reader with the following open problem to find a necessary and sufficient condition for the EMFL problem to have a unique optimal solution.

Open Problem 1 Find a necessary and sufficient condition for the EMFL problem to have a unique optimal solution.

Following Calamai and Conn [4], the objective function $f(x)$ in 13 can be written as

$$f(x) = \sum_{l\in M} f_l(x), \tag{15}$$

where M is a finite index set, $f_l(x) = ||r_l(x)||, r_l(x) = A_l^T x - b_l$, $b_l \in R^d$, A_l is an $n \times d$ by d matrix.

Let $x \in R^{n\times d}$. If $r_l(x) \neq 0$, then $f_l(x)$ is differentiable at this point and the gradient is given by $\nabla f_l(x) = A_l \frac{r_l(x)}{||r_l(x)||}$; otherwise $f_l(x)$ is not differentiable at this point.

Let $M_0(x) = \{l \in M | r_l(x) = 0\}$. For any $x \in R^{n\times d}$, define the following subprogram.

$$P(x): \qquad \min || \sum_{l\in M-M_0} \nabla f_l(x) + \sum_{l\in M_0} A_l u_l ||^2 \tag{16}$$
$$\text{s.t. } ||u_l||^2 \leq 1, l \in M_0.$$

Theorem 3.2 (Calamai and Conn [4], Dax [9]) $P(x)$ is well defined and its optimal solution set is a nonempty convex compact set. Let $u_l, l \in M_0(x)$ be an optimal solution of $P(x)$. Let $r = \sum_{l\in M-M_0} \nabla f_l(x) + \sum_{l\in M_0} A_l u_l$ be the residual. If $r = 0$ then x is an optimal solution of EMFL, otherwise $-r$ is a descent direction of $f(\bullet)$ at point x with $f'(x; -r) = -||r||^2 < 0$ where $f'(x; -r)$ is the directional derivative of f at point x along direction $-r$. □

3.3. The HAP and Its Convergence

For the ESFL problem, the objective function is nondifferentiable only at a finite number of points (where the new facility coincides with one of the existing facilities). Therefore, the ESFL problem is smooth in nature. For the EMFL problem, the objective function is nondifferentiable on one or more linear manifolds (where some facilities coincide). This makes the EMFL problem more difficult because we cannot avoid nondifferentiability easily. In order to avoid nondifferentiability in the EMFL problem, Eyster et al introduced a small positive number ϵ to the original problem, getting the following *smooth* perturbed objective function.

$$f_\epsilon(x) = \sum_{j=1}^{n}\sum_{i=1}^{m} w_{ji}\sqrt{||x_j - a_i||^2 + \epsilon} + \sum_{1\leq j<k\leq n} v_{jk}\sqrt{||x_j - x_k||^2 + \epsilon}. \tag{17}$$

A minimizer of $f_\epsilon(x)$ is called an ***ϵ-optimal solution*** to the EMFL problem.

Theorem 3.3 For any positive number ϵ, the function $f_\epsilon(x)$ is strictly convex and is continuously differentiable. The minimum objective function value of $f_\epsilon(x)$ converges to the minimum objective function value of $f(x)$ when ϵ converges to zero. □

The gradient of f_ϵ with respect to the jth new facility x_j is given by

$$\nabla_j f_\epsilon(x) = \sum_{i=1}^{m} w_{ji}\frac{x_j - a_i}{\sqrt{||x_j - a_i||^2 + \epsilon}} + \sum_{k\neq j}^{m} v_{jk}\frac{x_j - x_k}{\sqrt{||x_j - x_k||^2 + \epsilon}}. \tag{18}$$

The Miehle transformation $T_j : R^{n\times d} \to R^d, j = 1, 2, \cdots, n$ is defined by

$$T_j(x_1, \cdots, x_n) = \frac{\sum_{i=1}^{m} w_{ji}\frac{a_i}{\sqrt{||x_j-a_i||^2+\epsilon}} + \sum_{k\neq j}^{m} v_{jk}\frac{x_k}{\sqrt{||x_j-x_k||^2+\epsilon}}}{\sum_{i=1}^{m}\frac{w_{ji}}{\sqrt{||x_j-a_i||^2+\epsilon}} + \sum_{k\neq j}^{m}\frac{v_{jk}}{\sqrt{||x_j-x_k||^2+\epsilon}}}. \tag{19}$$

In 1973, Eyster, White and Wierwille [12] extended Weiszfeld's idea and proposed a Hyperboloid Approximation Procedure (HAP) for solving the Euclidean multifacility location problem.

Algorithm 3.1 (HAP)
Step 0. Choose any initial point $x^0 \in R^{n\times d}$. Set $s := 0$ and go to Step_1.
Step 1. For $j := 1, 2, \cdots, n$ do $x_j^{s+1} = T_j(x_1^s, \cdots, x_n^s)$.
Step 2. If $x^{s+1} = x^s$, stop; otherwise, replace s with $s+1$ and goto Step_1.

Note that the HAP is a closed form iteration formula. It can be implemented in parallel very easily because the $s + 1^{st}$ approximation of the new facilities can be computed simultaneously from the existing facilities and the sth approximation of the new facilities. [12] couldn't prove the convergence of the HAP, but conjectured that the HAP is a descent convergent algorithm, based on considerable computational experience. In 1977, Ostresh proved that the HAP is a descent algorithm under certain conditions. In 1981, Morris [25] proved that a variant of the HAP always converges. In 1991, Xue resolved this open problem in his thesis [39]. The following two theorems summarizes his solution to this open problem.

Theorem 3.4 (Xue [39]) Let $x = (x_1, x_2, \cdots, x_n)$ be any point in $R^{n\times d}$. Let $y = (y_1, y_2, \cdots, y_n)$ be the point generated by

$$y_j = T_j(x_1, x_2, \cdots, x_n), j = 1, 2, \cdots, n. \tag{20}$$

Then the following descent property for the HAP holds as long as $\epsilon > 0$.

$$2f_\epsilon(x) - 2f_\epsilon(y) \geq \sum_{j=1}^{n}\sum_{i=1}^{m} w_{ji}\frac{\left(\sqrt{||y_j - a_i||^2 + \epsilon} - \sqrt{||x_j - a_i||^2 + \epsilon}\right)^2 + ||y_j - x_j||^2}{\sqrt{||x_j - a_i||^2 + \epsilon}} \tag{21}$$
$$+ \sum_{j=1}^{n}\sum_{k\neq j} v_{jk}\frac{\left(\sqrt{||y_j - y_k||^2 + \epsilon} - \sqrt{||x_j - x_k||^2 + \epsilon}\right)^2}{\sqrt{||x_j - x_k||^2 + \epsilon}}$$

□

Theorem 3.5 (Xue [39]) From any initial point $x^0 \in R^{n\times d}$, the HAP either stops at the ϵ-optimal solution of the EMFL or generates an infinite sequence $\{x^s\}$ converging to the ϵ-optimal solution of the EMFL. □

3.4. Second-Order Methods

While some researchers are trying to extend Weiszfeld's algorithm to solve the EMFL problem, others are using the idea of projected Newton directions to design more efficient algorithms for the EMFL problem. In 1980, Calamai and Conn [2] proposed a projected gradient algorithm. In 1982, they proposed a projected Newton algorithm [3]. In 1983, Overton [29] proposed a projected Newton algorithm and proved that the algorithm has a quadratic rate of convergence provided the sequence of points generated by the algorithm converges to a strong minimizer plus some other conditions. In 1987, Calamai and Conn [4] proved that their projected Newton algorithm converges globally and quadratically under a nondegeneracy assumption and a strong minimizer assumption.

Since the nondegeneracy assumption and the strong minimizer assumption are nontrivial, one would ask whether the EMFL problem possess a globally convergent algorithm without these assumptions. In [33] Rosen and Xue proposed an algorithm for EMFL which, from any initial point, generates a sequence of points which converges to the closed convex set of optimal solutions of EMFL.

Since these algorithm require many additional notations to describe, we refer interested readers to those specific references for details. We would like to point out here that the algorithms of [4] and [29] are so far the most successful algorithms for solving the EMFL problems.

3.5. Polynomial Time Algorithms

In 1992 paper [40], Xue, Rosen and Pardalos observed that the dual of the EMFL problem studied by Francis and Cabot [13] is equivalent to the maximization of a linear function subject to convex quadratic inequality constraints and therefore can be solved in polynomial time by interior point methods recently studied by Jarre [16], Mehrotra and Sun [23], and Nesterov and Nemirovsky [26] for nonlinear convex programming. They also establish a theorem on the duality gap and present a procedure for recovering the primal solution from an interior dual feasible solution. However, no implementation of this algorithm is carried out. We believe that it is worth while to implement this algorithm and compare with the second-order algorithms.

4. The Facilities Location-Allocation Problem

4.1. The Problem

Let $a_1, a_2, \cdots, a_m$ be m points in R^d, the d-dimensional Euclidean space, which represent the respective locations of m clients in a given region. Let $c_1, c_2, \cdots, c_m$ be m positive weights, which represent the respective amount of service requests of the clients. Now, instead of setting up one service center to provide service to the m clients, we want to set up n service centers so that each client gets its service from one and only one of the n centers. The objective is to find the locations of the n service centers and assign the m clients to the n centers so that the summation of the weighted distances from each center to its clients is a minimum. In this case, we have a location-allocation problem.

In the location-allocation problem, there are essentially two phases: the allocation phase and the location phase. These two phases are defined and discussed in detail in the following paragraphs.

In the allocation phase, we need to allocate the m clients to the n centers. For this purpose, we introduce $m \times n$ zero-one variables w_{ij} with the meaning that $w_{ij} = 1$ if and only if the ith client gets its service from the jth center. We call these zero-one variables the `allocation variables`. It should be noticed that an arbitrary set of zero-one values for the allocation variables does not necessary corresponds to a feasible allocation, for it is required that every clients gets its service from one and only one center. Therefore we have the following definition for feasible allocations.

Definition 4.1 A location variable $w = \{w_{ij}\} \in \{0,1\}^{m \times n}$ is said to be feasible if and only if

$$\sum_{j=1}^{n} w_{ij} = 1, \forall i \in \{1, 2, \cdots, m\}. \tag{22}$$

The allocation corresponding to a feasible allocation variable is said to be a feasible allocation.

Given a feasible allocation, we need to find the locations for each of the n centers so that the summation of weighted distances form the center to its clients achieves a minimum. Therefore the location phase is equivalent to n independent ESFL problems. For this reason, we will call the n variables $x_j \in R^d, j = 1, 2, \cdots, n$ `location variables`.

For all $w \in [0,1]^{mn}$ and $x \in R^{n \times d}$, define

$$f(w, x) = \sum_{i=1}^{m} \sum_{j=1}^{n} c_i w_{ij} ||x_j - a_i||. \tag{23}$$

Then for a feasible allocation $w \in \{0,1\}^{mn}$ and location $x \in R^{n \times d}$, $f(w,x)$ defines the total operational cost associated with (w,x), which is the sum of weighted Euclidean distances from each client to its service center. The so-called *Euclidean facilities location-allocation* (EFLA) problem is to find a feasible allocation w and a location x such that the function $f(w,x)$ is globally minimized.

$$\min \ f(w,x), \tag{24}$$
$$\text{s. t. } w \in \{0,1\}^{mn}, x \in R^{d \times n}, \sum_{j=1}^{n} w_{ij} = 1, \forall i \in \{1, 2, \cdots, m\}.$$

4.2. Optimality Conditions

Unlike the problems discussed in the previous sections, the EFLA problem is a global optimization problem in the sense that its objective function is not convex. In this subsection, we will present some necessary conditions for optimality. We will also prove some interesting properties of the objective function.

Theorem 4.1 The function $f(w, \bullet)$ is a convex function on $R^{n \times d}$ for any fixed values of $w \in [0,1]^{mn}$. The function $f(\bullet, x)$ is a linear function on $[0,1]^{mn}$ for any fixed values of $x \in R^{n \times d}$.

Proof. For any fixed values of $w \in \{0,1\}^{mn}$, $f(w, \bullet)$ is the sum of the objective functions of n ESFL problems. Therefore it is convex. The second part of the theorem is clear from (23). □

Definition 4.2 For any given $w \in [0,1]^{mn}$, define

$$F(w) = \min\{f(w,x) | x \in R^{n \times d}\}. \tag{25}$$

In the case where w is a feasible allocation, it is clear that $F(w)$ is the minimum possible cost associated with allocation w. We will call $F(w)$ the cost of allocation w when w is a feasible allocation.

Theorem 4.2 The function $F(\bullet)$ is a concave function on $[0,1]^{mn}$.

Proof. Let w^1 and w^2 be two points in $[0,1]^{mn}$. Let $w^3 = \lambda w^1 + \mu w^2 \in [0,1]^{mn}$, where $\lambda + \mu = 1, \lambda \geq 0, \mu \geq 0$. We want to prove that

$$F(w^3) \geq \lambda F(w^1) + \mu F(w^2). \tag{26}$$

From Definition 4.?, there are three points $x^1, x^2, x^3 \in R^{n\times d}$ such that

$$F(w^k) = f(w^k, x^k) = \min\{f(w^k, x) | x \in R^{n\times d}\}, k = 1,2,3. \tag{27}$$

Therefore we have

$$f(w^1, x^3) \geq f(w^1, x^1) = F(w^1), \tag{28}$$

and

$$f(w^2, x^3) \geq f(w^2, x^2) = F(w^2). \tag{29}$$

From Theorem 4.?, we have

$$F(w^3) = f(\lambda w^1 + \mu w^2, x^3) = \lambda f(w^1, x^3) + \mu f(w^2, x^3) \geq \lambda F(w^1) + \lambda F(w^2). \tag{30}$$

This completes the proof. □

The following theorem provides a necessary condition for optimality.

Theorem 4.3 Suppose that (w, x) is a global minimizer of (24). Then

1. $\sum_{i=1}^m w_{ij} > 0, \forall j \in \{1,2,\cdots,n\}$;
2. $w_{ij} = 1 \Rightarrow ||x_j - a_i|| \leq ||x_k - a_i|| \forall k \in \{1,2,\cdots,n\}$;
3. x_j is the optimal solution of $\min\{\sum_{i=1}^m c_i w_{ij}||x - a_i||, x \in R^d\}$ $\forall j \in \{1,2,\cdots,n\}$.

Proof. Suppose that $\sum_{i=1}^m w_{ij_0} = 0$ for some $j_0 \in \{1,2,\cdots,n\}$. Since $m > n$, there must be a client a_{i_1} whose distance from its service center x_{j_1} is nonzero. Now define $(\overline{w}, \overline{x})$ as follows.

$$\overline{w}_{ij} = \begin{cases} w_{ij} & \text{if } i \neq i_1 \\ 0 & \text{if } i = i_1, j \neq j_0 \\ 1 & \text{if } i = i_1, j = j_0 \end{cases} \tag{31}$$

$$\overline{x}_j = \begin{cases} x_j & \text{if } j \neq j_0 \\ a_{i_1} & \text{if } j = j_0 \end{cases} \tag{32}$$

Then

$$f(\overline{w}, \overline{x}) = f(w, x) - c_{i_0}||x_{j_1} - a_{i_0}|| < f(w, x). \tag{33}$$

This contradiction proof the first claim of the theorem.

Now suppose that $w_{i_0 j_0} = 1$ but $||x_{j_0} - a_{i_0}|| > ||x_k - a_{i_0}||$ for some $k \in \{1, 2, \cdots, n\}$. Let $(\overline{w}, \overline{x})$ be defined as

$$\overline{w}_{ij} = \begin{cases} w_{ij} & \text{if } i \neq i_1 \\ 0 & \text{if } i = i_0, j \neq k \\ 1 & \text{if } i = i_0, j = k \end{cases} \tag{34}$$

$$\overline{x} = x. \tag{35}$$

Then

$$f(\overline{w}, \overline{x}) = f(w, x) - c_{i_0}[||x_{j_0} - a_{i_0}|| - ||x_k - a_{i_0}||] < f(w, x). \tag{36}$$

This contradiction proof the second claim of the theorem.

The last claim of the theorem is clear because x_j must be the optimal location of its clients. □

Theorem 4.3 has been the basis for heuristic algorithms for finding a *good* solution to (24). Any feasible solution (w, x) which satisfies the conditions in Theorem 4.3 is called a *local minimizer* for problem (24).

We will now prove that we can treat the allocation variables as continuous variables on $[0, 1]^{mn}$ instead of as discrete variables on $\{0, 1\}^{mn}$ without affecting the solution of the location-allocation problem.

Theorem 4.4 The minimum function value of the EFLA problem 24 is the same as the minimum function value of the following minimization problem.

$$\min \; f(w, x) = \sum_{i=1}^{m} \sum_{j=1}^{n} c_i w_{ij} ||x_j - a_i||, \tag{37}$$

$$\text{s. t. } x \in R^{d \times n}, w_{ij} \geq 0, i \in \{1, 2, \cdots, m\}, j \in \{1, 2, \cdots, n\}, \tag{38}$$

$$\sum_{j=1}^{n} w_{ij} = 1, \forall i \in \{1, 2, \cdots, m\}.$$

Proof. Since the feasible region of 37 is a superset of the feasible region of 24, the minimum objective function value of 37 is less than or equal to the minimum objective function value of 24.

Let (w, x) be a global minimizer of 37. We want to prove that there is a feasible solution $(\overline{w}, x)$ of 24 such that $f(w, x) = f(\overline{w}, x)$. It follows from Theorem 4.1 that $f(\bullet, x)$ is a linear function on $[0, 1]^{mn}$, therefore the minimum of $f(\bullet, x)$ can be achieved on an extreme point of $\{w \in [0, 1]^{mn}, \sum_{j=1}^{n} w_{ij} = 1 \forall i \in \{1, 2, \cdots, m\}\}$, which in turn must be an extreme point of $\{w \geq 0, \sum_{j=1}^{n} w_{ij} = 1 \forall i \in \{1, 2, \cdots, m\}\}$. Let $\overline{w}$ be such an extreme point. Then $\overline{w}$ corresponds to a feasible allocation and has the property that $f(\overline{w}, x) \leq f(w, x)$. This completes the proof. □

4.3. The Cooper Heuristic Algorithm

One of the earliest algorithms for computing a good local minimizer to problem 24 is the heuristic algorithm proposed by Cooper [7, 8].

Algorithm 4.1{A heuristic algorithm}

Step_1. Randomly Choose an initial allocation $w \in \{0,1\}^{mn}$ which satisfies $\sum_{j=1}^{n} w_{ji} > 0, \forall i \in \{1,2,\cdots,n\}$ and $\sum_{i=1}^{m} w_{ji} = 1, \forall j \in \{1,2,\cdots,m\}$.
Step_2. Compute the centers $x_1, x_2, \cdots, x_n$ by $\min\{f(w,x) | x \in R^{n\times d}\}$.
Step_3. If the conditions of Theorem 4.3 are satisfied, stop. Otherwise, we change the allocation w by assigning a_j to its closest center for all $j \in \{1,2,\cdots,m\}$ and goto Step_2.

The result of Algorithm 4.1 depends on the initial guess of the allocation variables. It is not guaranteed to find a global minimizer of problem (24). However, Algorithm 4.1 is usually called many times as a subroutine in a global optimization algorithm. The best solutions generated by Algorithm 4.1 is then treated as the *best possible* solution to problem (24).

Although branch-and-bound algorithms can be used to compute the global minimizer of (24), they are usually too time comsuming to be of any practical use. Therefore the Cooper heuristic algorithm is still in use.

4.4. The Ostresh Exact Algorithm for Two Centers

In this subsection, we will discuss an efficient exact algorithm for solving a special case of problem (24) when $n = 2$ and $d = 2$.

Theorem 4.5 An allocation $w \in \{0,1\}^{mn}$ may correspond to a local minimizer of (24) only if the point set $\{a_j | w_{j1} = 1\}$ and the point set $\{a_j | w_{j2} = 1\}$ can be separated by a straight line. Therefore the number of allocations which may correspond to a local minimizer of (24) is at most $\frac{n(n-1)}{2}$. □

Based on this fact, Ostresh proposed an exact algorithm for the two-center location-allocation problem. That is, for each allocation w which may correspond to a local minimizer of (24), compute the locations $x_1, x_2, \cdots, x_n$ and the corresponding cost. The allocation (and the corresponding locations of the centers) which has the minimum cost is the global minimizer of (24).

The Ostresh algorithm requires the solution of $O(m^2)$ ESFL problems each with $O(m)$ clients. So far, this has been the most efficient algorithm for solving the two-center location-allocation problem.

We conclude this section with the following open problem.

Open Problem 2 Find an algorithm which finds a global minimizer of (24) by solving a polynomial (in both n and m) number of ESFL problems with $O(m+n)$ existing facilities.

5. Spherical Facility Location

In this section, we will study spherical facility location problems where the distance is no longer Euclidean. This problem model is proposed because the distances on surface of the earth is geodesic, rather than Euclidean. However, as we will see in this section, there is a close connection between the spherical facility location problem and the planar Euclidean facility location problem. Therefore it is worth while to discuss this topic here.

5.1. The Problem

Let $a_1, a_2, \ldots, a_m$ be m points on the 3-dimensional sphere $S = \{x | x \in R^3, ||x|| = 1\}$, where $|| \bullet ||$ is the 2-norm. Let c_j, $j = 1, 2, \ldots, m$, be given positive numbers. We want to find a point $x \in S$ that minimizes the summation of weighted geodesic distances from x to all of the m given points, i.e.,

$$\min_{x \in S} f(x) = \sum_{j=1}^{m} c_j \cos^{-1}(a_j^T x). \tag{39}$$

This is called facility location on a sphere or the spherical facility location problem. Notice that $\cos^{-1}(a_j^T x)$ is the **geodesic distance** or **great circle distance** between x and a_j. We will call the shorter portion of the great circle passing through two points p_1 and p_2 on S the **great circle segment** connecting p_1 and p_2. We will also use $\widehat{p_1 p_2}$ to denote the great circle segment connecting p_1 and p_2 and use $d(p_1, p_2)$ to denote the **arc length** of $\widehat{p_1 p_2}$.

In the spherical facility location problem (39), $a_1, a_2, \ldots, a_m$ represent the locations of m existing facilities on the surface of the earth; c_j represent the importance of the jth existing facility; x represents the location of a new facility on the surface of the earth; the objective function $f(x)$ is the summation of weighted geodesic distances from the new facility to each existing facility. Our goal is to find an optimal location for the new facility, i.e., to minimize $f(x)$ subject to $x \in S$.

The spherical facility location problem is a generalization of the planar Euclidean facility location problem. It was first studied by Katz and Cooper and by Drezner and Wesolowsky where a Weiszfeld like algorithm was proposed. Xue [41, 42] proposed a globally convergent algorithm for solving this problem and proved some interesting property of the Katz-Cooper algorithm as well as some close connections between the spherical facility location problem and the planar Euclidean facility location problem.

5.2. Spherical Convexity

In the study of the spherical facility location problems, an interesting extension of convexity is introduced by Drezner and Wesolowsky [10]. It is called spherical convexity. In this subsection, we will present some basic concepts and properties on spherical convexity. So far, spherical convexity has only been used as a tool in solving the spherical facility location problem. It is worth while to do some extensive study on spherical convexity.

Definition 5.1 A **spherical circle** with a given **center** and **radius** is defined on a sphere by the locus of all points whose geodesic distance from the center is equal to that radius. A circle divides the sphere into two parts; a point is said to be **within** a circle only if the point and the center of the circle is included in the same part.

Definition 5.2 A **spherical convex set** is defined on the surface of a sphere as a set where for any two points of the set, the whole great circle segment connecting them is included in the set. The **spherical convex hull** of a set of points on a sphere is defined to be the smallest spherical convex set which contains the set of given points.

Definition 5.3 Let $\rho(r_1, r_2, \lambda)$ be the point on the great circle segment between r_1 and r_2 such that the distance between r_1 and $\rho(r_1, r_2, \lambda)$ is $\lambda d(r_1, r_2)$ for any number $\lambda \in [0, 1]$.

Definition 5.4 $f(r)$ is called a **spherical convex function** on a spherical convex set D on a sphere if for every $0 \leq \lambda \leq 1$ and $r_1, r_2 \in D$, we have

$$f(\rho(r_1, r_2, \lambda)) \leq (1 - \lambda)f(r_1) + \lambda f(r_2). \tag{40}$$

$f(r)$ is called a **strictly spherical convex function** if the inequality (40) is strict when $r_1 \neq r_2$ and $\lambda \in (0, 1)$.

Theorem 5.1 (Drezner and Wesolowsky [10]) The points within a circle on S with radius less than or equal to $\pi/2$ form a spherical convex set. The geodesic distance from a given point r on S is a spherical convex function within a circle of radius $\pi/2$ and center r. Every local minimizer of a spherical convex function on a spherical convex set is also a global minimizer. □

Theorem 5.2 (Drezner and Wesolowsky[10]) If all of the existing facilities are included within a circle of radius $\pi/4$, then the objective function $f(x)$ is a spherical convex function within that circle. Furthermore, if all of the existing facilities do not lie on a great circle segment, then the objective function is strictly spherical convex.
□

5.3. Optimality Conditions

Theorem 5.3 (Xue[41]) Suppose that all of the existing facilities are included within a circle of radius $\pi/4$. Then every global minimizer of (39) must lie within the spherical convex hull of the existing facilities. If all of the existing facilities lie on a great circle of length less than or equal to $\pi/2$, then one of the existing facilities is a global minimizer of the problem. If all of the existing facilities are included within a circle of radius $\pi/4$ but do not lie on a great circle, then the objective function of (39) is strictly spherical convex and there is a unique optimal solution to the spherical facility location problem (39). □

Theorem 5.4 (Xue[41]) An existing facility a_t is a global minimizer of (39) if and only if

$$\|\sum_{j=1, j\neq t}^{m} c_j \frac{a_t - \frac{a_j}{a_t^T a_j}}{\|a_t - \frac{a_j}{a_t^T a_j}\|}\| \leq c_t. \tag{41}$$

A smooth point x is a global minimizer of (39) if and only if

$$\sum_{j=1}^{m} c_j \frac{x - \frac{a_j}{(\frac{x}{\|x\|})^T a_j}}{\|x - \frac{a_j}{(\frac{x}{\|x\|})^T a_j}\|} = 0. \tag{42}$$

□

For any given $x \in S$, define

$$a_i^x = \frac{a_i}{x^T a_i}, i = 1, 2, \cdots, m. \tag{43}$$

Then all of the m points a_i^x, $i = 1, 2, \cdots, m$ lie on the plane which is tangent to S at point x. At point x, we can therefore define a corresponding Euclidean facility location as follows.

$$\min_{y \in R^3} \mathcal{F}_x(y) = \sum_{i=1}^{m} c_i ||y - a_i^x||. \tag{44}$$

Theorem 5.4 says that a point $x \in S$ is the optimal solution of the spherical facility location problem if and only if it is the optimal solution of the corresponding planar Euclidean single facility location problem.

5.4. Algorithm and Convergence

In this section, we will present two algorithms for solving the spherical facility location problem. The first algorithm was proposed Xue [41]. It first tries to see if any of the existing facilities is a global minimizer of the problem. If none of them is, the algorithm proceeds to generate a sequence of descent search directions and iteration points with strictly decreasing function values. This algorithm was proved to be globally convergent. The second algorithm is the Katz-Cooper algorithm which does not require a line search at each iteration. The current form of the algorithm is proposed by Xue [42] which was proved to be equivalent to the Katz-Cooper algorithm. However, global convergence of the second algorithm is still an open problem.

Algorithm 5.1(A Globally Convergent Algorithm)

Step 1. Find an existing facility a_t such that $f(a_t) \leq f(a_j)$ for all $j = 1, 2, \cdots, m$. Check the optimality condition for a_t. If a_t is an optimal solution, stop.

Step 2. Let $d = -\sum_{j=1, j\neq t}^{m} c_j \frac{a_t - a_j^{a_t}}{||a_t - a_j^{a_t}||}$. Find a small step size $\alpha > 0$ such that the point $a_t + \alpha d$ lies in the convex hull of $a_j^{a_t}$, for $j = 1, 2, 3, \cdots, m$, and that $x^1 = \frac{a_t + \alpha d}{||a_t + \alpha d||}$ has a function value less than $f(a_t)$. Let $k = 1$.

Step 3. Compute $a_j^{x^k}$ for $j = 1, 2, \cdots, m$. Compute $d^k = -\sum_{j=1}^{m} c_j \frac{x^k - a_j^{x^k}}{||x^k - a_j^{x^k}||}$. If $d^k = 0$, stop; otherwise compute $\alpha^k = \frac{1}{\sum_{j=1}^{m} \frac{c_j}{||x^k - a_j^{x^k}||}}$.

Step 4. Set $x^{k+1} = \frac{x^k + \alpha^k d^k}{||x^k + \alpha^k d^k||}$. If $f(x^{k+1}) \leq f(x^k) - 0.1\alpha^k ||d^k||^2$, then replace k with $k+1$ and goto step 3; otherwise, replace α^k with $0.5\alpha^k$ and goto step 4.

Remark 5.1 Steps 1 and 2 are used to eliminate nonsmooth points from further consideration. Let a_t be an existing facility whose function value is minimum among all the existing facilities. If a_t satisfies the optimality condition (41), then it is also a global minimizer of the problem. If a_t does not satisfy the optimality condition (41), then the optimal solution of the problem must be a smooth point and d computed in Step 2 is a descent direction of $f(x)$ at point a_t. Therefore the starting point x^1

does exist and can be computed using an inexact line search like the one proposed in [1]. Step 3 computes the search direction d^k, which is the negative of the gradient. If $d^k = 0$, then x^k satisfies the optimality condition (42) and is therefore a global minimizer. If $d^k \neq 0$, then it is a descent direction and an inexact line search [1] is carried out in Step 4. It is clear from the description of the algorithm that the whole iteration sequence $\{x^k\}$ lie in the spherical convex hull of the existing facilities.

Theorem 5.5 (Xue [41]) Algorithm 5.1 either stops at a global minimizer after a finite number of iterations; or generates an infinite sequence $\{x^k\}$ which converges to a global minimizer of the problem. □

Algorithm 5.2(The Katz-Cooper Algorithm)
Step 1. Starting with any point $x^1 \in S$, go to Step_2 with $k = 1$.

Step 2. $x^{k+1} = \dfrac{\sum_{i=1}^m \frac{c_i a_i^{x^k}}{\|x^k - a_i^{x^k}\|}}{\|\sum_{i=1}^m \frac{c_i a_i^{x^k}}{\|x^k - a_i^{x^k}\|}\|}$

Step 3. Replace k with $k+1$ and goto Step_2.

Remark 5.2 Just like the Weiszfeld algorithm for solving the Euclidean facility location problem, the Katz-Cooper algorithm for solving the spherical facility location problem is undefined when the current iteration point x^k coincides with one of the existing facilities. In this case, techniques that can move the the current iteration point to a smooth point with lower objective function value [34, 41] have to be used. This is not a serious problem however, since there are only a finite number of peak points in this model.

Theorem 5.6 (Xue [42]) If the oscillation of the sequence $\{x^k\}$ generated by Algorithm 5.2 converges to zero, then the whole sequence $\{x^k\}$ converges to the optimal solution of the spherical facility location problem. □

The global convergence of Algorithm 5.2 is still an open problem. The difficulty in proving the global convergence of the algorithm is to prove that the algorithm is a descent algorithm. We conclude this section with an open problem to the reader.

Open Problem 3 Prove that the Katz-Cooper algorithm is a descent algorithm by proving the following assertion: If $x \in S$ is not the optimal solution of the spherical facility location problem, and $y = \dfrac{\sum_{i=1}^m \frac{c_i a_i^x}{\|x - a_i^x\|}}{\|\sum_{i=1}^m \frac{c_i a_i^x}{\|x - a_i^x\|}\|}$, then $f(y) < f(x)$.

6. Conclusions

In this paper, we have discussed several problem models related to the Euclidean facilities location problem. These include the Euclidean single facility location problem, the Euclidean multifacility location problem, the Euclidean facilities location-allocation problem, and the spherical facility location problem. For each of these models, we have introduced the problem, addressed recent advances and discussed research directions. In particular, we have proposed three open problems in Sections 3, 4, and 5. We believe that study on these problems are fruitful.

Since there are so many papers published on this topic, it is very hard to cite all of them here. We would point out that the books [22, 15] contain a more complete reference.

Acknowledgment

The work of the first author was supported in part by the University of Vermont and by the Army Research Office contract number DAAL03-89-C-0038 with the University of Minnesota Army High Performance Computing Research Center. The work of the second author was supported in part by the Chinese Natural Science Foundation.

References

1. Armijo, Minimization of Functions Having Continuous Partial Derivatives, *Pacific Journal of Mathematics*, Vol.16 (1966), pp. 1-3.
2. P.H. Calamai and A.R. Conn, A Stable Algorithm for Solving the Multifacility Location Problem involving Euclidean Distances, *SIAM Journal on Scientific and Statistical Computing*, Vol. 1(1980), pp. 512-525.
3. P.H. Calamai and A.R. Conn, A Second-Order Method for Solving the Continuous Multifacility Location Problem, in Watson, G.A. ed., *Numerical Analysis: Proceedings of the Ninth Biennial Conference, Dundee, Scotland*, Lecture Notes in Mathematics 912, Springer-Verlag(1982), pp. 1-25.
4. P.H. Calamai and A.R. Conn, A Projected Newton Method for l_p Norm Location Problems, *Mathematical Programming*, Vol. 38(1987), pp. 75-109.
5. R. Chandrasekaran and A. Tamir, Open Questions Concerning Weiszfeld's Algorithm for the Fermat-Weber Location Problem, *Mathematical Programming*, Vol. 44(1989), pp. 293-295.
6. R. Chandrasekaran and A. Tamir, Algebraic Optimization: The Fermat-Weber Location Problem, *Mathematical Programming*, Vol. 46(1990), pp. 219-224.
7. L. Cooper, Location-allocation problems, *Operations Research*, Vol. 11(1963), pp. 331-343.
8. L. Cooper, Heuristic Methods for Location-allocation problems, *SIAM Review*, Vol. 6(1964), pp. 37-52.
9. A. Dax, A Note on Optimality Conditions for the Euclidean Multifacility Location Problem, *Mathematical Programming*, Vol. 36(1986), pp. 72-80.
10. Z. Drezner and G.O. Wesolowsky, Facility Location on a Sphere, *Journal of the Operational Research Society*, Vol. 29 (1978), pp. 997-1004.
11. U. Eckardt, Weber's Problem and Weiszfeld's Algorithm in General Spaces, *Mathematical Programming*, Vol. 18(1980), pp. 186-196.
12. J.W. Eyster, J.A. White and W.W. Wierwille, On Solving Multifacility Location Problems using a Hyperboloid Approximation Procedure, *AIIE Transactions*, Vol. 5(1973), pp. 1-6.
13. R.L. Francis and A.V. Cabot, Properties of a Multifacility Location Problem involving Euclidean Distances, *Naval Research Logistics Quarterly*, Vol. 19(1972), pp. 335-353.
14. R.L. Francis and J.M. Goldstein, Location Theory: A selected Bibliography, *Operations Research*, Vol. 22(1974), pp. 400-410.
15. R.L. Francis, Leon F. McGinnis, Jr. and John A. White, *Facility Layout and Location: An Analytical Approach*, Prentice Hall, 1991.
16. F. Jarre, On the Convergence of the Method of Analytic Centers when Applied to Convex Quadratic Programs, *Mathematical Programming*, Vol. 49(1991), pp. 341-358.
17. I.N. Katz, Local Convergence in Fermat's Problem, *Mathematical Programming*, Vol. 4(1974), pp. 98-107.

18. I.N. Katz and L. Cooper, Optimal Location on a Sphere, *Computers and Mathematics with Applications*, Vol. 6 (1980), pp. 175-196.
19. H.W. Kuhn, On a Pair of Dual Nonlinear Programs, in J. Abadie eds, *Nonlinear Programming*, pp. 39-54, North-Holland, 1967.
20. H.W. Kuhn, A Note on Fermat's Problem, *Mathematical Programming*, Vol. 4(1973), pp. 98-107.
21. R.F. Love, and J.G. Morris, Modelling Inter-City Road Distances by Mathematical Functions, *Operational Research Quarterly*, 23(1972), 61-71.
22. R.F. Love, J.G. Morris and G.O. Wesolowsky, *Facilities Location: Models & Methods*, North-Holland, 1988.
23. S. Mehrotra and J. Sun, A Method of Analytic Centers for Quadratically Constrained Convex Quadratic Programs, *SIAM Journal on Numerical Analysis*, Vol. 28(1991), pp. 529-544.
24. W. Miehle, Link Length Minimization in Networks, *Operations Research*, Vol. 6(1958), pp. 232-243.
25. J.G. Morris, Convergence of the Weiszfeld Algorithm for the Weber Problem using a Generalized Distance Function, *Operations Research*, Vol. 29(1981), pp. 37-48.
26. Y.E. Nesterov and A.S. Nemirovsky, Self-Concordant Functions and Polynomial Time Methods in Convex Programming, Moscow, 1990.
27. L.M. Ostresh, The Multifacility Location Problem: Applications and Descent Theorems, *Journal of Regional Science*, Vol. 17(1977), pp. 409-419.
28. L.M. Ostresh, On the Convergence of a Class of Iterative Methods for Solving the Weber Location Problem, *Operations Research*, Vol. 26(1978), pp. 597-609.
29. M.L. Overton, A Quadratically Convergent Method for Minimizing a Sum of Euclidean Norms, *Mathematical Programming*, Vol. 27(1983), pp. 34-63.
30. J.B. Rosen and G.L. Xue, A Computational Comparison of Two Algorithms for the Euclidean Single Facility Location Problem, *ORSA Journal on Computing*, Vol. 4(1991), pp. 207-212.
31. J.B. Rosen and G.L. Xue, On the Convergence of a Hyperbolold Approximation Procedure for the Perturbed Euclidean Multifacility Location Problem, *Operations Research*, accepted for publication.
32. J.B. Rosen and G.L. Xue, On the Convergence of Miehle's Algorithm for the Euclidean Multifacility Location Problem, *Operations Research*. Vol. 40(1992), pp. 188-191.
33. J.B. Rosen and G.L. Xue, A Globally Convergent Algorithm for the Euclidean Multifacility Location Problem, *Acta Mathematicae Applicatae Sinica*.
34. C.Y. Wang et al., On the Convergence and Rate of Convergence of an Iterative Algorithm for the Plant Location Problem (in Chinese), *Qufu Shiyun Xuebao*, Vol. 2(1975), pp. 14-25.
35. Wang, C.Y., Minimizing $\sum_{i=1}^{m} c_i||x - a_i||$ on a Closed Convex Set (in Chinese), *Acta Mathematicae Applicatae Sinica*, Vol. 1(1978), pp. 145-150.
36. E. Weiszfeld, Sur le Point par Lequel le Somme des Distances de n Points donnes est Minimum, *Tohoku Mathematical Journal*, Vol. 43(1937), pp. 355-386.
37. G.L. Xue, A Fast Convergent Algorithm for min $\sum_{i=1}^{m} c_i||x - a_i||$ on a Closed Convex Set (in Chinese), *Journal of Qufu Normal University*, Vol. 13, No. 3(1987), pp. 15-20.
38. G.L. Xue, A Globally and Quadratically Convergent Algorithm for min $\sum_{i=1}^{m} c_i||x - a_i||$ Type Plant Location Problem (in Chinese), *Acta Mathematicae Applicatae Sinica*, Vol. 12(1989), pp. 65-72.
39. G.L. Xue, *Algorithms for Computing Extreme Points of Convex Hulls and the Euclidean Facilities Location Problem*, Ph.D Thesis, Computer Science Department, University of Minnesota, Minneapolis, MN 55455, 1991.
40. G.L. Xue, J.B. Rosen and P.M. Pardalos, A Polynomial Time Dual Algorithm for the Euclidean Multifacility Location Problem, accepted for publication in in Proceedings of *Second Conference on Integer Programming and Combinatorial Optimization*, Pittsburgh, 1992, pp. 227-236.
41. G.L. Xue, A Globally Convergent Algorithm for Facility Location on a Sphere, *Computers and Mathematics with Applications*, accepted for publication.
42. G.L. Xue, On an Open Problem in Spherical Facility Location, submitted to *Numerical Algorithms*.

OPTIMAL DESIGN OF LARGE-SCALE OPENCUT COAL MINE SYSTEM

DEZHUANG YANG
Graduate School of Chinese Academy of Sciences, Beijing, China

1. Introduction

Energy is an important material base for sustained social and economic development. To a great extent, the realization of China's modernization lies on the production, supply and efficient utilization of energy. Therefore, the development of energy production has been listed as one of the focal points in the strategy for China to achieve a sustained social and economic development.

Coal holds an important position in China's energy production and consumption, amounting to about 70 percent of the total. In China, coal now makes up 75 percent of the industrial fuel and power, 65 percent of the raw materials of the chemical industry, and 85 percent of the fuel for urban use. Such a situation will remain for a long time. In China, the era during which petroleum becomes the major energy source will never come. Even in the world, coal will also be the major source of energy in the future. In order to develop China's economy it is necessary to develop the power industry and in order to develop the power industry, it is imperative to develop the coal industry. The major strategic measures for increasing coal output lies in the development of opencut coal mines. There are now five large-scale opencut coal mines in China. One of them is the Jungar Opencut Coal Mine in Inner Mongolia. The mine is located at the juncture of the two provinces of Shanxi and Shannxi as well as Inner Mongolia Autonomous Region. The Yellow River runs through the mine area. On the surface of the mine is a plain of grassland. The mine has big reserves and the coal is of high quality. The overburden is thin and the coal bed is thick. Besides, the mine also has the geographical advantage. Its coal has access to north China (including Beijing, Tianjin and Tangshan), northeast China and East China. It is of great significance to open up the mine. The Chinese government has decided to build at Jungar a large-scale opencut coal mine with an annual output of 60 million tons of coal.

The State Council organized some groups of experts in various fields to study how to open up the Jungar opencut coal mine. By applying the ideas and techniques of Operations Research and other fields many of which are interrelated to each other, we carried out system analysis of the big system and various subsystems, and developed mathematical models and a computing software of the optimal design for opencut coal mines. These models and software have played an important role in construction of opencut mines and produced great economic and social effects.

Ding-Zhu Du and Jie Sun (eds.), Advances in Optimization and Approximation, 332–346.

This paper discusses the problems related to these mathematical models.

2. System Analysis

Designing is a crucial link of the construction chain of the mine. It not only is, to a great extent, decisive to the technological level and economic efficiency of the mine, but also has impacts on the whole national economic and social system. In order to modernize the designing of mine construction, Operations Research should be used and the theory and technology of optimal design should be combined with computer so as to establish a set of mathematical models and methods for the optimal design of mines which are in accordance with the actual situation in China.

In establishing a mathematical model of the optimal design of a mine, first of all it is necessary to carry out system analysis in both the branch systems consisting of the projects of the mine construction and the system of the macro economy of the whole country and of the localities. The system, which consists of various subsystems of the mine construction such as the boundary of the mining area, the urban construction system of the mining area, the excavating machinery and equipment system of the mine, the arrangement for the division of the mining zones and the mining tempo, and the system for transporting excavated coal of the mine, is merely a partial and small system. In order to optimize the small system, it is necessary to start with the analysis of the whole system of the mine and the big system of the whole country. This means, in terms of scope, it is necessary to analyze the big system while analyzing the small system, and in terms of time, it is imperative to judge the issue from a long-term point of view. Only such a system analysis is effective. Only based on such a system analysis the established mathematical models are reliable and practical. The relations between the logical structure of the system analysis and the models are demonstrated in Diagram 1.

According to system analysis and the technology for decomposing big problems, it is clear that in the optimal design of an opencut coal mine, the following key models should be established:

1. The model of ore bodies be established on the basis of the prospecting data. This is the preparatory stage of the optimal design of a mine.

2. The model of optimal mining boundary of the opencut coal mine should be established.

Since the method of opencast mining is adopted, all the overburden should be removed in accordance with the model of the ore bodies and the stability of the geological structure of mine. The boundary of mine area whose overburden has been removed and which is considered feasible should meet the following two requirements: (1) Technologically, it should be kept within the geometric bounding of the angle of the stable geological structure. (2) Economically, it should be reasonable. First of all, an optimal boundary should be a feasible boundary. Meanwhile, it should be a boundary with optimal economic efficiency.

3. The model for the optimal investment in excavating machinery and equipment should be established.

Excavating machinery is the "tooth" of the mine and the crux of the mining equipment. It is the focal point in mine construction to choose suitable excavating

Diagram 1

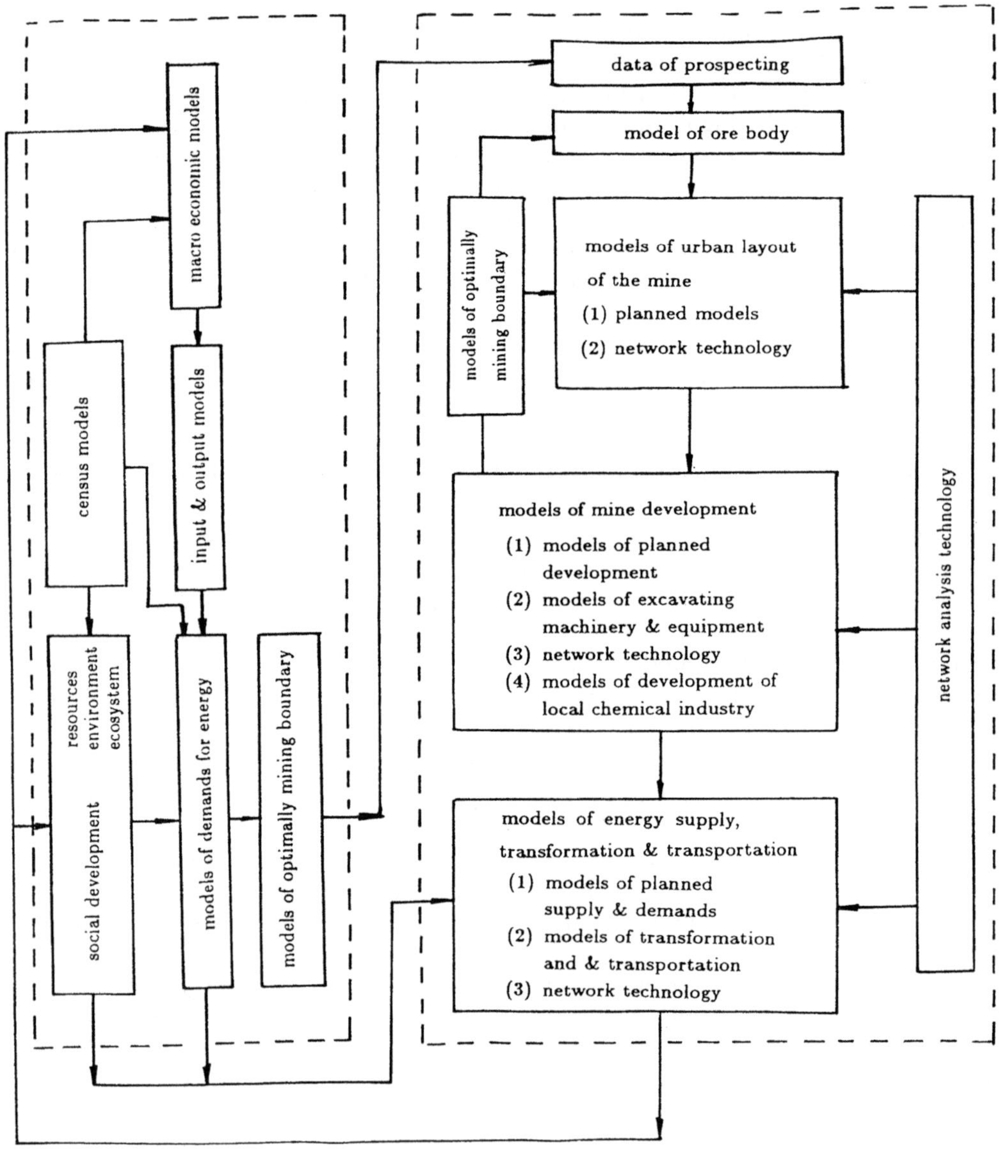
data of prospecting
model of ore body
macro economic models
census models
input & output models
models of optimally mining boundary
models of urban layout of the mine
(1) planned models
(2) network technology
network analysis technology
models of mine development
(1) models of planned development
(2) models of excavating machinery & equipment
(3) network technology
(4) models of development of local chemical industry
resources environment ecosystem
social development
models of demands for energy
models of optimally mining boundary
models of energy supply, transformation & transportation
(1) models of planned supply & demands
(2) models of transformation and & transportation
(3) network technology

equipment and mining technological process according to the local geological and mining conditions, and to equip the mines in reasonable steps according to different demands for coal in different period and to different scales of construction in different period. The machinery absorbs the major investment in the mine construction. The policy towards the investment in the machinery will be decisive to the production capacity of the mine and to the level of the mine's technology and equipment. This is a kind of dynamic investment, for which it is necessary to have in mind not only the advancement of the equipment, but also the rationality of the long-term operation cost and the maintenance fees. The development of such an optimal model is of great importance.

4. The model of power supply, transformation and transmission should be established.

In coal industry, transmission of coal energy is an important issue. Because of the characteristics of the distribution of coal resources in China, the country's pattern of energy transmission is featured by long-distance transmission of coal energy (or the so-called long-distance transportation of coal from the west to the east and from the north to the south).

Consumers mainly need the secondary energy or electric power. The primary energy, or coal, needs to be transformed by the power plant into the secondary energy, or electric power. Therefore, there are the following two ways for transmitting the coal energy: The first is to build power plants near the coal mining pits and then to transmit the electricity to consumers through ultrahigh-voltage technology. The second one is to transport the coal through transportation network to power plants at the area of consumers. The problem is: which way of the two is better?

First, by tackling this problem, we established a mathematical model, and made a conclusion.

Then, we came to the more practical problem. Since many factors should be taken into account in selecting the site of a power plant, including transportation network, geological conditions, hydrological and water-supply situation, the discharge ground, environment protection, ecological balance, and the economic development of the whole country and the localities, the qualified sites are rare. In China, some other departments concerned have nominated several sites on the basis of prospecting, analysis and research. So, we established another mathematical model, which integrates transportation of coal, selection of power plant sites, and transmission

5. The model of engineering network analysis should be established.

In China, the analysis method for networks has been widely applied. This is largely attributed to the work of Professor Hua Luogeng (Hau Loo-keng), my teacher, who had made great efforts to popularize applied mathematics. There are network analysis models in everyone of our subsystems. Moreover, we established network analysis models of the whole big system, which played an important role in coordinating the construction tempo of all the subsystems and the whole big system, and in allocating the funds and manpowers rationally so as to make the construction of the whole system both economic and of high quality.

Furthermore, we established the models of urban layout of the mine area and the models for the development of the local coal-chemical industry and the models of the environment protection, etc..

In China, the history of coal mine design can be roughly divided into three stages.

	Methods and Means	Scope	Research personnel
Stage I	Ordinary engineering design (Conventional single-mine design)	mine system	Coal mine design engineer
Stage II	1. Ordinary engineering design 2. System analysis for some subsystems of mine 3. OR application for some subsystem of mine	mine system and its subsystem	1.Coal mine design engineer 2. OR experts
Stage III	1. Ordinary engineering design 2. Large system (the mine, transportation, power plant, etc) analysis and all its subsystems analysis. 3. OR applied to (1) Coordination in large system and subsystems. (2) Selection of locations related to the large systems. (3) Design within systems. 4. Network analysis and technique.	1. National economic and society system 2. Mine system 3. Coal transportation system 4. Energy transformation system. 5. Electricity transmission 6. Other related system	1. Coal mine design engineer 2. OR experts 3. Experts in related fields

The present stage of China in coal mine design is the stage III. Our work is the first one at the stage III in China. It was begun in 1983.

3. Mathematical Models

This paper depicts merely three major models.

1. The model of optimum final pit contour.

A proper three dimensional euclidean space was established at the opencut coal mines. To introduce the mathematical model, let us first give some symbols and notations:

(1) Ω: Aggregate of three dimensional ore bodies.

(2) $H(x,y)$: The final surface of the pits of an opencut coal mine, which is just the surface we are going to find.

(3) $A(x,y,H(x,y),\alpha)$: The angle of the stabilized slope in the direction of angle α at point $(x,y,H(x,y))$ within Ω.

(4) D : The projection zone of Ω on the x,y plane. The boundary of the zone is denoted by S.

(5) $L(x,y)$: The function of the topographical surface of the mine.

(6) $f(x,y,z)$: The characteristic function of the value of the ore bodies at the point (x,y,z) of Ω.

The following is the general mathematical model of the optimum final pit contour.

Mathematical model:

$$\max \iint_D dxdy \int_{L(x,y)}^{H(x,y)} f(x,y,z)dz$$

subject to the following:

(i) slope bound

$$\frac{\partial H(x,y)}{\partial x}\cos\alpha + \frac{\partial H(x,y)}{\partial y}\sin\alpha \geq tgA(x,y,H(x,y),\alpha) \quad (x,y)\in D,\ 0\leq\alpha\leq 2\pi$$

(ii) topographical bounds

$$H(x,y) \geq L(x,y), \quad (x,y)\in D$$

(iii) boundary bounds

$$H(x,y) = L(x,y), \quad (x,y)\in S$$

The actual mining process is discrete, the ore body is divided into small cuboids whose length, width and height are in the proportion demanded by the slope angle in the x, y direction. $f(x,y,z)$ is defined on such cuboid. Its value is the value of this piece of coal after deducting the mining costs and transportation costs.

The discrete mathematical model is

$$\max \sum_{i=1}^{N}\sum_{j=1}^{M} f(i,j,k(i,j))$$

subject to the following: $\forall i = 1,2,\cdots,N, \forall j = 1,2,\cdots,M$,

(i) $|k(i,j+1) - k(i,j)| \geq 1$

(ii) $|k(i+1,j) - k(i,j)| \geq 1$

(iii) $0 \leq k(i,j) \leq L$

(iv) $k(1,j) = 1$ or $k(N,j) = 1$ or $k(i,M) = 1$ or $k(i,1) = 1$ (at least one holds)

(v) $k(i,j) \geq 0$ and be an integer.

The construction of such a mathematical model itself is based on the method of dynamic programming. So, it is called the dynamic programming method, which is a quick and fairly good method with little memory capacity requirement. However, the final results of the method cannot meet the requirements of the slope angle in some directions. We use the rolling-taper method to make revisions.

2. The optimization model for selecting the type of excavating equipment for the mine.

We suppose:

(a) The mining process is divided into m periods.

(b) The excavating equipment is divided into n types.

(c) x_{ij} : The amount of the i-type excavating equipment purchased at the jth period of the mining. $1 \leq i \leq n,\ 1 \leq j \leq m$.

(d) λ_{ij} : The cost of purchasing one piece of the i-type excavating equipment during the jth period of the mining. $1 \le i \le n,\ 1 \le j \le m$. (e) μ_{ij} : The value of one piece of the i-type excavating equipment purchased during the jth period of the mining. $1 \le i \le n,\ 1 \le j \le m$.

(f) ρ_{ij} : The damage coefficient in purchasing the i-type excavating equipment during the jth period of the mining. $1 \le i \le n,\ 1 \le j \le m$.

(g) α_{ij} : The manufacturer's minimum production size of the i-type excavating equipment during the jth period of the mining. $1 \le i \le n,\ 1 \le j \le m$.

(h) β_{ij} : The manufacturer's maximum capacity of producing the i-type excavating equipment during the jth period of the mining. $1 \le i \le n,\ 1 \le j \le m$.

(i) a_j : The minimum total amount of excavating equipment purchased during the time from the 1st period to the jth period, $1 \le j \le m$.

(j) b : The total expenditure on the purchase of the equipment during m periods.

$$\begin{aligned}
&\forall i = 1, 2, \cdots, n, \forall j = 1, 2, \cdots, m, \\
&\lambda_{ij} > 0, \mu_{ij} > 0, \alpha_{ij} \ge 0, \beta_{ij} \ge 0. \\
&b \ge 0.
\end{aligned}$$

Mathematical model:

$$\max \mu X^T$$

subject to

$$\begin{array}{ll}
\text{Quantities Constraints} & AX^T \ge a^T \\
\text{Expenditure Constraints} & \Lambda X^T \le b \\
\text{Manufacture Constraints} & x_{ij} \in K_{ij} \quad \text{and integer.}
\end{array}$$

where

(*i*) $X = (X_1, X_2, \cdots, X_n)$ where $\forall i = 1, 2, \cdots, n, \quad X_i = (x_{i1}, x_{i2}, \cdots, x_{im})$

(*ii*) $\mu = (\mu_1, \mu_2, \cdots, \mu_n)$ where $\forall i = 1, 2, \cdots, n, \quad \mu_i = (\mu_{i1}, \mu_{i2}, \cdots, \mu_{im})$.

(*iii*) $\Lambda = (\Lambda_1, \Lambda_2, \cdots, \Lambda_n)$ where $\forall i = 1, 2, \cdots, n, \quad \Lambda_i = (\lambda_{i1}, \lambda_{i2}, \cdots, \lambda_{im})$

(*iv*) $A = (A_1, A_2, \cdots, A_n)$ where $\forall i = 1, 2, \cdots, n, A_i = (a_{lj})$ and $\forall j = 1, 2, \cdots, m$,

$$a_{lj} = \begin{cases} \rho_{ij}^{l-j}, & j \le l; \\ 0, & j > l. \end{cases}$$

(*v*) $K_{ij} = \{0\} \cup [\alpha_{ij}, \beta_{ij}]$

(*vi*) $a = (a_1, a_2, \cdots, a_n)$.

This is a special problem of integer linear programming. The Matrix A is a block triangular matrix which can be computed with the method of decomposition. However, the difficulty is that the values of x_{ij} are taken generally in two or more closed intervals. In accordance with the speciality of the question, we gave a very efficient special algorithm, which is simple in programming and quick in computation, convenient for man-machine interactions and thus a good DSS method for policy-makers.

3. Transportation and transmission of coal energy.

3.1. Transportation of coal and transmission of electric power.

By solving a linear programming problem, we obtained an inequality:

$$(\alpha + \beta \cdot c(x)) \cdot t(x) \cdot E \leq \frac{r(1+r)^N}{(1+r)^N - 1}(a_1 - a_2) + \lambda dE + (\mu - u)$$

If the inequality holds, transporting coal is more economic than transmitting electricity. Otherwise, transmitting electricity is more economic than transporting coal. In the inequality,

α: The average cost of transporting coal per ton × Km;

β: The average rate of loss for transporting coal for one Km;

$C(x)$: The average price per-ton of coal with calorific value x;

$t(x)$: The amount of coal with calorific value x needed for generating one unit of electricity;

a_1: The average investment in one kilometer of electricity transmission line;

μ: The maintenance cost in one kilometer of electricity transmission line;

a_2: The average investment in one kilometer of coal transportation line;

u: The maintenance cost in one kilometer of coal transportation line;

λ: The average rate of loss for electricity transmission per Km;

d: The average price of per-unit electricity energy;

r: The interest rate.

In a certain historical period of a country (or region), the value of the right side of the inequality is specificed. The value is a criterion for judging which way of energy transmission is better. So, the value is called the critical value and is denoted by R. $f(x) = (\alpha + \beta \cdot c(x))t(x) \cdot E$. which is a decreasing function of x.

For a certain historical period in a country, the decision on whether to transport coal or to transmit electricity depends on x, the calorific value of coal. In China, x_0 or the critical value is about 3600 calorie per kg of coal.

3.2 Supply, transformation and transportation of coal energy.

According to the demands of the national and local macro economic development for coal energy, we drew up a coal mine development program, which lays the stress on making the policy towards the selection of equipment types. By doing so, the output of the coal mine in a certain period is fixed. The supply of coal energy is given by a supply-demand balancing programming.

In establishing the model of supply, transformation and transportation of the energy of the mine, the following problem should be considered. Suppose the output of the coal mine at different period (supply) and various consumer demands for electricity are known. Coal is transported through the transportation network to the power plant (whose site is to be decided). The power plant is a unit to transform the energy. It transforms coal energy into electric power and then the transmission network transmits the electric power to the consumers. This is a multi-period (from the longitudinal point of view) and a double-level (from a transverse point of view) complicated policy-making system with the energy transformation facilities. (The first level is the decision making transportation network while the second is the decision making of electric power transmission network). How to make the policy on the transformation facilities, and how to construct the coal transportation network and electric power transmission network so as to minimize the total investment of the multi-period and double-level decision-making system together with the operational costs is the decision-making problem for us to solve.

We introduce the following symbols and notations

$J_1 = (1, 2, \cdots, k_1)$: Index set of coal mine;

$J_2 = (k_1 + 1, \cdots, k_2)$: Index set of the nodes of the transportation network;

$J_3 = (k_2 + 1, \cdots, k_3)$: Index set of the power plant whose site is to be decided;

$J_4 = (k_3 + 1, \cdots, k_4)$: Index set of the source points of the power transmission network;

$J_5 = (k_4 + 1, \cdots, k_5)$: Index set of the nodes of the transmission network;

$J_6 = (k_5 + 1, \cdots, k_6)$: Index set of consumers.

l_{ij} : The length of the arc (i, j) of the transportation network;

d_{ij} : The maximum capacity of the arc (i, j) of the transportation network;

c_{ij} : The per-km cost for transporting one ton of coal through the arc (i, j) of the transportation network;

$\bar{l}_{ij}$ The length of the arc (i, j) of the electricity transmission network;

$\bar{d}_{ij}$: The maximum capacity of the arc (i, j) of the transmission network;

$\bar{c}_{ij}$: The per-km cost for transmitting per-unit of electricity through the arc (i, j) of the transmission network.

a_{ij} : The amount of jth coal energy source in the ith period;

$\bar{a}_{ij}$: The amount of the electricity of jth power plant in the ith period;

b_{ij} : The maximum capacity of jth power plant in the ith period;

$t(x)$: Transformation factor: the amount of coal with calorific value x needed for generating one unit of electricity;

x_{ijl} : The flow on the transportation network (i, j) in the lth period;

$\bar{x}_{ijl}$: The flow on the transmission network (i, j) in the lth period;

m_{itl} : The number of t-model generators put into operation at ith power plant in the lth period.

D : The set of the newly added arcs in the transportation network;

Ω : Set of all the arcs in the transportation network.

$\overline{D}$: The set of the newly added arcs in the transmission network;

$\overline{\Omega}$: Set of all the arcs in the transmission network.

S_{ij} : Construction time of the transportation network (i, j);

$\overline{S}_{ij}$: Construction time of the transmission network (i, j);

$$t_{ij} = \begin{cases} \min\{l | x_{ijl} > 0, & (i,j) \in D\} \\ \infty, \ \text{for all} & x_{ijl} = 0 \end{cases}$$

$$\bar{t}_{ij} = \begin{cases} \min\{l | \bar{x}_{ijl} > 0, & (i,j) \in \overline{D}\} \\ \infty, \ \text{for all} & \bar{x}_{ijl} = 0 \end{cases}$$

P_{ijl} : Investment cost in the lth period for $(i, j), (i, j) \in D$;

$\overline{P}_{ijl}$: Investment cost in the lth period for $(i, j), (i, j) \in \overline{D}$;

u_{ij} : Maintenance cost for (i, j) per-unit length, $(i, j) \in \Omega$; $\bar{u}_{ij}$: Maintenance cost for (i, j) per-unit length, $(i, j) \in \overline{\Omega}$;

r_{ij} : Loss rate of (i, j), $(i, j) \in \Omega$;

$\bar{r}_{ij}$: Loss rate of (i, j), $(i, j) \in \overline{\Omega}$;

g_{jtl} : Investment cost for each generator of model-t at jth power plant in the lth period;

$\lambda(m_{jtl})$: Installation time of m_{jtl} generators of model-t;
$\delta(x)$: Price of per-unit coal with calorific value x;
v_{jt} : Maintenance cost for each generator of model-t at jth power plant;
B_t : Amount of coal consumed by each generator of model-t per-period;
D_{jt} : Amount of electricity needed by jth point in lth period, $j \in J_6$.

Let the vector composed of x_{ijl} be X, the vector composed of $\overline{x}_{ijl}$ be $\overline{X}$, and the vector composed of m_{itl} be M. To give the mathematical model, let us introduce some functions.

Objective function: $f(X, \overline{X}, M) = \sum_{i=1}^{10} f(X, \overline{X}, M)$
Investment cost in D : $f_1(X, \overline{X}, M) = \sum_{(i,j)\in D} \sum_{l=t_{ij}-s_{ij}}^{t_{ij}} \frac{p_{ijl}}{(1+r)^l}$
Operational cost of Ω network: $f_2(X, \overline{X}, M) = \sum_{(i,j)\in\Omega} \sum_{l=1}^{m} \frac{x_{ijl} l_{ij} c_{ij}}{(1+r)^l}$
Maintenance cost of Ω network:

$$f_3(X, \overline{X}, M) = \sum_{(i,j)\notin D} \sum_{l=1}^{m} \frac{l_{ij} u_{ij}}{(1+r)^l} + \sum_{(i,j)\in D} \sum_{l=t_{ij}}^{m} \frac{l_{ij} u_{ij}}{(1+r)^l}$$

Compensation cost for losses of Ω network:

$$f_4(X, \overline{X}, M) = \sum_{(i,j)\in\Omega} \sum_{l=1}^{m} \frac{x_{ijl} r_{ij} \delta(x)}{(1+r)^l}$$

Investment fund in the power plant:

$$f_5(X, \overline{X}, M) = \sum_{j} \sum_{l=1}^{m} \sum_{k=l-\lambda(m_{jtl})} \sum_{t} \frac{m_{jtl} g_{jtk}}{(1+r)^k}$$

Maintenance cost of the power plant:

$$f_6(X, \overline{X}, M) = \sum_{j} \sum_{l=1}^{m} \sum_{t} \frac{m_{jtl} v_{jt}}{(1+r)^l}$$

Investment fund in $\overline{D}$:

$$f_7(X, \overline{X}, M) = \sum_{(i,j)\in\overline{D}} \sum_{l=\overline{t}_{ij}-\overline{s}_{ij}}^{\overline{t}_{ij}} \frac{\overline{p}_{ijl}}{(1+r)^l}$$

Operational cost of $\overline{\Omega}$ network:

$$f_8(X, \overline{X}, M) = \sum_{(i,j)\in\overline{\Omega}} \sum_{l=1}^{m} \frac{\overline{x}_{ijl} \overline{l}_{ij} \overline{c}_{ij}}{(1+r)^l}$$

Maintenance cost of $\overline{\Omega}$ network:

$$f_9(X, \overline{X}, M) = \sum_{l=1}^{m} \frac{\overline{l}_{ij} \overline{u}_{ij}}{(1+r)^l} + \sum_{(i,j)\in\overline{D}} \sum_{l=\overline{t}_{ij}}^{m} \frac{\overline{l}_{ij} \overline{u}_{ij}}{(1+r)^l}$$

Compensation cost for lossed of $\overline{\Omega}$ network:

$$f_{10}(X, \overline{X}, M) = \sum_{(i,j)\in\overline{\Omega}} \sum_{l=1}^{m} \frac{\overline{x}_{ijl}\overline{r}_{ij}\overline{c}_{ij}}{(1+r)^l}$$

Mathematical model:

$$\min \sum_{i=1}^{10} f_i(X, \overline{X}, M)$$

subject to the following constraints from (a) to (m).

(a) Constraints of the source points of Ω network:

$$\forall\ l = 1, 2, \cdots, m, \forall\ j \in J_1, \quad a_{lj} = \sum_i x_{jil}$$

(b) Requirements for the nodes of Ω network:

$$\forall l = 1, 2, \cdots, m, \forall j \in J_2, \quad \sum_i x_{ijl} = \sum_k x_{jkl}$$

(c) Requirements for the site of power plant of Ω network:

$$\forall\ l = 1, 2, \cdots, m, \forall\ j \in J_3, \quad \sum_i x_{ijl} = \sum_k x_{jkl} + \sum_{h=1}^{l} \sum_k m_{jkh} B_k$$

(d) Constraints to every arc of Ω network:

$$\forall (i, j) \in \Omega, \forall l = 1, 2, \cdots, m, \quad 0 \leq x_{ijl} \leq d_{ij}$$

(e) Constraints to the capacity of power plants:

$$\forall l = 1, 2, \cdots, m, \forall\ j \in J_3, \quad \sum_{h=1}^{l} \sum_k m_{jkh} B_k \leq b_{jl} \cdot t(x)$$

(f) Requirements for maintaining the first balance:

$$\forall\ l = 1, 2, \cdots, m, \quad \sum_{j\in J_3} \sum_{h=1}^{l} \sum_k m_{jkh} B_k = \sum_{k\in J_1} a_{lk}$$

(g) Requirements for maintaining the second balance:

$$\forall\ l = 1, 2, \cdots, m, \quad \sum_{j\in J_3} (\sum_{h=1}^{l} \sum_k m_{jkh} B_k) = \sum_{j\in J_3} \overline{a}_{lj} \cdot t(x)$$

(h) Requirements for the minimum electricity amount:

$$\forall \ l = 1, 2, \cdots, m, \quad \sum_{j \in J_3} \sum_{h=1}^{l} \sum_{k} m_{jkh} B_k \geq A_l$$

(i) Requirements for the source points of $\overline{\Omega}$ network:

$$\forall \ l = 1, 2, \cdots, m, \forall \ j \in J_3 \cup J_4, \quad \sum_{i} \overline{x}_{jil} = \overline{a}_{lj}$$

(j) Constraints of the branch points of $\overline{\Omega}$ network:

$$\forall \ l = 1, 2, \cdots, m, \forall \ j \in J_5, \quad \sum_{i} \overline{r}_{ij} \overline{x}_{ijl} = \sum_{k} \overline{x}_{jkl}$$

(k)

$$\forall \ l = 1, 2, \cdots, m, \forall \ j \in J_6, \quad \sum_{i} \overline{r}_{ij} \overline{x}_{ijl} = \sum_{k} \overline{x}_{jkl} + D_{jl}$$

(l) $\forall (i, j) \in \overline{\Omega}, \forall \ l = 1, 2, \cdots, m,$

$$0 \leq \overline{x}_{ijl} \leq \overline{d}_{ij}$$

(m)

$$m_{ijl} \geq 0, \quad \text{integer}.$$

Let us use an example to explain essential points for resolution.

Suppose that there are two power plants in the ith period ($i = 1, 2, 3$). Every power plant has extensive capacity minimum unit 0.1 M kw. The maximum capacity of 1st power plant is 2 units; the maximum capacity of 2nd power plant is 5 units. They transform coal into electricity: 3 units in the 1st period; 4 units in the 2nd period; 6 units in the 3rd period. Let S denote the set of mine sites. Let A_{ij} denote the jth feasible combinations of power-capacity distribution among the power plants in ith period. We construct a graph as shown in Figure 1. The vertex set consists of S, all A_{ij}, and a sink T. The arc set consists of all SA_{1j} and $A_{3j}T$ and all possible $A_{ij}A_{i+1,k}$. ($A_{ij}A_{i+1,k}$ is possible if it is possible to change A_{ij} into $A_{i+1,k}$.)

The length of each arc is assigned in the following way.

(1) For every j, there is an arc (S, A_{1j}) from S to A_{1j} of the first period. The length of the arc is defined as follows $d(S, A_{1j}) =$ the investment and optimal transportation cost for transporting the coal produced during the first period to the power plant selected by A_{1j}, plus the investment and operational cost of the power plant selected by A_{1j}, plus the investment and operational cost of the optimal transmission network from A_{1j} to the electricity consumer.

(2) For every k and j, there is an arc $(A_{kj}, A_{k+1,l})$ from A_{kj} of the kth period to $A_{k+1,l}$ of the $(k+1)$th period. The length of the arc is defined as that $d(A_{kj}, A_{k+1,l}) =$ the newly added investment and optimal transportation cost for transporting the coal produced during the $(k+1)$th Period to the power plants selected by $A_{k+1,l}$, plus the newly added investment and operational cost of the power

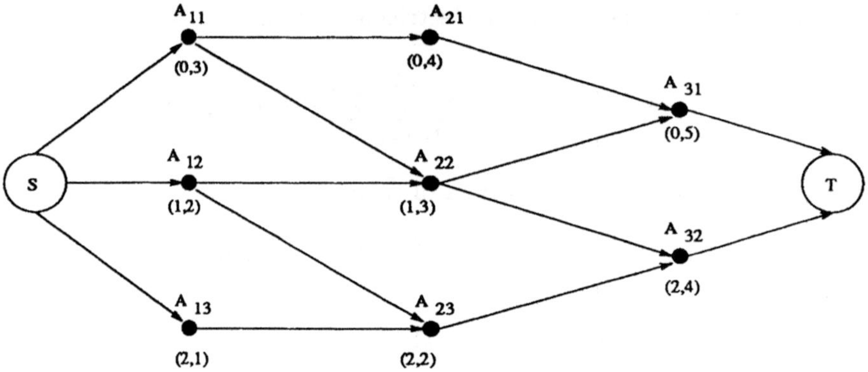

Fig. 1.

plants selected by $A_{k+1,l}$ expanded from power plants by A_{kj}, plus the investment for expanding the transmission network and to consumer.

(3) T is the sink. Set $d(A_{3j}, T) = 0, \ \ \forall\ j$.

The shortest path from S to T is the solution.

4. Flexible OR

The resolution of a large decision-making problem lies in a deep understanding of the real problem, the analysis of the large system and various subsystems, the comprehensiveness of the analysis, and the concerted efforts of the researchers in multi-discipline.

It is an important step to formulate the real problem into a resolvable model. We emphasize on the idea of integrating the establishment of the model with the algorithm of resolving the model. But, mathematical OR scientists are more interested in abstraction theory and algorithms with universal significance. It is no doubt that this is an important way. Besides, the computing software package with universal significance is of important value, which may also be used directly to solve practical problems. We hold that the vitality of real problems coming from practices (including economic system, military system and social system) lies in their particularity, or that any real problem is a particular problem. If we notice fully their particularity, it is possible to establish particular models and find particular solutions. Perhaps, it is more geared to actual circumstances, quicker to find the solution, and more acceptable for the decision-makers because it is more convenient for them to interact with the computer, easier for them to make revisions. Thus, it is more elegant in the sense of applicability and is just as noble as the pure mathematics.

In the research of applications, we put forward a new idea and a technique of applied mathematical programming, which we called the change-object method, or change-constraint method, or object and constraint-changing method. The fundamental idea of the change-object method is as follows: for a programming problem, objective and constraints are the two parts of the model. When the set of constraints is fixed, there will be many objective functions with the same point of the feasible set as their optimal solution. The object of our practical problem is only one of them. It might be one of the problems that are most difficult to resolve when our real objective function cooperates with the same set of constraints. We change the objective function, but with the same optimal solution, so that the new objective function together with the same constraint set form a special problem which can be readily solved by some special algorithm. In our research, this idea and the technique are very effective and have helped us to resolve some very difficult problems. We called it flexible OR.

The essence of the flexible OR lies in the fact that its theory and method are results of fully utilizing the existing bank of theories and methods accumulated by the human being and applying these theories and methods, ingeniously and comprehensively. It must be stressed that a particular problem should be resolved by a particular method or even by changing the objective function or the constraint set.

The flexible OR is quite different from the sensitivity analysis and simplifying-model method. The following table give a simplified explanation of their differences.

	Idea	Feature	Notes
Flexible OR	Change the objectives and/or constraints in order to find particular method for a particular problem.	After that the problem may become bigger, but it becomes essentially easier to solve	(1) pre-optimal analysis (2) equivalent model
Sensitivity analysis	Analyze the changes in the optimal solution due to changes in the data.	No change in problem but deal with only the sensitivity of the solution	post-optimal analysis
Simplify model	Simplifying the objective and constraints so as to make a difficult problem easier to solve	Problem seemingly become easier but may be oversimplified.	Simplied model

The ideas and techniques of the flexible OR were utilized in the optimal design of opencut coal mine systems. All the three major models listed emphasize on the particular methods, which can lead to computing softwares for special usage. Of course, the existing software package were also used in resolving some of the problems (such as the problem of coal transportation). These results have played an important role in construction of Jungar opencut coal mine and produced great economic benefit and social effect.

We also stress multi-level modeling: The first level is to make initial analysis and establish an initial model so as to make it clear what should be done and how to do it. The second level is to revise the analysis, establish a model and then repeat the process many times until the real problem is resolved satisfactorily. Our final aim is to solve the realistic problem.

Our research project brings about the following outputs for the Jungar mine:

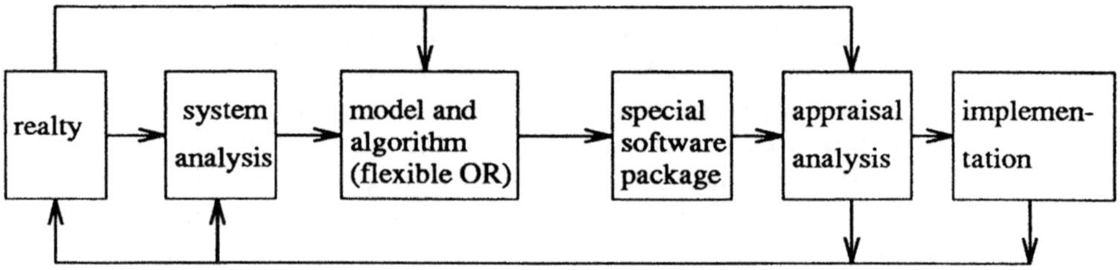

Fig. 2.

· Saving about 7% of the investment fund.

· Shortening the construction time by more than 2 years.

Now the Chinese government and local governments have started to apply our idea and technique OR to other projects. A new enterprise consisting of mine system, railway system, electric power system and others has established to coordinate all the parts involved. This is the first one in the coal industry in China. We believe that it is a very important step in the course of reform in our country.

References

1. Bellman, R.E. and Dreyfus, S.E., *Applied Dynamic Programming*, (Princeton University Press, 1962).
2. Dantzig, G.B., *Linear Programming and Extension*, (Princeton University Press, 1963).
3. Gass, Saul I., *Linear Programming: Methods and Application*, (McGraw-Hill Co., New York, 1985).
4. Gass, Saul I., Decision-aiding models: validation, assessment, and related issues for policy analysis, *Operation Research*, Vol. 31, No. 4, (1983).
5. Garfinkel. R.S. and Nemhauser, G.L., *Integer Programming*, (John Wiley & Sons, New York, 1972).
6. Hua Loo-Keng and Wang Yuan, *Popularizing Mathematical Methods in the People's Republic of China: Some Personal Experiences*, (World Publishing Co., Beijing, 1989).
7. *Proceeding of IIASA Conference 76*, a Conference Sponsored by International Institute for Applied System Analysis, Laxenburg, 1976.
8. Lipkewich, M.P. and Borggman, L., *Two and Three Dimensional pit Design Optimization Techniques*, (New York, AIME 1969).
9. [9] Song Chien and Yu, Jing-Yuan, *Population Control in China: Theory and Application*, (New York: Praeger Special Studies, 1985).

ON THE STRICTLY COMPLEMENTARY SLACKNESS RELATION IN LINEAR PROGRAMMING

SHUZHONG ZHANG
Econometric Institute
Erasmus University Rotterdam
P.O. Box 1738
3000 DR Rotterdam
The Netherlands

Abstract. Balinski and Tucker introduced in 1969 a special form of optimal tableaus for LP, which can be used to construct primal and dual optimal solutions such that the complementary slackness relation holds strictly. In this paper, first we note that using a polynomial time algorithm for LP Balinski and Tucker's tableaus are obtainable in polynomial time. Furthermore, we show that, given a pair of primal and dual optimal solutions satisfying the complementary slackness relation strictly, it is possible to find a Balinski and Tucker's optimal tableau in *strongly* polynomial time. This establishes the equivalence between Balinski and Tucker's format of optimal tableaus and a pair of primal and dual solutions to satisfy the complementary slackness relation strictly. The new algorithm is related to Megiddo's strongly polynomial algorithm that finds an optimal tableau based on a pair of primal and dual optimal solutions.

1. Introduction

Consider the following linear program

$$\begin{array}{lll} (P) & \min & c^T x \\ & \text{s.t.} & Ax = b \\ & & x \geq 0 \end{array}$$

where A is an $m \times n$ matrix, $b \in \mathcal{R}^m$ and $c \in \mathcal{R}^n$, and its dual problem

$$\begin{array}{lll} (D) & \max & b^T y \\ & \text{s.t.} & y^T A \leq c^T. \end{array}$$

Assume that both (P) and (D) have optimal solutions, and $rank\{A\} = m$.

It is well known that for optimal solutions of (P) and (D), the complementary slackness relation holds (see Chvátal [2]). Moreover, the following stronger statement can be proved.

Lemma 1 *There exist optimal solutions to (P) and to (D) such that*

$$x_i^*(c_i - y^{*T} A_i) = 0 \textit{ and } x_i^* + (c_i - y^{*T} A_i) > 0$$

for every $i = 1, 2, \cdots, n$.

Ding-Zhu Du and Jie Sun (eds.), Advances in Optimization and Approximation, 347–361.

Proof. See Goldman and Tucker [4], Dantzig [3] and Schrijver [9]. □

To distinguish from the normal complementary slackness relation, which does not exclude the possibility that both x_i^* and $(c_i - y^{*T} A_i)$ are zero, we call the relation stated in Lemma 1 the ***strictly complementary slackness relation***. Clearly, the concept of the strictly complementary slackness relation is interesting only for degenerate problems. Although the existence of a strictly complementary primal-dual pair is easily derived using Farkas' Lemma (cf. [9]), it is important to know how such a pair of solutions can be constructed. In 1969, Balinski and Tucker [1] showed a constructive way to get such strictly complementary pair in finite time. In their approach, a special format of the optimal tableau for linear programming was investigated. We will refer to this type of optimal tableau as ***the Balinski-Tucker tableau*** in this paper.

The recent development of the interior point methods for linear programming has opened new areas for research. In [6] an attempt was made to base the duality theory and the sensitivity analysis entirely on the interior point methodology. Typically, solutions produced by primal-dual interior point algorithms converge to the analytical center of the relative interior of the optimal face. Compared to the classical simplex method, where only basic feasible solutions are searched, this non-vertex property of the interior point methods was regarded as a disadvantage at early stages. To get a true vertex optimal solution from an approximative solution, a purification procedure can be used (cf. [7] and [12]). However, there are differences between a vertex optimal solution and an optimal basic solution, for there can be many bases corresponding to the same vertex. In [8], Megiddo showed that knowing only a primal optimal solution does not help in general to find an optimal basis which should be both primal and dual feasible. However, in the same paper he proved that if a primal optimal solution and a dual optimal solution are available simultaneously, then finding an optimal basis can be done in strongly polynomial time.

In this paper, we will show that knowing non-vertex primal and dual optimal solutions (in the degenerate case) actually we get more information. More precisely, we present strongly polynomial-time algorithms which can be used to construct "lexicographically feasible" extremal directions around the purified vertices. As a consequence, if we have both primal- and dual optimal solutions belonging to the relative interior of the respective optimal face, then a Balinski-Tucker tableau can be constructed in strongly polynomial time. In some sense, this result establishes that a Balinski-Tucker tableau and a pair of primal- and dual- optimal solutions in the relative interior of the optimal faces are equivalent. We remark also that a Balinski-Tucker tableau can be obtained in polynomial time in the ***weak*** sense, using a polynomial algorithm for LP. Our analysis is closely related to Megiddo's algorithm that finds an optimal basis using a pair of primal-dual solutions.

We organize this paper as follows. In the next section we discuss the properties of solutions satisfying the strictly complementary slackness relation. In the same section, the Balinski-Tucker tableau will be introduced, and its basic properties discussed. In Section 3, Megiddo's algorithm for finding an optimal basis will be included. New algorithms for finding Balinski-Tucker's tableau and the correctness proofs will be presented in Section 4. Finally, we conclude the paper in Section 5.

Before proceeding we mention the notations we use in this paper. We denote matrices by capital letters, and vectors by lower case letters with subscript denoting the coordinate. Index sets are denoted either by capital letters or by Greek letters. For a matrix $A = (a_1, \cdots, a_n)$ and an index set I, A_I denotes the submatrix of A whose columns belong to I, i.e. $A_I = \{a_j : j \in I\}$. The same rule applies to vectors. To ease the notation we do not distinguish a_j and A_j. A basis for the problem (P), denoted by B, is a maximal subset of $\{1, 2, \cdots, n\}$ such that the corresponding columns in A_B are independent. Finally, a tableau $T(B)$ associated with a basis B is a matrix given as follows:

$$T(B) = \begin{pmatrix} c^T - c_B^T A_B^{-1} A & -z_0 \\ A_B^{-1} A & A_B^{-1} b \end{pmatrix}$$

where $z_0 = c_B^T A_B^{-1} b$. If $A_B^{-1} b \geq 0$ we call the tableau *primal feasible* and we call it *dual feasible* if $c^T - c_B^T A_B^{-1} A \geq 0^T$. We call in this paper the matrix $A_B^{-1} A$ a *sub-tableau.*

2. Strictly Complementary Slackness and Balinski and Tucker's Tableau

Solutions for the primal and for the dual problem satisfying the strictly complementary slackness relation can be characterized by their topological position in the optimal faces. Let

$$\mathcal{F}_P = \text{ optimal solution set of (P)}$$

$$\mathcal{F}_D = \text{ optimal solution set of (D)}$$

and

$$\overset{o}{\mathcal{F}}_P = \text{ relative interior of } \mathcal{F}_P$$

$$\overset{o}{\mathcal{F}}_D = \text{ relative interior of } \mathcal{F}_D.$$

The definition of relative interiors can be found, e.g., in [9].

We have the following result.

Lemma 2 *Solutions x^* and y^* satisfy Lemma 1 if and only if $x^* \in \overset{o}{\mathcal{F}}_P$ and $y^* \in \overset{o}{\mathcal{F}}_D$.*

Proof. See Corollary 2.1 of [5].

□

Remark 1 By the definition, a single point also belongs to its 0-dimensional relative interior.

To see how solutions satisfying Lemma 1 can be obtained we introduce a special tableau form used by Balinski and Tucker in [1]. First we note the following result proved in [1].

Lemma 3 *Starting from an arbitrary tableau (matrix) it is always possible to get one of the following two tableau forms using pivot operations:*

$$(i)\quad \begin{array}{|ccccc|}\hline \cdot & \cdot & \cdots & \cdot \\ \cdot & \cdot & \cdots & \cdot \\ \oplus & \oplus & \cdots & \oplus \\ \cdot & \cdot & \cdots & \cdot \\ \cdot & \cdot & \cdots & \cdot \\ \hline \end{array}$$

$$(ii)\quad \begin{array}{|ccccc|}\hline \cdot & \cdot & \ominus & \cdot & \cdot \\ \cdot & \cdot & \ominus & \cdot & \cdot \\ \vdots & \vdots & \vdots & \vdots & \vdots \\ \cdot & \cdot & \ominus & \cdot & \cdot \\ \hline \end{array}$$

where $\oplus$ stands for either a positive number or zero, and $\ominus$ for a negative number or zero.

Proof. See Corollary 1 of [1]. For an alternative proof we can also use the criss-cross method as follows (cf. [10] or [11]). Fix one row and one column. If there is no negative elements in the row or no positive elements in the column, then stop. Otherwise, among the negative elements in the row and the positive elements in the column, select the one with the minimal index. Suppose it is a row element (for the other case follow the similar procedure). Then for the column corresponding to the chosen element find positive elements. If there is none, stop. Otherwise select the one with the smallest index. Pivot on this position. Since this criss-cross method guarantees no repetition of the bases, after a finite amount of pivoting steps, we will have to terminate with a tableau belonging to one of the two above mentioned formats.

□

Using this lemma we are now able to present the following result of Balinski and Tucker [1].

Theorem 1 *There exists an optimal tableau in the following format*

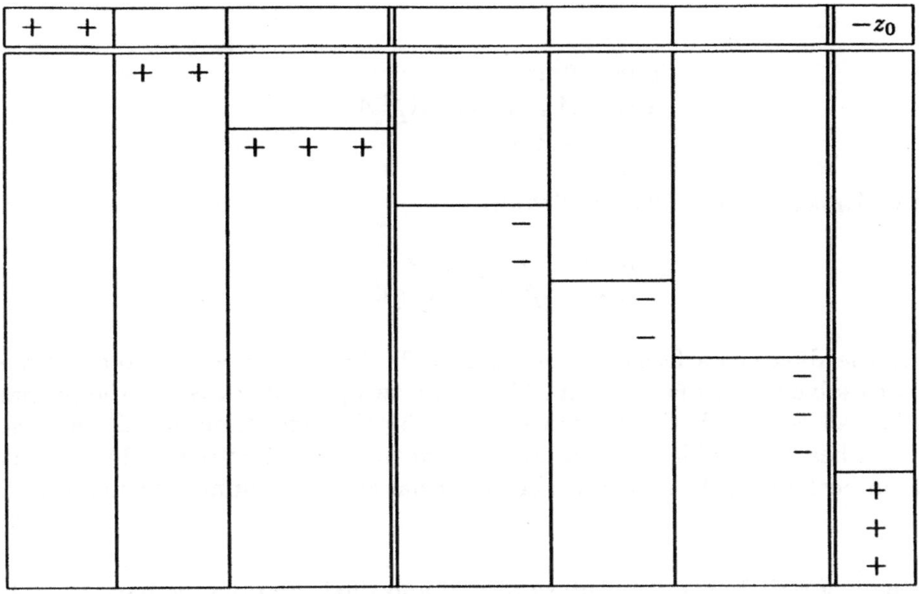

where + stands for a positive number and − for a negative number, and all the unspecified numbers above the staircase are zeros.

Proof. Applying Lemma 3 recursively on the north-east corner of the matrix which is not yet covered by positive rows or negative columns or the positive elements of the right-hand-side vector, the theorem easily follows.

□

Remark 2 Balinski-Tucker type tableaus are not unique in general.

Remark 3 A Balinski-Tucker tableau can also be called a *lexicographically optimal tableau.* Information about the dimension of the optimal faces is easily seen from a Balinski-Tucker tableau. The number of columns corresponding to the negative part of the staircase (columns between the two vertical double-lines in the displayed tableau), except for the basic ones, is equal to the dimension of the primal optimal face. The number of the rows corresponding to the positive part of the staircase equals the dimension of the dual optimal face.

Interestingly, using a polynomial algorithm for LP we have the following result.

Lemma 4 *It is possible to construct a Balinski-Tucker tableau in polynomial time.*

Proof. We need only to show that the tableau forms displayed in Lemma 3 can indeed be obtained in polynomial time. To see this, for a given sub-tableau $A_B^{-1}A$ we fix one arbitrary column, say $A_B^{-1}A_n$. Consider now the following linear program:

$$\begin{array}{ll} \min & 0^T x \\ \text{s.t.} & A_B^{-1}\bar{A}x = -A_B^{-1}A_n \\ & x \geq 0 \end{array}$$

where $\bar{A}$ is $A_{B\cup N\setminus\{n\}}$. The dual problem is:

$$\begin{array}{ll} \max & -(A_B^{-1}A_n)^T y \\ \text{s.t.} & y^T A_B^{-1}\bar{A} \leq 0^T. \end{array}$$

Clearly the dual has a feasible solution $y = 0$. Now apply a polynomial time algorithm to solve the primal problem. There are two possible cases: 1) the primal problem has an optimal basic solution reported by the algorithm; and 2) the primal problem has no feasible solution, and a dual unbounded extremal direction is reported. In both cases, the required tableau format given in Lemma 3 would follow. □

For a pair of strictly complementary primal and dual solutions we have the following important property.

Lemma 5 *For any $x^* \in \overset{o}{\mathcal{F}}_P$ and $y^* \in \overset{o}{\mathcal{F}}_D$, the index sets $I := \{i : x_i^* > 0\}$ and $J := \{j : c_j - {y^*}^T A_j > 0\}$ form a unique partition of $\{1, 2, \cdots, n\}$ which is independent of x^* and y^*.*

Proof. See, e.g., [6]. □

It is easily seen that if a Balinski-Tucker tableau is available, then a pair of optimal solutions for (P) and (D) satisfying the strictly complementary slackness relation can be obtained as shown in the following theorem. Remark that due to Lemma 2 this pair of optimal solutions belong to the relative interior of the corresponding optimal faces.

Theorem 2 *If a Balinski-Tucker tableau is given, then a pair of strictly complementary optimal solutions for (P) and (D) can be found in strongly polynomial time.*

Proof. See Section 4 of Balinski and Tucker [1]. □

The main purpose of this paper is to show that the reverse of the above theorem is also true. Namely, we are going to show that if such a pair of strictly complementary solutions for (P) and (D) are known, then we can construct a Balinski-Tucker tableau in strongly polynomial time. Before doing this, we first introduce in the next section Megiddo's algorithm which is useful for our analysis.

3. Megiddo's Algorithm

Megiddo [8] showed the following. To find an optimal tableau (basis), given an optimal solution x^* of (P), is in general as difficult as solving (P) from scratch. However, if *a pair* of optimal solutions x^* (of (P)) and y^* (of (D)) are available, it is possible to construct an optimal tableau (basis) in strongly polynomial time. We shall now present in the following Megiddo's strongly polynomial-time algorithm for finding an optimal basis.

Megiddo's Algorithm

- Input: $\bar{x} \in \mathcal{F}_P$ and $\bar{y} \in \mathcal{F}_D$.
- Output: An optimal basis B and the tableau $T(B)$.

Step 1 Let $X := \{i : \bar{x}_i > 0\}$ and $Y := \{j : c_j - \bar{y}^T A_j = 0\}$ (hence $X \subseteq Y$ due to the complementarity).

Step 2 If columns of A_X are linearly independent, go to Step 4.

Step 3 Choose $j \in X$ such that A_j is linearly dependent on $A_{X\setminus j}$. Find w such that $A_j = A_{X\setminus j} w$ and let

$$d_k := \begin{cases} 1, & \text{for } k = j \\ -w_k, & \text{for } k \in X \setminus j \\ 0, & \text{otherwise.} \end{cases}$$

Let

$$t := \min\{\frac{\bar{x}_i}{d_i} : \text{ for } i \text{ such that } d_i > 0\}$$

and $\bar{x} := \bar{x} - td$ and $X := \{i : \bar{x}_i > 0\}$. Go to Step 2.

Step 4 If there is $j \in Y$ such that A_j is linearly independent of A_X, then $X := X \cup j$ and repeat Step 4.

Step 5 If $rank\{A_X\} = m$, let $B := X$; construct the corresponding tableau $T(B)$ and stop.

Step 6 Let $j \notin X$ and A_j is linearly independent of A_X. Solve

$$\begin{aligned} z^T A_Y &= 0^T \\ z^T A_j &= 1. \end{aligned}$$

Let

$$t := \min\{\frac{c_k - \bar{y}^T A_k}{z^T A_k} : \text{ for } k \text{ such that } z^T A_k > 0\}.$$

Let $\bar{y} := \bar{y} + tz$ and $Y := \{j : c_j - \bar{y}^T A_j = 0\}$. Go to Step 4.

Theorem 3 *Megiddo's algorithm is correct and runs in strongly polynomial time.*

Proof. Clearly, during the procedure, $\bar{x}$ and $\bar{y}$ stay feasible and satisfy the complementary slackness relation and therefore remain optimal. Upon termination, the set X is a basis with $\bar{x}$ the corresponding optimal basic solution and $\bar{y}$ the corresponding dual optimal basic solution. The strong polynomiality follows immediately after noticing that each time at Step 3, $|X|$ strictly decreases, and at Step 6, $|Y|$ strictly increases.

□

4. The New Algorithms

In this section we shall present three new algorithms. The last algorithm is to construct a Balinski-Tucker tableau, and it is actually a summary of the first two algorithms and Megiddo's algorithm. We start our discussion by introducing some notations.

For a vector v in $\mathcal{R}^n$, we denote its positive index set by $\sigma(v) := \{i : v_i > 0\}$, and its negative index set by $\chi(v) := \{i : v_i < 0\}$, and the nonzero index set $\pi(v) := \sigma(v) \cup \chi(v)$.

For a feasible basis B, let the corresponding basic solution of (P) be $\bar{x}$. Clearly, $\bar{x}_B = A_B^{-1}b$ and $\bar{x}_N = 0$. We call a direction $r \in \mathcal{R}^n$ a *primal extremal direction w.r.t.* B if and only if $Ar = 0$ and there is $j \in N$ with $r_j > 0$ and $\pi(r) \subseteq B \cup j$. Similarly, if we let the corresponding dual basic solution be $\bar{y}$, then $\bar{y}^T A_B = c_B^T$. We call a direction $d \in \mathcal{R}^m$ a *dual extremal direction w.r.t.* B if and only if there is $i \in B$ such that $d^T A_i < 0$ and $\pi(d^T A) \subseteq N \cup i$. It is well known that the primal extremal directions are extended nonbasic column vectors in the sub-tableau and the dual extremal directions are the row vectors in the sub-tableau. In case of degeneracy, however, not all extremal directions are feasible. Note that extremal directions are related to the basis. In fact, for degenerate problems, finding a tableau with feasible extremal direction is as difficult as solving a linear program. To link extremal directions with Balinski and Tucker's tableau, we note the following lemma.

Lemma 6 *Suppose we have an optimal tableau $T(B)$ with primal optimal basic solution $\bar{x}$ and dual optimal basic solution $\bar{y}$. Moreover, suppose that there exist some primal extremal directions $r^{(1)}, r^{(2)}, \cdots, r^{(\nu)}$ w.r.t. B, and some dual extremal directions $d^{(1)}, d^{(2)}, \cdots, d^{(\tau)}$ w.r.t. B, in such a way that:*
1). $\bar{x}+\sum_{j=1}^{t} \delta_j r^{(j)} \in \mathcal{F}_P$ for $t = 1, 2, \cdots, \tau-1$, where δ_j are some positive numbers;
2). $\bar{y}+\sum_{j=1}^{s} \epsilon_j d^{(j)} \in \mathcal{F}_D$ for $s = 1, 2, \cdots, \nu-1$, where ϵ_j are some positive numbers;
3). $\bar{x} + \sum_{j=1}^{\tau} \delta_j r^{(j)} \in \overset{o}{\mathcal{F}}_P$;
4). $\bar{y} + \sum_{j=1}^{\nu} \epsilon_j d^{(j)} \in \overset{o}{\mathcal{F}}_D$.
Then the tableau is a Balinski-Tucker tableau.

Proof. From 1) and 2) we know that

$$\chi(r^{(t)}) \subseteq \sigma(\bar{x} + \sum_{j=1}^{t-1} \delta_j r^{(j)})$$

for $t = 1, 2, \cdots, \tau$ and,

$$\sigma({d^{(s)}}^T A) \subseteq \sigma(c^T - (\bar{y} + \sum_{j=1}^{s-1} \delta_j d^{(j)})^T A)$$

for $s = 1, 2, \cdots, \nu$.

By definition, it follows from 3) and 4) that

$$\sigma(\bar{x} + \sum_{j=1}^{\tau} \delta_j r^{(j)}) \text{ and } \sigma(c^T - (\bar{y} + \sum_{j=1}^{\nu} \delta_j d^{(j)})^T A)$$

form the optimal partition for strictly complementary pairs. Now put columns in the sub-tableau corresponding to $r^{(1)}, r^{(2)}, \cdots, r^{(\tau)}$ in the most right part of the sub-tableau with the right-to-left order, and arrange rows in $T(B)$ corresponding to $d^{(1)}, d^{(2)}, \cdots, d^{(\nu)}$ in the upper part of the sub-tableau from top to bottom.

With this construction it is clear that $T(B)$ is a Balinski-Tucker tableau.

□

The goal of the next two algorithms is in fact to find a basis set and extremal directions which can fulfill conditions of Lemma 6. Observe that if we have a vertex optimal solution and another optimal solution, then by connecting them we get a feasible direction for the vertex solution. Purify this direction to the extremal ones we obtain at least one feasible extremal direction. Applying this procedure repeatedly for a sequence of shrinking subproblems we finally get a set of needed extremal directions. Since we are interested in extremal directions induced in one tableau, the basic index must be recorded in the procedure in a consistent way. Another important observation is that finding primal extremal directions and finding dual ones can be done separately, due to the complementarity. The algorithms below are based on these observations. The primal part and the dual part are separately treated in two algorithms to ease the presentation. In the primal algorithm we call an index set I a *circuit* if the columns in A_I form a minimal dependent set, namely by deleting a certain column amongst them, the remaining ones will become independent.

The Primal Algorithm (PA)

- Input: $\bar{x} \in \overset{o}{\mathcal{F}}_P$.
- Output: A classification of indices in $\sigma(\bar{x})$ to either B or N; and a set of primal extremal directions $r^{(l)}$.

Step 0 Use Megiddo's algorithm to obtain a primal vertex optimal solution x^*.

Denote $I := \sigma(\bar{x})$ and $d := \bar{x} - x^*$.

Let $B := \{i : i \in \sigma(x^*)\}$, $N := \emptyset$ and $l := 1$.

Step 1 If $I = B \cup N$, stop.

Step 2 Let $K := I \setminus (B \cup N)$ and $d' := d$.

Step 3 If $B \cup K$ is not a circuit, go to Step 4.

Let

$$\lambda := \min\{\frac{d_i}{d'_i} : i \in K\}$$

and

$$\begin{aligned} d &:= d - \lambda d' \\ K' &:= \{i : d_i = 0 \text{ and } i \in K\}. \end{aligned}$$

Take any $i_l \in K'$ and let

$$\begin{aligned} r^{(l)} &:= d', \\ N &:= N \cup i_l, \\ B &:= B \cup K \setminus i_l, \\ l &:= l + 1. \end{aligned}$$

Go to Step 1.

Step 4 Find $f \in \mathcal{R}^n$ such that

- $Af = 0$;
- $\pi(f) \subseteq B \cup K$;
- $f_K \not\leq 0$ and f_K independent of d'_K.

Let

$$\lambda := \min\{\frac{d'_i}{f_i} : f_i > 0 \text{ and } i \in K\}$$

and

$$\begin{aligned} d' &:= d' - \lambda f \\ K &:= K \setminus \{i : d'_i = 0\}. \end{aligned}$$

Go back to Step 3.

Remark 4 We may call a cycle "Step 3 – Step 4 – Step 3" an ***inner-loop*** and a cycle from Step 1 to the next Step 1 an *outer-loop*. It is clear that the nonbasis set N is monotonically expanding after performing one outer-loop, and the working set K is decreasing during each inner-loop. In any loop, for $i \notin \sigma(x^*)$ we have $d_i \geq 0$, $\sigma(d') \subseteq \sigma(d)$ and $\pi(d') \subseteq B \cup K$ which means that at the beginning of an outer-loop, $d_N = d'_N = 0$. During an inner-loop, $d'_K > 0$. At the end of an inner-loop we have $d_{K'} = 0$. Because at Step 4 f_K is chosen to be independent of d'_K, and so when finishing Step 4 we always have $K \neq \emptyset$. Moreover, because at the end of Step 4 we have $A_{B\cup K} d'_{B\cup K} = 0$ and $d'_K > 0$, so we conclude that columns in the matrix $A_{B\cup K}$ are not independent. At the end of an inner-loop (Step 3) we have $d'_{i_l} > 0$ and this shows that by removing A_{i_l} the remaining columns in $A_{B\cup K}$ are independent. Finally, we remark that when the algorithm is terminated we have $\pi(d) \subseteq \sigma(x^*)$.

Now we prove the following result.

Theorem 4 *For the algorithm (PA) the following hold:*
1). The algorithm runs in strongly polynomial time;
2). All the produced directions $r^{(l)}$ $(1 \le l \le \tau)$ are primal extremal directions;
3). There exist positive numbers δ_l $(1 \le l \le \tau)$ with $1 \gg \delta_1 \gg \delta_2 \gg \cdots \gg \delta_\tau > 0$ such that

$$x^* + \sum_{l=1}^{t} \delta_l r^{(l)} \in \mathcal{F}_P$$

for $1 \le t \le \tau - 1$ *and*

$$x^* + \sum_{l=1}^{\tau} \delta_l r^{(l)} \in \overset{o}{\mathcal{F}}_P .$$

Proof. The strong polynomiality of the algorithm is easy to see. Because after each inner-loop, the set K strictly expands, and after each outer-loop, the set $B \cup N$ strictly expands. To prove that the directions $r^{(l)}$ $(1 \le l \le \tau)$ are extremal directions defined on the same tableau, we will have to check that for every $r^{(l)}$, $\pi(r^{(l)}) \subseteq B \cup i_l$. Notice that we have $d_N = 0$ after performing one outer-loop, where N is the set of nonbasic variables so far, and so the elements in N will not contribute as nonzero elements in the later constructed extremal directions. This proves the above statement. The last part of the theorem follows from the relation

$$\chi(r^{(t)}) \subseteq \sigma(x^*) \cup \bigcup_{l=1}^{t-1} \sigma(r^{(l)})$$

which can be proved by induction on t for $t = 1, 2, \cdots, \tau$. Remark that at the end of the t-th outer-loop we have $r_K^{(t)} > 0$ and $\pi(r^{(t)}) \subseteq B \cup K$.

Moreover,

$$\sigma(x^*) \cup \bigcup_{l=1}^{\tau} \sigma(r^{(l)}) = I,$$

and so the theorem is proved.

□

The second algorithm treats the dual part. It works in a similar way as its primal counterpart.

The Dual Algorithm (DA)

- Input: $\bar{y} \in \overset{o}{\mathcal{F}}_D$.
- Output: A classification of indices in $\sigma(c^T - \bar{y}^T A)$ to either B or N; and a set of dual extremal directions $d^{(l)}$.

Step 0 Use Megiddo's algorithm to obtain a dual vertex optimal solution y^*.

Denote $J := \sigma(c^T - \bar{y}^T A)$, $I := \{1, 2, \cdots, n\} \setminus J$ and $z := y^* - \bar{y}$.

Let $N := \{j : j \in \sigma(c^T - y^{*T} A)\}$, $B := \emptyset$ and $l := 1$.

Step 1 If $rank\{A_{I\cup B}\} = m$, let $N := J \setminus B$, stop.

Step 2 Let $K := \emptyset$ and $z' := z$.

Step 3 If $rank\{A_{I\cup B\cup K}\} < m - 1$, go to Step 4.

Let

$$\lambda := \min\{\frac{z^T A_j}{z'^T A_j} : j \in J \setminus (B \cup N \cup K)\}$$

and

$$\begin{aligned} z &:= z - \lambda z' \\ K' &:= \{i : z^T A_j = 0 \text{ and } i \in J \setminus (B \cup N \cup K)\}. \end{aligned}$$

Take any $j_l \in K'$ and let

$$\begin{aligned} d^{(l)} &:= -z', \\ B &:= B \cup j_l, \\ N &:= N \cup K \setminus j_l, \\ l &:= l + 1. \end{aligned}$$

Go to Step 1.

Step 4 Find $u \in \mathcal{R}^m$ such that

- $u^T A_{I\cup B\cup K} = 0$;
- $u^T A_{J\setminus N} \not\leq 0^T$ and $u^T A_{J\setminus N}$ independent of $z'^T A_{J\setminus N}$.

Let

$$\lambda := \min\{\frac{z'^T A_j}{u^T A_j} : u^T A_j > 0 \text{ for } j \in J \setminus N\}$$

and

$$\begin{aligned} z' &:= z' - \lambda u \\ K &:= K \setminus \{j : z'^T A_j = 0\}. \end{aligned}$$

Go back to Step 3.

Remark 5 Similar to Algorithms (PA), now the basis set B is strictly expanding in an outer-loop and K is strictly expanding in an inner-loop. Because $u \neq 0$ and $u^T A_{I\cup B\cup K} = 0$ at the end of one inner-loop, so that $rank\{A_{I\cup B\cup K}\} < m$ after one inner-loop. Other properties of Algorithm (PA) also hold here in a parallel way (see Remark 4).

Now we have the similar result as for Algorithm (PA).

Theorem 5 *For the algorithm (DA) the following hold:*
1). The algorithm runs in strongly polynomial time;
2). All the produced directions $d^{(l)}$ $(1 \leq l \leq \nu)$ are dual extremal directions;
3). There exist positive numbers ϵ_l $(1 \leq l \leq \nu)$ with $1 \gg \epsilon_1 \gg \epsilon_2 \gg \cdots \gg \epsilon_\nu > 0$ such that

$$y^* + \sum_{l=1}^{s} \epsilon_l d^{(l)} \in \mathcal{F}_D$$

for $1 \leq s \leq \nu - 1$ and

$$y^* + \sum_{l=1}^{\nu} \epsilon_l d^{(l)} \in \overset{o}{\mathcal{F}}_D .$$

Proof. Similar to the proof of Theorem 4, and is omitted here.

□

Remark 6 We remark here that the new algorithms can be applied to find "lexicographically feasible" extremal rays starting from any optimal pair of solutions (i.e. not necessary strictly complememtary pairs). The advantage of having a pair of primal- and dual- optimal points in the relative interiors is that a complete set of extremal directions will be found.

Finally, we present the combined algorithm to get a Balinski-Tucker tableau.

The Primal-Dual Algorithm (P&D)

– Input: $\bar{x} \in \overset{o}{\mathcal{F}}_P$ and $\bar{y} \in \overset{o}{\mathcal{F}}_D$.
– Output: A Balinski-Tucker Tableau.

Step 1 Apply Megiddo's algorithm to get the primal basic optimal solution x^* and the dual basic optimal solution y^*.

Step 2 Apply the primal algorithm (PA) to get a partition of variables in $\sigma(\bar{x})$ to either B or N, and to get a set of extremal directions $r^{(l)}$ $(1 \leq l \leq \tau)$.

Step 3 Apply the dual algorithm (DA) to get a partition of variables in $\sigma(c^T - \bar{y}^T A)$ to either B or N, and to get a set of extremal directions $d^{(l)}$ $(1 \leq l \leq \nu)$.

Step 4 Construct the tableau based on the basis B obtained at Step 2 and 3, and using the procedure decsribed in the proof of Lemma 6 to arrange and get a Balinski-Tucker type tableau.

Summarizing Theorem 3, Theorem 4, Theorem 5 and Lemma 6, we have the following main theorem of this paper.

Theorem 6 *The algorithm (P&D) yields a Balinski-Tucker tableau in strongly polynomial time.*

Proof. The theorem follows immediately from Theorem 3, Theorem 4, Theorem 5 and Lemma 6. Notice also that B is indeed a basis, because columns in A_B are independent according to the primal algorithm, and $rank\{A_B\} = m$ according to the dual algorithm. The extremal directions so derived satisfy the conditions required by Lemma 6.

□

5. Conclusions

In this paper we showed that if we have a pair of optimal primal and dual solutions satisfying the complementary slackness relation strictly, then we can construct a certain kind of optimal tableau which contains at least the information about the dimension of the optimal faces and the lexicographically feasible extremal directions. The construction is done in strongly polynomial time. Since linear programming is not yet known to be solvable in strongly polynomial time, this shows that optimal solutions can be further classified according to the information carried.

The algorithms can also be used for other purposes, e.g., to find the dimension of the optimal faces.

Acknowledgement: I like to thank Gert Tijssen for our extensive discussions on linear programming. Gert realized too that for finding a Balinski-Tucker type tableau, a proper projection should be used. I also like to thank Jos Sturm for commenting on an earlier version of this paper.

References

1. M.L. Balinski and A.W. Tucker, Duality theory of linear programming: A constructive approach with applications, *SIAM Review* 11 (1969) 347-377.
2. V. Chvátal, *Linear Programming,* W.H. Freeman and Company, New York, 1983.
3. G.B. Dantzig, *Linear Programming and Extensions,* Princeton University Press, Princeton, New Jersey, 1963.
4. A.J. Goldman and A.W. Tucker, Theory of linear programming, in *Linear Inequalities and Related Systems* (H.W. Kuhn and A.W. Tucker eds.), Annuals of Mathematical Studies, No. 38, Princeton University Press, Princeton, New Jersey, 1956.
5. O. Güller, C. Roos, T. Terlaky and J.-Ph. Vial, Interior point approach to the theory of linear programming, Technical Report No. 1992.3, Université de Genève, Switerland.
6. B. Jansen, C. Roos, T. Terlaky and J.-Ph. Vial, Interior-point methodology for linear programming: Duality, sensitivity analysis and computational aspects, Report 93-28, Faculty of Technical Mathematics and Informatics, Delft University of Technology, The Netherlands, 1993.
7. K.O. Kortanek and J. Zhu, New purification algorithms for linear programming, *Naval Research Logistics Quarterly* 35 (1988) 571-583.
8. N. Megiddo, On finding primal- and dual optimal bases, *ORSA Journal on Computing* 3 (1991) 63-65.

9. A. Schrijver, *Theory of Linear and Integer Programming,* John Wiley & Sons, New York, 1986.
10. T. Terlaky, A convergent criss-cross method, *Math. Oper. und Stat., ser. Optimization* 16 (1985) 683-690.
11. T. Terlaky and S. Zhang, Pivoting rules for linear programming: A survey on recent theoretical developments, to appear in *Annuals of Operations Research: Special Volume on Degeneracy Problems.*
12. L. Zhang and S. Zhang, Convex exact penalty functions, space dilation and linear programming, Research Memorandum Nr. 524, Institute of Economics Research, University of Groningen, The Netherlands, 1993.

ANALYTICAL PROPERTIES OF THE CENTRAL TRAJECTORY IN INTERIOR POINT METHODS

GONGYUN ZHAO
Department of Mathematics, National University of Singapore, Singapore 0511.

and

JISHAN ZHU
Department of Decision Sciences, National University of Singapore, Singapore 0511.

Abstract. In this paper, we study some analytical properties of the central trajectory for a class of optimization problems. We establish the bounds for the higher order derivatives and prove the relative Lipschitz continuity as well as the convergence radius of the Taylor series. We also relate the performance of a conceptual algorithm, which follows the central trajectory, to a curvature integral.

Key words: Central Trajectory, Interior Point Methods, Analytical Property, Relative Lipschitz Continuity, Condition of acute angle, Higher order derivatives

1. Introduction

In the last couple of years, interior point methods has been studied intensively and provided an efficient solution to many optimization problems, (for a survey, cf [3]). Following the socalled central trajectory has been showed an effective way towards the optimal. While there are numerous proposals of how to follow the central trajectory in designing an efficient algorithm, only a few papers have studied the analytical property of the central trajectory in the literature. The present work studies the analytical property of the central trajectory for a class of optimization problems satisfying the *Condition of Acute Angle.* We are interested in finding some analytical properties related to the higher order derivatives of the central trajectory. Although some of the results has been obtained within the content of linear programing (LP) and linear complementarity problems (LCP), [8],[10], the study provides a unified point of view on this subject and, hence, generizes some results in the literature.

The problem considered in this paper is the following: (P) Find $x \in R^n, s \in R^n$ such that

$$x^T s = 0, \tag{1.1}$$

$$(x, s) \in S \tag{1.2}$$

where S is the *feasible region* of the problem. Eventually, (P) represents the Karush-Kuhn-Tucker conditions of many optimization problems. We list a few as the examples:

Ding-Zhu Du and Jie Sun (eds.), Advances in Optimization and Approximation, 362–375.

1) Linear Programming (LP): min $c^T x$, s.t. $Ax = b, x \geq 0$. The feasible region is:

$$S = \{(x, s) \mid Ax = b, A^T y + s = c, y \in R^m, x \geq 0, s \geq 0\}, \tag{1.3}$$

2) Linear Complementarity Problem (LCP). The feasible region is:

$$S = \{(x, s) \mid s = Mx + b, x \geq 0, s \geq 0\}, \tag{1.4}$$

3) Linearly Constrained Convex Programming ($LCCP$): min $f(x)$, s.t. $Ax = b, x \geq 0$. The feasible region is:

$$S = \{(x, s) \mid Ax = b, A^T y + s = \nabla f(x), y \in R^m, x \geq 0, s \geq 0\}. \tag{1.5}$$

For simplicity, throughout the paper, we define operation "$\circ$" as follows:
i) if both x and s are vectors, then

$$x \circ s = Xs,$$

ii) if x is a vector and M is a matrix, then

$$x \circ M = XM$$

$$M \circ x = MX,$$

where X is a diagonal matrix with x as its diagonal elements, i.e., $X = diag(x)$.

The interior point methods define a family of trajectories in the interior of the feasible region S by relaxing the complementary slackness (1.1) to

$$x \circ s = r\xi, \tag{1.6}$$

where ξ is a positive vector and $r \geq 0$ is a parameter. For any fixed $\xi > 0$, we refer to the solution $z(r, \xi) = (x(r, \xi), s(r, \xi))$ of (1.6) and (1.2) as a *ξ-trajectory*. Within the family, the member with $\xi = e$, where e is the vector of ones, is called *central trajectory*, and has been paid special attension in the literature.

Many interior point algorithms explicitly or implicitly follow one of the ξ-trajectory, most of them follow the central trajectory. The efficiency of the algorithm heavily depends on the analytical properties of the trajectory especially when it use the tangent of the trajectory as the improving direction.

The remaining of the paper is orgnized as follows. In section 2, we establish the bounds for the higher order derivatives. In section 3, we prove the relative Lipschitz continuity of the derivatives. In section 4, we study the convergence radius of the Taylor series. In section 5, we indicate that the performance of algorithms (which follows the central trajectory) depends on the integration of a curvature over the central trajecory. We conclude our study in section 6.

2. Bounds for Derivatives of ξ-Trajectory

We first establish a recursive formular for calculating the higher order derivatives.

Lemma 2.1: *For given $\xi > 0$ and $r > 0$, the derivatives (with respect to r) $x_i := x_i(r,\xi) := x^{(i)}(r,\xi)/i!$, $s_i := s_i(r,\xi) := s^{(i)}(r,\xi)/i!$ satisfy the following equations,*

$$s \circ x_1 + x \circ s_1 = \xi$$

$$s \circ x_k + x \circ s_k = -\sum_{i=1}^{k-1} x_i \circ s_{k-i} \qquad k \geq 2 \tag{2.1}$$

where $(x, s) = (x_0, s_0)$ is the solution of (1.6) *and* (1.2). **Proof**: For any $r > 0$ and $r + \epsilon > 0$ set $x(r+\epsilon) = \sum_0^\infty x_i\epsilon^i$, $s(r+\epsilon) = \sum_0^\infty s_i\epsilon^i$ in the left hand side of (1.6),

$$\Big(\sum_{i=0}^{\infty} x_i\epsilon^i\Big) \circ \Big(\sum_{i=0}^{\infty} s_i\epsilon^i\Big)$$

$$= x \circ s + \sum_{k=1}^{\infty}(s \circ x_k + x \circ s_k)\epsilon^k + \sum_{k=2}^{\infty}\sum_{i=1}^{k-1} x_i \circ s_{k-i}\epsilon^k.$$

A comparison with the right hand side $(r+\epsilon)\xi$ gives the equations of (2.1). #

We introduce the *weighted derivatives*

$$p_i := r\xi^{1/2} \circ x^{-1} \circ x_i = \xi^{-1/2} \circ s \circ x_i \tag{2.2a}$$

$$q_i := r\xi^{1/2} \circ s^{-1} \circ s_i = \xi^{-1/2} \circ x \circ s_i \tag{2.2b}$$

By using the notation of p_i and q_i, we can further simplify the formulae of Lemma 2.1,

$$p_1 + q_1 = \xi^{1/2}$$

$$p_k + q_k = -\frac{1}{r}\xi^{-1/2} \circ \sum_{i=1}^{k-1} p_i \circ q_{k-i} \qquad k \geq 2. \tag{2.3}$$

The main result in this section shows an important property of ξ-trajectories $z(r) = z(r,\xi)$: The weighted derivatives of $z(r)$ grow geometrically, that is,

$$\max\{\|p_k\|_2, \|q_k\|_2\} \leq \|p_k + q_k\|_2 \leq C \cdot \alpha^k,$$

for all $k = 1, 2, \ldots$, with constants C and α. To derive this property we impose a condition on ξ-trajectory $(x(r), s(r))$.

Condition of Acute Angle:

$$x_i^T s_i \geq 0 \quad \text{for all} \quad i = 1, 2, \ldots. \tag{2.4}$$

This condition is equivalent to

$$p_i^T q_i \geq 0, \qquad \forall i = 1, 2, \ldots \tag{2.5}$$

since $p_i^T q_i = r x_i^T s_i$.

Remark: The reader can easily verify that linear programming problems, linear complementarity problems and convex quadratic problems satisfy this condition.

By the condition of acute angle we have

$$\max\{\|p_k\|_2, \|q_k\|_2\} \leq (\|p_k\|_2^2 + \|q_k\|_2^2)^{1/2} \leq \|p_k + q_k\|_2. \tag{2.6}$$

Before we present the theorem, we need one more lemma which plays the key role in the proof of the above property.

Lemma 2.2: *The solution of the following difference equation*

$$v_k = \rho \sum_{i=1}^{k-1} v_i v_{k-i}, \qquad k = 2, 3, \ldots$$

is given by

$$v_k = 2(-1)^{k-1} \binom{1/2}{k} (4\rho)^{k-1} v_1^k = 2\frac{(2k-3)!!}{(2k)!!}(4\rho)^{k-1} v_1^k, \qquad k \geq 2, \tag{2.7}$$

resulting in the estimate $|v_k| \leq \frac{1}{k}(4\rho)^{k-1}|v_1|^k$.

Proof: Let $\epsilon > 0$ be so small, that $|4\rho v_1 \epsilon| < 1$. Consider the generating function

$$f(\epsilon) = \sum_{k=1}^{\infty} v_k \epsilon^k \ .$$

Then from

$$\sum_{k=2}^{\infty} v_k \epsilon^k = \rho \sum_{k=2}^{\infty} \sum_{i=1}^{k-1} v_i v_{k-i} \epsilon^k$$

follows

$$f(\epsilon) - v_1 \epsilon = \rho f(\epsilon)^2 \ ,$$

this gives

$$f(\epsilon) = \frac{1 - \sqrt{1 - 4\rho\epsilon v_1}}{2\rho} \ .$$

Taylor expansion of the right hand side then gives v_k of (2.7). #

Corollary 2.1: *If* $0 \leq u_k \leq \rho \sum_{i=1}^{k-1} u_i u_{k-i}, \forall k \geq 2$ *then*

$$u_k \leq v_k \qquad \text{for } k = 1, 2, \ldots,$$

where v_k *are as* (2.7) *and* $v_1 := u_1$.

Proof: Use induction. If $u_i \le v_i$ for all $1 \le i \le k-1$, then

$$u_k \le \rho \sum_{i=1}^{k-1} u_i u_{k-i} \le \rho \sum_{i=1}^{k-1} v_i v_{k-i} = v_k \qquad \#$$

Denote

$$\sigma_k := p_k + q_k, \tag{2.8}$$

$$\psi := \psi(r,\xi) := \frac{1}{r^2} p_1 \circ q_1(r,\xi), \qquad \kappa := \kappa(r,\xi) := \|\psi(r,\xi)\|_2^{1/2}. \tag{2.9}$$

The function κ has the following several equivalent expressions:

$$\kappa^2(r,\xi) = \frac{1}{r}\|x_1 \circ s_1\|_2 = \frac{1}{r^2}\|p_1 \circ q_1\|_2 = \frac{1}{r}\|\xi^{1/2} \circ \sigma_2(r)\|_2. \tag{2.10}$$

We call κ the curvature of a ξ-trajectory since it is a measure on the magnitute of the second order derivative of the curve (trajectory).

From

$$p_1 \circ (\xi^{1/2} - p_1) = q_1 \circ (\xi^{1/2} - q_1) = r^2 \psi$$

follows by solving a quadratic inequality

$$\left.\begin{matrix} \|p_1\|_\infty \\ \|q_1\|_\infty \end{matrix}\right\} \le r\kappa + \|\xi\|_\infty^{1/2}. \tag{2.11}$$

Now we present one of the main theorems in this paper.

Theorem 2.1: *For any $\xi > 0$ and for all $k = 2, 3, \ldots$, if the ξ-trajectory satisfies the condition of acute angle, then we have*

$$\max\{\|p_k\|_2, \|q_k\|_2\} \le \|\sigma_k\|_2 \le v_k \le \frac{1}{k}(4\rho)^{k-1}\beta^k ,$$

where $v_1 = \beta$, v_k is as (2.7) and

$$\rho := \rho(r,\xi) := \|\xi^{-1/2}\|_\infty / r, \quad \beta := \beta(r,\xi) := r\kappa + \|\xi\|_\infty^{1/2}. \tag{2.12}$$

Proof: For $k = 2$ the inequality can be checked directly. For $k \ge 3$ we have by (2.3)

$$\begin{aligned} \|\sigma_k\|_2 &\le \frac{1}{r}\|\xi^{-1}\|_\infty^{\frac{1}{2}} \sum_{i=1}^{k-1} \|p_i \circ q_{k-i}\|_2 \\ &\le \frac{1}{r}\|\xi^{-1}\|_\infty^{\frac{1}{2}} \Big(\sum_{i=2}^{k-2} \|\sigma_i\|_2 \cdot \|\sigma_{k-i}\|_2 + (\|p_1\|_\infty + \|q_1\|_\infty)\|\sigma_{k-1}\|_2 \Big). \end{aligned}$$

Let $u_1 := \beta$ and $u_k := \|\sigma_k\|_2$ for $k \ge 2$. Then

$$u_k \le \rho \sum_{i=1}^{k-1} u_i u_{k-i}$$

and we complete the proof by using Corollary 2.1 and (2.11). #

This theorem essentially describes the growth of the derivatives. For practical uses we still need the following theorem about the higher order derivatives of σ_k. Theorem 2.1 eventually turns out to be the special case of the following more general result by setting $u = 0$.

Theorem 2.2: *For any $\xi > 0$ and for all $k = 2, 3, \ldots$ and $u = 0, 1, \ldots$, if the ξ-trajectory satisfies the condition of acute angle, then we have*

$$\frac{1}{u!}\|\sigma_k^{(u)}\|_2 \le \binom{k+u}{k} v_{k+u} \le \frac{(k+u-1)!}{k!u!}(4\rho)^{k+u-1}\beta^{k+u},$$

where v_k, ρ and β are as in Theorem 1.1.

Proof: Applying Leibniz formula to

$$\sigma_k = p_k + q_k = -\xi^{-1/2} \circ \sum_{i=1}^{k-1} \frac{1}{i!(k-i)!} x^{(i)} \circ s^{(k-i)}$$

we have

$$\begin{aligned}
\sigma_k^{(u)} &= -\xi^{-1/2} \circ \sum_{i=1}^{k-1}\sum_{j=0}^{u} \frac{\binom{u}{j}}{i!(k-i)!} x^{(i+j)} \circ s^{(k+u-i-j)} \\
&= -\frac{(k+u)!}{k! \cdot r}\xi^{-1/2} \circ \sum_{i=1}^{k-1}\sum_{j=0}^{u} P(k,u;i,j) p_{i+j} \circ q_{k+u-i-j} \qquad \text{by (2.2)},
\end{aligned}$$

where

$$P(k,u;i,j) := \binom{k}{i}\binom{u}{j} \Big/ \binom{k+u}{i+j}.$$

By Theorem 2.1,

$$\begin{aligned}
\frac{1}{u!}\|\sigma_k^{(u)}\|_2 &\le \binom{k+u}{k} \rho \sum_{i=1}^{k-1}\sum_{j=0}^{u} P(k,u;i,j) v_{i+j} v_{k+u-i-j} \\
&= \binom{k+u}{k} \rho \sum_{l=1}^{k+u-1} \sum_{i+j=l, 1\le i\le k-1, 0\le j\le u} P(k,u;i,j) v_l v_{k+u-l}.
\end{aligned}$$

For fixed l $(0 \le l \le k+u)$

$$\sum_{i+j=l, 0\le i\le k, 0\le j\le u} P(k,u;i,j) = 1.$$

Hence

$$\begin{aligned}
\frac{1}{u!}\|\sigma_k^{(u)}\|_2 &\le \binom{k+u}{k} \rho \sum_{l=1}^{k+u-1} v_l v_{k+u-l} \\
&= \binom{k+u}{k} v_{k+u} \qquad \text{(by Lemma 2.2)} \\
&\le \frac{(k+u-1)!}{k!u!}(4\rho)^{k+u-1}\beta^{k+u}.
\end{aligned}$$

3. Relative Lipschitz Continuity of the Weighted Derivatives

In the remaining of the paper, we only consider the case that $\xi = e$. The results are, however, true for any ξ-trajectory, besides a factor $\|\xi^{\pm 1}\|_\infty$. To simplify the notations we write $z(r) := z(r,e)$, $\sigma_k(r) := \sigma_k(r,e)$, $\kappa(r) := \kappa(r,e)$, etc. by omitting the notation of vector e.

Now we make use of the results developed in Section 2 to prove the following lemma, a *relative Lipschitz continuity* of function $\sigma_k(r)$.

Theorem 3.1: *Suppose that the central trajectory satisfies the condition of acute angle. Then for any* $0 < T < 1/6$ *and all* $r_1 > r_2 > 0$ *satisfying*

$$r_1\kappa(r_1) \geq 2, \qquad \kappa(r_1)(r_1 - r_2) = T \tag{3.1}$$

the following inequality holds

$$\|\sigma_k(r_1) - \sigma_k(r)\|_2 \leq \delta_k r_1 \kappa^{k+1}(r_1)(r_1 - r) \qquad \forall r \in [r_2, r_1] \text{ and } k = 2,3,\ldots,$$

where

$$\delta_k := \frac{6^{k+1}}{4(1-6T)^k}.$$

Proof: Using Taylor expansion and Theorem 2.2, we have for any $r \in [r_2, r_1]$

$$\begin{aligned}\|\sigma_k(r_1) - \sigma_k(r)\|_2 &\leq \sum_{u=1}^{\infty} \frac{1}{u!}\|\sigma_k^{(u)}(r_1)\|_2 (r_1 - r)^u \\ &\leq \frac{(4\rho\beta)^k}{4\rho}\sum_{u=1}^{\infty}\frac{(k+u-1)!}{k!u!}(4\rho\beta)^u(r_1-r)^u,\end{aligned} \tag{3.2}$$

where, from (2.12), $\rho = 1/r_1$ and $\beta = r_1\kappa(r_1) + 1$. The ($k$-1)-th order derivative of

$$\sum_{u=0}^{\infty} t^u = \frac{1}{1-t}$$

with respect to $t \in (0,1)$ gives

$$\sum_{u=1}^{\infty}\frac{(k+u-1)!}{(k-1)!u!}t^u = \frac{1}{(1-t)^k} - 1 \leq \frac{kt}{(1-t)^k}. \tag{3.3}$$

By condition (3.1), we have

$$4\rho\beta \leq 6\kappa(r_1)$$

and

$$4\rho\beta(r_1 - r) \leq 6T < 1.$$

Hence by (3.2) and (3.3) with $t = 4\rho\beta(r_1 - r)$

$$\|\sigma_k(r_1) - \sigma_k(r)\|_2 \leq \frac{(6\kappa(r_1))^{k+1}(r_1 - r)}{4\rho(1-6T)^k}. \qquad \#$$

In the case $k = 2$, noting that $r\kappa^2(r) = \|\sigma_2(r)\|_2$, we have

$$\|\sigma_2(r_1) - \sigma_2(r)\|_2 \leq \delta_2 T \|\sigma_2(r_1)\|_2 \frac{(r_1 - r)}{(r_1 - r_2)} \qquad \forall r \in [r_2, r_1] \tag{3.4}$$

where $\delta_2 = 6^3/(4(1-6T)^2)$.

4. On the Taylor Expansion of the Weighted Derivatives

As an analytical curve, the central trajectory can be expanned in Taylor series. In this section, we investigate the convergence radius of the Taylor series. Finding such a radius may helps us to estimate the residual in Taylor approximation.

Let's observe Taylor series expanned at r. We denote $x_i = x_i(r)$. Then we have

$$x_j(r') = \sum_{i=0}^{\infty} \frac{(j+i)!}{i!j!} x_{j+i}(r'-r)^i. \quad j = 1, 2, ... \tag{4.1}$$

Consider the function $p_j(r') = r\xi^{1/2} \circ x^{-1} x_j(r')$. By (2.3) $p_j(r')$ has the following expression directly follows (4.1),

$$p_j(r') = \sum_{i=0}^{\infty} \frac{(j+i)!}{i!j!} p_{j+i}(r'-r)^i. \tag{4.2}$$

By Theorem 2.1 in Section 2, we have

$$\|p_{j+i}\|_2 \leq \frac{r}{4}\left(\frac{4}{r}(r\kappa + 1)\right)^{j+i} = \frac{r}{4}(4(1 + \frac{1}{r\kappa})\kappa)^{j+i},$$

The bound is also true for $\|q_{j+i}\|_2$.

if $r\kappa \neq 0$. Now the redius of convergence of the Taylor series is

$$d = \frac{1}{\limsup_{i\to\infty}(\|p_{i+j}\|_2(j+i)!/(i!j!))^{1/i}} \geq \frac{1}{4(1+1/(r\kappa))\kappa}.$$

So we have the following lemma:

Lemma 4.1: *Suppose that the central trajectory satisfy the condition of acute angle. Then at any $r > 0$, if $\kappa = \kappa(r) \neq 0$, the radius of convergence of Taylor series of the central trajectory is at least as large as*

$$\left(4(1 + 1/(r\kappa))\kappa\right)^{-1}. \qquad \#$$

What will the central trajectory be if $\kappa(r) = 0$ at some point r? By the definition of κ (2.10), $\kappa = 0$ if and only if $p_1 \circ q_1 = 0$. By the recursive formula (2.3),

$$p_2 + q_2 = -r^{-1} p_1 \circ q_1 = 0.$$

So by the condition of acute angle, we have

$$p_2 = q_2 = 0.$$

Otherwise,

$$0 = \|p_2 + q_2\|_2 \geq \|p_2\|_2 + \|q_2\|_2 > 0,$$

is a contradiction. Now suppose that $p_i = q_i = 0$, for all $i = 2, 3, \ldots, k-1$. Then by (2.3),

$$p_k + q_k = 0.$$

So by the condition of acute angle, we have $p_k = q_k = 0$. Therefore, by induction we have the following lemma.

Lemma 4.2: *Suppose that the central trajectory satisfy the condition of acute angle. If the weighted curvature $\kappa(r) = 0$ at some point r, then the whole central trajectory is a straight line.*

Remark: Indeed, for any $v > 0$ and any analytic function $f : R^{v+1} \to R$, the function $g(r) := f(z(r), z_1(r), \ldots, z_v(r))$ is also analytic. The method we used above for determining the radius of convergence of Taylor series can also be applied to the function $g(r)$.

5. The Integration of a Curvature over the Trajectory

To avoid technical details, we consider a conceptual algorithm, and outline the relationship between a curvature integral and the number of iterations needed by the predictor-step of the following algorithm.

Algorithm 1: Given $r_0 > \epsilon > 0$, $\alpha > 0$, and a starting point $(x(r_0), s(r_0))$ on the central trajectory. For $k = 0, 1, 2, \ldots$,

Predictor-Step: Compute the derivetive $(x_1(r_k), s_1(r_k))$, and consider the tangent line $(\bar{x}(r), \bar{s}(r)) = (x(r_k), s(r_k)) + (r - r_k)(x_1(r_k), s_1(r_k))$. Determine $r_{k+1} := r$ so that

$$d(\bar{x}(r), \bar{s}(r), r) = \alpha, \tag{5.1}$$

which is explicitly given by

$$\|x_1(r_k) \circ s_1(r_k)\|_2 \frac{(r - r_k)^2}{r} = \alpha. \tag{5.2}$$

Corrector-Step: Find point $(x(r_{k+1}), s(r_{k+1}))$ on the central trajectory. If $r_{k+1} \leq \epsilon$, stop. Otherwise, go to Predictor-Step. #

We will relate the number of iterations needed by the above algorithm to a curvature integral. There are some advantages of representing the complexity as a form of curvature integral. First, the curvature, $\kappa(r, e)$, is continous function with respect to r and hence may be easier to analize. And second, the curvature can better reflect the trajectory for each individual problem. So for some 'nice' problems, the complexity may be estimated lower. The reader can find some interesting examples in [6].

N. Karmarkar [2] and G. Sonnevend [6] first independently pointed out the relationship between the complexity of interior point methods and curvature integrals. Encouraged by G. Sonnevend, the first auther and Stoer [8] proved this relationship for a primal-dual path-following algorithm. In the following we present a similar theorem for Algorithm 1.

Theorem 5.1: *Suppose that the central trajectory satisfies the condition of acute angle. Then for any* $r_0 > \epsilon > 0$, *the number* $N := N(r_0, \epsilon)$ *of steps* $r_k \to r_{k+1}$ *of Algorithm 1 with*

$$r_0 > r_1 > \ldots > r_N > \epsilon \geq r_{N+1}$$

is bounded above by

$$N \leq C_1 \int_{\epsilon}^{r_0} \kappa(r, e)dr + C_2 \ln \frac{r_0}{\epsilon}. \tag{5.3}$$

Here C_1 *and* C_2 *are constants depending only on the choice of* α. We first prove a lemma and then the theorem. The numbers, such as 2, 1/66, used below as well as those in the proof of Theorem 5.1 are not essential, they are chosen only for convenience.

Lemma 5.1: *Suppose that the central trajectory satisfies the condition of acute angle. Then for any* $r_1 > r_2 > 0$ *satisfying*

$$r_1\kappa(r_1) \geq 2, \qquad \kappa(r_1)(r_1 - r_2) \geq \frac{1}{66},$$

the following inequality holds

$$\int_{r_2}^{r_1} \kappa(r)dr \geq \frac{1}{99}$$

Proof: Let $T = 1/66$ in Lemma 3.1, then $T < 1/6$ and $\delta_2 T \leq 1$. Let r_2' satisfy $\kappa(r_1)(r_1 - r_2') = T$, then we have by (3.4) that

$$\begin{aligned}
\int_{r_2}^{r_1} \kappa(r)dr &\geq \int_{r_2'}^{r_1} \kappa(r)dr \\
&\geq \int_{r_2'}^{r_1} \sqrt{\frac{1}{r_1}(\|\sigma_2(r_1)\|_2 - \|\sigma_2(r_1) - \sigma_2(r)\|_2)}\,dr \\
&\geq \kappa(r_1) \int_{r_2'}^{r_1} \sqrt{1 - \delta_2 T \frac{r_1 - r}{r_1 - r_2'}}\,dr \\
&\geq \kappa(r_1) \int_{r_2'}^{r_1} \sqrt{\frac{r - r_2'}{r_1 - r_2'}}\,dr \\
&= \frac{2}{3}T = \frac{1}{99}.
\end{aligned}$$

Proof of Theorem 5.1 According to whether

$$r_k\kappa(r_k,e)\geq 2 \qquad \text{and} \qquad \frac{r_{k+1}}{r_k}\geq\frac{1}{2} \tag{5.4}$$

holds or not, the iteration steps $r_k \to r_{k+1}$, $k=0,1,\ldots,N-1$, will be divided in two parts:

$$I_1 := \{k \mid (5.4) \text{ is satisfied for } k\}, \quad I_2 := \{k \mid (5.4) \text{ is not satisfied for } k\}.$$

Denote by $N_1 = |I_1|$ the number of iterations in I_1, and $N_2 = |I_2|$. Obviously $N = N_1 + N_2$. We will show that

$$N_1 \leq 99\int_\epsilon^{r_0}\kappa(r,e)dr \tag{5.5}$$

and

$$N_2 \leq 16\ln(\frac{r_0}{\epsilon}), \tag{5.6}$$

by selecting $1\geq\alpha\geq 1/16$ in Algorithm 1. (The selection of α is quite arbitrary. In practice, a large α can reduce duality gap fast, however, if it is too large, then the Corrector-Step demands more efforts).

To prove (5.5) let $k\in I_1$. Then by (5.2) and (2.10),

$$\kappa(r_k,e)(r_k-r_{k+1})\sqrt{\frac{r_k}{r_{k+1}}}=\sqrt{\alpha}$$

and, hence, by (5.4),

$$\kappa(r_k,e)(r_k-r_{k+1})\geq\sqrt{\frac{\alpha}{2}}>\frac{1}{66}.$$

By Lemma 5.1, we have

$$\int_{r_{k+1}}^{r_k}\kappa(r,e)dr\geq\frac{1}{99}.$$

Hence

$$\int_\epsilon^{r_0}\kappa(r,e)dr\geq\sum_{k=0}^{N-1}\int_{r_{k+1}}^{r_k}\kappa(r,e)dr\geq\sum_{k\in I_1}\int_{r_{k+1}}^{r_k}\kappa(r,e)dr\geq\frac{N_1}{99},$$

this shows (5.5).

To prove (5.6) let $k\in I_2$. Then either

$$r_k\kappa(r_k,e)<2$$

or

$$\frac{r_k}{r_{k+1}}>2.$$

In the first case, we get by (5.2)

$$r_k\kappa(r_k,e)\frac{r_k - r_{k+1}}{\sqrt{r_k r_{k+1}}} = \sqrt{\alpha},$$

By (5.4) and the fact that $\alpha \geq 1/4$, we have

$$\frac{r_k}{r_{k+1}} - \frac{1}{8}\sqrt{\frac{r_k}{r_{k+1}}} \geq 1,$$

which implies that

$$\frac{r_k}{r_{k+1}} \geq 1 + \frac{1}{8}.$$

In either case $r_k/r_{k+1} \geq 1 + 1/8$. It follows

$$(1+\frac{1}{8})^{N_2} \leq \prod_{k\in I_2} \frac{r_k}{r_{k+1}} \leq \prod_{k=0}^{N-1} \frac{r_k}{r_{k+1}} \leq \frac{r_0}{\epsilon},$$

which gives

$$N_2 \leq \frac{1}{\ln(1+1/8)} \ln\frac{r_0}{\epsilon} \leq 16\ln\frac{r_0}{\epsilon}.$$

This shows (5.6) and completes the proof of Theorem 5.1. #

The results developed above enable us to represent the complexity of a predictor-corrector algorithm in a form of curvature integral along the central trajecorty. It is easy to show that an upper bound of the curvature integral is $O(\sqrt{n}\ln(\frac{r_0}{\epsilon}))$, since for any $r > 0$,

$$\max\{\|p_1(r)\|_2, \|q_1(r)\|_2\} \leq \|e\|_2 = \sqrt{n} \tag{5.7}$$

and by (2.10),

$$\kappa(r) = \frac{1}{r}\|p_1(r)\circ q_1(r)\|_2^{1/2} \leq \frac{1}{r}(\|p_1(r)\|_2\|q_1(r)\|_2)^{1/2} \leq \frac{\sqrt{n}}{r}.$$

So we have

$$\int_\epsilon^{r_0} \kappa(r)dr \leq \sqrt{n}\ln(\frac{r_0}{\epsilon}) \tag{5.8}$$

Observe that the equality for (5.8) to hold if and only if $r\kappa(r) = \sqrt{n}$. Indeed, $r\kappa(r) \to 0$ when (x,s) converge to the optimal solution, cf. [1] and [7]. So under certain assumptions, there exists a constant (which may problem dependant), λ, such that $r\kappa(r) \leq O(1)$ other than $O(\sqrt{n})$ when $r < \lambda$. Then the number of iterations needed is

$$O\left(\sqrt{n}\ln\left(\frac{r_0}{\lambda}\right) + \ln\left(\frac{\lambda}{\epsilon}\right)\right).$$

In theory, such a constant, λ, may be greater than any given ϵ specified in the algorithm. Also, for any $r > 0$, it may be true that $r\kappa(r) << \sqrt{n}$. Mizuno, Todd and Ye [4], [5] provided some heuristic reasoning that the expected value of

$r\kappa(r) \approx n^{0.25}$. Hence, by representing the complexity by a curvature integral, we provide some possibility to improve the worst-case complexity estimation. At least, we feel that it could be done for certain class of problems satisfying the condition of acute angle.

In practice, the resulting point of the corrector step may not be on the central trajectory. A practical algorithm stop executing the corrector step when the solution is near center. Theorem 5.1 remains true for the practical algorithm with careful parameter choosing. We direct the interested reader to [8],[9],[10] for more details.

6. Conclusions

This paper serves two purposes. One is to *fill* the gap in the literature. Central trajectory has been widely used but its analytical properties (higher order derivatives) lack investigation. The other purpose is to provide a more general study for a class of optimization problems. A further study on such a class of optimization problems can depart and benefit from the results of this paper.

References

1. J. Ji, F. Potra and S. Huang, "A predictor-corrector method for linear complementarity problems with polynomial complexity and superlinear convergence," Report No. 18, Department of mathematics, The University of Iowa (Iowa City, IA), 1991.
2. N. Karmarkar, "Riemannian geometry underlying interior point methods", Lecture and preprint presented at the 13th Int. Symp. on math. programming, Tokyo (Tokyo, 1988).
3. M. Kojima, N. Megiddo, T. Noma and A. Yoshise, "A unified approach to interior point algorithms for linear complementarity problems," Research Report RJ 7493 (70008), IBM Almaden Research Division, San Jose, CA 95120-6099, USA, 1990.
4. S. Mizuno, M.J. Todd and Y. Ye, "Anticipated behaviour of path-following algorithm for linear programming," Technical Report No. 878, School of Operations Research and Industrial Engineering, Cornell University (Ithaca, NY, 1989).
5. S. Mizuno, M.J. Todd and Y. Ye, "On adaptive-step primal-dual interior-point algorithm for linear programming," Technical Report No. 944, School of Operations Research and Industrial Engineering, Cornell University (Ithaca, NY, 1990), to appear in *Mathematics of Operations Research*.
6. G. Sonnevend, J. Stoer and G. Zhao, "On the complexity of following the central path of linear programs by linear extrapolation II," *Mathematical Programming* 52 (1991) 527–553.
7. Y. Ye, O. Güler, R.A. Tapia and Y. Zhang, "A quadratically convergent $O(\sqrt{n}L)$-iteration algorithm for linear programming," Working Paper No 91-14, College of Business Administration, The University of Iowa (Iowa City, IA), 1991. Also TR91-26, Department of mathematical Sciences, Rice University (Houston, TX), 1991.
8. G. Zhao, "Relationship between the curvature integral and the complexity of path-following methods in linear programming," Research Report No. 556, Department of mathematics, National University of Singapore, Singapore 0511, 1993.
9. G. Zhao and J. Stoer, "Estimating the complexity of a class of path-following programs by curvature integrals," *Appl. Math. Optim.* 27 (1993) 27-85.

10. G. Zhao and J. Zhu, "The Curvature Integral and the Complexity of Linear Complementarity Problems," Research Report No. 558, Department of mathematics, National University of Singapore, Singapore 0511, 1993.

THE APPROXIMATION OF FIXED POINTS OF ROBUST MAPPINGS

QUAN ZHENG
Department of Mathematics
Shanghai University of Science and Technology
Shanghai, China

and

DEMING ZHUANG*
Department of Mathematics and Computer Studies
Mount Saint Vincent University
Halifax, Nova Scotia , Canada B3M 2J6

1. Introduction

The study of the existence results of fixed points of discontinuous mappings was started in 1976 by Caristi [2]. However, to the best of our knowledge, serious effort has yet been devoted to the numerical computations of fixed points of discontinuous mappings. In this work we propose a method to find fixed points of a discontinuous robust mapping using the integral global minimization approach. In Section 2, we explain the reason for introducing the concept of robustness for global minimization. In Section 3, we define robust mappings and upper robust functions and study basic properties of robust mappings. In Section 4, we consider the existence results of fixed points of robust mappings. The computation of the fixed points of an robust mapping is deduced to finding the set of global minima of an robust function. Thus, we can use the integral global optimization method to find them. Numerical examples are also presented in this section to demonstrate that this approach is competitive with other algorithms.

2. Robust Sets and Robust Minimizers

Let X be a Hausdorff topological space, $f : X \rightarrow R^1$ a function and D a subset of X. Consider the following minimization problem:

$$c^* = \inf_{x \in D} f(x). \tag{1}$$

Usually, the study of a minimization problem consists of three parts: i) the existence of minimizers; ii) optimality conditions for describing the properties of minimizers and iii) methods for finding minimizers.

* Research supported partially by NSERC grant and Mount St Vincent University internal grant.

Ding-Zhu Du and Jie Sun (eds.), Advances in Optimization and Approximation, 376–389.

If f is lower semicontinuous and there is a real number $c > c^*$ such that the level set

$$H_c := \{x \in D : f(x) \leq c\} \tag{2}$$

is a nonempty compact set, then the set of global minima is nonempty:

$$H^* = \{x \in D : f(x) = c^*\} \neq \emptyset.$$

The above condition of the existence of minimizers is not very restrictive. The objective function f is not even required to be continuous as the following example shows.

Example 2.1 Let $D = [-\pi, \pi]$ and

$$f(x) = \begin{cases} 1 + \sin x, & x \neq 0, \\ 0, & x = 0. \end{cases}$$

For such a discontinuous function f, the existence condition is satisfied. The function f has two minimizers: $x_1 = 0$ and $x_2 = -\pi/2$.

The classical optimality conditions for describing properties of a minimizer are often quite demanding. Most of them are based on the concept of gradient or its generalizations. Usually, these are necessary but not sufficient conditions. The numerical algorithms designed based on these classical optimality conditions approximate a local minimizer but not a global one. Quite often, extra assumptions must be imposed in order to guarantee the convergence of the algorithms. Furthermore, because the classical optimality conditions are based on gradients, it is difficult to design a "direct" minimization algorithm that corresponds to the optimality conditions; the proofs of the convergence of the algorithms can sometimes be complex.

We intend to develop a unified theory and methodology (robust analysis and the integral approach of global minimization) to deal with the global minimum of a robust function over a robust constraint set. This approach studies the properties of robust sets and functions, characterizes the optimality conditions of global minima and develops numerical methods to find global minimizers in a direct and uniform fashion. The reader is referred to [3, 12, 13] for details of integral optimization and robust analysis.

Let us recall some basic definitions and facts of robust sets.

Definition 2.1 *Let X be a topological space, D a subset of X. D is said to be* **robust** *if*

$$cl\ D = cl\ int\ D. \tag{3}$$

where clA is the topological closure of A and intA is the topological interior of A. A point $x \in D$ is said to be a **robust point** *of D if $x \in$ cl intD, When the point $x \in D$ is a robust point of D, then D is said to be a* **semi-neighbourhood** *of x.*

It is clear that $x \in D$ is a robust point of D if and only if there is a net $\{x_\lambda\} \subset$ int D such that $x_\lambda \to x$; the set D is robust if and only if every point of D is a robust point of D. If A is a semi-neighbourhood of x and $A \subset B$, then B is also a semi-neighbourhood of x.

The concepts of robust sets, robust points and semi- neighbourhoods are extensions of those of open sets, interior points and neighbourhoods. Any open set, including the empty set, is robust. Any interior point of D is a robust point of D. Any neighbourhood of x is a semi-neighbourhood of x.

A union of robust sets is robust; but an intersection of two robust sets may not be robust. An intersection of a robust set and an open set is robust.

Now, let us look at Example 2.1 again. We are not interested in the minimizer $x_1 = 0$ because, unlike the other minimizer $x_2 = -\pi/2$, the function value $f(0) = 0$ is not associated with the points nearby. We say such minimizers are *unstable solutions* of (1) in the following sense: a small perturbation near this point causes a large deviation of the value of the mapping. We want to exclude such solutions. It is for this reason that we introduce the following concept of robust solutions of a minimization problem.

Definition 2.2 *Let X be topological space. A point $x^* \in D$ is said to be a* **robust minimizer** *of (1) if*

1. *x^* is a global minimizer of (1); and*
2. *x^* is robust point of the level set F_c for all $c > c^* = f(x^*)$ where*

$$F_c = \{x \in D : f(x) < c\}. \tag{4}$$

It is clear from the definitions that, if x^* is a global minimizer of (1), then x^* is a *robust* minimizer of (1) if and only if for every $c > c^* = f(x^*)$ there is a net of points $\{x_\lambda\} \subset \text{int } F_c$, such that

$$x_\lambda \to x^* \text{ and } f(x_\lambda) \to f(x^*) \tag{5}$$

When X is a metric space, the property (5) provides a possibility of finding the global minimum point x^* of f by selecting a sequence of points $\{x_k\}$ in the open set int F_c, and the corresponding function values $f(x_k)$ converge to $c^* = f(x^*)$. Note that if f is continuous at x^*, then for every sequence of points $\{x_k\} \subset \text{int } F_c (c > c^*)$, such that $x_k \to x^*$, we always have $f(x_k) \to f(x^*)$ as $k \to \infty$. x^* being a robust global minimizer ensures the existence of such a sequence even if f is not continuous.

If the point x^* is not a global minimizer of f over D, then (5) usually does not hold. as the following example shows.

Example 2.2 Let $X = R^1, D = [-10, 10]$ and

$$f(x) = \begin{cases} x - 1, & x < 0, \\ 0, & x = 0, \\ x + 1 & x > 0. \end{cases}$$

Take $c = 0.1$, even though $x = 0$ is a robust point of $F_c = \{x \in D : f(x) < c\}$, it is clear that for every sequence $\{x_k\} \subset$ int F_c $x_k \to 0$ $f(x_k) \to -1 \neq f(0) = 0$.

The above definition of global robust minimizer does exclude the minimizer $x_1 = 0$ in Example 2.1 from being robust minimizer. Indeed, if we take $c = 0.1$, then

$$F_c = \{0\} \cup (-\pi/2 - \arcsin 0.1, -\pi/2 + \arcsin 0.1)$$

and $\mathrm{int}(F_c) = (-\pi/2 - \arcsin 0.1, -\pi/2 + \arcsin 0.1)$. We cannot find a sequence $\{x_k\}$ such that (5) holds.

Example 2.3 Let

$$f(x) = \begin{cases} 1, & x \text{ is a irrational number,} \\ |x|, & x \text{ is a rational number} \end{cases}$$

and $D = (-\infty, \infty)$. The point $x^* = 0$ is the unique minimizer. Let $x_k = 1/k$, $k = 1, 2, \ldots$. Then $f(x_k) = 1/k, f(0) = 0$, so that (8) is satisfied. However, the point $x^* = 0$ is not a robust minimizer. Indeed, if we take $c = 0.5$, then $F_c = \{x : f(x) < 0.5\}$ and $\mathrm{int}\, F_c = \emptyset$. This kind of minimizers is also excluded by our definition of robustness.

3. Robust Mappings, Upper Robust Functiona and Their Properties

In this section we introduce the concept of robust mappings and upper robust functions. We also study their fundamental properties.

Definition 3.1 *Let X and Y be topological spaces. $T : X \to Y$ is said to be* **robust** *at $x \in X$ if for every neighbourhood $U_Y(y)$ of $y = T(x)$, $T^{-1}(U_Y(y))$ is a semi-neighbourhood of x, T is said to be a* **robust** *mapping if T is robust at every x in X.*

It is clear from the definitions that T is robust at x if x is a robust point of $T^{-1}(U_Y(y))$; T is a robust mapping if for each open set G in Y the preimage $T^{-1}(G)$ is a robust set in X.

Definition 3.2 *Let $f : X \to R^1$ be a real-valued function, x_0 a point of X, f is said to be* **upper robust** *at x_0 if $x_0 \in F_c$ implies that x_0 is a robust point of F_c. f is said to be* **upper robust** *on X if for each real number c, the set $F_c = \{x \in X : f(x) < c\}$ is robust.*

It is clear that f is upper robust on X if and only if f is upper robust at each point of X.

Remark 3.1 We define upper robustness of a function when considering a minimization problem. If one wishes to consider a maximization problem with the integral approach, one has to require that $-f$ be upper robust. In this case f is said to be **lower robust**.

The concepts of robust mappings, upper robust functions and lower robust functions are natural extensions of that of continuous mappings, upper semi-continuous functions and lower semi- continuous functions. A continuous mapping T is robust since for each open set in Y the preimage $T^{-1}(G)$ is an open set.

For similar reasons, an upper (respectively, lower) semi- continous function is upper (respectively, lower) robust. However, a robust mapping may not be continuous and an upper (respectively, lower) robust function may not be upper (respectively, lower) robust.

Example 3.1 Let $X = Y = R^1$, and

$$Tx = \begin{cases} x-1, & x<0, \\ x+1, & x \geq 0. \end{cases}$$

Then T is robust but it is discontinuous.

Remark 3.2 If $Y = R^1$, then a robust mapping T is an upper robust function. Indeed, for each real number c, the set $\{x : Tx < c\} = T^{-1}((-\infty, c))$ is robust in X, since $G = (-\infty, c)$ is an open set in R^1. However, an upper robust function, as a mapping from X to R^1, may not be robust as the following example shows.

Example 3.2 Let $X = Y = R^1$, and

$$Tx = \begin{cases} x-1, & x<0, \\ 0, & x=0, \\ x+1, & x>0. \end{cases}$$

T is a monotone function on R^1, so it is upper robust. However, T is not robust: if we take the open set $G = (-0.2, 0.2)$ in Y, then $T^{-1}(G) = \{0\}$; this is not a robust set in Y.

The above example also shows that a robust mapping $f : X \to R^1$ may not be lower semi-continous.

As we mentioned in the previous section, when we study a minimization problem the objective function is usually assumed to be lower semicontinuous to ensure the existence of global minimum points. The following proposition shows that a lower semicontinuous upper robust function from X to R^1 is robust.

Proposition 3.1 *A lower semi-continuous upper robust function $f : X \to R^1$ is robust.*

Proof. An open set G in R^1 can be represented as a union of a finite or denumerable number of disjoint intervals:

$$G = \bigcup_{i=1}^{\infty}(a_i, b_i), \quad (a_i, b_i) \cap (a_j, b_j) \neq \emptyset, \quad \text{if } i \neq j,$$

and

$$f^{-1}(a_i, b_i) = \{x : f(x) < b_i\} \cap \{x : f(x) > a_i\},$$

where $\{x : f(x) < b_i\}$ is a robust set since f is assumed to be upper robust, and $\{x : f(x) > a_i\}$ is an open set since f is lower semicontinuous. As the intersection of a robust set and an open set, $f^{-1}((a_i, b_i))$ is robust. But,

$$f^{-1}(G) = \bigcup_{i=1}^{\infty} f^{-1}((a_i, b_i))$$

is a union of robust sets, it is also robust. Hence, f is a robust mapping.

Proposition 3.2 *Let $T : X \to Y$ be a mapping, then T is a robust mapping if and only if for every point in X, there exists a neighbourhood $O(x)$ of x, such that for any open set G in Y, $O(x) \cap T^{-1}(G)$ is a robust set of X.*

Proof. Assume that T is robust, take a point $x \in X$ and a neighbourhood $O(x)$ of x. For a given open set $G \subset Y$, $T^{-1}(G)$ is a robust set of X. Then, as the intersection of an open set and a robust set, the set $O(x) \cap T^{-1}(G)$ is a robust set of X.

Now, take an open set G in Y and take $x \in T^{-1}(G)$. Let $O(x)$ be the neighbourhood of x such that $O(x) \cap T^{-1}(G)$ is a robust set. For any neighbourhood $O_1(x)$ of x, the set $O_1(x) \cap O(x) \cap T^{-1}(G)$ is the intersection of the open set $O_1(x)$ and the robust set $O(x) \cap T^{-1}(G)$; thus it is a robust set. Moreover, since x belongs to all of these sets, their intersection is nonempty. Therefore, $\operatorname{int}(O_1(x) \cap O(x) \cap T^{-1}(G)) \neq \emptyset$. This implies that

$$O_1(x) \cap \operatorname{int} T^{-1}(G) \neq \emptyset.$$

Thus we have proved that x is a robust point of $T^{-1}(G)$. Since x is arbitrarily choosen, this implies that $T^{-1}(G)$ is a robust set. Therefore, T is a robust mapping.

We now consider the robustness of composition and product and sum of mappings. The results will be applied in the next setion.

Let X, Y and Z be topological spaces, $T_1 : X \to Y$ and $T_2 : Y \to Z$ mappings. Recall the composition of T_1 and T_2 is the mapping $T = T_1T_2 : X \to Z$ defined as follows:

$$Tx = T_2(T_1x), \qquad \text{for all } x \in X;$$

The product of T_1 and T_2 is the mapping $T = (T_1, T_2) : X \to Y \times Z$ defined as follows:

$$Tx = (T_1x, T_2x), \qquad \text{for all } x \in X.$$

Theorem 3.1 *Let X, Y and Z be topological spaces, $T_1 : X \to Y$ a robust mapping and $T_2 : Y \to Z$ a continuous mapping. Then the composition $T = T_1T_2 : X \to Z$ of T_1 and T_2 is a robust mapping.*

Proof. For each open set G in Z, we have

$$T^{-1}(G) = (T_1T_2)^{-1}(G) = T_1^{-1}(T_2^{-1}(G)).$$

Because T_2 is continuous, $T_2^{-1}(G)$ is an open set in Y. Furthermore, $T_1^{-1}(T_2^{-1}(G))$ is a T_1-preimage of an open set, the robustness of $T_1^{-1}(T_2^{-1}(G))$ follows because of the robustness of T_1. Hence, T is a robust mapping from X to Z.

Theorem 3.2 *Let X, Y and Z be topological spaces, $T_1 : X \to Y$ and $T_2 : X \to Z$ mappings, and $T = (T_1, T_2) : X \to Y \times Z$ the product of T_1 and T_2. If T is robust, then both T_1 and T_2 are robust; if one of T_1 and T_2 is robust and other one is continuous, then T is robust.*

Proof. Let $\pi_1 : Y \times Z \to Y$ and $\pi_2 : Y \times Z \to Z$ be projections on Y and Z, respectively. They are continuous mappings. Indeed, for each open set G in $Y, \pi_1^{-1}(G) = G \times Z$, is obviously open in $Y \times Z$. Thus π_1 is continuous. Similarly, π_2 is continuous. If T is robust, then so are $T_1 = \pi_1 T$ and $T_2 = \pi_2 T$ by Theorem 3.1.

Suppose T_1 is a robust mapping and T_2 is a continuous mapping. Take a product open set $G \times U$ in $Y \times Z$, we have

$$T^{-1}(G \times U) = T_1^{-1}(G) \cap T_2^{-1}(U).$$

Because $T_1^{-1}(G)$ is a robust set and $T_2^{-1}(U)$ is an open set, $T_1^{-1}(G) \cap T_2^{-1}(U)$ is a robust set. In a product topological space, each open set can be represented as a union of product open sets, i.e., $O = \bigcup_{\alpha \in \Lambda} O_\alpha$, where O_α is a product open set, and $T^{-1}(O_\alpha)$ is a robust set. Moreover,

$$T^{-1}(O) = \bigcup_{\alpha \in \Lambda} T^{-1}(O_\alpha). \tag{6}$$

As a union of robust sets, $T^{-1}(O)$ is robust. Therefore T is a robust mapping.

If both T_1 and T_2 are only assumed to be robust, the product mapping $T = (T_1, T_2)$ may not be robust as the following example illustrates.

Example 3.3 Let $X = Y = R^1$,

$$T_1 x = \begin{cases} x, & x \leq 0, \\ x + 1, & x < 0 \end{cases}$$

and

$$T_2 x = \begin{cases} x - 1, & x > 0 \\ x, & x \geq 0 \end{cases}$$

and take $G \times U = (-2, 0.5) \times (-0.5, 2)$. Then, $T_1^{-1}(G) = (-2, 0]$ and $T_2^{-1}(U) = [0, 2)$. Thus,

$$T^{-1}(G \times U) = T_1^{-1}(G) \cap T_2^{-1}(U) = \{0\},$$

which is a nonrobust set.

The above example may also be used to illustrate that the composition of two robust mappings need not be robust. Indeed, let $T_1 : X \to Y$ and $T_2 : Y \to Z$. Take $G = (-0.5, 0.5)$, then $T_2^{-1}(G) = [0, 0.5)$, but, $T_1^{-1}(T_2^{-1}(G)) = T_1^{-1}([0, 0.5)) = \{0\}$. It is not a robust set.

We now suppose Y is a topological linear space, T, T_1 and $T_2 : X \to Y$ mappings, and α a scalar. Then we define αT and $T_1 + T_2$ as follows:

$$(\alpha T)(x) = \alpha T(x), (T_1 + T_2)(x) = T_1(x) + T_2(x), \text{ for all } x \in X.$$

Theorem 3.3 *If T is robust at x_0, then αT also robust at x_0. If T_1 is robust at x_0 and T_2 is continuous at x_0, then $T_1 + T_2$ is robust at x_0.*

Proof. If $\alpha = 0$, then αT is a continuous mapping. Thus αT is robust at every point of X. Now, suppose that $\alpha \neq 0$. Let G be an arbitrarily given open set such that $x_0 \in (\alpha T)^{-1}(G)$. Thus, $\alpha T(x_0) \in G$ and then $T(x_0) \in \frac{1}{\alpha}G$. However, $\frac{1}{\alpha}G$ is also an open set in Y. Furthermore, $x_0 \in T^{-1}(\frac{1}{\alpha}G)$, thus, x_0 is robust to $T^{-1}(\frac{1}{\alpha}G)$. We also have $T^{-1}(\frac{1}{\alpha}G) = (\alpha T)^{-1}(G)$, i.e., x_0 is robust to $(\alpha T)^{-1}(G)$. Hence, we have proved that the mapping αT is robust at x_0.

To prove T_1+T_2 is robust at x_0, we first assume that T_1 is continuous at x_0. Then, T_1+T_2 is continuous at x_0, it is robust at x_0. Suppose now that T_1 is robust but not continuous at x_0. Let $N(y_1 + y_2)$ be a neighbourhood of $y_1 + y_2 = (T_1 + T2)(x_0)$. By the continuity of the addition operation in a linear metric space, there is a neighbourhood $N_1(y_1)$ and a neighbourhood $N_2(y_2)$ such that

$$N_1(y_1) + N_2(y_2) \subset N(y_1 + y_2).$$

we can also prove that if D and E are subsets of Y then we have

$$T_1^{-1}(D) \cap T_2^{-1}(E) \subset (T_1 + T_2)^{-1}(E + D). \tag{7}$$

Indeed, suppose $z \in T_1^{-1}(D) \cap T_2^{-1}(E)$, then

$$z \in T_1^{-1}(D) \text{ and } z \in T_2^{-1}(E), \text{ or } T_1(z) \in D \text{ and } T_2(z) \in E.$$

It follows that

$$(T_1 + T_2)(z) = T_1(z) + T_2(z) \in E + D.$$

That is $z \in (T_1 + T_2)^{-1}(E + D)$, thus (7) holds.

Since T_1 is robust at x_0, $T_1^{-1}(N_1(y_1))$ is a robust set. Since T_2 is continuous at x_0, $T_2^{-1}(N_2(y_2))$ is an open set. Hence, $T_1^{-1}(N_1(y_1)) \cap T_2^{-1}(N_2(y_2))$ is a robust set. It is clear that x_0 is in the set. So, it is a semi-neighbourhood of x_0. By the above and note that if $A \subset B$ then $f^{-1}(A) \subset f^{-1}(B)$ for any mapping f, we have $T_1^{-1}(N_1(y_1)) \cap T_2^{-1}(N_2(y_2)) \subset (T_1 + T_2)^{-1}(N_1(y_1) + N_2(y_2)) \subset (T_1 + T_2)^{-1}(N(y_1 + y_2))$. Therefore, $(T_1 + T_2)^{-1}(N(y_1 + y_2))$ is also a semi-neighbourhood of x_0. That is, the mapping $T_1 + T_2$ is robust at x_0.

Corollary 3.1 *If T is robust, Then so is αT for all α in R. If T_1 is robust and T_2 is continuous on X, then $T_1 + T_2$ is robust.*

Remark 3.3 Note that for an upper robust function f, the robustness of f at a point x_0 does not imply the robustness of αf for $\alpha < 0$. The following example shows that the sum of two robust mappings may not be robust.

Example 3.4 Let $X = Y = R^1$, and

$$T_1(x) = \begin{cases} 1, & \text{if } x < 0, \\ 0, & \text{if } x \geq 0 \end{cases}$$

$$T_2(x) = \begin{cases} 0, & \text{if } x \leq 0, \\ 1, & \text{if } x > 0. \end{cases}$$

Mappings T_1 and T_2 are robust; i.e., T_1 and T_2 are robust at each point $x \in X$. However, the sum

$$T_1(x) + T_2(x) = \begin{cases} 1, & \text{if } x \neq 0, \\ 0, & \text{if } x = 0. \end{cases}$$

is nonrobust at $x = 0$.

4. Existence and Computation of Fixed Points of Robust Mappings

The study of the existence of fixed points of a discontinuous mapping started in 1976, see [2, 4]. The following theorem, due to Caristi, is of the simplest form.

Theorem 4.1 *Let (X, d) be a complete metric space and $T : X \to X$ a given mapping. If there is a lower semi-continuous function ϕ, which is bounded from below, such that*

$$d(x, Tx) \leq \phi(x) - \phi(Tx), \text{ for all } x \in X, \tag{8}$$

then T has fixed points.

The following example shows that T, satisfying all assumptions in Theorem 4.1, may be discontinuous.

Example 4.1 Let $X = [-2, 2]$ and $T : X \to X$ be defined as follows

$$Tx = \begin{cases} \frac{1}{2}x - 1, & -2 \leq x < 0, \\ 0, & x = 0 \\ \frac{1}{2}x + 1, & 0 < x \leq 2. \end{cases}$$

And let $d(x, y) = \mid x - y \mid$. This mapping has three fixed points: $x = 0, x = -2$ and $x = 2$, where $x = 0$ is a discontinuous point of T. We have,

$$d(x, Tx) = \begin{cases} 0, & \text{if } x = 0 \\ 1 - \frac{1}{2} \mid x \mid, & \text{if } 0 < \mid x \mid \leq 2. \end{cases}$$

We take the lower semicontinuous function $\phi(x)$,

$$\phi(x) = \begin{cases} 0, & \text{if } x = 0 \\ 2 - \mid x \mid, & \text{if } 0 < \mid x \mid \leq 2. \end{cases}$$

Then $\phi(x) - \phi(Tx) = d(x, Tx)$, for all $x \in [-2, 2]$. The conditions of Theorem 4.1 are satisfied.

Note that the lower semicontinuous function $\phi(x)$ in the example is not robust at $x = 0$.

Example 4.2 Let $X = [-2, 2]$ and let $T : X \to X$ be defined as follows

$$Tx = \begin{cases} \frac{1}{2}x, & -2 \le x \le 0 \\ \frac{1}{2}x + 1, & 0 < x \le 2. \end{cases}$$

Let $d(x, y) = | x - y |$. This mapping has two fixed points: $x = 0$ and $x = 2$, where $x = 0$ is a discontinuous point of T. We have,

$$d(x, Tx) = \begin{cases} \frac{1}{2} | x |, & \text{if } -2 \le x \le 0 \\ 1 - \frac{1}{2} | x |, & \text{if } 0 < x \le 2. \end{cases}$$

We take the lower semicontinuous function $\phi(x)$ as,

$$\phi(x) = \begin{cases} | x |, & \text{if } -2 \le x \le 0 \\ 2 - x, & \text{if } 0 < x \le 2. \end{cases}$$

Then we have $\phi(x) - \phi(Tx) = d(x, Tx)$, for all $x \in [-2, 2]$. The conditions of Theorem 4.1 are satisfied.

Note that, in this example, the lower semicontinuous function $\phi(x)$ is robust at $x = 0$.

Like minimization problems, a nonrobust solution of the equation $Tx = x$ is *unstable*. Therefore, we wish to exclude such a solution. In other words, we wish to find robust solutions of $Tx = x$ or ***robust fixed points*** of the mapping T. Note that discontinuous solutions of $Tx = x$ such as the point $x = 0$ in Example 4.2 are not excluded by the following definition of robust fixed points.

Definition 4.1 *Let X be topological space, $T : X \to X$ a mapping. A point $x \in X$ is said to be a* **robust fixed point** *of T if $Tx = x$ and the mapping T is robust at x.*

It is clear from the definition that the fixed point $x = 0$ in Example 4.1 is not robust while the fixed point $x = 0$ in Example 4.2 is.

It follows from definistions and our discussion in the previous sections that if $T : X \to X$ is a robust mapping then all fixed points of T are robust.

Theorem 4.2 *Let (X, d) be a complete metric space and $T : X \to X$ a given robust mapping. If there is a lower semicontinuous function ϕ, which is bounded from below, such that*

$$d(x, Tx) \le \phi(x) - \phi(Tx), \text{ for all } x \in X,$$

then T has robust fixed points.

We can use the algorithms of the integral global minimization to compute the fixed points of a robust mapping, thanks to the following theorem.

Theorem 4.3 *Let (X, d) be a metric space and $T : X \to X$ a given robust mapping. Then the set of fixed points of T is nonempty if and only if* $\min d(x, Tx) = 0$*; in this case, the set of fixed points of the mapping T is coincident with the set of global minima of $d(x, Tx)$:*

$$\{x \in X : Tx = x\} = \{x \in X : d(x, Tx) = 0\}. \tag{9}$$

Remark 4.1 Most of the existing optimization methods can only find local minimizers of a function. Such methods cannot be applied to find the fixed points of a mapping. This is because the local minimum value of $d(x, Tx)$ may not be equal to zero, therefore local minimum points may not be fixed points of T.

The following theorem ensures that the objective function $d(x, Tx)$ is upper robust if T is robust. This is precisely what the integral global minimization method requires.

Theorem 4.4 *Let (X, d) be a metric space and $T : X \to X$ a given mapping. If T is a robust mapping, then $d(x, Tx)$ is an upper robust function on X.*

Proof. Let $T_1 = (Ix, Tx) : X \to X \times X$ and $T_2 = d(x, Tx) : X \times X \to R^1$, where $Ix = x$ is a continuous mapping and T is assumed to be robust. Thus, as a product mapping T_1 is robust (Theorem 3.3). It is also clear that T_2 is a continuous mapping; thus, as the composition of the robust mapping T_1 and the continuous mapping T_2, $d(x, Tx) = T_2T_1 : X \to R^1$ is a robust mapping.

Therefore, $d(x, Tx)$ is (upper) robust.

We now consider the numerical tests for finding fixed points of a discontinuous mapping. The reader is again referred to [3, 11, 12] for the full account of the numerical implementation of the integral global optimization.

Example 4.3 Let

$$g_i(x) = \frac{\pi}{n}\{|\sin(\pi x_1)|^{\alpha_i} + \sum_{r=1}^{n-1} |x_r|^{\beta_i}(1 + |\sin(\pi x_{r+1})|^{\alpha_i}) + |x_n|^{\beta_i}\},$$
$$\alpha_i = 1 + i/n,\ \beta_i = 3 - i/n,\ i = 1, 2, \ldots, n. \tag{10}$$

The functions $g_i(x), i = 1, 2, \ldots, n$ are continuous but nonsmooth functions; We further let

$$t_i(x) = x_i + g_i(x) + [g_i(x)]/n,\ i = 1, 2, \ldots, n,$$

where $[g_i(x)]$ is the integer part of $g_i(x)$. Then $t_i(x)$, $i = 1, 2, \ldots, n$ are discontinuous. The mapping

$$Tx = (t_1(x), t_2(x), \ldots, t_n(x))$$

is a robust mapping and has a fixed point $x^* = (0, 0, \ldots, 0)$ on $D = \prod_{i=1}^{n}[-9.0, 11.0]$. Let

$$f(x) = d(x, Tx) = ||x - Tx||,$$

where $||.||$ is a norm in R^m. We use the following three metrics:

$$l_1 - \text{metric} \ : d_1(x, 0) = \sum_{i=1}^{m} |x_i|,$$

$$l_2 - \text{metric} \ : d_2(x, 0) = \sum_{i=1}^{m} |x_i|^2,$$

$$l_\infty - \text{metric} \ : d_\infty(x,0) = \max_{1\le i\le m} |x_i|.$$

Numerical tests have been done on a PC computer for $n = 10, 20, 50$. For simplicity, we only given the results in $n = 10$ case. The stopping criterion used in the integral minimization is

$$V_1(f,c) = \frac{1}{\mu(H_c)} \int_{H_c} (f(x) - c)^2 d\mu < 10^{-12},$$

where $V(f,c)$ is the variance of f over the level set $H_c = \{x : f(x) < c\}$, see [13]. In the table below we give the test results for three metrics,where f^* is the function value,$V_1(f,c)$ is the modified variance and N_f is the number of function evaluation. From these results, one sees that see that the integral global minimization algorithm can handle a complicated problem with a high accuracy in a reasonable computing effort.

It is obvious that for different metrics we should use different stopping criterions. For example, for l_2-metric it is better to require $V(f,c) < 10^{-24}$ in order to compare with other metric.

We now consider the numerical computation of equilibrium prices. Let n be the number of commodities in economy and m the total number of economic agents. The j-th agents is assumed to respond to a vector of nonnegative prices $p = (p_1,\ldots,p_n)$ by a vector of excess demands

$$g_1^j(p),\ldots,g_n^j(p) \text{ for the } n \text{ commodities },$$

where (for the constant elasticity of substitution utility function)

$$g_i^j(p) = \frac{a_{ji}\sum_k w_{jk}p_k}{p_i^{bj}\sum_k a_{jk}p_k^{1-bj}} - w_{ji}.$$

$W = (w_{jk})$ and $A = (a_{jk})$ are $m \times n$ matrices with positive entries, and $b = (b_1,\ldots,b_m)$ is a positive vector. For each commodity i,

$$g_i(p) = \sum_{j=1}^{m} g_i^j(p)$$

is defined as the market excess demand for that commodity. It is a function of the price vector p when the matrices W and A, and the vector b are given.

A vector of prices $p = (p_1,\ldots,p_n)$ is in equilibrium if at these prices the market excess demand for each commodity is less than or equal to zero. Thus we have a minimization problem:

$$\min_{p\ge 0} \|g(p)^+\|,$$

where

$$g(p)^+ = (\max(g_1(p),0),\ldots,\max(g_m(p),0))$$

and $\|.\|$ is the l_∞-metric. We consider three examples discussed in Scarf's paper [8]. We could not verify the results of $(g_i(\hat{p}))$ with the price vectors given by [8].

For instance, in his Example 1, if we change $\hat{p} = (104.9, 12.3, 5.2, 23.6, 14.1)$ to $\hat{p} = (104.9, 12.3, 5.2, 19.0, 14.1)$, then we obtain a better approximations (f^* from 2.163063 to 1.399183×10^{-2}. There the change of the fourth component from 23.6 to 19.0 is suggested by comparing with our computations (see Example 1 below and notice that $g(\alpha p) = g(p)$ for any positive constant α).

Now, suppose we accept the fact that the data of $W = (W_{jk}), A = (a_{jk})$ and $b = (b_j)$ given in these three examples of Scarf's paper [8] are correct, then we obtain the following results:

Example 4.4 (Example 1 of [8]) ($n = 5$, $m = 3$, $f^* = 1.430512 \times 10^{-6}$)

Example 4.5 (Example 2 of [8]) ($n = 8$, $m = 5$, $f^* = 2.905727 \times 10^{-6}$)

Example 4.6 (Example 3 of [8]) ($n = 10$, $m = 5$, $f^* = 4.768372 \times 10^{-6}$)

We can see from the above results that each component of $(g_i(\hat{p}))$ converges to zero uniformly. If the approximations is less accurate, as showing in Scarf's computation [8], some components of $(g_i(\hat{p}))$ may have large negative values. Even though such accuracy might not be essential in the computation of economic equilibria, they do demonstrate the convergence process of the algorithm.

Note that one of the advantage of computing economic prices by the integral global minimization approach is that we can start a computation in a large search domain and obtain accurate results. For instance, in Example 4.5, the search domain is

$$0 \le p_1 \le 100, \quad 0 \le p_2 \le 50, \quad 0 \le p_3 \le 50, \quad 0 \le p_4 \le 50, \quad 0 \le p_5 \le 50,$$

$$0 \le p_6 \le 150, \quad 0 \le p_7 \le 150, \quad 0 \le p_8 \le 150$$

After 43 iterations, we obtain a solution $f^* = 2.03836 \times 10^{-2}$. One needs 74 iterations for $f^* = 2.0843 \times 10^{-4}$ and 100 iterations for $f^* = 8.463859 \times 10^{-6}$.

References

1. C. Berge, *Espaces Topologique, Fonctions Multivoques*, Dunod, Paris, 1966.
2. J. Caristi, Fixed point theorems for mappings satisfying inwardness conditions, *Trans. Amer. Math. Soc.*, **215** (1976), 186-192.
3. S. Chew and Q. Zheng, *Integral Global Optimization*, Lecture Notes in Economics and Mathematical Systems, No. **298**, Springer-Verlag, 1988.
4. I. Ekeland, Nonconvex minimization problems, *Bull.(new) Amer. Math. Soc.* **1** (1979), 443-474.
5. Ky Fan, A generalization of Tucher's combinatorial lemma with topological applications, *Ann. of Math.*, **56** (1952), 431-437.

6. Ky Fan, Combinatorial properties of certain simplicial and cubical vertex maps, *Arch. Math.*, **11** (1960), 368-377.
7. Ky Fan, Simplicial maps from an orientable n-pseudomanifold into S^m with the octahedral triangulation, *J. Combinatorial Theory*, **2** (1967), 588-602.
8. H. Scarf, The approximation of fixed point of a continuous mapping, *SIAM J. Appl. Math.*, **15** (1967), 1328-1343.
9. H. Scarf (with the collaboration of T. Hansen), *Computation of Economic Equilibria*, New Haven, Yale University Press, 1973.
10. S. Shi, Q. Zheng and D. Zhuang, Discontinuous robust mappings are approximatable, preprint.
11. Zheng Quan, Global optimization of a class of discontinuous functions, *Numerical Mathematics, A Journal of Chinese Universities*, **8:1** (1985), 31-43, in chinese.
12. Zheng Quan, Robust Analysis and global minimization of a class of discontinuous function (I), *Acta Mathematicae Applicatae Sinica* (English Series), **6:3** (1990), 205-223.
13. Zheng Quan, Robust Analysis and global minimization of a class of discontinuous function (II), *Acta Mathematicae Applicatae Sinica* (English Series), **6:4** (1990), 317-337.
14. Q. Zheng, Measurability and discontinuity of robust functions, *Computers and Mathematics with Applications*, 1993.

Ky Fan, Combinatorial properties of certain simplicial and cubical vertex maps, Arch. Math. 11 (1960), 368-377.

GPSR Compliance

The European Union's (EU) General Product Safety Regulation (GPSR) is a set of rules that requires consumer products to be safe and our obligations to ensure this.

If you have any concerns about our products, you can contact us on ProductSafety@springernature.com

In case Publisher is established outside the EU, the EU authorized representative is:

Springer Nature Customer Service Center GmbH
Europaplatz 3
69115 Heidelberg, Germany

Batch number: 10755284

Printed by Printforce, the Netherlands